Wireless Communications

Wireless Communications

Principles, Theory and Methodology

KEITH Q.T. ZHANG

WILEY

This edition first published 2016
© 2016 John Wiley & Sons Ltd

Registered office
John Wiley & Sons Ltd, The Atrium, Southern Gate, Chichester, West Sussex, PO19 8SQ, United Kingdom

For details of our global editorial offices, for customer services and for information about how to apply for permission to reuse the copyright material in this book please see our website at www.wiley.com.

Library of Congress Cataloging-in-Publication Data

Zhang, Keith Q.T.
  Wireless communications : principles, theory and methodology / Keith Q.T. Zhang.
    pages cm
  Includes bibliographical references and index.
  ISBN 978-1-119-97867-1 (hardback)
1. Wireless communication systems. I. Title.
  TK5103.2.Z5245 2015
  621.384–dc23
                              2015021967

A catalogue record for this book is available from the British Library.

Set in 10/12pt, TimesLTStd by SPi Global Private Limited, Chennai, India.
Printed and bound in Singapore by Markono Print Media Pte Ltd

1   2016

In loving memory of my parents

Contents

Preface

This book is intended for graduate students and senior undergraduates majoring in wireless communications, but can be also used as a reference book for practicing engineers. Human society is now in an information era, and knowledge has been explosively accumulating, especially in the last three decades. It is, therefore, impossible to embody all new developments in a single textbook. In selecting the contents of this book, we have tried to cover the most representative achievements since C.E. Shannon established his revolutionary foundation for information theory in 1948, with emphasis on the thoughts, philosophy, and methodology behind them. Indeed, the existing knowledge is undoubtedly important. It is the thoughts and methodology that help create new knowledge for the future.

The transportation of information requires energy and channel resources, the latter of which can be further subdivided into frequency resource and spatial resource. The purpose of wireless communications is to fully exploit these resources to implement satisfactory transmission of information over a wireless channel with a desired data rate and a reasonable transceiver complexity. The satisfaction is usually measured in terms of three metrics: reliability, spectral efficiency, and system complexity. The impedance to reliable and spectrally efficient transmission stems from the impairments of a physical channel, which may include additive white Gaussian noise (AWGN), interference, and multipath propagation.

Wireless communication is a legend. In early 1980s, people talked about data rates of 9.6 kbps on telephone lines, but today cellular phones are everywhere with data rates exceeding 100 Mbps on wireless channels. Wireless communication evolves from generation to generation by developing various key technologies, strategies, and principles to handle challenges arising therein.

The first strategy is related to coding and turbo principle. As revealed by Shannon, channel coding is an effective means to implement reliable communications in AWGN. Digital modulation provides a basic data rate for communications. It converts the problem of analog waveform reception to the one of discrete statistical decision, thereby significantly increasing the transmission reliability. However, digital modulation, when used alone, leaves a big gap from the maximum possible rate that can be offered by a physical channel. This gap can be shortened by error-correction encoding. It is, thus, fair to say that without coding there is no modern communications. Through coding, certain algebraic structures are embedded into an information-bearing sequence. Once embedded, such code structures belong not only to the coder/decoder subsystem but also to the entire communications link. Thus, symbol estimation at the receiver end is, in essence, a global optimization problem. Global decoding, though optimal, is computationally prohibitive in practice. A much less complex strategy moving along this way is the *turbo principle*. By combining code structures with other functional blocks of communications for joint processing in an iterative manner, the turbo principle represents a great leap toward the globally optimized symbol estimation. The consequence is a significant improvement in reliability and spectral efficiency.

The second principle is the orthogonality strategy used in interference management. Traditional channel equalization represents a defensive strategy for combating inter-symbol interference (ISI), aiming to minimize the influence of ISI under certain criteria, the details of which are addressed in Chapter 9. In its nature, ISI has a similar generating mechanism as multi-user interference (MUI). The only difference is that the former results from a set of sequentially transmitted symbols that compete for the use of a temporally dispersive channel, whereas the latter stems from sharing the same frequency band by multiple users. Thus, a proactive strategy is to remove the source that generates interference by assigning an orthogonal or approximately orthogonal subspace to each symbol or each user to avoid their mutual collision. This orthogonality principle is a widely used paradigm in modern wireless communications. The application of this paradigm to a temporally dispersive ISI channel of Toeplitz structure leads to the multi-carrier or the orthogonal frequency division multiplexing (OFDM) technique, as detailed in Chapter 10. Using this orthogonality paradigm to multiple-access channels, the resulting schemes are code division multiple access (CDMA), time division multiple access (TDMA), and OFDMA, which are treated in Chapter 12. The orthogonality principle is also successfully applied to adjacent-interference management in cellular systems.

The third principle is multiple strategies for exploiting spatial electromagnetic resources. Mobile communication is characterized by multipath propagation, a phenomenon that causes a significant drop in the system error performance. Therefore, multipath fading is historically treated as a harmful effect, as clearly indicated in the phrase "combating multipath fading." However, random scattering and random motion that create multipath propagation constitute a two-dimensional random field. Such a random field embodies an abnormal channel capacity, representing a spatially distributive resource for wireless communications. Powerful techniques to exploit such a spatial resource include diversity combining, multi-input multi-output (MIMO) antenna systems, and cooperative relay communications, which are addressed in Chapter 7, Chapter 13, and Chapter 14, respectively.

Apart from the aforementioned principles, energy allocation between modulation/coding, over different frequencies, and along spatial beamformers to approach the channel capacity is also very important. The discussion of this issue is scattered over different chapters. Fully understanding the above principles represents a grasp of thoughts behind modern wireless communications, at least at a philosophical level. This is one effort made in this book.

In addition to providing the reader with an overall picture of wireless communications, we also carefully expound its technical details, not only covering a variety of main results and conclusions but also revealing the methodology used for their derivations. The solution to a communication problem is often not unique, but rather allows different formulations. This flexibility is demonstrated in this book, wherever possible, to provide a platform for students to foster their ability of divergent thinking.

Many problems at the end of each chapter are adapted from IEEE Transactions papers, aiming to give students a smooth transition from the course study to frontier research. As an aid to the teacher of the course, a solution manual for all the problems in the book is available from the publisher. The book covers enough topics and materials for two one-semester courses. For example, Chapters 3–7 and Chapter 9 can be used for a course on wireless communications, while Chapter 8 and Chapters 10–15 can be used for a course on advanced topics in wireless communications.

Acknowledgments

I am indebted to Dr. Shenghui Song for preparing an excellent draft of Chapter 14, which was partly edited by Professor Ranjan Mallik, to whom I am much grateful. I wish to express my deep gratitude to Dr. Jiayi Chen for plotting most of figures in the book and providing solutions to many problems at the end of each chapter. I am also grateful to Professor Kai Niu for creating Figures 11.7–11.9, and providing useful suggestions on turbo codes. I am indebted to Dr. Kai Chen and Dr. Qian Wen for providing all the figures and solutions to many problems in Chapters 6 and 11. Thanks also go to Dr. Guangchi Zhang who created Figures 13.8 and 13.9 and drew all the block diagrams in Chapter 9. I appreciate Dr. Xiaowei Cui for correcting the derivation of one property of CAZAC sequences. I am also grateful to Professor Xiangming Li for sharing his vision on random codes and algebraic codes. Dr. Dai Lin carefully proofread Chapters 9 and 13 providing many useful feedbacks, for which I deeply appreciate. I am also grateful to Dr. Jiayi Chen, Professor Danpu Liu, and Dr. Jun Wang for their valuable helps in proofreading Chapters 6, 10 and 15, respectively. I would like to extend my thanks to Mr. Fangchao Yuan for drawing Figure 1.2 and some other technical assistances.

I am particularly grateful to my project editors Liz Wingett and Teresa Netzler for their strong support and help throughout the writing of this book. I am indebted to Teresa Netzler, Content Capture Manager at Wiley, for her tireless effort in overseeing the production of the book in its various stages. My appreciation goes to Sandra Grayson for her elegant book-covers design. I am very grateful to Lavanya Prasannam, the Project Manager, and her team of SPi Global, Chennai, for their excellent jobs in copy editing.

Finally, I would like to express my sincere gratitude to my wife, Huisi, for her understanding and support during the years of writing.

1

Introduction

Abstract

The main goal of modern wireless communications is reliable transmission over an imperfect channel with the data rate to approach the channel capacity as much as possible. A physical channel often introduces additive white Gaussian noise, interference of various natures, and probably also multipath fading. The latter two impairments are often the sources that eventually limit the performance of a wireless system. Efforts of combating interference and multipath fading constitute an important part of the history of communications. The mathematical nature of interference is the collision between multiple symbols or multiple users in a *low-dimensional* space. Today, the paradigm of orthogonality has become a basic thought for combating different types of interference, while multi-antenna technology is a powerful means to exploit the inherent capacity of multipath fading channels.

James Clerk Maxwell's electromagnetic theory, established in 1864, uncovered the field nature of electromagnetic waves, marking a transition in our understanding of electromagnetism from phenomenology to physical theory. It is a prelude to the two far-reaching and revolutionary events in contemporary sciences: quantum physics and Einstein's theory of relativity. In engineering aspect, Maxwell's electromagnetic theory essentially becomes a physical foundation for telecommunications, stimulating the invention of the radio by Guglielmo Marconi in 1901, the invention of television broadcasting by Philo Farnsworth in 1928, and the invention of frequency modulation by Edwin Armstrong in 1933. These inventions and their practical applications, in turn, spurred the revolutionary advances in electronic devices, as exemplified by the invention of the vacuum tube in 1904 by John Fleming, transistors in 1948 by Walter Brattain, John Bardeen, and William Stockley at Bell Labs, and digital computers in 1946 by a team led by von Neumann. Further development of telecommunications required various key technologies for transmission, among which we can list the most representatives as follows:

- Nyquist's sampling theorem in 1928 by Harry Nyquist;
- Pulse-code modulation (PCM) in 1937 by Alec Reeves;
- Matched filters in 1943 by D.O. North.

Indeed, the electromagnetic theory and the emergence of various electronic devices had prepared the physical stage for communications. However, communication is of a different nature and is aimed at transporting information over an impaired channel; it is a process always associated with uncertainty and randomness. Clearly, we need a rigorous theory to fully reveal the nature of communications before the possibility to implement it reliably. Such a theory occurred in 1948 when Claude Shannon published his celebrated paper entitled "A mathematical theory of communication." Shannon proved, for the first time, that the maximum data rate achievable by a communication system was upper-bounded by the channel capacity and that error-free communication was possible as long as the transmission rate did not exceed the Shannon limit. It is fair to say that, while Maxwell's electromagnetic theory laid the physical foundation for communications, Shannon's theory provided it with a rigorous information-theoretic framework. Although various communication experiments had been conducted well before Shannon published his seminal papers, it was Shannon who made communication a rigorous science, casting dawn light on the horizon of modern communications.

With the advances in theories and technology, cellular mobile communications gradually developed as an industry. The idea of mobile communications dated back to 1946, when the Federal Communications Commission (FCC) granted a license for the first frequency-modulation (FM)-based land-mobile telephone operating at 150 MHz. But, a successful improved mobile-telephone service (IMTS) did not become a reality until the early 1960s when semiconductors became a matured technology. The flourishing of mobile phones, however, relied on frequency reuse over different geographical areas. That was the concept of cellular systems initiated by the Bell Laboratories in late 1940s as it requested the frequency band 470–890 MHz from the FCC for cellular telephony. Unfortunately, the request was declined twice due to other spectrum arrangements. The conflict between the industry and the FCC continued for many years. Ultimately, in 1981, the FCC finalized a bandwidth of 50 MHz from 800 to 900 MHz for cellular mobile communications, giving birth to the first-generation (1G) cellular mobile systems.

Since its first deployment in the early 1980s, cellular mobile communication technology evolved, shortly in two decades, from its first generation to the second employing digital voice transmission in 1985–1988, as represented by the GSM and IS-95, to meet the ever-increasing global market. It then evolved into its third generation (3G), which employed the wide-band CDMA technology and offered multimedia services. All the 3G standards were developed and released by an international organization called the Third Generation Partnership Project (3GPP). On the way to the fourth-generation (4G) cellular technology, the 3GPP took the strategy of long-term evolution (LTE) and released its LTE-8 standard in 2008. The standard LTE-8 and the subsequently released LTE-9 represent the transition from 3G to 4G. The true 4G is defined by the standard LTE-Advanced, which was released on December 6, 2010. 4G cellular technology can support data rate up to 1 Gb/s, and allows for variable bandwidth assignments to meet the requirements of different users.

Today, cellular phones, the Internet, and information exchange are ubiquitous, bringing us into the era of a knowledge explosion. In the short span of the past five decades, wireless communication has evolved from its inception to 4G, and is now moving to 5G, leaving a variety of dazzling achievements behind. We may give a long list of various events and activities that have had far-reaching impacts upon the human society and technological evolution, but it is impossible and unnecessary. What we really need is the fundamentals as well as the thoughts and philosophy behind them. Knowledge can be easily found by a google search. Only thoughts and philosophy that dictate the development of modern wireless communication can inspire us to further create novel knowledge and technology for the future. Indeed, as stated by Will Durant, "Every science begins as philosophy and ends as arts." Searching for the trajectory of thoughts, methodology, and philosophy from the history of wireless communications is interesting and challenging, and it is the direction of this book to endeavor.

We need to comb through the aforementioned dazzling events to uncover the underlying philosophy behind them. We identify four key issues, namely Shannon's theory and channel coding, the principle of orthogonality, diversity, and the turbo principle.

1.1   Resources for wireless communications

A typical communication system consists of transmitters, receivers and physical channels. By a physical channel we mean a medium connecting the transmitter to the receiver, which can be an optical fiber, a cable, twisted lines, or open air as encountered in wireless communications. Transmitters and receivers can be configured as a point-to-point communication link, a point-to-multipoint broadcast system, or a communication network. Communication networks are not in the scope of this book. A physical channel is usually imperfect, introducing noise, distortion, and interference. The objective of communication is the reliable transmission of information-bearing messages from the transmitter to the destination.

The three basic resources available for wireless communication are frequency resource, energy resource, and spatial resource. The first is usually called *frequency bandwidth*, the second is called *the transmit power*, and the third takes the form of *random fields* created when a wireless signal propagates through a spatial channel with a multitude of scatterers in random motion relative to the transceiver. The central issue to modern wireless communications is to fully exploit these resources to implement reliable communication between transmitters and receivers to satisfy a certain optimal criterion in terms of, for instance, spectral efficiency, energy efficiency, or error performance.

1.2   Shannon's theory

A fundamental challenge to communications is to look for a rigorous theoretical answer to the question of what is the maximum data rate that can be reliably supported by a given physical channel. Shannon answered this question by establishing three theorems, from the information-theoretic point of view. In his celebrated paper published in 1948 [1], Shannon showed that when a Gaussian random signal of power P watts is transmitted over an additive white Gaussian noise (AWGN) channel of one-sided power spectral density N_0 W/Hz and frequency bandwidth B Hz, the reliable communication data rate is upper-bounded by the channel capacity

$$C = B\log_2(1 + \rho) \text{ bits/s,} \tag{1.1}$$

where $\rho = P/\sigma_n^2$ is the signal-to-noise ratio (SNR) and $\sigma_n^2 = N_0 B$ is the noise power. Channel capacity is an inherent feature of a physical channel. We use a white Gaussian signal to test it, in much the same way as a delta function is used to test the impulse response of a linear system. This theorem clarifies that the resources of system bandwidth and signal power are convertible to reliable transmission data rate which, however, has a ceiling.

Shannon further asserted in his second theorem that error-free communication at a capacity-approaching rate was possible through coding. To this end, coding should be made as random and as long as possible. Before Shannon, a plausible belief was that the transmission reliability decreased with increased data rate. Shannon laid a solid foundation for information theory, igniting the dawn of modern communications.

Next, let us show how to determine the maximum reliable data rate on an AWGN channel with a given SNR. In (1.1), note that the transmitted power P is related to the bit energy E_b, the bit rate R_b, and the bit duration T_b by $P = E_b/T_b = E_b R_b$. Error-free transmission requires $R_b \leq C$. Thus, on the capacity boundary defined by the Shannon theorem, we have $R_b = C$, so (1.1) becomes

$$C/B = \log_2\left(1 + \frac{E_b C}{N_0 B}\right), \tag{1.2}$$

or equivalently,

$$\frac{E_b}{N_0} = \frac{2^{C/B} - 1}{C/B}. \tag{1.3}$$

Clearly, on the Shannon bound, the SNR is a function only of the spectral efficiency. It uncovers the conversion rule between the transmitted power (through SNR) and the normalized data rate (through the spectral efficiency). We are interested in the extreme case when $C/B \to 0$. Using the L'Hopital's rule, we obtain the asymptotic minimum SNR as

$$E_b/N_0 \to \log_e 2 = 0.693 = -1.6 \text{ dB}. \tag{1.4}$$

This indicates that, if we accept a very low spectral efficiency using, for example, coding, we can implement reliable reception as long as $E_b/N_0 \geq -1.6\,\text{dB}$.

✍ **Example 1.1** _____

Consider a binary phase shift keying (BPSK) system operating in an AWGN channel, and use $P_b = 10^{-5}$ as the measure of reliable reception. Without coding, we need SNR $= 11.39\,\text{dB}$ to achieve 1 b/s/Hz and $P_b = 10^{-5}$. As comparison, two rate-$\frac{1}{2}$ codes are used along with BPSK. One is a convolutional code with generators $G_1 = (11111)$ and $G_2 = (10001)$, alongside a 256×256 interleaver and decoding length $65,536$ bits. Another is a turbo code with two recursive systematic convolutional (RSC) coders. The results are summarized in Table 1.1.

Table 1.1 *Gaps of practical systems from their Shannon limit*

Scheme	Spectral efficiency (b/s/Hz)	SNR needed (dB)	Shannon SNR (dB)	Gap from Shannon (dB)
Modulation alone	1	11.39	0	11.39
Convolutional code	0.5	0.7	−0.82	1.52
Turbo	0.5	0.5	−0.82	1.32

1.3 Three challenges

Shannon's theorems uncovered the existence of good codes to approach the capacity of AWGN channels but did not provide any practical schemes for its implementation, leaving a rugged way for communications engineers to explore. Furthermore, Shannon only addressed the case of an AWGN channel. Its extension to a communication network is still unknown [2]. All these relevant issues constitute a challenge.

Other than AWGN, many practical communication systems suffer interference of different types. Even worse, it is often the interference, rather than the AWGN, that dominates the system performance. For example, inter-symbol interference (ISI) occurs when a sequence of symbols is transmitted through a bandlimited channel; inter-user interference (IUI) arises when multiple users share the same physical channel; and inter-antenna interference (IAI) happens when a number of parallel data streams share the same spatial multi-input multi-output (MIMO) channel. Regardless of their different appearances, they share a similar generating mechanism in which multiple symbols, multiple users, or multiple data streams compete for the use of the same resource. With this additional interference, the signal-to-interference plus noise-ratio (SINR) quickly deteriorates into a significant drop in the channel capacity, or equivalently, in the error performance, in accordance with (1.1). Thus, combating interference is another major challenge to the system designer.

For wireless communications, the third challenge arises from multipath propagation, which causes random fluctuation of the received signal, a phenomenon usually referred to as *multipath fading*. The consequence of multipath fading is the degradation in the system performance, and therefore multipath fading is usually considered a harmful effect. The traditional strategy is to mitigate it as much as possible. Then a philosophical question arises: is multipath fading an angel or a devil? Recent research work pioneered by Telator [3] and Foschini [4, 5] has changed our notion, uncovering that multipath fading is indeed an angel. There is huge capacity inherent in multipath random fields, which can be exploited by using multiple antennas.

Clearly, three issues central to wireless communications are coding, anti-interference, and the exploitation of spatial resource in random fields.

1.4 Digital modulation versus coding

Channel orthogonality is just like highway scheduling, aiming to eliminate or reduce IAI or IUI. Even with a dedicated channel, transmitted signals still suffer from the corruption of AWGN. Therefore at the source end, we still need to mark the "caravan" (i.e., the message sequence) with a special pattern so as to increase their recognition from the background noise. This is the job of channel coding.

Modern communication adopts the digital format, typically starting from digital modulation where information messages are represented in terms of discrete symbols. Each symbol is defined in a two-dimensional (2-D) Euclidean space, consistent with the fact that communication signals consist of in-phase and quadrature components. Digital modulation has a number of advantages. First, the retrieval of information messages from the received noisy data is, in essence, an estimation problem. According to statistical theory, the estimation accuracy is inversely proportional to the square-root of the available sample size. The difficulty with signal recovery in a communication system is that estimation must be done based on a single sample, implying a poor estimation accuracy even with a very large SNR. The use of discrete symbols converts the estimation problem to a decision one, in which estimation errors approach zero as the SNR tends to infinity. Second, digital modulation is a convenient way to provide the basic data rate. This data rate, obtained by digital modulation alone, shows the big gap from the Shannon's capacity bound. But it creates an easy structure to add algebraic structures (coding), usually in the Galois field, for spectral efficiency improvement. Finally, the digital format enables the exploitation of various advanced digital processing techniques to improve the system's overall performance.

Viewing the close relationship between coding and modulation, Urgerboeck asserts that "digital modulation and coding are two aspects of the same entity." Modulation provides the necessary data rate which, when used alone however, is far from the Shannon bound. The philosophy is to sacrifice part of the data rate, through adding some algebraic structure, to trade for a coding gain so that, at a lower rate than provided by modulation, the spectral efficiency approaches the Shannon bound. Stated in another way, we invest part of the symbol energy in code structures for better reliability. For example, a k-bit message is binary Hamming coded as an n-bit codeword to produce κ-bit correction capability. Suppose the message bit's energy is E_b. Then, the coded bit's energy reduces to $E_c = (k/n)E_b$, and the error performance of the coded system in AWGN with two-sided power spectral density $N_0/2$ is upper-bounded by

$$P_e = \sum_{i=\kappa+1}^{n} \binom{n}{i} \left[Q \left(\sqrt{\frac{2(k/n)E_b}{N_0}} \right) \right]^i \left[1 - Q \left(\sqrt{\frac{2(k/n)E_b}{N_0}} \right) \right]^{n-i}$$

$$= f \left(\frac{(k/n)E_b}{N_0} \right). \tag{1.5}$$

The uncoded system invests all of its available energy E_b into modulation, resulting in an error performance that follows the rule of the Q-function, as shown by $P_e = Q(\sqrt{2E_b/N_0})$. With coding, the system invests part of its energy, $(n - k)E_b/n$, into coding, producing a much faster dropping error probability defined above by the function $f(\cdot)$. A more powerful code can do even better than the fall-off function defined above by $f(\cdot)$.

On one hand, codes should be made as random and as long as possible, according to Shannon. Indeed, the probability of the decoding error can be made exponentially decreasing as the block size of a random-like code approaches infinity. This performance is achieved, however, at the cost of an exponential increase in decoding complexity. On the other hand, codes should have certain algebraic structures for ease of decoding. The task is therefore to seek well-structured codes with random behavior. Two classes of codes meet these requirements. They are concatenated codes and low-density parity-check (LDPC) codes. The former is constructed from two constituent encoders connected with a random interleaver, the idea originated by Dave Forney in his 1965 Ph.D. thesis. The constituent encoders are just as the usual ones, having certain algebraic structures. The random interleaver makes the synthesized code of random appearance. The LDPC codes are sparse codes with an LDPC matrix, the scheme proposed in 1960 by Robert G. Gallager in his doctoral dissertation. LDPC codes can be intuitively represented in terms of a Tanner graph, and the random behavior of the LDPC codes reflects in their random connectivity in the Tanner graph.

The use of powerful codes provides only a potential capability to approach the channel capacity, and the arrival at this goal must also rely on efficient decoding techniques. Turbo processing is one such powerful technique invented by Claude Berrou in 1993. The LDPC and Turbo codes do not eliminate codewords of small weights but make them occur with negligible probability. A turbo code consists of two constituent encoders, in series or in parallel. Its receiver correspondingly consists of two decoders forming a feedback loop with information exchange. Each decoder employs the BCJR algorithm to generate LLR soft symbol information, which is fed back to another decoder as extrinsic information to improve its detection performance. Thus, the turbo receiver can be viewed as an information-exchange learning machine.

Coding may take different forms compatible with a particular operating environment. Typically, algebraic structures can be embedded in a string as a coded sequence. It can also be embedded across different subcarriers to make even the error performance along them, as encountered in OFDM, or embedded across different transmit antennas forming a 2-D pattern, as encountered in MIMO systems.

1.5   Philosophy to combat interference

Competition for a limited frequency resource often happens in wireless communications. Interference is the consequence of competitive use of an imperfect channel, despite its occurrence in different forms and diverse applications. It appears in the form of ISI when a signal is transmitted over a bandlimited channel, in the form of IAI when a parallel data streams are transmitted over a MIMO antenna channel, and in the form of multiuser interference in multi-access environments. Though they have different appearances, these interferences have a similar generating mechanism and share the same philosophy in their avoidance.

There exist two different strategies in tackling interference, each following a distinct philosophy. The first strategy has no intent to prevent the generation of interference but, rather, tries to remedy it once it happens. An example is the use of channel equalizers to suppress ISI on a temporally dispersive channel. The second strategy aims to eliminate the source of interference before its generation, representing an active philosophy. The basic tool to implement the second strategy is the paradigm of *orthogonality*, which has formed the foundation for multiple access techniques, multicarrier theory, and adjacent-cell interference suppression. It is fair to say that the paradigm of orthogonality is one of the pillars that support the architecture of modern wireless communications.

Many functional blocks we see today in a typical wireless communication systems result from the use of these two strategies for combating interference. The examples include precoder, equalizer, spreading and despreading, OFDM and its inversion, and so on. They remain in use because of their simplicity. Their basic thoughts are further demonstrated as we go through Chapters 9–13.

1.6   Evolution of processing strategy

As described above, the conventional strategy for combating interference mainly addresses the flaws of channels that are responsible for the generation of interference. Processing is done on a block-by-block basis, isolating itself from the coding and decoding process. According to Shannon, coding is an indispensable ingredient to reliable communications. It is usually done in the Galois field, thereby possessing certain good structures of finite groups or polynomial rings, which endow a communication system with the capability for noise/interference resistance. For example, Andrew Viterbi showed how to exploit the Markovian properties of convolutional codes for their efficient decoding based on a trellis diagram. The coded structure, however, is not isolated from the overall system. If we use $\mathbf{b}$ to denote the information-bearing vector, $\mathbf{C}$ to denote the linear coding operation, and the operator $\Psi(\cdot)$ to denote the overall function of the remaining blocks such as modulation, channel matrix, spreading, and despreading, then the received signal vector is expressible in the form

$$\mathbf{y} = \Psi(\mathbf{C}\mathbf{b}) + \mathbf{n},$$

where $\mathbf{n}$ is the additive noise component. The task of reliable communications includes two aspects: first, given the channel matrix, design the encoder $\mathbf{C}$ and the overall system function $\Psi(\cdot)$; second, once the system is designed and upon receipt of the vector $\mathbf{y}$, retrieve the message vector $\mathbf{b}$. Directly retrieving $\mathbf{b}$ is extremely involved and is, in essence, a nonlinear integer programming problem. Faced with this intractable issue, the question is: what is a feasible strategy for a practical wireless system? Three lines of thought are

1. block-by-block isolative design and processing;
2. local joint optimization; and
3. overall system design and global optimization.

The first line of thought represents the traditional philosophy of local optimization, and has been widely adopted in practice. The result is obviously a suboptimal solution. The second reflects a current trend toward joint/global processing, rooted in the following insight. Namely, an algebraic structure, once introduced into a coded sequence, belongs not only to the coding and decoding subsystem but also to the overall system. The coded structure penetrates all the functional blocks, integrating itself with the structures in modulation, OFDM, MIMO channels, ISI channels, multiuser spreading codes, or another encoder to form an even more abundant superstructure. Exploiting such a superstructure, even partly, can significantly improve the overall system performance. Such joint processing is made possible today thanks to the invention of the turbo principle by Claude Berrou [6, 7], and to the advance in the computational power of IC chips. We will revisit these important strategies in more detail in Chapter 8.

The third philosophy, though leading to a global optimal solution, is still intractable today, due to the limited computational signal processing power. However, it represents the efforts for the future. Its implementation awaits breakthroughs in notion, methodology, and technology.

1.7   Philosophy to exploit two-dimensional random fields

As previously mentioned, random fields represent a *spatial resource* available to wireless communications. When a mobile unit moves around a complex propagation environment, random scattering from distributive

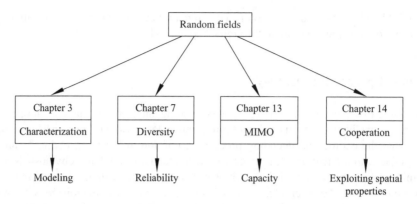

Figure 1.1   *Exploiting the spatial electromagnetic resource in random fields*

objects constitutes a random field. Random fields have many important characteristics, causing, for example, propagation loss and multipath fading of the received signals.

Multipath fading is a two-sided sword. In AWGN channels, the error probability of a coherent communications system drops exponentially with increased SNR through the relationship of a Q-function. Rayleigh fading makes the average error performance fall off much more slowly, only on the order of magnitude inversely proportional to the SNR. As such, multipath fading is traditionally treated as a harmful effect and, thus, must be suppressed as much as possible. Research in the recent two decades has changed our vision on fading, that is, random fields possess abnormal channel capacity. Exploring such capacity is increasingly important to high-data-rate wireless communications.

A powerful means to exploit fading resources is inserting multiple antennas in a 2-D random field. When placing a set of collocated antennas at a receiver, the resulting system is diversity combining, which makes the system's average error performance, in the dB scale, improve linearly with the number of antennas.

Placing an antenna array at each side of a transceiver constitutes a MIMO system, an enabling device to exploit the capacity of a random field. The average mutual information of a MIMO system is roughly proportional to the minimum number of transmit and receive antennas. A MIMO system can be configured in a multiplexing mode for high-data-rate wireless transmission or in a diversity mode for reliability enhancement. There is a tradeoff between the two. If antennas are distributively placed in a random field, they form a distributive MIMO system, which can be used, for example, for collaborative communications.

The characterization of random fields and various techniques for their exploitation are elucidated in several chapters of the present book, as outlined in Figure 1.1.

1.8   Cellular: Concept, Evolution, and 5G

The idea central to cellular systems is frequency reuse over different geographical locations, usually known as cells, so that a limited frequency bandwidth can be used to support a huge user population over a service region. A direct consequence of frequency reuse is co-channel interference caused by co-channel users at different cells, which ultimately limits the performance of a cellular system. Co-channel interference management is one of the key issues to cellular system design, and its strategy continuously evolves from 1G cellular to 5G [32–34]. The traditional technique, as widely adopted in 1G and 2G cellular, is to implement frequency reuse over well-separated cells, usually through the strategy of an N-cell reuse pattern to share

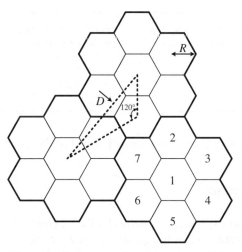

Figure 1.2   *Illustrating the seven-cell reuse pattern of hexagonal cellular*

a total of Ω frequency channels. Each cell possesses only Ω/N channels. The parameter N, known as the *cluster size*, is chosen to balance the signal to co-channel interference ratio (SIR) and the channels available in each cell. A typical N-cell reuse pattern is illustrated in Figure 1.2 where $N = 7$. A cellular system with hexagonal geometry and an N-cell reuse pattern has the following property.

$$N = i^2 + ij + j^2.$$

Let R denote the outer radius of a cellular hexagon, and let D denote the center-to-center separation between two nearest co-channel cells. The ratio $q = D/R$ is called the *reuse ratio*. For a cellular system with hexagonal geometry, the reuse ratio meets the following condition:

$$q = \frac{D}{R} = \sqrt{3N}.$$

Clearly, the idea behind the 1G and 2G systems for frequency reuse is to simply exploit the power-decay law of electromagnetic propagation with distance. This simple idea, however, fails to implement 100% frequency reuse. To further increase the user population, CDMA cellular in 2G and 3G adopts a totally different strategy for frequency reuse by separating each cell with a particular pair of short pseudo-noise (PN) sequences long scrambling code, such that interference from unintended cells is substantially blocked after decorrelation. In so doing, all the frequency channels are 100% reused over each cell. The 4G cellular employs OFDM-based access (OFDMA) schemes in which the technique based on PN sequences is no longer applicable. Thus, the suppression of adjacent-cell interference has to rely on joint frequency managements in both physical and higher layers.

Wireless communications today is on the way to its fifth generation. The 5G cellular should be a heterogeneous network, supporting cellular networks, internet of things, and others. In 5G, multiple antennas will be intensively distributed over the entire service area to fully exploit the channel capacity in a random field to support very high data rates. In the next two decades, the total global data volume of transmission is predicted to be on the order of magnitude of $E18$ with a main feature of big data and low information. The wireless capacity to be explored to meet this challenge includes efforts along three aspects. They are frequency spectrum extension, spectral efficiency improvement, and network density increase. The potential contributions of the three aspects are summarized as follows [8]:

Predicting agent	Frequency spectrum extension (times)	Spectrum efficiency improvement (times)	Network density (times)
Nokia	10	10	10
NTT	2.8	24	15
Com Mag [9]	3	5	66

Before concluding this chapter, references of relevance [16–34] are included for additional reading.

1.9   The structure of this book

Wireless communication is a still-expanding wonderland full of engineering miracles and elegant philosophy that await exploration. Some mathematical background is briefly reviewed in Chapter 2. Chapters 4–5 are dedicated to digital modulation, while Chapter 6 is devoted to channel coding. Interference is a major channel impairment that limits the error performance of wireless communications. The philosophy of combating ISI caused by temporally dispersive channels can be classified into defensive and proactive strategies, which are investigated in Chapters 9 and 10, respectively. Interference arising in multiuser channels is studied in Chapter 12. Another important issue to wireless communications is the investigation and exploitation of the spatial electromagnetic resource of a multipath channel, and these cover four chapters. Among them, Chapter 3 investigates channel modeling and characterization, Chapter 7 investigates diversity reception, Chapter 13 is devoted to MIMO wireless systems, and Chapter 14 describes cooperative wireless communications. Coding usually takes the form of an embedded algebraic structure that belongs to the entire communication system. An efficient tool towards globally exploiting such a structure is the turbo processing principle, which is studied in Chapter 11.

1.10   Repeatedly used abbreviations and math symbols

Some nomenclatures and mathematical symbols used throughout the book are tabulated below for ease of reference.

AWGN	Additive white Gaussian noise
CDF	Cumulative distribution function
CDMA	Code-division multiple access
CHF	Characteristic function
CR	Cognitive radio
CSIR	Channel state information available at receiver
CSIT	Channel state information available at transmitter
DMT	Diversity and multiplexing tradeoff
EGC	Equal gain combining
FMDA	Frequency-division multiple access
LDPC codes	Low-density parity-check codes
MASK	M-ary amplitude shift keying
MFSK	M-ary frequency shift keying
MPSK	M-ary phase shift keying

QAM	Quadrature amplitude modulation
QPSK	Quadrature phase shift keying
MIMO	Multiple-input multiple-output
ML	Maximum likelihood
MRC	Maximal ratio combining
MSK	Minimum shift keying
OFMDA	Orthogonal frequency-division multiple access
PDF	Probability density function
PN sequences	Pseudo-noise sequences
RSC codes	Recursive systematic convolutional codes
SC	Selection combining
SNR	Signal-to-noise ratio
TMDA	Time-division multiple access
ZFE	Zero-forcing equalization
i.i.d.	Independent and identically distributed

E_s	Symbol energy
N_0	One-sided power spectral density of AWGN in W/Hz
$N_0/2$	Two-sided power spectral density of AWGN in W/Hz
$\sim$	Distributed as
$\otimes$	Kronecker product; namely, $\mathbf{A} \otimes \mathbf{B} = [a_{ij}\mathbf{B}]$
$\mathbf{A}^\dagger$	Hermitian transpose of complex matrix $\mathbf{A}$
$\mathbf{A}^T$	Transpose of a real matrix $\mathbf{A}$
$\mathbb{E}[\cdot]$	Expectation operator
$\mathrm{var}[x]$	Variance of random variable x
$\mathrm{Cov}[\mathbf{x}]$	Covariance of random vector $\mathbf{x}$
$\Re(z)$	The real part of z
$\Im(z)$	The imaginary part of z
$\langle x(t), y(t) \rangle$	Inner product of two functions or two vectors
$\|x(t)\|$	Norm of $x(t)$; namely $\|x(t)\| = [\langle x(t), x(t) \rangle]^{1/2}$
$Q(x)$	Q-function; namely, $Q(x) = \int_x^\infty \frac{1}{2\pi} e^{-\xi^2/2} d\xi$
$\Pr\{A\}$ or $\Pr(A)$	Probability of event A
$\mathcal{N}(\mathbf{m}, \mathbf{R})$	Gaussian distribution with mean $\mathbf{m}$ and covariance matrix $\mathbf{R}$
$\mathcal{CN}(\mathbf{m}, \mathbf{R})$	Complex Gaussian distribution with mean $\mathbf{m}$ and covariance matrix $\mathbf{R}$ [10].
$\mathcal{CW}_m(N, \mathbf{R})$	m-Dimensional complex Wishart distribution with sample size N and covariance matrix $\mathbf{R}$ [11, 12]
$\mathcal{NK}(m, \Omega)$	Nakagami distribution with fading parameter m and power Ω
$\mathcal{LN}(\mathbf{m}, \mathbf{R})$	Joint lognormal distribution with mean $\mathbf{m}$ and covariance matrix $\mathbf{R}$
$\chi(m)$	Chi distribution with m degrees of freedom. ([13], p. 421)
$\chi^2(m)$	Chi-square distribution with m degrees of freedom [14]
$\Gamma(x)$	Gamma function
$\tilde{\Gamma}(x)$	Complex gamma function
$_1F_1(a; b; z)$	Confluent (Kummer's) hypergeometric function. ([15], p. 504)
$_2F_1(a, b; c; z)$	Gauss hypergeometric function. ([15], p. 556)

As a convention throughout the book, an equation referred to in the text is indicated only by its number for simplicity. For example, Equation (9.16) is simply written as (9.16). Whenever necessary, a subscript will be added to a distribution symbol to indicate the dimension of the distribution, for example, $C\mathcal{N}_m(\mathbf{v}, \mathbf{R})$ and $\mathcal{N}_m(\mathbf{v}, \mathbf{R})$. The subscript m for dimension is dropped when no ambiguity is introduced.

Problems

Problem 1.1   Name five breakthroughs you consider the most significant in communications theory in past six decades, and justify your assertions.

Problem 1.2   Enumerate five technological breakthroughs you consider the most significant in communications in past six decades, and justify your assertions.

References

1. C.E. Shannon, "A mathematical theory of communications," *Bell Tech. J.*, vol. 27, pp. 379–423, 623–656, 1948.
2. P. Gupta and P.R. Kumar, "The capacity of wireless networks," *IEEE Trans. Inf. Theory*, vol. 46, no. 2, pp. 388–404, 2000.
3. I.E. Telatar, "Capacity of multi-antenna Gaussian chaneels," *Bell Labs Technical Memorandum*, June 1995.
4. G.J. Foschini, "Layered space-time architecture for wireless communication in a fading environment when using multi-element antennas," *Bell Labs Tech. J.*, vol. 1, no. 2, pp. 41–59, Autumn 1996.
5. G.J. Foschini and M.J. Gan, "On the limits of wireless communication in a fading environment when using multiple antennas," *Wireless Pers. Commun.*, vol. 6, no. 3, pp. 311–335, 1998.
6. C. Berrou, A. Glavieux, and P. Thitimajshima, "Near Shannon limit error correcting coding and decoding: turbo codes," *Proceedings of the 1993 IEEE International Conference on Communications*, pp. 1064–1070, Geneva, Switzerland, May 1993.
7. C. Berrou, "The ten-year-old Turbo codes are entering into service," *IEEE Commun. Mag.*, vol. 41, no. 8, pp. 110–116, 2003.
8. J.H. Lu, "Thoughts on problems of future wireless communications," in *National Conference on Wireless Applications and Managements*, Nov. 23, 2013, Tianjin, China.
9. J. Zander and P. Mahonen, "Riding the data tsunami in the cloud: myths and challenges in future wireless access," *Commun. Mag.*, vol. 51, no. 3, pp. 145–151, 2013.
10. K. Miller, *Complex Stochastic Processes: An Introduction to Theory and Application*, Reading, MA: Addison-Wesley, 1974.
11. M.L. Mehta, *Random Matrices*, 3rd edn, Elsevier Academic Press, 2004.
12. A.T. James, "Distributions of matrix variates and latent roots derived from normal samples," *Ann. Math. Stat.*, vol. 35, no. 2, pp. 475–501, 1964.
13. M.D. Springer, *The Algebra of Random Variables*, New York: John Wiley & Sons, Inc., 1976.
14. N.L. Johnson and S. Kotz, *Continuous Univariate Distributions*, vol. 2, New York: John Wiley & Sons, Inc., 1976.
15. M. Abramowitz and I.A. Stegun, *Handbook of Mathematical Functions*, New York: Dover Publications, 1970, p. 932.
16. L. Brandenburgh and A. Wyner, "Capacity of the Gaussian channel with memory: a multivariate case," *Bell Syst. Tech. J.*, vol. 53, pp. 745–779, 1974.
17. T.S. Rappaport, *Wireless Communications*, Prentice-Hall 1996, Chapter 2, Parts of Chapters 3–4, and Chapter 5.
18. A. Goldsmith, *Wireless Communications*, Cambridge University Press, 2005.
19. A.F. Molisch, *Wireless Communications*, Wiley/IEEE Press, 2007.
20. G.L. Stuber, *Principles of Mobile Communication*, 2nd edn, Springer-Verlag.
21. David Tse, *Fundamentals of Wireless Communication*, Cambridge University Press, 2005.
22. R.E. Ziemer and R.L. Peterson, *Introduction to Digital Communication*, 2nd edn, Chapter 10, Prentice-Hall, 2001, pp. 650–763.
23. R.L. Peterson, R.E. Ziemer, and D.E. Borth, *Introduction to Spread Spectrum Communications*, Prentice-Hall, 1995, Chapters 11–3, Chapters 9 and 11.

24. V.K. Garg, *IS-95 CDMA and CDMA2000*, Prentice-Hall, 2000, Chapter 2 (good introduction to CDMA) and Chapter 7 (overall system).
25. M.D. Yacoub, *Foundations of Mobile Radio Engineering*, Boca Raton, FL: CRC Press, 1993.
26. W.C. Lee, "Overview of cellular CDMA," *IEEE Trans. Veh. Technol.*, vol. 40, no. 2, pp. 291–302, 1991.
27. W.C. Lee, "Applying the intelligent cell concept to PCS," *IEEE Trans. Veh. Technol.*, vol. 43, no. 3, pp. 672–679, 1994.
28. To gain newest information about 3G systems, visit the website–http://www.itu.int/imt/.
29. F. Adachi, D. Garg, S. Takaoka, and K. Takeda, "Broadband CDMA techniques," *IEEE Wireless Commun. Mag.*, vol. 12, no. 2, pp. 8–18, 2005.
30. A. Perotti and S. Benedetto, "A new upper bound on the minimum distance of turbo codes," *IEEE Trans. Inf. Theory*, vol. 50, no. 12, pp. 2985–2997, 2004.
31. H. El Gammal and A.R. Hammons, "Analyzing the turbo decoder using the Gaussian approximation," *IEEE Trans. Inf. Theory*, vol. 47, no. 2, pp. 671–686, 2001.
32. Y. Wang, J. Li, L. Huang, A. Georgakopoulos, and P. Demestichas, "5G mobile," *IEEE Technol. Mag.*, vol. 9, no. 3, pp. 39–46, 2014.
33. Ericsson, '5G radio access: research and vision, white paper. [Online] available: http://www.ericsson.com/res/docs/white-paper/wp-5g.pdf.
34. P. Demestichas, A. Georgakopoulos, D. Karvounas, K. Tsagkaris, V. Stavroulaki, L. Jianmin, C. Xiong, and J. Yao, "5G on the horizon: key challenges for the radio-access network," *IEEE Technol. Mag.*, vol. 8, no. 3, pp. 47–53, 2013.

2

Mathematical Background

Abstract

This chapter briefly outlines some mathematical skills, techniques, and methodology to be used in the subsequent study of the book, with emphasis on their applications. Readers with a good mathematical background only need to go through the section on congruence mapping and signal spaces, which supplies a powerful means for understanding many communication systems and relevant signal processing schemes involved therein. Other material in this chapter can be skipped for the first time of reading and revisited whenever necessary.

♣

2.1 Introduction

It is impossible, and not intended, to systematically describe the mathematical background for communications in a limited space. But rather, the purpose of this chapter is to present mathematical results and skills of relevance.

2.2 Congruence mapping and signal spaces

Congruence mapping is a transformation that maps elements from one metric space to another such that distance between the elements is preserved in the new space. This implies that the shape and size of a geometric figure or object remain unchanged before and after mapping.

As a powerful tool, congruence mapping is widely used in wireless communications. In wireless systems, message data to be handled is often in discrete form. They must be converted into appropriate waveforms to acquire sufficient energy and to match a given physical channel for efficient transmission. At the receiver end, continuous waveforms are converted back to the discrete domain to facilitate signal processing. The equivalence between the signals in discrete and continuous forms is usually ensured by a congruence mapping.

Wireless Communications: Principles, Theory and Methodology, First Edition. Keith Q.T. Zhang.
© 2016 John Wiley & Sons, Ltd. Published 2016 by John Wiley & Sons, Ltd.
Companion Website: www.wiley.com/go/zhang7749

Suppose we are given m discrete data $\mathbf{x} = [x_1, \cdots, x_m]$, which are converted into a continuous waveform by virtue of a set of orthonormal functions $\{\phi_1(t), \cdots, \phi_m(t)\}, a \leq t \leq b$, such that

$$x(t) = x_1 \phi_1(t) + \cdots + x_m \phi_m(t). \tag{2.1}$$

The basis functions $\{\phi_i(t)\}$ satisfy the following inner-product conditions:

$$\langle \phi_k(t), \phi_\ell(t) \rangle = \begin{cases} 0, & \ell \neq k \\ 1, & \ell = k \end{cases} \tag{2.2}$$

with the inner product defined by

$$\langle \phi_\ell(t), \phi_k(t) \rangle = \int_a^b \phi_\ell^*(t) \phi_k(t) dt. \tag{2.3}$$

Equation (2.1) establishes a mapping from a discrete to a continuous space:

$$\underbrace{[x_1, \cdots, x_m]}_{\mathbf{x}} \rightarrow \underbrace{x_1 \phi_1(t) + \cdots + x_m \phi_m(t)}_{x(t)}. \tag{2.4}$$

The vector $\mathbf{x}$ is a point in the m-dimensional Euclidean space spanned by the bases $\{\mathbf{e}_i, i = 1, \cdots, m\}$, where $\mathbf{e}_i$ is an all-zero vector except entry i of unit value. The geometric significance of (2.9) is that a signal can be represented either in a Euclidean space or in a functional space. The two spaces are congruent. The *only difference* is that they adopt different basis functions. From the mathematical point of view, using different sets of basis functions does not change the nature of the space. However, it does have different implications in physical implementation, as will be clear in the subsequent discussion of various multiple access schemes in Chapter 12. The two representations are equivalent in the sense that they comply with the property of invariance of inner product. Suppose two m-vectors, $\mathbf{v}_1$ and $\mathbf{v}_2$, are mapped into the functional space producing $v_1(t)$ and $v_2(t)$, respectively, in the same way as shown in (2.4). Then, we have

$$\mathbf{v}_2^\dagger \mathbf{v}_1 = \langle v_1(t), v_2(t) \rangle. \tag{2.5}$$

It is, in fact, a restatement of Parseval's theorem. If we set $\mathbf{v}_2 = \mathbf{v}_1 = \mathbf{v}$, the above expression simply asserts the length invariance under the mapping. Note that dividing the left side by the length of two vectors, $\|\mathbf{v}_1\| \|\mathbf{v}_2\|$, geometrically represents the cosine of the angle between the two vectors. Likewise, the right side, divided by $\|v_1(t)\| \|v_2(t)\|$, has a similar interpretation. Combining these arguments enables another assertion that the inner-product, or the angle, is invariant under the mapping. Clearly, a geometry in the two spaces is invariant under mapping, thereby maintaining the same lengths and same angles, and thus the same shape. Accordingly, the mapping is an isometric mapping.

Regardless of their equivalence, the representation in the Euclidean space is more intuitive and easier to visualize, and is, thus, widely adopted in signal design and performance analysis. The resulting Euclidean space is referred to as the *signal space*, the term coined by a Russian student in his Ph.D. dissertation and popularized through the textbook by Wozencraft and Jacobs [39].

The waveform $x(t)$ is often corrupted by AWGN, $n(t)$, producing [39] a noisy observation

$$y(t) = x(t) + n(t) = \sum_{i=1}^{m} x_i \phi_i(t) + n(t). \tag{2.6}$$

This received signal is then projected back onto the signal space to facilitate signal detection, signal processing, performance analysis, or other purposes, via a bank of parallel correlators to obtain

$$\langle y(t), \phi_k(t) \rangle = x_k + n_k \tag{2.7}$$

for $k = 1, \cdots, m$. This is the mapping from the continuous-time functional space to the Euclidean space. The correlation can be done alternatively, and more efficiently, through a matched filter. Though important to radar or sonar signal detection in the sense of maximizing the output SNR, matched filters are only a means to implement projection on the signal space but not at the core of congruence mapping. The local reference in the correlators can be any scaled version of $\phi_k(t)$, which has no influence on the system performance since the resulting SNRs remain unchanged. However, the use of basis functions as local references ensures the noise component n_k to have the same variances as $n(t)$, thus eliminating the need for recalculation and simplifying the analysis.

For the purpose of system design or performance analysis, there is no need to work out the projections, but one can write the received vector from (2.6), yielding

$$\mathbf{y} = \mathbf{x} + \mathbf{n} \tag{2.8}$$

where $\mathbf{n} = [n_1, \cdots, n_m]$ of i.i.d. Gaussian entries each of zero mean and the same variance as $n(t)$.

In communications, the choice of an appropriate set of basis functions depends very much on the nature of the signals and channels under consideration; more details will be presented as we proceed to the issues of modulation, OFDM, and multiple access. In what follows, we give two examples for illustration.

✍ Example 2.1

As the first illustrating example, we consider a typical digital modulation scheme, QAM, as sketched in Figure 2.1. Without loss of the generality, we focus on the message data pair (a_k, b_k) transmitted over the kth symbol interval $(k-1)T_s \leq t < kT_s$, since the treatment of other periods is exactly the same. The message data pair is used to modulate the two orthogonal carriers to generate a modulated signal

$$s(t) = a_k \underbrace{(\sqrt{2/T_s} \, \cos \omega_c t)}_{\phi_1(t)} + b_k \underbrace{(\sqrt{2/T_s} \, \sin \omega_c t)}_{\phi_2(t)} \tag{2.9}$$

for $(k-1)T_s \leq t < kT_s$. This process is of duality in understanding. As a physical process, it is called *modulation* in communications, an indispensable step in order to obtain energy and to match the channel for transmission. Its mathematical nature is an isometric mapping from a Euclidean space to a continuous-time functional space. Sine and cosine functions are a natural choice to form a basis for various problems in communications, given that electromagnetic waves often take the form of sinusoids with a phase.

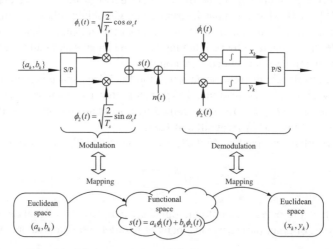

Figure 2.1 *Modulation and demodulation versus congruence mapping*

The process of demodulating the noisy received signal $x(t) = s(t) + n(t)$ is a reverse operation, namely projecting $x(t)$ onto the space spanned by $\phi_1(t)$ and $\phi_2(t)$. To determine the demodulated output, there is no need to conduct integration again, but one can directly collect the coefficients of the normalized bases $\phi_1(t)$ and $\phi_2(t)$ in (2.9) to write the result as

$$\mathbf{z} = [a_k, b_k] + [n_c(k), n_s(k)] \tag{2.10}$$

where $n_c(k)$ and $n_s(k)$ are projections of noise $n(t)$ onto $\phi_1(t)$ and $\phi_2(t)$ over the interval $(k-1)T_s \le t < kT_s$. They are independent and have the same property as $n(t)$. For AWGN channels at any given instant t, $n(t) \sim \mathcal{N}(0, \sigma_n^2)$. Then, $n_c(k)$ and $n_s(k)$ have the same distribution. A pair of variables defined in the 2-D vector $\mathbf{z}$ is isomorphic to the complex number defined by

$$z = (a_k + jb_k) + \underbrace{(n_c(k) + jn_s(k))}_{w_k} \tag{2.11}$$

where $j = \sqrt{-1}$ and w_k is the complex Gaussian noise distributed as $w_k \sim \mathcal{CN}(0, 2\sigma_n^2)$. ○

In the above example, the bases consist only of two functions, sine and cosine, defined over the finite symbol interval. In the next example, the bases are an infinite set of orthonormal functions defined over $-\infty < t < \infty$.

✍ **Example 2.2**

Determine the signal space for a bandlimited signal.

Solution

Consider a bandlimited baseband signal $x(t)$ of bandwidth W Hz. Denote $T_s = \frac{1}{2W}$ and denote

$$\phi_k(t) = \sqrt{2W}\ \text{sinc}2W(t - kT_s), k = \cdots, -1, 0, 1, \cdots \tag{2.12}$$

whose Fourier transform is given by

$$\Phi_k(f) \overset{FT}{\to} \frac{1}{\sqrt{2W}}\text{rect}\left(\frac{f}{2W}\right)e^{-j2\pi\,kT_s}.$$

It is easy to examine that $\{\phi_k(t)\}$ constitutes a complete set of bases over $-\infty < t < \infty$ by showing that

$$\int_{-\infty}^{\infty} \phi_m(t)\phi_\ell^*(t)dt = \int_{f=-\infty}^{\infty} \Phi_m(f)\Phi_\ell^*(f)df$$

$$= \frac{1}{2W}\int_{-\infty}^{\infty}\text{rect}^2\left(\frac{f}{2W}\right)e^{j2\pi(\ell-m)fT_s}\,df$$

$$= \frac{\sin\pi(\ell - m)}{\pi(\ell - m)} = \begin{cases}1, & m = \ell \\ 0, & m \neq \ell.\end{cases} \tag{2.13}$$

Next, we show that $\{\phi_k(t)\}$ spans a signal space for all bandlimited signals $x(t)$ of finite energy. Let us represent the sampled version of $x(t)$ by

$$x_s(t) = x(t)\sum_{k=-\infty}^{\infty}\delta(t - kT_s) = x(t)\sum_{m=-\infty}^{\infty}T_s^{-1}e^{j2\pi m/T_s},$$

where the sum of impulses is a periodic function and thus can be expressed as a Fourier series leading to the result shown on the right. Hence, by taking the Fourier transform, we obtain the Fourier spectrum of $x_s(t)$ as $X_s(f) = T_s^{-1} \sum_{m=-\infty}^{\infty} X\left(f - \frac{m}{T_s}\right)$. Since Nyquist sampling rate is used, there is no aliasing in $X_s(f)$. Thus, we can completely recover $x(t)$ by passing $x_s(t)$ through an ideal low-pass filter with $H(f) = T_s \text{rect}\left(\frac{f}{2W}\right)$. The filter output is then equal to $X(f) = H(f)X_s(f)$, which, in the time domain, represents the convolution

$$x(t) = \underbrace{2WT_s\text{sinc}(2Wt)}_{=1} * \left(\sum_{k=-\infty}^{\infty} x(t)\delta(t - kT_s)\right)$$

$$= \sum_{k=-\infty}^{\infty} x(kT_s)\text{sinc}[2W(t - kT_s)]$$

$$= \sum_{k=-\infty}^{\infty} \sqrt{T_s}\, x(kT_s)\, \phi_k(t). \tag{2.14}$$

From the last line, we can write $x(t)$, in vector form, to obtain its representation in the signal space, yielding

$$x(t) \rightarrow \mathbf{x} = \sqrt{T_s}\, [\cdots, x(-T_s), x(0), x(T_s), \cdots].$$

This is a congruence mapping for bandlimited signals. ◯

〰〰✍

As an application of orthogonal expansion, let us take mutual information as an example. Suppose a zero-mean Gaussian random signal $x(t)$, with power P_s and frequency bandlimited to $(-B, B)$ Hz, is transmitted through an AWGN channel, characterized by $n(t)$, with two-sided power spectral density σ_n^2 watt/Hz, producing the received signal

$$y(t) = x(t) + n(t).$$

We want to determine the mutual information between $x(t)$ and $y(t)$ over the time span $(0, T)$ in seconds. Recall that we are only able to evaluate the entropy of a random sample but not a continuous signal. The first step is, therefore, to represent $y(t)$ in vector form. Denoting the Nyquist sampling interval as $T_s = 1/(2B)$ and expanding $x(t)$ in terms of the sinc function, we obtain

$$y(t) = \sum_{i=-\infty}^{\infty} \sqrt{T_s}x(iT_s)\phi_i(t) + n(t). \tag{2.15}$$

Since we are only concerned with signal components within $(0, T)$, denote $m = T/T_s$. Further, denote $x(iT_s)$ and $y(iT_s)$ by x_i and y_i, respectively, and denote the projection of $n(t)$ onto $\phi_i(t)$ by n_i. It yields the discrete representation of $y(t)$ over $(0, T)$ in Euclidean space:

$$y_i = \sqrt{T_s}\, x_i + n_i, \; i = 1, \cdots, m,$$

where $x_i \sim \mathcal{CN}(0, P_s)$ and $n_i \sim \mathcal{CN}(0, \sigma_n^2)$. It is clear that the transmission of a bandlimited Gaussian waveform is equivalent to transmitting Gaussian symbols with sinc-shaping. The samples $\{y_i\}$ are independent, and thus their mutual information is additive.

The mutual information of y_i equals $H(y_i) - H(y_i|x_i)$, where $H(y) = \mathbb{E}_y[\log f(y)]$ and $H(y|x) = \mathbb{E}_y[f(y|x)]$, with $f(y)$ denoting the probability density of y and $f(y|x)$ defined in a similar manner. For

Gaussian signals in Gaussian noise, the entropy is simply equal to the logarithm of the corresponding variance. Performing simple calculation, we obtain the mutual information of $y(t)$ over $(0, T)$ as

$$I = \sum_{i=1}^{m} [\log(\sigma_n^2 + P_s T_s) - \log(\sigma_n^2)]$$

$$= \sum_{i=1}^{m} \log \left(1 + \frac{P_s T_s}{\sigma_n^2}\right) = m \log \left(1 + \frac{P_s T_s}{\sigma_n^2}\right). \tag{2.16}$$

For a noise-like signal with very wide frequency bandwidth $B \to \infty$, we have $T_s \to 0$ and thus

$$I = \lim_{T_s \to 0} \frac{T}{T_s} \log \left(1 + \frac{P_s T_s}{\sigma_n^2}\right) = P_s T / \sigma_n^2 \quad \text{in nat.} \tag{2.17}$$

More generally, any physical variations carry mutual information, be it in the form of mechanical displacement, visible light, electromagnetic wave, or quantum entanglement. Mutual information is measured by the energy that is consumed to create such variations, normalized by the noise power spectral density.

Property 2.2.1 *The energy consumed in transmitting a noise-like Gaussian signal in AWGN is completely converted into mutual information.*

Jostein Garrder is correct, "The truth is a diamond with many facets." Equation (2.17) reveals, from one aspect, the energy nature of mutual information.

The result in (2.16) can be obtained alternatively through a physical argument. To this end, we pass the noisy signal $y(t)$ through an ideal band-pass filter to eliminate the noise component outside the frequency band of interest. The output noise power thus becomes $P_n = 2B\sigma_n^2 = \sigma_n^2/T_s$, which allows us to rewrite (2.16) as

$$I = m \log \left(1 + \frac{P_s}{P_n}\right). \tag{2.18}$$

The details are left to Problem 2.2 as an exercise.

2.3 Estimation methods

Given observations $\mathbf{x} = [x_1, \cdots, x_n]$, generated by a random mechanism, which is characterized by a set of parameters θ, we want to estimate the parameters θ. Let $f(\mathbf{x}|\theta)$ denote the PDF of the random samples $x_1, \cdots, x_n$, and let $G(\theta)$ denote the prior probability distribution of θ. Then, according the Bayes' rule, the *a posteriori* probability of θ after the receipt of the samples is given by

$$p(\theta|\mathbf{x}) = \begin{cases} \frac{f(\mathbf{x}|\theta)G(\theta)}{\sum_\theta f(\mathbf{x}|\theta)G(\theta)}, & \text{discrete parameter } \theta \\ \frac{f(\mathbf{x}|\theta)g(\theta)}{\int f(\mathbf{x}|\theta)g(\theta)d\theta}, & \text{continuous parameter } \theta \end{cases} \tag{2.19}$$

For continuous parameters θ, the only difference is to replace the PDF $g(\theta)$ by its discrete counterpart $G(\theta)$. Note that the denominators have no influence on the estimation of θ, and thus can be dropped. For parameter estimation, the PDF $f(\mathbf{x}|\theta)$ can be interpreted as the likelihood function of θ given a set of observations $\mathbf{x}$, and thus is often denoted as $L(\mathbf{x}|\theta)$. The two functions have the same form, but *different* interpretations. Thus, up to a scalar, the *a posteriori* PDF or the *a posteriori* probability can be rewritten as

$$p(\theta|\mathbf{x}) = \begin{cases} L(\mathbf{x}|\theta)G(\theta), & \text{discrete parameter } \theta \\ \\ L(\mathbf{x}|\theta)g(\theta), & \text{continuous parameter } \theta \end{cases} \quad (2.20)$$

The *a posteriori* probability provides a platform to incorporate the prior information into the samples for a more accurate estimation of θ, which is given by

$$\hat{\theta}_{\text{MAP}} = \max_{\theta} p(\theta|\mathbf{x}). \quad (2.21)$$

The maximum a posteriori (MAP) estimation procedure provides a natural way to incorporate the past experience or knowledge into the estimation process and, thus, offers better results than its maximum likelihood maximum likelihood (ML) counterpart. The key to MAP estimation is the determination of the prior probability, which is usually done in two ways. In the first, the prior is obtained from some empirical evidences or experiments. The second is based on the Jeffreys' invariance principle [24], in which the prior probability is shown to be $p(\theta) = [\det(I(\theta))]^{1/2}$, with $I(\theta)$ denoting the Fisher information matrix of the parameters θ. In wireless communication, however, prior probability plays a different role and is obtained in a different manner, as is evident in its elegant use in turbo decoding where the prior probability serves as a natural tie between the two decoders for the same message sequence. The soft output from one decoder is applied to anther decoder as the prior.

2.3.1   Maximum likelihood estimation (MLE)

When no prior information about θ is available, we simply assume θ to be uniformly distributed over the region it is defined. In this case, all the information about θ is contained in $L(\mathbf{x}|\theta)$, leading to the ML estimator, as shown by

$$\hat{\theta}_{ML} = \max_{\theta} L(\mathbf{x}|\theta). \quad (2.22)$$

The second case that leads to ML estimation is when the sample size n is very large, such that the likelihood factor dominates and the influence of $g(\theta)$ is negligible. The methodology of ML estimation was first advocated and pioneered by Ronald Fisher, and since then has dominated statistical theory for many decades. Its popularity is strongly supported by its three elegant asymptotic properties that are achievable as the sample size tends to infinity. These properties are asymptotic consistency, asymptotic efficiency, and asymptotic normality. By asymptotic consistency we mean that the MLE is asymptotically unbiased, that is,

$$\lim_{n \to \infty} \hat{\theta} \to \theta. \quad (2.23)$$

The MLE is said to be asymptotically efficient, in the sense that it achieves the Cramer–Rao bound, that is, the covariance matrix of $\hat{\theta}$

$$\lim_{n \to \infty} \text{Cov}(\hat{\theta}) = \frac{1}{n} \left(\mathbb{E} \left[\frac{\partial^2 \ln L}{\partial \theta \partial \theta^T} \right] \right)^{-1} \quad (2.24)$$

The reciprocal of this asymptotic covariance matrix, denoted by $\mathbf{I}(\theta)$, is often known as the Fisher information matrix. The Cramer–Rao bound defines the minimum lower bound that can be achieved by the MLE. For any finite sample size, the difference matrix $\text{Cov}(\hat{\theta}) - [nI(\theta)]^{-1}$ is always semidefinite. Given the asymptotic mean and covariance matrix, it is easy to represent the third property of asymptotic normality as

$$\lim_{n \to \infty} \hat{\theta} \to \mathcal{N}(\theta, [nI(\theta)]^{-1}). \quad (2.25)$$

In many applications, however, the sample size is moderate or even small, so that the influence of estimation bias is no longer negligible. As common practice, a correction factor is often introduced to an MLE to remove the bias.

2.3.2   Maximum a posteriori estimation

For a finite set of samples, the incorporation of *a priori* information can improve the estimation accuracy, thereby justifying MAP estimation. The debate on the use of use prior information $g(\theta)$ has a long history, tracing back to the era of Ronald Fisher and Egon Pearson. A main objection is that prior information is often subjective.

In wireless communication, symbol decisions are made usually based on a very limited number of observations. When the strategy of symbol-by-symbol processing is adopted, the problem falls into the framework of binary or multiple decision, which is done via hypothesis test, often relying on a single received sample. To enhance the transmission reliability, a certain algebraic structure is embedded into the transmitted data to form a coded sequence for transmission. At the receiver, the original message symbols are treated as a single entity that needs to be retrieved from the noisy (and probably also faded) received data, by exploiting the coding structure. This is an estimation problem with coding constraints. Given that symbols take discrete values, such an estimation problem naturally falls into the framework of integer programming which, alongside the constraints imposed by coding structures, make it extremely difficult to solve.

2.4   Commonly used distributions in wireless

In this section, we introduce some commonly used distributions in wireless communications, which include chi-square distributions, correlated joint gamma distributions, Wishart distributions, and distributions of the maximum and minimum eigenvalues of Gaussian MIMO channels.

2.4.1   Chi-square distributions

The chi-square distribution is widely encountered in wireless communications, and is often derivable from a number of i.i.d. random Gaussian variables. Chi-square distributions can be classified into central chi-square and non-central chi-square which, in communication, arise naturally from Rayleigh and Rician fading channels, respectively.

2.4.1.1   Central-χ^2 distribution

A good example to illustrate central chi-square distributions is a Rayleigh fading channel whose channel gain, in complex baseband form, is given as

$$h = x_1 + jx_2. \tag{2.26}$$

If $h \sim \mathcal{CN}(0, 1)$, then x_1 and x_2 are i.i.d. Gaussian random variables of zero mean and variance of $1/2$, that is, $x_i \sim \mathcal{N}(0, 1/2), i = 1, 2$. The magnitude of h follows the Rayleigh distribution, and its power follows the $\chi^2(2)$ distribution. Specifically, we have

$$2|h|^2 = \underbrace{2x_1^2}_{\chi^2(1)} + \underbrace{2x_2^2}_{\chi^2(1)} \sim \chi^2(2). \tag{2.27}$$

We normalize each real Gaussian variable so that $\sqrt{2}x_i$ has unit variance, enabling the assertion that its power is $\chi^2(1)$. In general, given m i.i.d. Gaussian variables $x_i \sim \mathcal{N}(0, \sigma_x^2)$, we have

$$\sum_{i=1}^{m} \left(\frac{x_i}{\sigma_x}\right)^2 \sim \chi^2(m) \tag{2.28}$$

or equivalently

$$\sum_{i=1}^{m} x_i^2 \sim \sigma_x^2 \chi^2(m). \tag{2.29}$$

The symbol $\chi^2(m)$ means a chi-square variable of m degrees of freedom (DFs). Therefore, a chi-square variable of an integral DF m can be considered to have been generated from m i.i.d. Gaussian variables.

In the above, the DF m is assumed to be an integer in order to show how a chi-square variable is related to the Gaussian variables. In fact, we can extend m to an arbitrary real number $\nu > 0$. Suppose $x \sim \chi^2(\nu)$. Then, its PDF is given by

$$\chi^2(\nu) \sim \frac{1}{2\Gamma(\nu/2)} \left(\frac{x}{2}\right)^{\frac{\nu}{2}-1} e^{-x/2}, \; x \geq 0, \tag{2.30}$$

where $\Gamma(x)$ is the gamma function. Sometimes, it is more convenient to work on its Fourier transform, referred to as the *characteristic function* (CHF). For $x \sim \chi^2(\nu)$, it is straightforward to work out its CHF $\Phi_x(jt)$ by definition, yielding

$$\Phi_x(jt) \triangleq \mathbb{E}[e^{jtx}] = (1 - 2jt)^{-\nu/2}. \tag{2.31}$$

2.4.1.2 *Non-central χ^2 distribution*

A natural way to introduce non-central chi-square distributions is to consider a Riciain fading channel where there exists a line-of-sight path of constant gain μ, in addition to a diffuse random component x, so that the channel gain is equal to

$$h = x + \mu \tag{2.32}$$

where $x \sim \mathcal{CN}(0, \sigma_x^2)$. Similar to the treatment of Rayleigh fading, we normalize h by σ_x to have unit variance, resulting in $h/\sigma_x \sim \mathcal{CN}(\mu/\sigma_x, 1)$. It follows that

$$2|h|^2 \sim \sigma_x^2 \, \chi^2(2, \lambda), \tag{2.33}$$

where the non-centrality parameter is given by $\lambda = 2|\mu|^2/\sigma_x^2$.

The DF of non-central chi-square distributions can be also extended to any arbitrary positive real number ν. In general, the PDF and CHF of $\chi^2(\nu, \lambda)$ are given by Johnson and Kotz [1]

$$f(x) = \frac{1}{2}\left(\frac{x}{\lambda}\right)^{\frac{1}{4}(\nu-2)} I_{\frac{1}{2}(\nu-2)}(\sqrt{\lambda x}\,) \exp\left[-\frac{1}{2}(\lambda + x)\right], x > 0$$

$$\phi_x(jt) \triangleq \mathbb{E}[e^{jtx}] = (1 - j2t)^{-\frac{\nu}{2}} \exp\left(\frac{jt\lambda}{1 - j2t}\right). \tag{2.34}$$

Here, $I_n(z)$ is a modified Bessel function of the first kind with order n. By setting $\lambda = 0$, these expressions reduce to the case of the central chi-square distribution.

2.4.1.3 *Properties of chi-square distribution*

The chi-square distribution has a number of good properties that make it very popular in practical applications. Recall that the Gaussian distribution is the only distribution that enjoys the closure property, in the sense that any linear operation on a Gaussian variable remains Gaussian. The chi-square distribution partly shares such a property.

Property 2.4.1 *An equally weighted combination of independent chi-square variables is still a chi-square variable. Specifically, for m independent chi-square variables $x_i \sim \chi^2(\nu_i, \lambda_i)$, their equally weight sum is distributed as*

$$x = x_1 + \cdots + x_m \sim \chi^2(\nu, \lambda)$$

with the new degrees of freedom (DF) and non-centrality parameter given by

$$\nu = \nu_1 + \cdots + \nu_m$$
$$\lambda = \lambda_1 + \cdots + \lambda_m. \tag{2.35}$$

In the above, the equal weighting factor is set to unity for notational simplicity. This property is not applied to the general case with unequal weighting, and hence the term semiclosure property.

Property 2.4.2 *The mean, variance, and the first four central moments (i.e., the moment about the mean) of $x \sim \chi^2(\nu, \lambda)$ are given by Kendall and Stuart [2]*

$$\mathbb{E}[x] = \nu + \lambda$$
$$\mathrm{var}[x] = \sigma_x^2 = 2(\nu + 2\lambda)$$
$$\mathbb{E}[\|x - \mathbb{E}[x]\|^3] = 8(\nu + 3\lambda)$$
$$\mathbb{E}[\|x - \mathbb{E}[x]\|^4] = 48(\nu + 4\lambda) + 12(\nu + 2\lambda)^2 \tag{2.36}$$

In the previous study of chi-square distributions, we assumed a set of independent Gaussian variables. We now investigate the properties of chi-square variables derived from a correlated Gaussian vector. A direct, but useful, extension of (2.28), due to Giri [3], is stated below.

Property 2.4.3 *If $x \sim \mathcal{CN}_m(a, R_x)$, then $2x^\dagger R_x^{-1} x \sim \chi^2(2m, 2a^\dagger R_x^{-1} a)$.*

This result is easy to understand since $R_x^{-1/2}$ is used to normalize the vector x to produce m i.i.d. complex variables each of unit variance. Each complex variable possesses two DFs and each DF has variance $1/2$, resulting in the factor of 2 in the above expression.

Property 2.4.4 *If $x \sim \mathcal{CN}(v, R)$ and Q is a Hermitian matrix, the CHF of the quadratic form $z = x^\dagger Q x$ is given by Turin [4]*

$$\phi_z(jt) = \mathbb{E}[e^{jtz}] = \det(I - jtRQ)^{-1} \exp(jt v^\dagger Q (I - jtRQ)^{-1} v). \tag{2.37}$$

Property 2.4.5 *The PDF of the power sum of independent Rayleigh variables is always expressible as a linear combination of chi-square PDFs.*

The power of each Rayleigh faded variable follows a $\chi^2(2)$ distribution. Consider the sum of L independent $\chi^2(2)$ variables x_i, namely, $x = x_1 + \cdots + x_L$, where $f_i(x) = \frac{1}{\alpha_i} e^{-x/\alpha_i}$. For simplicity, we assume all α_i's are different. The treatment for the case with identical α_i is similar, except that we need to take derivatives to determine the higher order terms. To determine the PDF of x, we use the CHF method by calculating

$$\phi_x(jt) = \prod_{k=1}^{L} (1 - jt\alpha_k)^{-1}$$

$$= \sum_{k=1}^{L} \frac{b_k}{1 - jt\alpha_k}, \tag{2.38}$$

where

$$b_k = \prod_{i=1, i \neq k}^{L} (1 - jt\alpha_i)^{-1} \big|_{t:1 - jt\alpha_k = 0}$$

$$= \prod_{i=1, i \neq k}^{L} \left(1 - \frac{\alpha_i}{\alpha_k}\right)^{-1}. \tag{2.39}$$

Note $\phi_x(0) = \sum_{k=1}^{L} b_k = 1$. The inverse transform the CHF for the PDF of x yields

$$f(x) = \sum_{k=1}^{L} \frac{b_k}{\alpha_k} e^{-x/\alpha_k}. \tag{2.40}$$

The semiclosure property of a chi-square distribution asserts that an equally weighted sum of independent chi-square variables is still a chi-square variable. In wireless communication, however, we are often confronted with an *unequally* weighted sum of chi-square variables, and a natural question is whether we can accurately approximate this sum by a single chi-square variable. The answer is in the affirmative, and the approximation method is widely adopted by the statistical community.

Property 2.4.6   Approximation of a χ^2 Sum
The first three central moments of a chi-square variable $\chi^2(v, \lambda)$ are given by $\mu_1 = v + \lambda, \mu_2 = \sigma^2 = 2(v + 2\lambda), \mu_3 = 8(v + 3\lambda)$ (Pearson, 1959). Suppose we want to approximate

$$\sum_{i=1}^{k} c_i \chi^2(v_i) \sim c\chi^2(v),$$

with the first two moments. Then, the fitting coefficients are given by

$$v = \frac{\left(\sum_{i=1}^{k} c_i v_i\right)^2}{\sum_{i=1}^{k} c_i^2 v_i} \quad c = \frac{\sum_{i=1}^{k} c_i^2 v_i}{\sum_{i=1}^{k} c_i v_i}.$$

Porteous's lemma, stated below, often finds applications in communication.

Property 2.4.7   *If $u \sim \chi^2(k)$, then*

$$\mathbb{E}[\ln u] = \ln k - \frac{1}{k} - \frac{1}{3k^2} + \frac{2}{15k^4} + O(k^{-6}). \tag{2.41}$$

The extension of this property to non-central chi-square distributions, due to [5], is given below:

Property 2.4.8 Expected Logarithm of chi-square variable
Let $x \sim \chi^2(2m, 2\mu)$. We assert that [5]

$$\mathbb{E}[\ln x] = g_m(\mu)$$

$$\mathbb{E}[x^{-n}] = \frac{(-1)^{n-1}}{(n-1)!} \, g_{m-n}^{(n)}(\mu), \; m > n, \tag{2.42}$$

and for integer $m \in N$,

$$g_m(\xi) = \begin{cases} \ln \xi - Ei(-\xi) + \sum_{j=1}^{m-1} \left(-\frac{1}{\xi}\right)^j \left[e^{-\xi}(j-1)! - \frac{(m-1)!}{j(m-1-j)!} \right], & \xi > 0 \\ \psi(m), & \xi = 0 \end{cases} \tag{2.43}$$

where the exponential integral function $Ei(\cdot)$ and Euler's function $\psi(\cdot)$ are defined by

$$Ei(-\xi) = -\int_\xi^\infty \frac{e^{-t}}{t} dt, \; \xi > 0,$$

$$\psi(m) = -\gamma + \sum_{j=1}^{m-1} \frac{1}{j}, \; m \in N, \tag{2.44}$$

with $\gamma \approx 0.577$ denoting Euler's constant. Note that $g_m(\xi)$ is a continuous, monotonically increasing concave function over $0 \le \xi < \infty$. In particular

$$g_m^{(1)}(\xi) = e^{-\xi} \sum_{i=0}^\infty \frac{1}{k!} \frac{\xi^k}{k+m},$$

$$g_m^{(2)}(\xi) = -e^{-\xi} \sum_{i=0}^\infty \frac{1}{k!} \frac{\xi^k}{(k+m)(k+m+1)},$$

$$g_m^{(3)}(\xi) = e^{-\xi} \sum_{i=0}^\infty \frac{1}{k!} \frac{2\xi^k}{(k+m)(k+m+1)(k+m+2)}. \tag{2.45}$$

2.4.2 Gamma distribution

A gamma random variable is closely related to the corresponding central chi-square by a scalar. All the properties of the chi-square distribution are shared by the gamma distribution with PDF:

$$f(x) = \frac{a}{\Gamma(m)} (ax)^{m-1} e^{-ax}, x > 0$$

where $a > 0$ and $m > 0$. It has a form similar to that of the chi-square distribution. In fact, if we simply denote it by $G(m, a)$, it can be readily related to the latter by

$$G(m, a) \sim \frac{1}{2a} \chi^2(2\nu),$$

which indicates that up to a scale, the gamma distribution is the same as the central chi-square distribution.

For a random variable with range over $(0, \infty)$, we can take a gamma-type expansion [6] in terms of the Laguerre polynomials generated from the gamma distribution.

2.4.3   Nakagami distribution

The Nakagami distribution, in nature, arises from taking the square-root of the corresponding chi-square or gamma variable. In terms of distribution, it is nothing but a chi-variable. Its popularity lies in its good fit to many empirical data over certain fading channels, a discovery and subsequent in-depth statistical analysis made by Nakagami. Such channels are thus known as *Nakagami fading*, to recognize his contributions.

To be specific, if z follows the Nakagami distribution with *fading parameter* m and power $\mathbb{E}[z^2] = \Omega$, or simply denoted as $z \sim \mathcal{NK}(m, \Omega)$, then it is related to a chi variable [7] by

$$\mathcal{NK}(m, \Omega) \sim \sqrt{\frac{\Omega}{2m}} \chi(2m), \tag{2.46}$$

whose PDF, when written explicitly, is given by

$$f(z) = \frac{2z^{2m-1}}{\Gamma(m)} \left(\frac{m}{\Omega}\right)^m \exp\left(-\frac{mz^2}{\Omega}\right), \; z > 0. \tag{2.47}$$

Each fading parameter counts on the number of complex variables that would have been used to generate the Nakagami distribution. Thus, the fading parameter m corresponds to the DF of $2m$. The quantity $\Omega/(2m)$ physically represents the power per DF. The correctness of (2.46) can be easily examined by evaluating the expected power of the two sides. The PDF of $\xi \sim \chi(v)$ is given by

$$f_{\chi(v)}(\xi) = \frac{\xi^{v-1} e^{-\xi^2/2}}{\Gamma(v/2) \, 2^{(v/2)-1}}.$$

Property 2.4.9 *If $\xi \sim \chi(v)$, then $\xi^2 \sim \chi^2(v)$.*

Property 2.4.10 *If $\xi \sim \chi(v)$, its non-central moments (i.e., about the origin) is given by Wikipedia [8]*

$$\mathbb{E}[\xi^k] = 2^{k/2} \frac{\Gamma(\frac{v+k}{2})}{\Gamma(\frac{v}{2})}. \tag{2.48}$$

Various useful properties of Nakagami distributions have been derived by its originator, and the interested reader is referred to his seminal paper [9] for details. Rayleigh distribution is a special case of the Nakagami distribution with $m = 1$. Namely, if $z = x + jy \sim \mathcal{CN}(0, \Omega)$, then $|z|$ is Rayleigh-distributed and is related to $\chi(2)$ by

$$|z| \sim \sqrt{(\Omega/2)} \, \chi(2) = \sigma \chi(2), \tag{2.49}$$

where σ^2 denotes the variance of x or y.

2.4.4   Wishart distribution

In the history of statistics, the chi-square distribution is typically derived from the estimation of the variance from i.i.d. samples generated by a Gaussian population, a direct result of the semiclosure property.

The Wishart distribution is a multivariate extension of the chi-square distribution, arising from the estimation of the covariance matrix based on i.i.d. vector samples from a multivariate Gaussian population. Given N independent samples $\mathbf{x}_i, i = 1, \cdots, N$ from a vector complex Gaussian population $\mathcal{CN}_p(\mu, \boldsymbol{\Sigma})$, and denoting $\bar{\mathbf{x}} = (1/N) \sum_{i=1}^{N} \mathbf{x}_i$, then the Hermitian sample covariance

$$\mathbf{A} = \sum_{i=1}^{N} (\mathbf{x}_i - \bar{\mathbf{x}})(\mathbf{x}_i - \overline{\mathbf{x}_i})^{\dagger} \tag{2.50}$$

is complex and Wishart-distributed. More specifically, by denoting $n = N - 1$, the PDF of $\mathbf{A}$ is given by

$$f(\mathbf{A}) = \frac{|\mathbf{A}|^{n-p}}{\pi^{\frac{1}{2}p(p-1)} \Gamma(n) \cdots \Gamma(n-p+1) |\boldsymbol{\Sigma}|^n} \exp(-\mathrm{tr}(\boldsymbol{\Sigma}^{-1}\mathbf{A})), \tag{2.51}$$

which is often briefly denoted as $\mathbf{A} \sim \mathcal{CW}_n(p, \boldsymbol{\Sigma})$.

Property 2.4.11 *If i.i.d. variables* $c_k \sim \mathcal{CN}_L(\mathbf{0}, \mathbf{R}_c)$, *then* $\mathbf{A} = \sum_{k=1}^{M} \mathbf{c}_k \mathbf{c}_k^{\dagger} \sim \mathcal{CW}_L(M, \mathbf{R}_c)$. *We assert*

$$\xi = (\mathbf{s}^{\dagger}\mathbf{A}^{-1}\mathbf{s})^{-1} \sim \mathcal{CW}_1(M - L + 1, (\mathbf{s}^{\dagger}\mathbf{R}_c^{-1}\mathbf{s})^{-1}) \tag{2.52}$$

or, more specifically

$$f_{\xi|\mathbf{s}}(z|\mathbf{s}) = \frac{\alpha^{p+1}}{p!} z^p e^{-\alpha z} = \frac{(-1)^{p+1}}{p!} z^p \frac{d^{p+1}}{dz^{p+1}}(e^{-\alpha z}) \tag{2.53}$$

with $p = M - L$ *and* $\alpha = \mathbf{s}^{\dagger}\mathbf{R}_c^{-1}\mathbf{s}$.

This useful property is due to A.T. James [10, 11].

Property 2.4.12 *If* $\mathbf{A} \sim \mathcal{CW}_m(N, \mathbf{R})$, *then*

$$\det(\mathbf{A}) \sim 2^{-m} \det(\mathbf{R}) \chi^2(2N) \chi^2(2N - 2) \cdots \chi^2(2N - 2m + 2). \tag{2.54}$$

For a proof, see Problem 2.3.

2.4.4.1 PDF of sample eigenvalues

The joint density function of the eigenvalues $\ell_1 > \ell_2 > \cdots > \ell_p > 0$ of a sample covariance matrix $\mathbf{S} \sim \mathcal{CCW}_p(n, \boldsymbol{\Sigma}), (n > p)$ is given by

$$\frac{\pi^{p^2/2} \, 2^{-np/2} (\det\boldsymbol{\Sigma})^{-n/2}}{\Gamma_p\left(\frac{n}{2}\right) \Gamma_p\left(\frac{p}{2}\right)} \prod_{i=1}^{p} \ell_i^{(n-p-1)/2} \prod_{j:j>i}^{p} (\ell_i - \ell_j) \int_{\mathcal{O}_p} \mathrm{etr}\left(-\frac{1}{2} n \boldsymbol{\Sigma}^{-1} \mathbf{H} \mathbf{L} \mathbf{H}^T\right) (d\mathbf{H}). \tag{2.55}$$

For $\boldsymbol{\Sigma} = \lambda \mathbf{I}$, the integral over the orthonormal group $\mathcal{O}_p$ is simplified to $\exp(-\lambda^{-1} \sum_{i=1}^{p} \ell_i)$ and, thus, the above expression reduces to

$$f(\ell_1, \cdots, \ell_p) = \frac{\pi^{p^2/2} \, 2^{-np/2}}{\Gamma_p\left(\frac{n}{2}\right) \Gamma_p\left(\frac{p}{2}\right)} \exp\left(-\frac{1}{\lambda} \sum_{i=1}^{p} \ell_i\right) \prod_{i=1}^{p} \ell_i^{(n-p-1)/2} \prod_{j:j>i}^{p} (\ell_i - \ell_j). \tag{2.56}$$

2.5   The calculus of variations

Calculus of variations is a systematic approach to solving extremal problems for a functional. It traces back to the Brachistochrone problem, initiated by an optical optimization problem introduced in 1662 by Piere de Fermat. The problem was subsequently formulated in 1696 as one of finding the quickest descent path, and was solved by Johann Bernoulli, among the others including Newton and Jacob Beroulli. All these solutions, however, lacked the generality in methodology. In 1774, Leonhard Euler extended the most descending problem to a more general one of functional optimization involving an integral objective function, and provided a solution. The term *calculus of variations* was coined by Euler. The method of Euler was significantly improved in 1755 by a young French mathematician J.-L. Lagrange. Euler employed Lagrange's viewpoint to summarize similar optimization problems, resulting in the first textbook on calculus of variations, published in 1744 [11, 12].

In communication, we only come across the simplest class of variational problems, in which the unknown is a continuously differentiable scalar function, and the functional to be minimized depends upon at most its first derivative. The basic minimization problem, then, is to determine a suitable function $y = y(x) \in C^1[a, b]$ that minimizes the objective functional

$$J[y] = \int_a^b L(x, y(x), y'(x))dx. \tag{2.57}$$

The upper and lower bounds a and b are usually determined by the boundary conditions on $y(x)$. The integrand L is known as the *Lagrangian* for the variational problem, in honor of Lagrange, one of the main founders of the subject. We usually assume that the Lagrangian $L(x, y, y')$ is a reasonably smooth function of all three of its (scalar) arguments x, y, and y'. Unlike traditional optimization problems, we seek a functional $y(x)$ to minimize or maximize the objective function $J[y]$. Lagrange modified the method by Euler by wisely introducing a perturbation function and exploiting the fact that the objective functional must have an extremum when the perturbation becomes zero [29]. It turns out to be a very elegant solution called the *Euler–Lagrange equation*, given by

$$\frac{d}{dx}\left[\frac{\partial L}{\partial y'}\right] - \frac{\partial L}{\partial y} = 0. \tag{2.58}$$

In differentiation, $y(x)$ and $y'(x)$ are treated just as variables.

For illustration of the Euler–Lagrange equation, consider the famous problem of power allocation to maximize the capacity of a dispersive channel in wireless communications.

✍ **Example 2.3** ──

The capacity of a dispersive channel with frequency response $H(f)$ and transmit power allocation filter $P(f)$ is given by

$$C = \int_{-\infty}^{\infty} \ln(1 + \rho|P(f)|^2|H(f)|^2)df,$$

where ρ denotes the SNR at the transmitter. We need to optimize the transmit filter $P(f)$ such that

$$\max_{|P(f)|^2} C$$

$$\text{s.t.} \int_{-\infty}^{\infty} |P(f)|^2 df$$

To be in the same form as required by the Euler–Lagrange equation, we combine the capacity and the constraint using the Lagrange multiplier λ to construct a single objective function

$$J[|P(f)|^2] = \int_{-\infty}^{\infty} \underbrace{[\ln(1 + \rho|P(f)|^2|H(f)|^2) + \lambda|P(f)|^2]}_{L} df.$$

Denoting $x = f$, $y(f) = |P(f)|^2$, and $z = |H(f)|^2$, $L = \ln(1 + \rho \, yz) + \lambda \, y$. It follows from (2.58) that

$$\frac{\partial L}{\partial y} - \frac{d}{dx}\frac{\partial L}{\partial y'} = 0$$

which, after simplification, leads to

$$\frac{\rho z}{1 + \rho yz} - \lambda = 0. \tag{2.59}$$

Or equivalently,

$$|P(f)|^2 = \max\left\{0, \left[\frac{1}{\lambda} - \frac{1}{\rho|H(f)|^2}\right]\right\} \triangleq \left[\frac{1}{\lambda} - \frac{1}{\rho|H(f)|^2}\right]^+. \tag{2.60}$$

This is the popular water-filling formula in a continuous frequency version.   ◯

2.6   Two inequalities for optimization

Two inequalities are often useful in system optimization for communication. One pertains to Rayleigh quotients and the other to determinants.

2.6.1   Inequality for Rayleigh quotient

Let $\mathbf{A}$ be an $m \times m$ Hermitian matrix with minimum eigenvalue $\lambda_{\min}$ and maximum eigenvalue $\lambda_{\max}$. Then, for an arbitrary $m \times 1$ nonzero vector $\mathbf{x}$, the Rayleigh quotient is bounded by

$$\lambda_{\min} \le \frac{\mathbf{x}^\dagger \mathbf{A} \mathbf{x}}{\mathbf{x}^\dagger \mathbf{x}} \le \lambda_{\max}. \tag{2.61}$$

The upper and lower bounds are achieved when $\mathbf{x}$ is equal to, up to a scale factor, the eigenvector corresponding to $\lambda_{\max}$ and $\lambda_{\min}$, respectively.

2.6.2   Hadamard inequality

Calculus of variations is a powerful means to handle optimization problems involving continuous functions. In discrete optimization, we may need the Hardamard inequality, which is stated below:

For any positive semidefinite Hermitian matrix $\mathbf{A} = [a_{ij}]$ of $M \times M$ dimensions, we have

$$\det(\mathbf{A}) \le \prod_{i=1}^{M} a_{ii}. \tag{2.62}$$

The equality holds if and only if $\mathbf{A}$ is diagonal.

✍ **Example 2.4** _____

Given an $M \times M$ positive semidefinite Hermitian matrix $\mathbf{A} = [a_{ij}]$, find a square matrix $\mathbf{X}$ such that $\text{tr}(\mathbf{X}) = 1$ for which

$$C = \log \, \det(\mathbf{I} + \mathbf{AX}) \tag{2.63}$$

is maximized.

Solution

Let us see how to use the Hadamard inequality to simplify the problem. We eigen-decompose the Hermitian matrix $\mathbf{A}$ to give $\mathbf{A} = \mathbf{U}^\dagger \mathbf{D} \mathbf{U}$, where $\mathbf{U}$ is the unitary matrix of eigenvectors, and $\mathbf{D}$ is the diagonal matrix formed by the eigenvalues $d_i, i = 1, \cdots, M$ of $\mathbf{A}$. Hence, we can write

$$\det(\mathbf{I} + \mathbf{AX}) = \det(\mathbf{I} + \mathbf{UDU}^\dagger \mathbf{X}) = \det(\mathbf{I} + \mathbf{DU}^\dagger \mathbf{XU}). \tag{2.64}$$

Denote $\mathbf{Y} = \mathbf{U}^\dagger \mathbf{XU}$. Then, using the property of trace operations, we have $\text{tr}(\mathbf{Y}) = \text{tr}(\mathbf{XUU}^\dagger) = \text{tr}(\mathbf{X}) = 1$. Since $\mathbf{D}$ is diagonal, it follows from the Hadamard inequality that $\mathbf{Y}$ must also be diagonal. In other words, the optimizer $\mathbf{X}$ must have the same eigenvectors as $\mathbf{A}$. It yields

$$\det(\mathbf{I} + \mathbf{DY}) \le \prod_{i=1}^{M}(1 + d_i y_{ii}),$$

whereby the original optimization problem is simplified to

$$\max \, \log \prod_{i=1}^{M}(1 + d_i y_{ii})$$

$$\text{s.t.} \sum_{i=1}^{M} y_{ii} = 1, y_{ii} \ge 0. \tag{2.65}$$

In its present form, the problem can be easily solved by using the method of Lagrange multiplier, and is therefore left to the reader as an exercise.   ○

〜〜✍

2.7   Q-function

The Q-function is an indispensable tool widely used in wireless systems design and their performance evaluation on AWGN channels. It is defined by a Gaussian integral

$$Q(x) = \int_x^\infty \frac{1}{\sqrt{2\pi}} e^{-\frac{x^2}{2}} dx. \tag{2.66}$$

However, it is found that many analyses can be done equally well, or even better, in the Fourier transform domain involving CHFs. Another equally popular function is the complementary error function, defined as

$$\text{erfc}(x) \stackrel{\Delta}{=} 1 - \int_{-x}^{x} \frac{1}{\sqrt{\pi}} e^{-u^2} du. \tag{2.67}$$

The two functions are closely related, as shown by

$$\text{erfc}(x) = 2Q(\sqrt{2}\,x), \text{ or } Q(x) = (1/2)\text{erfc}(x/\sqrt{2}). \tag{2.68}$$

The Q-function is a special function, and various approximate polynomials are available allowing its simple calculation to achieve an accuracy of up to 10^{-8} ([13], p. 932).

A discovery by Weinstein in 1974 builds a bridge, facilitating the use of the CHF method. In particular, he relates the Q-function to an exponential integral, as shown by Craig [15–17]

$$Q(x) = \begin{cases} \dfrac{1}{\pi}\displaystyle\int_0^{\pi/2} \exp\left(-\dfrac{x^2}{2\sin^2\theta}\right) d\theta, \ x \geq 0; \\[3mm] 1 - \dfrac{1}{\pi}\displaystyle\int_0^{\pi/2} \exp\left(-\dfrac{x^2}{2\sin^2\theta}\right) d\theta, \ x \leq 0. \end{cases} \tag{2.69}$$

$$Q^2(x) = \frac{1}{\pi}\int_0^{\pi/4} \exp\left(-\frac{x^2}{2\sin^2\theta}\right) d\theta, \ x \geq 0. \tag{2.70}$$

From (2.69) for $x > 0$, the integrand is an increasing function over $(0, \frac{\pi}{2})$. Thus, the Q-function for $x > 0$ can be bounded in different ways.

Property 2.7.1 *If we upper-bound the integrand its value at $\theta = \pi/2$, it yields*

$$Q(x) < \frac{1}{2}e^{-x^2/2}, \text{for } x > 0. \tag{2.71}$$

If, instead, the integrand is bounded segmentally to improve the bounding tightness, we obtain the following results:

Property 2.7.2 *For $x > 0$, we have*

$$Q(x) < \frac{1}{4}e^{-x^2} + \frac{1}{4}e^{\{-x^2/2\}}, \ (\theta_1 = \pi/4)$$

$$Q(x) < \frac{1}{6}e^{-2x^2} + \frac{1}{6}e^{-2x^2/3} + \frac{1}{6}e^{\{-x^2/2\}}, \ (\theta_1 = \pi/6, \theta_2 = \pi/3)$$

$$Q(x) \approx \frac{1}{12}e^{-x^2/2} + \frac{1}{4}e^{-2x^2/3}. \tag{2.72}$$

The approximation on the third line is quite accurate over the range 0–13 dB.

The Q-function has a special relationship with the chi-square distribution. To see this, set $x = \sqrt{\gamma}$ in (2.66) and change the integration variable $z = u^2$. It turns out that

$$Q(\sqrt{\gamma}) = \frac{1}{2}\int_\gamma^\infty \frac{1}{\sqrt{2\pi}} z^{-1/2} e^{-z/2} \, dz. \tag{2.73}$$

Recognize that the integrand above is the PDF of $\chi^2(1)$, enabling us to further write

$$Q(\sqrt{\gamma}) = \frac{1}{2}(1 - F_{\chi^2(1)}(\gamma)) \tag{2.74}$$

with $F_{\chi^2(1)}$ denoting the CDF of $\chi^2(1)$ variable.

Property 2.7.3 *Suppose now $\gamma \sim \gamma_0 \chi^2(2)$; namely, its PDF equals $f(\gamma) = \frac{1}{2\gamma_0} e^{-\frac{\gamma}{2\gamma_0}}$. Then*

$$\mathbb{E}[Q(\sqrt{\gamma}\,)] = \frac{1}{2}\left[1 - \int_0^\infty F_{\chi^2(1)}(\gamma) f(\gamma) d\gamma\right]$$

$$= \frac{1}{2}\left[1 - \int_0^\infty f_{\chi^2(1)}(\gamma) e^{-\frac{\gamma}{2\gamma_0}} d\gamma\right], \tag{2.75}$$

where the second line is obtained by using integration by parts. We recognize that the integral is essentially the CHF of the $\chi^2(1)$ variable with the transform variable $z = -\frac{1}{2\gamma_0}$, and is thus equal to $(1 - 2z)^{-1/2}|_{z=-\frac{1}{2\gamma_0}}$. After simplification, it yields

$$\mathbb{E}[Q(\sqrt{\gamma}\,)] = \frac{1}{2}\left(1 - \sqrt{\frac{\gamma_0}{1 + \gamma_0}}\right). \tag{2.76}$$

This provides a novel perspective to visualize the popular diversity formula, which will be addressed in Chapter 7.

2.8   The CHF method and its skilful applications

In the analysis of wireless communication systems, we often deal with the PDF of a sum of random variables. Three tools offer a link to the CHF domain.

2.8.1   Gil-Pelaez's lemma

In wireless communication, it is often required to determine the probability of the type $\Pr\{D < 0\}$, where D is a decision variable with CHF $\phi_D(jt)$. Such a probabilistic evaluation can be converted to a much simpler calculation on the CHF domain via the Gil-Pelaez lemma [18], given by

$$\Pr\{D < 0\} = \frac{1}{2} - \frac{1}{2\pi}\int_{-\infty}^\infty \frac{\Im\{\phi_D(jt)\}}{t} dt. \tag{2.77}$$

Here and throughout the book, $\Re\{z\}$ and $\Im\{z\}$ are used to denote the real and imaginary parts of a complex number z. It is straightforward to verify this assertion by invoking the Fourier transform of the unit step function. In many applications with Rayleigh fading, one can work out the integral for a closed-form solution by virtue of the Residue theorem.

2.8.2   Random variables in denominators

Consider the determination of the expected value of y, which is a function of random variables $x_i, i = 1, \cdots, m$, such that

$$y = \frac{\prod_{i=1}^m g_i(x_i)}{\left(\sum_{i=1}^m x_i\right)^m} \tag{2.78}$$

where $g_i(\cdot)$ are algebraic functions. The random variables appearing in the denominator constitute a major difficulty to the treatment of expectation. We must remove them from the denominator. By invoking the identity [19]

$$\left(\sum_{i=1}^n x_i\right)^{-m} = \frac{1}{\Gamma(m)}\int_0^\infty z^{m-1} e^{-z\sum_{i=1}^n x_i} dz, m > 0,$$

we can rewrite

$$y = \frac{1}{\Gamma(m)} \int_0^\infty z^{m-1} \left[\prod_{i=1}^n g_i(x_i) e^{-z x_i} \right] dz. \tag{2.79}$$

The expectation of y is now simplified to finding the CHFs of $g_i(x_i)$. This formula and the methodology behind it are particularly useful in the outage performance analysis of various systems.

2.8.3 Parseval's theorem

Parseval's theorem simply reveals the invariance of the inner product after a congruence mapping. In wireless communication, we are particularly interested in a set of basis functions defined by complex exponentials, which leads to the Fourier transform. Suppose that two temporal functions $f(t)$ and $g(t)$ have Fourier transforms $F(\omega)$ and $G(\omega)$, respectively. Then, we have

$$\int_{-\infty}^\infty f(t) g^*(t) dt = \frac{1}{2\pi} \int_{-\infty}^\infty F(\omega) G^*(\omega) d\omega. \tag{2.80}$$

In the special case with $g(t) = f(t)$, the above expression simply states that the energy evaluated in the frequency domain is the same as that evaluated in the time domain. This is the law of energy preservation over different congruence spaces.

The above formula will be used in the error performance analysis of equal-gain combining systems.

2.9 Matrix operations and differentiation

The following results are useful in the development and evaluation of various multiple antenna systems.

2.9.1 Decomposition of a special determinant

Suppose that matrix $\mathbf{S}_k$ takes the form $\mathbf{S}_k = \mathbf{I} + \rho \sum_{i=1}^k \mathbf{x}_i \mathbf{x}_i^\dagger$ with $\mathbf{x}_i$ denoting a column vector and ρ a constant. We have [20]

$$\det(\mathbf{S}_k) = (1 + \rho \mathbf{x}_1^\dagger \mathbf{x}_1)(1 + \rho \mathbf{x}_2^\dagger \mathbf{S}_1^{-1} \mathbf{x}_2) \cdots (1 + \rho \mathbf{x}_k^\dagger \mathbf{S}_k^{-1} \mathbf{x}_k),$$

which lays a solid foundation for the well-known V-BLAST receiver for MIMO channels, besides its many other applications.

2.9.2 Higher order derivations

Let $u(x)$ and $v(x)$ be n times differentiable. The Leibniz rule for higher order differentiation is as follows:

$$\frac{d^n(uv)}{dx^n} = \sum_{i=0}^n \binom{n}{i} \frac{d^i v}{dx^i} \frac{d^{n-i} u}{dx^{n-i}}, \tag{2.81}$$

where

$$\binom{n}{0} = 1, \quad \frac{d^0 u}{dx^0} = u.$$

This rule for differentiation is useful in the performance analysis of many diversity systems.

2.9.3   Kronecker product

Let $\mathbf{A}$ be $n \times p$ and $\mathbf{B}$ be $m \times q$. Then, the $mn \times pq$ matrix

$$\mathbf{A} \otimes \mathbf{B} = \begin{pmatrix} a_{11}\mathbf{B} & a_{12}\mathbf{B} & \cdots & a_{1p}\mathbf{B} \\ a_{21}\mathbf{B} & a_{22}\mathbf{B} & \cdots & a_{2p}\mathbf{B} \\ \vdots & & & \vdots \\ a_{n1}\mathbf{B} & a_{n2}\mathbf{B} & \cdots & a_{np}\mathbf{B} \end{pmatrix}$$

is called the Kronecker product of $\mathbf{A}$ and $\mathbf{B}$. The Kronecker product has the following properties:

$$\mathbf{A} \otimes (\mathbf{B} \otimes \mathbf{C}) = (\mathbf{A} \otimes \mathbf{B}) \otimes \mathbf{C}$$

$$\mathbf{A} \otimes (\mathbf{B} + \mathbf{C}) = (\mathbf{A} \otimes \mathbf{B}) + (\mathbf{B} \otimes \mathbf{C})$$

$$(\mathbf{A} + \mathbf{B}) \otimes \mathbf{C} = (\mathbf{A} \otimes \mathbf{C}) + (\mathbf{B} \otimes \mathbf{C})$$

$$(\mathbf{A} \otimes \mathbf{B})(\mathbf{C} \otimes \mathbf{D}) = (\mathbf{AC}) \otimes (\mathbf{BD})$$

$$(\mathbf{A} \otimes \mathbf{B})^{\dagger} = \mathbf{A}^{\dagger} \otimes \mathbf{B}^{\dagger}$$

$$(\mathbf{A} \otimes \mathbf{B})^{-1} = \mathbf{A}^{-1} \otimes \mathbf{B}^{-1}$$

$$|\mathbf{A} \otimes \mathbf{B}| = |\mathbf{A}|^{n}|\mathbf{B}|^{m} \quad (\mathbf{A} : m \times m, \ \mathbf{B} : n \times n)$$

$$tr(\mathbf{A} \otimes \mathbf{B}) = tr(\mathbf{A})tr(\mathbf{B})$$

$$rank(\mathbf{A} \otimes \mathbf{B}) = rank(\mathbf{A})rank(\mathbf{B})$$

$$vec(\mathbf{AXB}) = (\mathbf{B}^{T} \otimes \mathbf{A})vec(\mathbf{X})$$

$$tr(\mathbf{A}^{T}\mathbf{B}) = (vec\mathbf{A})^{T}vec\mathbf{B}$$

$$tr(\mathbf{AX}^{T}\mathbf{BXC}) = (vec\mathbf{X})^{T}(\mathbf{CA} \otimes \mathbf{B}^{T})vec\mathbf{X}.$$

For details, the interested reader is referred to [21].

2.10   Additional reading

So far, we have briefly reviewed mathematical results for subsequent use in this book. For more insightful understanding, readers are referred to other books and papers of relevance. For example, reader is referred to [24, 33–34] for early work on complex Gaussian distributions, referred to [9–10, 32, 33] for asymptotic distributions of sample eigenvalues and to [31, 35] for random matrices and more advanced modern treatment, referred to [20, 21] for vec operators and pattern matrices, referred to [29–30] for matrix theory, referred to [22] for distributions of ordered statistics, and referred to [25] for algebraic functions of random variables.

Problems

Problem 2.1   Assume that $\mathbf{A}$ and $\mathbf{B}$ are positive-definite Hermitian matrices. Find $\mathbf{x}$ that maximizes and minimizes the ratio

$$\rho = \frac{\mathbf{x}^{\dagger}\mathbf{AX}}{\mathbf{x}^{\dagger}\mathbf{Bx}},$$

respectively, and determine its corresponding maximum and minimum values.

Problem 2.2   Consider the issue of mutual information associated with transmission of a zero-mean Gaussian random signal $x(t)$, with power P_s and frequency-band limited to $(-B, B)$ Hz, through an AWGN channel $n(t)$ with two-sided spectral density $N_0/2$ W/Hz. In practical implementation, the received signal is applied to an ideal band-pass filter over $(-B, B)$ before sampling at the Nyquist rate of $\frac{1}{T_s} = 2B$ to produce a discrete output

$$y_i = x_i + v_i, \ i = 1, \cdots, m,$$

where $m = T/T_s$ and v_i denotes the output noise sample. Show that the system's mutual information over time period $(0, T)$ is given by $I = \log(1 + P_s/P_n)$ with P_n denoting the output power of the bandpass filter.

Problem 2.3   Assume $\mathbf{X} \sim \mathcal{CN}_{N,m}(\mathbf{0}, \mathbf{I} \otimes \mathbf{R})$ with $m \leq N$. Show that the determinant of $\mathbf{A} = \mathbf{X}^\dagger \mathbf{X}$ is distributed as the product of m independent chi-square variables, as shown by

$$\det(\mathbf{A}) \sim 2^{-m} \det(\mathbf{R}) \, \chi^2(2N) \chi^2(2N - 2) \cdots \chi^2(2N - 2m + 2).$$

Hint: use the Gram–Schmidt procedure to orthogonalize the column vectors of $\mathbf{X}$.

Problem 2.4   Suppose that random variable x is formed by the sum of three independent chi-square variables $x = 4.8\chi^2(2) + 1.5\chi^2(3.6) + 3.3\chi^2(5)$. Approximate x by a single chi-square variable.

Problem 2.5   Let $z = x + jy$, where $x \sim \mathcal{N}(0, \sigma^2)$ and $y \sim \mathcal{N}(0, \sigma^2)$ are independent. It is well known that $r = |z|$ follows the Rayleigh distribution, given by

$$f(r) = \frac{r}{\sigma^2} \exp\left(-\frac{r^2}{2\sigma^2}\right), \ r > 0$$

which is usually derived by using the method of Jacobian transform. Now, use the properties of chi-square distribution to show this assertion.
Hint: derive the PDF of $|z|^2$ first.

Problem 2.6   In wireless communication, a linear combination of m independent lognormal variables is often approximated by a single lognormal variate. But numerical results show a discrepancy, which increases quickly with m. Use a principle in statistics to argue that such an approximation is mathematically not well justified.

Problem 2.7   Use the knowledge you learnt in this chapter to quickly determine the following integral by observation.

$$I = \int_0^\infty x^2 e^{-x} dx.$$

Problem 2.8   Show that a linear combination of independent chi-square variables with equal weighting is still a chi-square variable.

Problem 2.9   Consider the transmission of a random signal x with mean μ_x and variance σ_x^2 over an AWGN channel, such that the received signal is given by $y = x + n$, where $n \sim \mathcal{N}(0, \sigma_n^2)$. Show that the optimal PDF of x that achieves the channel capacity is Gaussian.

References

1.   N.L. Johnson and S. Kotz, *Continuous Univariate Distributions*, vol. 2, New York: John Wiley & Sons, Inc., 1970.
2.   M. Kendall and A. Stuart, *The Advanced Theory of Statistics*, vol. 2, 4th edn, London: Griffin Publishing Group, 1979.
3.   N. Giri, "On the complex analogue of T^2- and R^2- tests," *Ann. Math. Stat.*, vol. 36, pp. 664–670, 1965.

4. G.L. Turin, "The characteristic function of Hermitian quadratic forms in complex normal variables," *Biometrika*, vol. 47, no 1–2, pp. 199–201, 1960.
5. S.M. Moser, "The fading number of memoryless multiple-input multiple-output fading channels," *IEEE Trans. Inf. Theory*, vol. 53, no. 7, pp. 652–2666, 2007.
6. N.L. Bowers Jr., "Expansion of probability density functions as a sum of gamma densities with applications in risk theory," *Trans. Soc. Actuar.*, vol. 18, Part 1, no. 52, pp. 125–147, 1966.
7. "The chi-distribution," http://www.wikipedia.com.
8. M. Nakagami, *"The m-distribution: a general formula of intensity distribution of rapid fading,"* in *Statistical Methods in Radio Wave Propagation*, W.C. Hoffman, Ed. New York: Pergamon, 1960.
9. R.J. Muirhead, *Aspects of Multivariate Statistical Theory*, New York: John Wiley & Sons, Inc., 1982.
10. A.T. James, "Distributions of matrix variates and latent roots derived from normal samples," *Ann. Math. Stat.*, vol. 35, no. 2, pp. 475–501, 1964.
11. F.J. Swetz and V.J. Katz, *Mathematical Treasures–Leonhard Euler's Integral Calculus*, 2011.
12. R.A. Nowlan, *A Chronicle of Mathematical People*, http://www.robertnowlan.com.
13. M. Abramowitz and I.A. Stegun, *Handbook of Mathematical Functions*, New York: Dover Publications, 1970, p. 932.
14. J.W. Craig, "A new, simple and exact result for calculating the probability of error for two-dimensional signal constellations," in *IEEE MILCOM Conference Record, Boston, MA*, 1991, pp. 25.5.1–25.5.5.
15. F.S. Weinstein, "Simplified relationships for the probability distribution of the phase of a sinewave in narrow-band normal noise," *IEEE Trans. Inf. Theory*, vol. 20, pp. 658–661, 1974.
16. R.F. Pawula, "Generic error probabilities," *IEEE Trans. Commun.*, vol. 47, no. 5, pp. 697–702, 1999.
17. J. Gil-Pelaez, "Notes on the inversion theorem," *Biometrika*, vol. 38, pp. 481–482, 1951.
18. S.B. Provost, "The exact density of a statistic related to the shape parameter of a gamma variate," *Metrika*, vol. 35, pp. 191–196, 1988.
19. E.J. Kelly, "An adaptive detection algorithm," *IEEE Trans. Aerosp. Electron. Syst.*, vol. 22, no. 1, pp. 115–127, 1986.
20. H.V. Henderson and S.R. Searle, "Vec and vech operators for matrices, with some uses in Jacobians and multivariate statistics," *Can. J. Stat.*, vol. 7, no. 1, pp. 65–81, 1979.
21. D.P. Wiens, "On some pattern-reduction matrices which appear in statistics," *Linear Algebra Appl.*, vol. 67, pp. 233–258, 1985.
22. E.B. Manoukian, *Mathematical Nonparametric Statistics*, New York: Gordon and Breach Science, 1986.
23. H. Jeffreys, *Theory of Probability*, Oxford: Clarendon Press, 1961.
24. K. Miller, *Complex Stochastic processes: An Introduction to Theory and Application*, Reading, MA: Addison-Wesley, 1974.
25. M.D. Springer, *The Algebra of Random Variables*, New York: John Wiley & Sons, Inc., 1976.
26. I. Todhunter, *A History of the Calculus of Variations*, New York: Chelsea.
27. I.M. Gelfand and S. Fomin, *Calculus of Variations*, Dover Publications, 2000.
28. R.M. Gray, *Toeplitz and Circulant Matrices: A Review*, Stanford, CA: Stanford University.
29. D.S. Berstein, *Matrix Mathematics: Theory, Facts, and Formulas*, Princeton, NJ: Princeton University Press, 2009.
30. D.A. Harville, *Matrix Algebra from a Statistician Perspective*, New York: Springer-Verlag, 2008.
31. M.L. Mehta, *Random Matrices*, 3rd edn, Amsterdam: Elsevier Academic Press, 2004.
32. E.P. Wigner, "Distribution laws for the roots of a random Hermitian matrix," pp. 446–462.
33. N.R. Goodman, "Statistical analysis based on a certain multivariate complex Gaussian distribution (An introduction)," *Ann. Math. Stat.*, vol. 34, pp. 152–176, 1963.
34. C.G. Khatri, "Classical statistical analysis based on a certain multivariate complex Gaussian distribution," *Ann. Math. Stat.*, vol. 36, pp. 98–114, 1965.
35. L.G. Ordonez, D.P. Palomar, and J.R. Fonollosa, "Ordered eigenvalues of a general class of Hermitian random matrices with application to the performance analysis of MIMO systems," *IEEE Trans. Signal Process.*, vol. 57, no. 2, pp. 672–689, 2009.
36. J.M. Wozencraft and I.R. Jacobs, *Principles of Communication Engineering*, New York: Wiley, 1965.
37. E.S. Pearson and H.O. Hartley, ed., *Biometrika Tables for Statisticians*, London: Biometrika Trust, 1959.

3

Channel Characterization

Abstract

Cellular mobile communication systems operate in the worst terrestrial environments characterized by mobility and severe random electromagnetic scattering, making the propagation behavior therein substantially different from its counterpart in the free space. Channel characterization is, therefore, an indispensable step in the design and testing of a cellular system, aiming to reveal how the scattering channels influence the power and frequency characteristics of a transmitted signal. Propagation loss behavior is of direct relevance to cellular power budget design and cellular planning. Large-scale and local fluctuation behaviors are critical to the assessment of the system's dynamic performance, whereas channel frequency characteristics affect the design of the wireless transceiver structure.

Channel characterization requires the incorporation of both physical and statistical descriptions. This chapter consists of three parts. Part 1 is devoted to the large-scale behavior, while Part 2 is devoted to the statistical modeling of the local behavior of a mobile signal. Finally, in Part 3, we present some algorithms for the generation of typical fading processes encountered in wireless communications.

♣

3.1 Introduction

Wireless mobile communication is characterized by the random mobility of mobile units relative to their transmitters. As a mobile unit moves around a scattering field, the electromagnetic propagation paths vary accordingly, and so do their channel gains. Modeling such propagation channels is indispensable for fully understanding the nature of wireless communication and its statistical performance.

A typical behavior of the signal strength received at a mobile unit is a function of distance, as illustrated in Figure 3.1. The fluctuating curve is the result of two mechanisms that dictate radio propagation over a wireless channel, namely diffraction and scattering [1]. First, we observe that the average signal strength experiences a *relatively smooth* drop with increasing propagation distance. This slow variation, called *shadowing*, results from secondary electromagnetic waves diffracted from large obstructions on radio paths between the transmitter and the receiver. We also observe that at each given location, the received signal strength fluctuates *rapidly*

Wireless Communications: Principles, Theory and Methodology, First Edition. Keith Q.T. Zhang.
© 2016 John Wiley & Sons, Ltd. Published 2016 by John Wiley & Sons, Ltd.
Companion Website: www.wiley.com/go/zhang7749

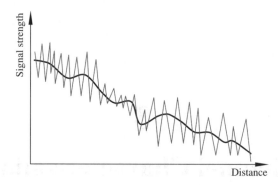

Figure 3.1   *Illustrating the variation of mobile signal strength with distance*

about its mean value because of local multipath propagation. The mechanism behind multipath propagation is scattering from small objects such as lamp poles and foliage.

Indeed, cellular mobile communication operates in the worst channel conditions among all the terrestrial communication systems, in the sense that the transmitted signals suffer from very large propagation loss and severe local multipath fluctuation, in addition to possible co-channel interference. Therefore, channel characterization of the operational environments for a wireless system is an indispensable step in the design and testing of cellular systems. To this end, we need an in-depth understanding of the way a wireless channel influences a transmit-through radio signal, at least in two aspects: namely the average propagation loss and local fluctuations around the average behavior. The average propagation loss, representing the large-scale behavior of wireless propagation, is required for system budget design (e.g., the minimum transmitted power), system coverage, and cell planning. On the other hand, local multipath behavior is required for the assessment of the system's dynamic performance. In the second aspect, we are concerned with the way the multipath channel influences the frequency characteristics of the mobile signals. A multipath channel behaves like a discrete filter with random tap coefficients. The question is how to determine whether a single-tap or a multitap filter should be used for channel characterization, and how to characterize such a filter so that we can differentiate a fast and a slow fading channel. Different attributes of a mobile channel require different tools for characterization, and the commonly used tools for characterizing the average propagation loss, channel filters and their random coefficients are briefly summarized as follows:

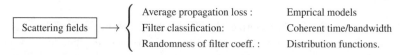

In general, channel characterization requires the combination of both physical and statistical descriptions. This chapter consists of three parts. Part 1 is devoted to the large-scale behavior, while Part 2 is devoted to modeling the local behavior of a mobile signal. Finally, in Part 3, we presents some algorithms for generating typical fading processes encountered in wireless communications.

3.2   Large-scale propagation loss

The first impact of cellular scattering fields upon transmitted signals is the propagation loss. Propagation loss in terrestrial environments is much larger than the theoretic prediction in the free space. It is influenced by various factors including the carrier frequency, the height of antennas, the nature of the terrain, the extent of urbanization, changes in weather, the speed of mobile units, changes in foliage conditions, and the surrounding buildings and other scatterers. Although the average propagation loss demonstrates certain deterministic behaviors similar to theoretic prediction, many of these factors are random in nature, thereby making it impossible to obtain an exact large-scale channel description. In this regard, both deterministic and random models can find their applications.

Various empirical models have been proposed for the deterministic prediction of average path loss, aiming to uncover the mechanism behind empirical data by means of statistical and electromagnetic theories. Two such models are summarized in the empirical formulas due to Okumura and Hata, and are the focus of this section. We first define the notations for use in this section. This section is focused on empirical deterministic models of large-scale loss, leaving its random modeling to the next section.

P_T: Tx power
P_R: Rx power
G_T: Tx antenna gain along T-R
G_R: Rx antenna gain along T-R
$P_t G_T$: EIRP (the effective isotropic radiated power)
d: Distance between Tx and Rx
λ: Radio wavelength
c: Light speed
f_c: Carrier frequency (RF)
D: Largest dimension of the antennas
h_t : Effective base TX antenna height between 30 and 200 m
h_r : Effective mobile antenna height

3.2.1 Free-space propagation

Since free-space propagation loss is often used as a benchmark for constructing empirical models, it is appropriate to provide a brief review before proceeding. The transmitted power produces the power density $W = P_T G_T / (4\pi d^2)$ at the receiver d meters apart, which, along with the fact that the receive antenna has the effective aperture of $A = \lambda^2 G_R / (4\pi)$, enables us to determine the received power as

$$P_R = WA = \frac{\lambda^2 P_T G_T G_R}{(4\pi)^2 d^2}. \tag{3.1}$$

The product $P_T G_T$ physically represents the effective isotropic radiated power, and hence often abbreviated as EIRP. Dividing both sides of the above expression by the transmitted power P_T leads to the channel gain

$$\frac{P_R}{P_T} = \frac{G_T G_R}{L_F}, \tag{3.2}$$

where

$$L_F = [4\pi \, (d/\lambda)]^2 \tag{3.3}$$

represents the propagation loss in the free space without taking the antenna gains into account. With L_F, we rewrite

$$P_R = G_R \underbrace{(P_T G_T)}_{EIRP} / L_F. \tag{3.4}$$

In practice, it is difficult to measure P_T at $d = 0$. This difficulty can be circumvented by representing the P_R at distance d with reference to the received power at a *close-in* reference distance d_0 in the far field. It yields

$$P_R(d) = P_R(d_0)(d_0/d)^2.$$

The reference distance d_0 must be in the far field, which is defined as a point exceeding the value of $\frac{2D^2}{\lambda}$. As a typical value, $d_0 = 1$ km is used for a cellular system of large coverage, whereas $d_0 = 100$ m or $d_0 = 1$ m for micro-cellular systems. In the log scale, we have

$$P_R(d) = \begin{cases} 10 \log_{10} P_R(d_0) + 20 \log_{10}(d_0/d) \text{ dB} \\ 10 \log_{10} P_R(d_0) + 20 \log_{10}(d_0/d) + 30 \text{ dBm}. \end{cases} \tag{3.5}$$

3.2.2   Average large-scale path loss in mobile

Because of its complicated radio propagation environments, terrestrial mobile communication experiences a much larger propagation loss compared to its free-space counterpart. Thus, the propagation loss in the free space must be modified accordingly. Modifications are made along two ways.

One is to extend the inverse-square law for the free space to a general case with a propagation loss exponent $n > 0$, so that the received power is rewritten as

$$P_R(d) \propto \frac{P_T}{d^n} \tag{3.6}$$

which, when measured in terms of a far field reference distance d_0, is given by

$$P_R(d)_{\text{dB}} = 10 \log_{10} P_R(d_0) + 20n \log_{10}(d_0/d) \text{ dB}. \tag{3.7}$$

Empirical values for n are obtained by data fitting, and typical values for cellular mobile communications range between 3 and 5. In cellular planning and outage performance analysis, the loss exponent n is often set to 4, as opposed to $n = 2$ in the free space. The advantage of the revised propagation model given in (3.6)–(3.7) lies in its simplicity, and is often incorporated into statistical models as a factor to account for the average loss behavior.

Another way of modification is to use the free-space formula (3.2) as a basic component by adding some terms to account for the physical and geometrical aspects of practical cellular operational environments. The results along this line include the well-known Okumura, Hada, and JTC air models.

3.2.3   Okumura's model

Okumura uses the propagation loss in the free space as a benchmark, denoted here by L_F, to build his model by adding a number of correction terms to account for various physical and topological factors surrounding the receiver. The Okumura model characterizes the median loss L_{50} for built-up areas, yielding [2]

$$L_{50} = L_F(d) + C_F(d, f) - G(h_t) - G(h_r) + G_{\text{area}} \quad \text{(in dB)}. \tag{3.8}$$

This model is built by using the free-space propagation loss L_F as a reference. Compared to L_F, the operation in a real mobile environment introduces additional loss, which is accounted for by the correction term $C_F(d, f)$. The term C_F is defined under three particular conditions: 200-m high transmit antenna, 3-m high receive antenna, and a built-up urban area. If the operational conditions are different, further correction is needed. The correction terms $G(h_t)$ and $G(h_r)$ signify the antenna gains due to the base station antenna height and the mobile antenna height, respectively, as shown by

$$G(h_t) = 20 \log_{10}(h_t/200), \quad 10 \text{ m} < h_t < 1000 \text{ m},$$

$$G(h_r) = \begin{cases} 10 \log_{10}(h_r/3) & h_r \leq 3 \text{ m} \\ 20 \log_{10}(h_r/3) & 3 \text{ m} < h_r < 10 \text{ m}. \end{cases}$$

Table 3.1   *Fitting parameters of $C_F(d, f)$ versus distance d*

d (km)	$C_F(d, f) = a_2 f^2 + a_1 f + a_0$ (dB), f in MHz		
	a_2	a_1	a_0
1	-5.0334×10^{-7}	5.0583×10^{-3}	14.782
2	-4.3600×10^{-7}	5.0296×10^{-3}	17.800
5	-8.8352×10^{-7}	6.4510×10^{-3}	20.908
10	-7.7598×10^{-7}	6.0528×10^{-3}	24.057
20	-6.7627×10^{-7}	5.9191×10^{-3}	26.611
30	-8.9176×10^{-7}	7.1336×10^{-3}	29.796
40	-1.1742×10^{-6}	8.0077×10^{-3}	33.461
50	-1.3100×10^{-6}	9.0236×10^{-3}	37.420
60	-1.5114×10^{-6}	1.0047×10^{-2}	41.166
70	-1.2142×10^{-6}	9.1819×10^{-3}	44.872
80	-1.5858×10^{-6}	1.0299×10^{-2}	47.495
100	-1.6837×10^{-6}	1.0973×10^{-2}	51.098

The term G_{area} represents the environment-dependent gain. The correction terms $C_F(d, f)$ and G_{area} were obtained via field tests by Okumura and his colleagues in 1968 [2] and are plotted in the form of look-up charts. It is sometimes inconvenient to accurately read C_F and G_{area} from these charts. Since the curves in these charts are quite smooth, we can fit them with polynomials of the form

$$C_F(d, f) = a_2 f^2 + a_1 f + a_0,$$

$$G_{\text{area}} = b_2 f^2 + b_1 f + b_0.$$

In the following fitting, f must be written in megahertz. In particular, we find $b_2 = 0$ and the result for G_{area} is given by

$$G_{\text{area}} = \begin{cases} 0.0051 f + 5.0536, & \text{suburban area} \\ 0.0079 f + 16.1297, & \text{quasi open area} \\ 0.0078 f + 21.6186, & \text{open area.} \end{cases} \tag{3.9}$$

The coefficients a's for C_F are given in Table 3.1.

✑ Example 3.1

Suppose a base station operates in a suburban area and employs an antenna of height 50 m to transmit 1.8-GHz signals to a receiver located $d = 1000$ m away from the transmitter. Determine the minimum EIRP if the receive antenna is isotropic with height 0.05 m, and the expected receiver power is -95 dB.

Solution

From the given condition $f = 1.8\,\text{GHz}$, $h_t = 50\,\text{m}$, $h_r = 0.05\,\text{m}$, and $d = 1000\,\text{m}$, we have $\lambda = 3 \times 10^8 / f = \frac{1}{6}\,\text{m}$. Thus, the free-space propagation loss (in dB) is equal to

$$L_F = 20 \log_{10}(4\pi d / \lambda) = 97.55\,\text{dB}.$$

The antenna correction factors are determined as

$$G(h_t) = 20 \log_{10}(50/200) = -12.04\,\text{dB}$$

$$G(h_r) = 10 \log_{10}(0.05/3) = -17.78\,\text{dB}.$$

Given $f = 1800$ MHz, use the coefficients on the first row for $d = 1$ km in Table 3.1 to determine C_F, to obtain

$$C_F(d, f) = -5.0334 \times 10^{-7}f^2 + 5.0583 \times 10^{-3}f + 14.782 = 22.56 \text{ dB},$$

and use the coefficients for suburban area in (3.9), yielding

$$G_{\text{area}} = 0.0051f + 5.0536 = 14.2336 \text{ dB}.$$

It follows that

$$L_{50} = L_F + C_F(d, f) + G(h_t) + G(h_r) + G_{\text{area}}$$
$$= 97.55 + 22.56 + 12.04 + 17.78 - 14.23$$
$$= 135.70 \text{ dB}$$

which, along with the condition of $G_R = 0$ dB, enables the use of (3.4) to obtain

$$\text{EIRP} = P_R + L_{50} = -95 \text{ dBm} + 135.70 = 40.70 \text{ dB}.$$

3.2.4   Hata's model

Okumura's model has its parameters expressed in terms of graphs. Hata parameterizes these graphs, obtaining the empirical formula

$$L_{50}(\text{urban})(\text{dB}) = 69.55 + 26.16 \log_{10}f_c - 13.82 \log_{10}h_t$$
$$-a(h_r) + (44.9 - 6.55 \log_{10}h_t) \log_{10}d$$
$$L_{50}(\text{suburban}) = L_{50}(\text{urban}) - 2[\log_{10}(f_c/28)]^2 - 5.4$$
$$L_{50}(\text{rural}) = L_{50}(\text{urban}) - 4.78[\log_{10}(f_c)]^2 - 18.33 \log_{10}(f_c) - 40.98$$

where $h_t \in [30, 200]$ m is the effective base TX antenna height in meters, $h_r \in [1, 10]$ m is the effective mobile antenna height, $a(h_r)$ is the correction factor for the mobile antenna height, d denotes the Tx and Rx separation in kilometers, and $f_c \in (150, 1500)$ MHz is the carrier frequency. For a small to medium city

$$a(h_r) = (1.1 \log f_c - 0.7)h_r - (1.56 \log f_c - 0.8) \text{ dB}.$$

For a larger city

$$a(h_r) = \begin{cases} 8.29[\log(1.54h_r)]^2 - 1.1 \text{ dB} & f_c \leq 300 \text{ MHz} \\ 3.2[\log(11.75h_r)]^2 - 4.97 \text{ dB} & f_c \geq 300 \text{ MHz}. \end{cases}$$

Here, $\log$ is the short-form of $\log_{10}$.

The empirical formulas of Okumura–Hata are recommended by ITU-R as rural wave propagation models, suitable for the propagation range $1 - 15$ km.

3.2.5   JTC air model

The joint technical committee (JTC) proposes three propagation loss models for the cellular interface standards, which are designed for indoor, outdoor micro-cellular, and outdoor macro-cellular, respectively [3]. Here we give only a brief description of the outdoor macro-cellular model. For a macro-cellular environment with antenna heights above the roof-top level, the gross path loss with distance d (in meters) is given by

$$L_{\text{total}} = \max\{A + B \log_{10}(d), 38.1 + 20 \log_{10}(d)\} + \varphi, \tag{3.10}$$

where

$$A = 20.62 - 5.83 \log_{10}(h_t) + C,$$
$$B = 44.9 - 6.55 \log_{10}(h_t),$$

where C is the clutter correction factor C (in dB) and is equal to 0 for urban high-rise, -6 for urban/suburban low-rise, -12 for residential, and -18 for rural settings. The term φ is a correction term for the building penetration loss (in dB); it is equal to 15 for urban and suburban areas, and equal to 10 for residential and rural areas.

3.3   Lognormal shadowing

As shown in Figure 3.1, the average path loss between the transmitter and receiver, though smooth, is slowly varying, indicating that the large-scale variation is a function not only on d but also on terrains and obstructions, such as foliage and buildings in the transmission path. Two strategies can be used to describe such a slow variation. The first strategy, as addressed in Section 3.2, was adopted by Okumura and Hata among others, and amends the free-space model, location by location, by adding correction terms to account for local particulars. This type of models is particularly suited for cellular planning.

In the second strategy, the large-scale slow variation is treated as a random variable, influenced by large-scale geography. It, therefore, suggests the incorporation of statistical models into the inverse power-n path attenuation for a better characterization, whereby the overall propagation attenuation can be written as the product of two factors

$$\ell(d, \xi) = d^n 10^{\xi/10} \tag{3.11}$$

Which (in dB scale) gives

$$\underbrace{10 \log_{10}\ell}_{L(d)} = \underbrace{10n \log_{10}d}_{\overline{L}(d)} + \xi. \tag{3.12}$$

The random variable ξ represents the large-scale variation caused by variations in terrains, hills, and so on, usually called *shadowing*. Empirical data show that a good model for shadowing is the lognormal distribution, given by

$$\xi \sim \mathcal{N}(0, \sigma_{\text{dB}}^2) \tag{3.13}$$

which enables us to write

$$L \sim \mathcal{N}(\overline{L}, \sigma_{\text{dB}}^2), \tag{3.14}$$

or more explicitly

$$f_L(x) = \frac{1}{\sqrt{2\pi\sigma_{\text{dB}}^2}} \exp\left[-\frac{(x - 10n \log_{10}d)^2}{2\sigma_{\text{dB}}^2}\right]. \tag{3.15}$$

The standard deviation σ_{dB} of the channel loss L is known as the *dB-spread*, which ranges between 5 and 12 dB depending on the operational environment. A typical value of σ_{dB} is 6 dB.

3.4   Multipath characterization for local behavior

We have studied the characteristics of propagation loss in terrestrial mobile communications, aiming to provide a theoretic basis to power-budget planning for cellular systems. From the perspective of system design, we are also concerned with the impact of wireless channels on the transmitted signals in terms of their spectra and statistical behaviors.

There are many discrete scatterers located between the wireless transmitter and the receiver. It is, therefore, natural to think the propagation channel as a discrete filter of isolated paths, in the form

$$h(t, \tau) = \sum_{\ell=0}^{L-1} a_\ell(t)\delta(\tau - \tau_\ell), \tag{3.16}$$

which is sometimes called *Turin's* model. This is a time-variant tap-delay-line filter of L taps, where t is the time index, τ is the time delay relative to t, and $a_\ell(t)$ is the complex time-variant tap coefficient. The tap coefficients $\{a_\ell(t)\}$ are used to capture the scattering and deflection features of each path. Since the receiver is often in motion relative to the transmitter, the tap coefficients are typically varying with time. This assertion is true even when both the transmitter and the receiver are static, because of the mobility of some scatterers. According to the duality theory of stochastic processes, $a_\ell(t)$ can be treated either as random or time-varying. In the former framework, $a_\ell(t)$ represents a random population, while in the latter it represents one temporal realization. Both treatments are useful in wireless communication. For example, $a_\ell(t)$ are treated as deterministic when dealing with channel equalization, whereas they are treated as stochastic when conducting system performance evaluation.

The function $h(t, \tau)$ represents the time-variant impulse response of the wireless channel. If we take the Fourier transform with respect to the time delay τ, the result is just the same as the traditional frequency transfer function at time t. If the Fourier transform is taken with respect to t, the result reflects the Doppler spectrum. For system analysis and simulations, however, we are more interested in the behavior of $h(t, \tau)$ in the time and time-delay domains, and proceed along this direction in what follows. We need to answer two questions. First, does the channel filter $h(t, \tau)$ behave like a multitap filter or a single-tap filter? Second, what is the behavior of each tap coefficient?

3.4.1   An equivalent bandwidth

For a wireless system to operate in a given environment with impulse response $h(t, \tau)$, it is of primary importance to practical applications to determine the equivalent bandwidth of a multipath channel so as to decide, for example, whether a channel equalizer is needed. Recall that in linear system theory, it is the time duration of the impulse response of a filter that determines its frequency bandwidth. We also know that the $-3\,\mathrm{dB}$ bandwidth of a rectangular pulse is equal to the reciprocal of its time duration. This estimation is roughly correct even for waveforms without a rectangular shape. The difference in wireless communication is that channels confronting us are discrete filters with time-varying or random tap coefficients and time delays. Intuitive analogy, suggests us to use the average time duration to describe their frequency characteristics. The difficulty is that mobile channels are usually time-varying. We therefore consider, instead, its average delay dispersion in the form of the root-mean-squared (rms) delay spread.

3.4.1.1   *Characterization for time dispersion*

If the tap coefficients $a_\ell(t)$ in (3.16) are treated as random variables, we need their joint probability density function (PDF) to determine the average delay dispersion of the time-varying channel, which, however, is not available and even impossible to acquire in practice. Thus, the more easily acquirable power delay profile is used instead. The power $P(\tau)$ as a function of excess time delay τ is called the *power delay profile* (PDP). For a given value of τ, it is obtained by time-averaging over all the instantaneous power $|h(t,\tau)|^2$ acquired at a given local area, as shown by

$$\overline{\tau} = \frac{\sum_k P(\tau_k)\tau_k}{\sum_k P(\tau_k)}, \quad \text{mean delay}$$

$$\overline{\tau^2} = \frac{\sum_k P(\tau_k)\tau_k^2}{\sum_k P(\tau_k)}, \quad \text{the second moment}$$

$$\sigma_\tau = \sqrt{\overline{\tau^2} - (\overline{\tau})^2}, \quad \text{rms delay spread} \tag{3.17}$$

where $P(\tau_k)$ denotes the average power at the time delay τ_k, and summation is taken over all possible k. The first parameter represents the mean time delay, and the third represents the rms delay spread. Their calculation is similar to that of moments in probability. The difference is that the latter uses the probabilities as weighting coefficients calculating the statistical average, whereas the former employs the normalized powers as weighting to reflect the relative importance of a particular delay.

3.4.1.2 *Coherence bandwidth*

In analogy to the way of defining the bandwidth of a rectangular pulse, we may use the rms delay spread to define the *coherence* bandwidth of a channel filter:

$$B_c = \begin{cases} \frac{1}{50\sigma_\tau}, & \text{if cross-correlation} \geq 0.9 \\ \frac{1}{5\sigma_\tau}, & \text{if cross-correlation} \geq 0.5. \end{cases} \tag{3.18}$$

This is an empirical expression. Since $h(t,\tau)$ is time-varying and not of a rectangular shape, correction factors of 5 and 50 are used. Physically, the coherence bandwidth stands for the maximum frequency difference for which two wireless signals are still strongly correlated in amplitude. There are different ways to define strong correlation. If we use a cross-correlation of 0.9 or above to define strong correlation between the two signals, the coherence bandwidth is estimated by the first line. If the cross-correlation of 0.5 or above is used, instead, the coherence bandwidth is given by the second line.

3.4.1.3 *Criterion for Filter-type decision*

The relative coherence bandwidth is more meaningful compared to the signal to be transmitted. In so doing, we can classify mobile channels into two types: flat fading and frequency-selective fading. The criterion for this partition is as follows:

$$B_c \gg B_s \text{ and } T_s \gg \sigma_\tau : \text{Flat}$$

$$B_c < B_s \text{ and } T_s < \sigma_\tau : \text{Frequency-selective} \tag{3.19}$$

where T_s and B_s denote the signal duration and bandwidth, respectively. The first condition on the first line means that the channel coherence bandwidth is much greater than the signal bandwidth. The second condition implies that the channel delay spread is much less than the symbol duration. The combination of the two conditions means that all signal components can go through the channel with equal gain and linear phase. Thus, for flat fading, the channel filter can be characterized by a single-tap filter. On the second line, the received signal is a superposition of multiple copies of well-separated signal waveforms, implying that the transmitted signal experiences severe distortion and its spectrum is not preserved. Thus, for frequency-selective fading, an appropriate model for the channel is a multitap filter with random or time-varying tap coefficients.

✍ **Example 3.2**

Consider the outdoor Channel A recommended in the JTC Standards for urban/suburban low-rise and low antenna with tap parameters given below [3]:

Tap	1	2	3	4	5	6
Delay (ns)	0	50	150	325	550	700
Average power (dB)	0	−1.6	−4.7	−10.1	−17.1	−21.7

(a) Determine the 90% correlation coherence bandwidth of the channel.
(b) Determine whether a channel equalizer is required if GSM is used.

Solution

Before using (3.17) for calculation, we must convert the power in dB into the linear scale, and the resulting power vector is $\mathbf{P} = [1; 0.6918; 0.3388; 0.0977; 0.0195; 0.0068]$ which, after normalization by the total power $\mathrm{sum}(\mathbf{P}) = 2.1547$, produces the weighting coefficients

$$\mathbf{p} = [0.4641; 0.3211; 0.1573; 0.0454; 0.0090; 0.0031].$$

Then, using (3.17) leads to

$$\bar{\tau} = \sum_{i=1}^{6} p_i \tau_i = 61.5574; \quad \overline{\tau^2} = \sum_{i=1}^{6} p_i \tau_i^2 = 1.3407 \times 10^4$$

whereby

$$\sigma_\tau = (\overline{\tau^2} - \bar{\tau}^2)^{1/2} = 98.068 \text{ ns}.$$

With 90% correlation, the coherence bandwidth is given by

$$B_c = (50\sigma_\tau)^{-1} = 222.05 \text{ kHz}.$$

Since B_c exceeds the 200-kHz bandwidth requirement by GSM, no channel equalization is needed.   ○

3.4.1.4   *Theoretic coherence bandwidth*

The empirical formula (3.18) offers a simple and feasible way to estimate the channel coherence bandwidth. It is worth spending some more time to understand its philosophy. Recall that the coefficients of the channel filter are random parameters, and so is its bandwidth. Suppose at time instant t the equivalent time duration of the channel response $h(t, \tau)$ is $\Delta\tau$; then the instantaneous bandwidth is determined as

$$B_c(t) = \frac{\kappa}{\Delta\tau} \tag{3.20}$$

where κ is a correction factor to account for the non-rectangular shape of $h(t, \tau)$. Since the coherence bandwidth $B_c(t)$ is randomly varying, what we are really concerned with is its mean value

$$\underbrace{B_c = \mathbb{E}\left[\frac{\kappa}{\Delta\tau}\right]}_{\text{definition}} \overset{?}{\rightarrow} \underbrace{\frac{\kappa}{\mathbb{E}[\Delta\tau]} = \frac{\kappa}{\sigma_\tau}}_{\text{current practice}}.$$

The left-hand side stands for the definition of coherence bandwidth, where the expected value of $\kappa/\Delta\tau$ is not easy to determine. The right-hand side is the empirical formula in current use, where the expectation operator is moved to the denominator and σ_τ is used as an estimator of $\mathbb{E}[\Delta\tau]$. However, moving the expectation operator to the denominator is *not justified* in mathematics, given that the nonlinear operation is involved there.

We, therefore, want to seek a more accurate method for coherence bandwidth evaluation [4]. To this end, let us consider a discrete multipath Rayleigh fading channel with tap weight coefficients $\{h_k\}$ and time delays $\{\tau_k\}$, such that its impulse response $h(\tau)$ and frequency transfer function $H(\omega)$ are given by

$$h(\tau) = \sum_{k=0}^{D-1} h_k \delta(\tau - \tau_k)$$

$$H(\omega) = \int_{-\infty}^{\infty} h(\tau)e^{-j\omega\tau} = \sum_{k=0}^{D-1} h_k e^{-j\omega\,\tau_k} \tag{3.21}$$

where we assumed that $\{h_k\}$ are independent with PDP $\{P_0, P_1, \cdots, P_{D-1}\}$. Coherence bandwidth is defined as the range of frequencies over which two frequency components show a strong amplitude correlation. Physically, it represents a frequency range over which the channel can be considered "flat." We, therefore, need to focus on the channel frequency response. For notational simplicity, denote

$$\mathbf{h} = [h_0, h_1, \cdots, h_{D-1}]^T$$

$$\mathbf{w}_i = [e^{j2\pi\,f_i\tau_0}, \cdots, e^{j2\pi\,f_i\tau_{D-1}}]^T, \tag{3.22}$$

and simply denote the frequency response at f_i (i.e., $H(\omega_i)$) and its magnitude by x_i and r_i, respectively. Namely,

$$x_i = \mathbf{w}_i^\dagger \mathbf{h}$$

$$r_i = |x_i|. \tag{3.23}$$

Under an independent Rayleigh fading assumption, we have

$$\mathbf{h} \sim \mathcal{CN}(0, \mathbf{R}_h) \tag{3.24}$$

with $\mathbf{R}_h = \text{diag}(P_0, \cdots, P_{D-1})$. The coherence bandwidth of a fading channel is probed by sending two sinusoids, separated by $\Delta f = f_1 - f_2$ Hz, through the channel. The coherence bandwidth is defined as Δf over which the cross correlation between r_1 and r_2 is greater than a preset threshold, say, $\eta_0 = 0.9$. Namely,

$$C_{r_1, r_2} = \frac{\text{cov}(r_1, r_2)}{\sqrt{\text{var}(r_1)\text{var}(r_2)}} = \eta_0. \tag{3.25}$$

where $\text{var}(r_1) = \text{var}(r_2)$.

Since various moments of r_i are expressible in terms of the covariance structure of the 2×1 frequency response vector $\mathbf{x} = [x_1, x_2]^T$, we start from the PDF of $\mathbf{x}$, as shown by

$$\mathbf{x} \sim \mathcal{CN}(0, \mathbf{R}_x), \tag{3.26}$$

where the (i, j)th entry of $\mathbf{R}_x$ is given by

$$\mathbf{R}_x(i, j) = \mathbf{w}_i^\dagger \mathbf{R}_h \mathbf{w}_j, \quad i.j = 1, 2. \tag{3.27}$$

To this end, we define

$$\mathbf{S} = \mathbf{R}_x^{-1}, \quad \lambda_{12}^2 = \frac{|S_{12}|^2}{S_{11} S_{22}} \tag{3.28}$$

and then invoke the results of Miller [5] to obtain

$$\mathbb{E}[r_1 r_2] = {}_2F_1(1.5, 1.5; 1; \lambda_{12}^2) \frac{\pi \det(\mathbf{S})}{4 S_{11}^{1.5} S_{22}^{1.5}}$$

$$\mathbb{E}[r_i^2] = \mathbb{E}[|x_i|^2] = P$$

$$\mathbb{E}[r_i] = \frac{1}{2}\sqrt{\pi P}. \tag{3.29}$$

Here, P denotes the total power, that is, $P = \sum_{i=0}^{D-1} P_i$. After some direct manipulation and denoting $p_i = P_i/P$, we can represent $\mathbb{E}[r_1 r_2]$ and λ_{12}^2 in terms of the parameters of the channel power delay profile, given by

$$\lambda_{12}^2 = \sum_{i=0}^{D-1} \sum_{j=0}^{D-1} p_i p_j \cos 2\pi (\tau_i - \tau_j) \Delta f,$$

$$\mathbb{E}[r_1 r_2] = {}_2F_1(1.5, 1.5; 1; \lambda_{12}^2) \frac{\pi P(1 - \lambda_{12}^2)^2}{4}, \tag{3.30}$$

which enables us to rewrite

$$C_{r_1, r_2} = \frac{{}_2F_1(-0.5, -0.5; 1; \lambda_{12}^2) - 1}{(4/\pi) - 1}. \tag{3.31}$$

Accordingly, the coherence bandwidth for a given threshold η_0 is defined by the value of Δf, such that

$$_2F_1\left(-\frac{1}{2}, -\frac{1}{2}; 1; \lambda_{12}^2\right) - \left(\frac{4}{\pi} - 1\right) \eta_0 - 1 = 0, \tag{3.32}$$

where the Gauss hypergeometric function can be calculated by directly calling the MATLAB function hypergeom.m, or using the following series representation:

$$_2F_1(a; b; c, z) = \sum_{n=0}^{\infty} \frac{(a)_n (b)_n}{(c)_n} \frac{z^n}{n!} \tag{3.33}$$

with $(x)_0 = 1$ and $(x)_n = x(x + 1) \cdots (x + n - 1)$. Note from (3.30) that λ_{12}^2 is essentially the Fourier transform of the sample correlation of PDP, thereby representing the PDP "spectrum" as a function of Δf. Accordingly, a rough conclusion is that the smaller the rms delay spread, the wider the coherence bandwidth. Such an explanation is intuitively appealing but not rigorous. In fact, λ_{12}^2 depends on the structure of the PDP pattern rather than simply on its average characteristic – the rms delay spread. In fact, in some application scenarios, the variation of the coherence bandwidth with η_0 can exhibit fluctuations, as shown Figures 6 and 9 in Ref. [6]. We, therefore, expect that simply using the simple reciprocal of rms delay spread for coherence bandwidth calculation will lead to inaccurate results.

To determine the coherence bandwidth for a given threshold η_0, we can solve (3.32) using the Newton–Raphson iteration. To this end, define

$$g(\Delta f) = {}_2F_1(-0.5, -0.5; 1; \lambda_{12}^2) - \left(\frac{4}{\pi} - 1\right) \eta_0 - 1, \tag{3.34}$$

whose derivative with respect to Δf is given by

$$\dot{g}(\Delta f) = -_2F_1(0.5, 0.5; 2; \lambda_{12}^2) \times \frac{\pi(\tau_i - \tau_j)}{2} \sum_{i=0}^{D-1} \sum_{j=0}^{D-1} p_i p_j \sin 2\pi(\tau_i - \tau_j)\Delta f. \tag{3.35}$$

The coherence bandwidth is the value of Δf such that $g(\Delta f) = 0$, which can be numerically calculated through a Newton–Raphson type of iteration:

$$\Delta f_{n+1} = \Delta f_n - \frac{g(\Delta f_n)}{\dot{g}(\Delta f_n)}. \tag{3.36}$$

✍ **Example 3.3** _____

Consider Example 5.5 of Ref. [7]. The PDP is defined by four components $[-20, -10, -10, 0]$ dB located at delays $0, 1, 2,$ and 5 μs, respectively. The coherence bandwidths calculated using the empirical and exact formulas are listed below.

Formula	90%-BW	50%-BW
Empirical	14.55 kHz	145.54 kHz
Theoretic	35.54 kHz	100.88 kHz

It is clear that the empirical formulas underestimate the 90% coherence bandwidth but overestimate the 50% one.   ○

〰✍

3.4.2   Temporal evolution of path coefficients

We have studied the coherence bandwidth of the multipath filter (3.16). Another important aspect of a channel filter is the evolution of its path coefficients in time, caused by the relative motion between a mobile unit and the base station. The rule that dictates such variations is essential in the performance evaluation of differential MPSK systems and in the determination of intervals of sounding signals for channel estimation.

Each path coefficient varies independently and, hence, we will focus on a single path. For simplicity, we drop the subscript and simply write $a_\ell(t)$ as $a(t)$. The motion of the mobile, or the motion of the surrounding objects, in a radio channel causes a Doppler shift. The maximum Doppler spread, denoted by f_d, is related to the speed v of the mobile unit by

$$f_d = \frac{v}{\lambda}, \tag{3.37}$$

where λ denotes the wavelength of the carrier. The Doppler spread tends to decorrelate the temporal correlation of $a(t)$.

3.4.2.1   *Coherence time to characterize filter evolution*

Coherence time is an important parameter that characterizes the decorrelation process of $a(t)$ caused by the Doppler shift. It is defined by

$$T_c = \begin{cases} \frac{1}{f_d} & \text{time correlation} = 0.9 \\ \frac{9}{16\pi f_d} & \text{time correlation} = 0.5 \end{cases}, \tag{3.38}$$

The cutoff temporal correlation for defining the coherence time is usually set to 0.9 or 0.5, as shown on the first and second lines. As a tradeoff, one may use their geometric mean to define T_c, yielding

$$T_c = 0.75/(\sqrt{\pi} \, f_d) = 0.423/f_d. \tag{3.39}$$

Having defined the coherence time, we can qualitatively classify the variation of a path coefficient into fast fading and slow fading, by virtue of the following criterion:

$$T_s > T_c \quad \text{and } B_s < B_D : \text{fast fading}$$

$$T_s \ll T_c \quad \text{and } B_D \ll B_s : \text{slow fading.}$$

The condition $T_s > T_c$ implies that, within a signal symbol duration, the signal experiences more than one period of Doppler variation, thus causing a fast amplitude fluctuation. The case of $T_s \ll T_c$ represents another extreme case of slow fading, implying that over one symbol duration, the signal experiences Doppler spread only during a small portion of a Doppler period. When the signal baseband width greatly exceeds B_D, the Doppler spread can be neglected.

3.4.3   Statistical description of local fluctuation

The classification of a path evolution into fast and slow fading provides only a rough characterization of the Doppler spread. In system design and performance analysis, we often need a complete probabilistic description. To this end, we treat each time-varying path coefficient as random, which is denoted here by $z(t)$. Assume that $z(t)$ remains unchanged over each transmitted symbol duration but can vary from one symbol to another. We sample $z(t)$ once a symbol period, and denote its ith sample as z_i. Further we write the n samples so obtained in vector form $\mathbf{z} = [z_1, z_2, \cdots, z_n]^T$. A probabilistic model for $\mathbf{z}$ should appropriately reflect the influence of the Doppler spread caused by the relative motion of a mobile unit.

3.4.4   Complex Gaussian distribution

The first model of practical interest is the complex Gaussian distribution, usually represented as $\mathbf{z} \sim \mathcal{CN}(\mathbf{c}, \mathbf{R})$ or more explicitly with its joint PDF

$$f(\mathbf{z}) = \frac{1}{\pi^n \det(\mathbf{R})} \exp[-(\mathbf{z} - \mathbf{c})^\dagger \mathbf{R}^{-1}(\mathbf{z} - \mathbf{c})]. \tag{3.40}$$

The complex representation stems from the quadratic reception widely adopted in communication. The in-phase and quadrature components produced therein can be naturally mapped into a complex variable. The complex Gaussian distribution provides a powerful tool to describe a Rayleigh or Rician fading channel. The complex Gaussian distribution of $\mathbf{z}$ implies that its real and imaginary parts are independent and jointly follow a real Gaussian distribution. To be specific, we denote the real and imaginary parts of $\mathbf{z}$ by $\mathbf{x} = \Re(\mathbf{z}), \mathbf{y} = \Im(\mathbf{z})$, and define

$$\mathbf{u} = \begin{bmatrix} \mathbf{x} \\ \mathbf{y} \end{bmatrix}, \ \mathbf{m} = \begin{bmatrix} \Re(\mathbf{c}) \\ \Im(\mathbf{c}) \end{bmatrix}, \ \boldsymbol{\Sigma} = \begin{bmatrix} \Re(\mathbf{R}) & -\Im(\mathbf{R}) \\ \Im(\mathbf{R}) & \Re(\mathbf{R}) \end{bmatrix}.$$

With this notation, we assert $\mathbf{u} \sim \mathcal{N}(\mathbf{m}, \boldsymbol{\Sigma})$. The treatment in complex vector is equivalent to its counterpart in real. The equivalence can be easily shown by checking that [5]

$$\det \boldsymbol{\Sigma} = 2^{-2n} (\det \mathbf{R})^2$$

$$(\mathbf{z} - \mathbf{c})^\dagger \mathbf{R}^{-1}(\mathbf{z} - \mathbf{c}) = \frac{1}{2}(\mathbf{u} - \mathbf{m})^T \boldsymbol{\Sigma}^{-1}(\mathbf{u} - \mathbf{m}), \tag{3.41}$$

which enables the assertion

$$f(\mathbf{u}) = \frac{1}{(2\pi)^n (\det \boldsymbol{\Sigma})^{1/2}} \exp\left[-\frac{1}{2}(\mathbf{u} - \mathbf{m})^T \boldsymbol{\Sigma}^{-1}(\mathbf{u} - \mathbf{m})\right] = f(\mathbf{z}).$$

The two representations are equivalent; but the treatment of the complex one is much simpler and, therefore, widely adopted in practice.

3.4.4.1 *Rayleigh fading*

In a practical propagation environment, the constant $\mathbf{c}$ in (3.40) can physically represent a line-of-sight (LOS) component. If there is no LOS component in propagation, $\mathbf{c}$ is zero and we say the channel gain amplitude suffers Rayleigh fading. Let us focus on a single entry of $\mathbf{z}$ and denote it as $z = x + jy$, where x and y are independent each of the distributions $\mathcal{N}(0, \sigma^2)$. Then, the amplitude of z is equal to

$$r = \sqrt{x^2 + y^2} \tag{3.42}$$

whose PDF can be determined through the Jocobian approach. The result is the Rayleigh distribution

$$f_r(r) = \frac{r}{\sigma^2} \exp\left(-\frac{r^2}{2\sigma^2}\right), \ 0 \leq r < \infty. \tag{3.43}$$

The Rayleigh distribution, up to a constant factor, is the $\chi(2)$ in the Pearson family of distributions. To be specific, $r \sim \sigma\chi(2)$. It is named after Lord Rayleigh to recognize his pioneering contributions.

3.4.4.2 *Rician fading*

When an LOS component $\mathbf{c}$ exists in (3.40), we say the channel amplitude experiences Rician fading to recognize the early jobs of Stephen O. Rice. To see the amplitude distribution, we focus, again, on a single entry of $\mathbf{z}$, denoted here by $z = x + jy$, where x and y are independent each of distribution $\mathcal{N}(c, \sigma^2)$. Then, the distribution of the amplitude $r = \sqrt{x^2 + y^2}$ can be shown to be

$$f_r(r|c, \sigma^2) = \frac{r}{\sigma^2} \exp\left[-\frac{(r^2 + |c|^2)}{2\sigma^2}\right] I_0\left(\frac{|c|r}{\sigma^2}\right), \ 0 \leq r < \infty, \tag{3.44}$$

where $I_0(\cdot)$ is the modified Bessel function of the first kind with order zero. For Rician fading, the total average power $\Omega = 2(\sigma^2 + |c|^2)$, with $|c|^2$ representing the power from the direct path (LOS) and σ^2 representing the power from the scattered component. In Rician fading channels, the ratio of the two power components, defined by $K = \frac{|c|^2}{\sigma^2}$, is often called the Rician factor.

3.4.5 Nakagami fading

In Rayleigh and Rician distributions, a fading channel is endowed with two statistical degrees of freedom corresponding to the two independent real and imaginary components. The Nakagami m-distribution is even more flexible, allowing a faded channel magnitude to have arbitrary degrees of freedom to characterize the various possible fading severities encountered in practice. The Nakagami model has been widely shown to provide a better fit to many urban multipath empirical data than its competitors. It is used for magnitude data. Let $\mathbf{z}$ denote the complex baseband sample data from a path coefficient, and denote its magnitude and power vectors, respectively, by

$$\mathbf{v} = [|z_1|, \cdots, |z_n|]^T,$$
$$\mathbf{w} = [|z_1|^2, \cdots, |z_n|^2]^T. \tag{3.45}$$

Suppose that $\mathbf{v}$ is a Nakagami vector with fading parameter m and covariance matrix $\mathbf{R}_v$. We are interested in its joint Nakagami PDF. However, such a joint distribution is not available in the literature. Recall that the power of a Nakagami variable follows the Gamma distribution. A simple statistical description does exist for jointly Gamma distributed variables in the form of a characteristic function. Denote the covariance matrix of $\mathbf{v}$ by $\mathbf{R}_v$, denote the transform variables of $w_i, i = 1, \ldots, n$ in the CHF domain by $t_i, i = 1, \ldots, n$, and denote and $\mathbf{T} = \text{diag}(t_1, \cdots, t_n)$. Then, the joint CHF of $\mathbf{w}$ is given by

$$\phi_{\mathbf{w}}(j\mathbf{t}) = \mathbb{E}[e^{jt_1 w_1 + \cdots + jt_n w_n}] = \det(\mathbf{I} - j\mathbf{TA})^{-m}, \tag{3.46}$$

where $\mathbf{A}$ is a positive-definite symmetrical matrix and its (p, q)th entry is obtainable from the corresponding one of $\mathbf{R}_v$ by

$$A(p, q) = \sqrt{R_v(p, q)/m}. \tag{3.47}$$

The fading parameter m ranges between $1/2$ and infinity, and is used to characterize the rate at which a path gain fluctuates. A smaller value of m corresponds to more severe fading. When $m = 1$, the channel reduces to the case of Rayleigh fading.

The PDF of a single Gamma variable with variance Ω and fading parameter m is explicitly expressible as

$$f_w(\xi) = \left(\frac{m}{\Omega}\right)^m \frac{\xi^{m-1}}{\Gamma(m)} \exp\left(-\frac{m_k \xi}{\Omega}\right), \quad \xi > 0. \tag{3.48}$$

One may imagine that w has m fading dimensions, and each dimension is of power Ω/m.

3.4.6 Clarke–Jakes model

In the above, we described various fading distributions for locally (i.e., short-term) fluctuating behaviors of a path-gain process. Yet, its correlation structure has not been specified. One of the most popular correlation structures for short-term fading is defined by the Clarke–Jakes model [8, 9], a model derived under the assumption of a homogeneous horizontal scattering field. Any signal transmitted therein is scattered by densely and uniformly distributed scatterers, forming a large number of horizontally traveling wave fronts to impinge upon the received antenna with random and independent angles and phases of arrival. Hence, a sinusoidal signal with carrier frequency ω_c and information phase θ_m, after propagating over a flat fading channel in such a random field, produces a received output of the form

$$x(t) = r(t) \cos(\omega_c t + \theta_m + \phi) = \Re\{z(t) \exp(j\omega_c t + \theta_m)\} \tag{3.49}$$

where $z(t) = r(t) \exp(j\phi)$ denotes the complex envelope. The use of the central limit theorem enables us to assert that $z(t)$ is a complex Gaussian process. The correlation structure of $z(t)$ is important in many applications such as differential demodulation. Assuming a homogeneous planar scattering field with scatterers uniformly distributed at angle $[0, 2\pi]$, Clarke [8] derives the following results:

Property 3.4.1 *Let $C(\tau)$ denote the correlation of $z(t)$ and let σ_z^2 denote its variance.*

$$C(\tau) = \mathbb{E}[z(t)z^*(t - \tau)]/\sigma_z^2 = J_0(2\pi f_d \tau), \tag{3.50}$$

where $J_0(\cdot)$ is the Bessel function of the first kind and order zero.

This property clearly uncovers the influence of the Doppler frequency shift f_d on the evolution of a path gain in time. The first zero-crossing point of $J_0(x)$ is $x = 2.40$, implying that two samples of $z(t)$ separated by $\tau = 0.382/f_d$ are totally uncorrelated. In this sense, we may say that the signal fades at a rate proportional to the Doppler frequency shift.

The Fourier transform of $C(\tau)$ provides the spectral structure of the fading process $z(t)$.

Property 3.4.2 *The baseband power spectral density is the Fourier transform of $C(\tau)$, given by*

$$S(f) \propto \begin{cases} \dfrac{1}{\pi f_d \sqrt{1 - \left(\frac{f}{f_d}\right)^2}}, & |f| \le f_d \\ 0, & \text{elsewhere} \end{cases} \tag{3.51}$$

The power spectrum is band-limited within the maximum Doppler shift $f_d = v/\lambda$, and has two poles at $f = \pm f_d$. The maximum Doppler shift can be determined by

$$f_d = \frac{v}{\lambda}, \tag{3.52}$$

where the wavelength λ of the carrier can be determined from the speed of light c and the carrier frequency f_c by $\lambda = c/f_c$.

Very often, $z(t)$ is sampled with an appropriate rate to generate a sequence of $\mathbf{z} = [z_1, z_2, \cdots, z_n]$ for the purpose of signal processing.

Property 3.4.3 *The complex envelope sequence $\mathbf{z}$ is jointly Gaussian distributed, as shown by*

$$\mathbf{z} \sim \mathcal{CN}(0, \sigma_z^2 \mathbf{C}), \tag{3.53}$$

with the (i, j)th component of $\mathbf{C}$ defined by $J_0(2\pi|i - j|f_d^\circ)$, where f_d° is the normalized Doppler frequency by the sampling rate.

The effect of transmission distance d on the propagation loss can be incorporated into the Jakes model in two ways. If the focus is only on the local behavior, we may write $\sigma_z^2 = P_T/d^n$ up to a constant factor, where P_T signifies the transmit power and n is the propagation loss exponent. If, on the other hand, the dynamic variation over a large scale is of major concern, we may consider σ_z^2 as a lognormal variable.

Property 3.4.4 *The magnitude of $z(t)$ is a Rayleigh variable, and the correlation between two arbitrary magnitude samples, say $|z(t)|$ and $|z(t - \tau)|$, is given by $J_0^2(2\pi f_d \tau)$.*

3.5   Composite model to incorporate multipath and shadowing

On carefully observing Figure 3.1, one can find that the received signal amplitude comprises two multiplicative components, which forms a composite channel model

$$\xi(t) = \ell(t)\, r(t) \tag{3.54}$$

with $r(t)$ representing the local fluctuation caused by multipath, and $\ell(t)$ characterizing the large-scale *shadowing* behavior. We want to derive the PDF of ξ in the composite model. Assume that $r(t)$ follows the Nakagami distribution $\mathcal{NK}(m, \Omega)$ with total power Ω and fading parameter m. As such, conditioned on ℓ or equivalently conditioned on the power $p = \ell^2$, ξ also follows the Nakagami distribution but with power $p\Omega$, with PDF

$$f_{\xi|p}(\xi|p) = \frac{2\xi^{2m-1}}{\Gamma(m)} \left(\frac{m}{p\Omega}\right)^m \exp\left(-\frac{m\xi^2}{p\Omega}\right), \ \xi > 0, \tag{3.55}$$

where $p = \ell^2$ follows the lognormal distribution, that is, $10 \log_{10}(p) \sim \mathcal{N}(0, \sigma_{dB}^2)$ with the dB spread typically falling into $\sigma_{dB} \in (5, 12)$ dB. In the linear scale, the PDF of p is given by

$$f_p(y) = \frac{10}{y(\ln 10)\sqrt{2\pi\sigma_{dB}^2}} \exp\left(-\frac{50 \log_{10}^2(y/d^n)}{\sigma_{dB}^2}\right), \tag{3.56}$$

which, when combined with (3.55), enables us the determine the PDF of ξ, yielding

$$f_\xi(\xi) = \int_0^\infty f_{\xi|p}(\xi|y) f_p(y) dy, \ \xi > 0. \tag{3.57}$$

Though focusing only on a single sample of a single path, the composite model (3.57) is easily extensible to cover the temporal evolution of a path gain by considering a sample vector of $\xi(t)$, denoted by **a**. Note that the large-scale shadowing $\ell(t)$ is slowly varying as compared to the local fluctuation $r(t)$, and is thus treated as a scalar lognormal random variable. As such, we can write $\mathbf{a} = \ell\mathbf{r}$. The generation of a joint Nakagami-distributed vector **r** will be addressed in Section 3.7.2. In fact, $r(t)$, or equivalently **r**, can be any other process such as a complex Gaussian. Further extension to the case with multiple paths is straightforward, given that different paths in the Turin model are independent.

3.6   Example to illustrate the use of various models

Each of the channel models we have developed so far has its own applications in wireless communication. The choice of a particular model depends on the application. As illustration, let us consider the issue of reuse pattern design and its outage performance evaluation.

3.6.1   Static design

Suppose a cellular system employs an N-cell reuse pattern with hexagonal geometry. Consider the worst case in which a mobile is located at a vertex of a hexagon and is simultaneously corrupted by $K = 6$ interfering stations on the first tier. Determine N such that the average signal to co-channel interference ratio (SINR) is not less than 17 dB. Since the average SINR is used as a criterion and the problem is focused on small-scale cellular geometry, it is appropriate to use the simple model $P_R = P_T/d^n$. As an approximation, assume that all the K interferers have the same distance to the receiver; namely, $D_1 = \cdots = D_K = D$. Thus, by assuming equal transmit power P_T for all $(K + 1)$ co-channel stations and denoting the cellular outer radius by R, we can write the SINR as

$$\frac{S}{I} = \frac{R^{-n}}{K \cdot D^{-n}} = \frac{1}{K}\left(\frac{D}{R}\right)^n.$$

Recall that for the hexagonal cell geometry, $D/R = \sqrt{3N}$. Note that 17 dB = 50.12. With $K = 6$ and propagation loss exponent $n = 4$, it yields

$$\frac{S}{I} = \frac{1}{6}(\sqrt{3N})^4 = 50.12 \rightarrow N = 5.78. \tag{3.58}$$

Since N must take the form $N = i^2 + ij + j^2$, the minimum value is $N = 7$. The corresponding SIR is $73.5 = 18.66$ dB, which satisfies the requirement.

3.6.2 Dynamic design

The above calculation represents a static design. In practice, both the wanted and interfering signals are randomly varying. We, therefore, need a dynamic evaluation of the resulting design of seven-cell reuse pattern in terms of its outage performance. The term *outage* represents the event of unsatisfactory reception over an intended service area caused by co-channel interference. The central idea of cellular systems is frequency reuse over different geographic locations. For this reason, the received signal comprises not only the desired signal $s_0(t)$ but also interferers $s_k(t)$ from the co-channel users, as shown by

$$z(t) = r_0 s_0(t) + \sum_{k=1}^{K} r_k s_k(t), \tag{3.59}$$

where r_i denote the corresponding channel gains. The noise component is dropped from $z(t)$ since it is negligible compared to the interference. Assume that all $s_k(t)$ have been normalized to have unit variance, so that the instantaneous received power from user k is equal to $\xi_k = |r_k|^2$. The desired signal power is ξ_0, and the total co-channel interference power is

$$\xi = \sum_{k=1}^{K} \xi_i. \tag{3.60}$$

Both ξ_0 and ξ are random variables. To ensure reliable communication, therandom SINR $\gamma = \xi_0/\xi$ must exceed a minimum protection threshold, say q. Otherwise, an outage event occurs. The outage probability is a fundamental measure of the service quality of a cellular system and is formally expressible as $P_{\text{out}} = \Pr\{\gamma < q\}$. The ratio γ involves nonlinear operation, and, furthermore, finding the PDF of the summation ξ implies an intractable multifold convolution. All these troubles can be removed if we work on the CHF domain, where a convolution in the original domain is simply transformed into a product. Following this philosophy, we rewrite the outage event as $\alpha = q\xi - \xi_0 > 0$ in terms of a linear combination of random variables, whereby

$$P_{\text{out}} = \Pr\{\alpha > 0\}. \tag{3.61}$$

To implement such a dynamic assessment, we select statistical models for all $|r_k|$. Assume that they are independent and follow the Nakagami distribution. In particular, $|r_k| \sim \mathcal{NK}(m_k, \Omega_k)$. The effect of the distance, P_T/d^n, is absorbed into the variance Ω_k. The CHF of α can be easily determined as

$$\phi_\alpha(t) = \left(1 + \frac{jt}{\lambda_0}\right)^{-m_0} \prod_{k=1}^{K} \left(1 - \frac{jqt}{\lambda_k}\right)^{-m_k}, \tag{3.62}$$

where $\lambda_k = m_k/\Omega_k$. Using the Gil-Palaez Lemma, we can directly calculate the outage probability from the CHF to obtain

$$P_{\text{out}} = \frac{1}{2} + \frac{1}{\pi} \int_0^\infty \frac{\Im\{\phi_\alpha(t)\}}{t} \, dt. \tag{3.63}$$

The integrand is smooth and well behaved, approaching $(q \sum_{k=1}^{K} \Omega_k) - \Omega_0$ as $t \to 0$ and tending to 0 as $t \to \infty$. For Rayleigh fading for which all $m_k = 1$, the above integral can be worked out to obtain exact results by using the residue theorem. See Problem 3.8 for details.

3.6.3   Large-scale design

Very often, it is required that the outage performance be satisfied over a large service area. The inclusion of the lognormal shadowing into the channel model provides a more dynamic environment to examine the system performance and, thus, often leads to more accurate results. Therefore, the composite model (3.54) is widely used in simulation-based system evaluations. The problem is that the use of the composite model often makes it very difficult to analyze, thereby failing to reveal a simple and elegant rule that dictates the system performance. As such, Viterbi et al. use only the pure lognormal model of the form (3.15) in their capacity evaluation of the CDMA cellular system [10]. In some application scenarios in which we are interested only in the local behavior of a wireless system, it suffices to incorporate the static exponentially decaying power model into (3.40) or (3.55) by setting the received power to $P_R = P_T/d^n$.

Theoretic analysis can provide useful guidelines for system design. More comprehensive performance assessment, however, must rely on system simulations. How to generate various typical channel environments for simulations is the subject of the next section.

3.7   Generation of correlated fading channels

In this section, we describe techniques for simulating several commonly used fading environments, including Rayleigh, Rician, Nakagami, and lognormal shadowing. Approximation of a lognormal sum by a single random variable is also addressed.

3.7.1   Rayleigh fading with given covariance structure

Rayleigh fading refers to the random behavior of an envelope process resulting from a complex Gaussian sequence. Consider a complex Gaussian L-vector

$$\mathbf{x} = \mathbf{u} + j\mathbf{v}, \tag{3.64}$$

the ith entry of which $x_i = u_i + jv_i$ has the magnitude

$$r_i = |x_i| = \sqrt{u_i^2 + v_i^2}, i = 1, \cdots, L. \tag{3.65}$$

The corresponding magnitude sequence

$$\mathbf{r} = [r_1, \cdots, r_L] \tag{3.66}$$

represents a random process subject to Rayleigh fading. Our task here is to construct a Rayleigh fading process **r** with a given correlation matrix

$$\mathbf{R}_r = \begin{pmatrix} R_r(1,1) & R_r(1,2) & R_r(1,3) \cdots & R_r(1,L) \\ R_r(2,1) & R_r(2,2) & R_r(2,3) \cdots & R_r(2,L) \\ \vdots & & & \vdots \\ R_r(L,1) & R_r(L,2) & R_r(L,3) \cdots & R_r(L,L) \end{pmatrix}, \tag{3.67}$$

where the off-diagonal entries $R_r(i,j), i \neq j$ can be written in terms of the correlation coefficient $\rho_r(i,j)$, such that

$$\rho_r(i,j) = \frac{R_r(i,j)}{\sqrt{R_r(i,i)R_r(j,j)}}. \tag{3.68}$$

For a stationary fading process, $\mathbf{R}_r$ is a Toeplitz matrix with $\rho_r(i,j) = \rho_r(i-j)$ and $\mathbf{R}_r^* = \mathbf{R}_r$.

Because of the closure property of the Gaussian distribution, we consider using a linear transform, defined by matrix $\mathbf{L}$, which operates on a white Gaussian vector $\mathbf{w} \sim \mathcal{CN}(0, \mathbf{I})$ to produce

$$\mathbf{x} = \mathbf{L}\mathbf{w} \tag{3.69}$$

such that the resulting Rayleigh sequence, defined by the magnitude of $\mathbf{x}$, is of the specified covariance matrix $\mathbf{R}_r$. If we know the covariance matrix $\mathbf{R}_x$ of $\mathbf{x}$, the transform matrix $\mathbf{L}$ is easy to determine by employing Cholesky decomposition. Thus, the key issue is to determine $\mathbf{R}_x$ or, more precisely, the correlation coefficients $\rho_x(i, j)$.

We may directly determine ρ_x from its counterpart ρ_r; the problem of so doing is that the resulting expression involves a quite complicated elliptic integral of the second kind [11, 12]. We note that the correlation coefficients ρ_y of the power sequence $\{y_i\} = \{x_i^2\}$ is much easier to handle, suggesting that ρ_y be used for the calculation of $\mathbf{R}_x$. More details are given in the subsequent discussion of the generation of Nakagami fading processes. Here, we only list the algorithm for generating Rayleigh fading channels.

Algorithm 3.1 *Generation of correlated Rayleigh fading channels. Denote $a = \pi/(4 - \pi)$ and $\xi = 1 - (\pi/4)$.*

(a) Determine ρ_y from ρ_r by using the relationship

$$\rho_r(i, j) = a \left[{}_2F_1 \left(-\frac{1}{2}, -\frac{1}{2}; 1, \rho_y(i, j) \right) - 1 \right].$$

　Efficient calculation can be implemented through Newton–Raphson type iteration.

(b) Employ ρ_y to determine $\mathbf{R}_x$ by using

$$R_x(i, j) = \frac{1}{2\xi} \sqrt{R_r(i, i) R_r(j, j) \rho_y(i, j)}$$

　for $i, j = 1, \cdots, L$.

(c) Cholesky-decompose $\mathbf{R}_x = \mathbf{L}\mathbf{L}^\dagger$ to obtain a lower triangular matrix $\mathbf{L}$.

(d) Generate an i.i.d. complex Gaussian L-vector $\mathbf{w} \sim \mathcal{CN}(0, \mathbf{I})$, which, when processed by the linear operator $\mathbf{L}$, produces the required Rayleigh fading vector $\mathbf{x} = \mathbf{L}\mathbf{w}$.

3.7.2　Correlated Nakagami fading

Two methods can be used to generate the Nakagami variable, that is, the decomposition method [13, 14] and the method by Sim [15].

It is well known that the power of a Nakagami variable, up to a scalar, follows the chi-square distribution, which can be directly synthesized from Gaussian variables. Thus, to generate an $L \times 1$ Nakagami sequence $\mathbf{z}$ with fading parameter m and covariance matrix $\mathbf{R}_z$, that is,

$$\mathbf{z} \sim \mathcal{NK}(m, \mathbf{R}_z), \tag{3.70}$$

we may start from $L \times 1$ Gaussian vectors $\mathbf{x}_k \sim \mathcal{N}(0, \mathbf{R}_x)$ and take the following steps to synthesize $\mathbf{z}$:

$$\{\mathbf{x}_k\} \xrightarrow{\Sigma} \mathbf{y} \xrightarrow{\sqrt{\ }} \mathbf{z}, \tag{3.71}$$

where $\mathbf{y} \sim \mathcal{GM}(m, \mathbf{R}_y)$ denotes the corresponding gamma distribution with covariance matrix $\mathbf{R}_y$. To generate $\mathbf{z}$ for a given $\mathbf{R}_z$, we need to determine $\mathbf{R}_x$ following the logical relation

$$\mathbf{R}_z \rightarrow \mathbf{R}_y \rightarrow \mathbf{R}_x. \tag{3.72}$$

The jointly gamma distributed vector $\mathbf{y}$ is uniquely characterized by its CHF

$$\phi_{\mathbf{y}}(\mathbf{s}) = \mathbb{E}[e^{\mathbf{y}^T \mathbf{s}}] = \det\left(\mathbf{I} - m^{-1/2}\mathbf{S}\mathbf{R}_y^{\odot\frac{1}{2}}\right)^{-m}, \tag{3.73}$$

where $\mathbf{s} = [s_1, \cdots, s_L]^T$ denotes the transform vector of $\mathbf{y}$ in the CHF domain, and $\mathbf{S} = \mathrm{diag}(s_1, \cdots, s_L)$. This CHF is the starting point of our derivations.

To proceed, denote

$$\mathbf{y}^{\odot\, r} = [y^r(1), \cdots, y^r(L)]^T$$

$$\xi = 1 - \frac{1}{m}\frac{\Gamma^2(m + \frac{1}{2})}{\Gamma^2(m)}$$

$$a = \frac{\Gamma^2(m + \frac{1}{2})}{\Gamma(m)\Gamma(m + 1) - \Gamma^2(m + \frac{1}{2})}. \tag{3.74}$$

We first determine the variance σ_y^2 of $y(i)$ from σ_z^2 of $z(i)$. Directly calculating the moments $\mathbb{E}[z^2]$ by using the Nakagami PDF and recognizing that $y = z^2$, we obtain

$$\sigma_y^2 = \frac{(\sigma_z^2)^2}{m\xi^2}. \tag{3.75}$$

We next show how to find the cross-correlation coefficient $\rho_y(i, j)$ between $y(i)$ and $y(j)$ from its counterpart $\rho_z(i, j)$ for $\mathbf{z}$. According to Nakagami [16], we have

$$\rho_z(i, j) = a\left[{}_2F_1\left(-\frac{1}{2}, -\frac{1}{2}; m; \rho_y(i, j)\right) - 1\right]. \tag{3.76}$$

Theoretically, given $\rho_z(i, j)$, we can find the corresponding $\rho_y(i, j)$ numerically via an iterative algorithm such as the Newton–Raphson method. Denote $\rho_y(i, j)$ by v, and denote its value after iteration n by v_n. Further, denote the iterative function

$$f(v) = a\left[{}_2F_1\left(-\frac{1}{2}, -\frac{1}{2}; m; v\right) - 1\right] - \rho_z(i, j). \tag{3.77}$$

Its derivative with respect to v can be easily determined as

$$\dot{f}(v) = \frac{a}{4m}\left[{}_2F_1\left(-\frac{1}{2}, -\frac{1}{2}; m; v\right)\right], \tag{3.78}$$

whereby the Newton–Raphson iteration for v is given by

$$v_{n+1} = v_n - \frac{f(v_n)}{\dot{f}(v_n)} \rightarrow \rho_y(i, j). \tag{3.79}$$

To initialize the iteration, we may set $v_0 = \rho_z(i, j)$. The iteration usually converges in a few steps. The results are used as $\rho_y(i, j)$ to form $\mathbf{R}_y$.

We next need to relate $\mathbf{R}_x$ to $\mathbf{R}_y$. Note that the CHF of $\mathbf{u} = \mathbf{x}^{\odot 2}$ is given by

$$\phi_{\mathbf{u}}(\mathbf{s}) = \mathbb{E}[e^{\mathbf{u}^T \mathbf{s}}] = \det(\mathbf{I} - 2\mathbf{S}\mathbf{R}_x)^{-1/2}, \tag{3.80}$$

which can be easily related to (3.73) if m is a multiple of half, as shown by

$$\phi_{\mathbf{y}}(\mathbf{s}) = \prod_{k=1}^{2m} \det(\mathbf{I} - 2\mathbf{S}\mathbf{R}_x)^{-1/2} = \det(\mathbf{I} - 2\mathbf{S}\mathbf{R}_x)^{-m}. \tag{3.81}$$

It follows that

$$\mathbf{R}_x = \frac{1}{2\sqrt{m}} \mathbf{R}_y^{\odot(1/2)}. \tag{3.82}$$

Recall that a product in CHF corresponds to the summation of independent variables in the original domain. Hence, we have

$$\mathbf{y} = \sum_{k=1}^{2m} \mathbf{x}_k^{\odot\,2}. \tag{3.83}$$

Summarizing (3.75)–(3.84), we can produce $\mathbf{R}_x$ directly from a given $\mathbf{R}_z$ as follows:

$$R_x(i,j) = \begin{cases} \frac{1}{2m\xi}\,\mathrm{var}[z(i)], & i = j \\[2mm] \frac{1}{2m\xi}\sqrt{\mathrm{var}[z(i)]\,\mathrm{var}[z(j)]\rho_y(i,j)}, & i \neq j \end{cases}, \tag{3.84}$$

where $\mathrm{var}[z(k)] = R_z(k,k)$.

To generate a Nakagami process with a general fading parameter $0.5 \leq m < \infty$, two techniques can be used. To proceed, denote

$$p = \lfloor 2m \rfloor, \tag{3.85}$$

where $\lfloor r \rfloor$ denotes the integer part of $r > 0$. The first technique is based on the observation in statistical theory that a linear sum of chi-square variables can be well approximated by a single chi-square variable. Application of this principle enables us to represent $\mathbf{y}$ as a linear sum of $p + 1$ independent variables. In our case, we can write

$$\mathbf{y} = \alpha \sum_{k=1}^{p} \mathbf{x}_k^{\odot\,2} + \beta\mathbf{x}_{p+1}^{\odot\,2}. \tag{3.86}$$

The first term on the right is a summation creating an order-p chi-square process, and the second term is a correction vector. The parameters α and β are determined by equating the first two moments of both sides of the above expression. The result is given by

$$\alpha = \frac{2pm + \sqrt{2pm(p + 1 - 2m)}}{p(p + 1)}$$

$$\beta = 2m - p\alpha. \tag{3.87}$$

The second technique is based on chi-square mixture [17]; namely, the desired sequence $\mathbf{y}$ is generated by probabilistically switching between two chi-square processes, as shown by

$$\mathbf{y} = \begin{cases} \sum_{k=1}^{p} \mathbf{x}_k^{\odot\,2}, & \text{with probability } q \\[2mm] \sum_{k=1}^{p+1} \mathbf{x}_{p+k}^{\odot\,2}, & \text{with probability } 1 - q \end{cases} \tag{3.88}$$

The probability q can be determined by equating one moment of the above expression. In particular, if the fourth moment is used, we obtain

$$q = \frac{p(p + 1 - 2m)}{2m}. \tag{3.89}$$

Algorithm 3.2 *To generate an $L \times 1$ Nakagami vector $\mathbf{z}$ of fading parameter m and covariance matrix $\mathbf{R}_z$, we may employ the following algorithm:*

(a) *Determine the coefficients a and ξ using (3.74).*
(b) *Generate the correlation coefficients $\rho_y(i,j)$ using (3.79).*
(c) *Generate $\mathbf{R}_x$ using (3.84).*

(d) Eigen-decompose $\mathbf{R}_x = \mathbf{U}\mathbf{\Lambda}\mathbf{U}^\dagger$, where $\mathbf{\Lambda} = diag(\lambda_1, \cdots, \lambda_L)$ is the diagonal matrix formed by the eigenvalues of $\mathbf{R}_x$. Denote $\mathbf{D} = diag(\sqrt{\lambda_1}, \cdots, \sqrt{\lambda_L})$.

(e) Generate the Gaussian vector $\mathbf{x}_k = \mathbf{U}\mathbf{D}\mathbf{e}_k$, where $\mathbf{e}_k \sim \mathcal{N}(0, \mathbf{I})$ for $k = 1, \cdots, p + 1$.

(f) Synthesize $\mathbf{y}$ through the decomposition technique (3.86) or the probabilistic switching model (3.88).

(g) Scale $\mathbf{y}$ to have a wanted average, and take the square root of the resulting vector to obtain $\mathbf{z} = \mathbf{y}^{\odot 1/2}$.

Let us say a few words on the two methods for the generation of $\mathbf{y}$. The first method directly produces a Nakagami z-sequence with the specified fading parameter m and covariance matrix $\mathbf{R}_z$. When using the second method for y-sequence generation, however, each realization of the resulting z-sequence, in general, does not possess the same fading parameter as specified. Therefore, it is more suitable for a simulative evaluation of the statistical performance of a system in correlated Nakagami fading. We also note that eigen decomposition, rather than its Cholesky counterpart, is used for $\mathbf{R}_x$ in the above algorithm. The reason is that the eigenvalues of $\mathbf{R}_x$ can be zero, especially for a Bessel-type correlation. In numerical calculation, these zero or near-zero eigenvalues can be negligibly negative, making the square-root operations involved in $\mathbf{D}$ unjustified. To avoid this situation, it is better to remove negative eigenvalues from $\mathbf{\Lambda}$ before evaluating $\mathbf{D}$. This goal can be achieved by using the MATLAB syntax: $\mathbf{K} = \mathbf{\Lambda} > 0; \mathbf{D} = \mathrm{sqrt}(\mathbf{K} \cdot *\mathbf{\Lambda})$.

Let us illustrate the use of the algorithm by an example.

✍ Example 3.4

We want to generate a correlated Nakagami process $\mathbf{z}$ with a specified correlation matrix. Since there is no commonly accepted covariance model for a general Nakagami process, we still assume $C_z(\tau) = J_0^2(2\pi f_d \tau)$ as its correlation structure for illustration. Fast or slow fading is a relative concept measured by the Doppler

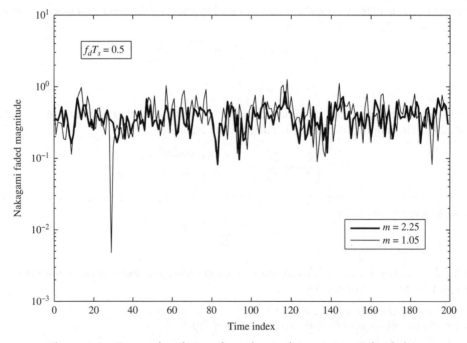

Figure 3.2 *Temporal evolution of a path gain that experiences fast fading*

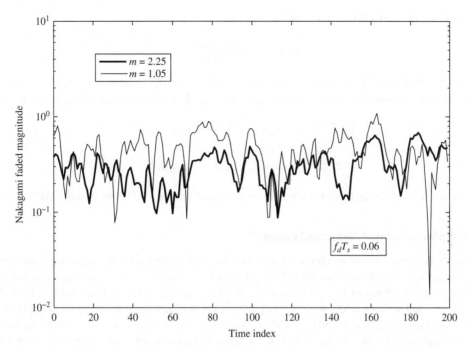

Figure 3.3 *Temporal evolution of a path gain that experiences slow fading*

shift f_d normalized by the signaling symbol rate $1/T_s$. Here, T_s is the symbol duration. To generate a Nakagami vector, we first need to discretize $C_z(\tau)$. For rectangular shaping, the system bandwidth is roughly equal to $1/T_s$, for which the Nyquist sampling rate is $f_{Nyq} = 2/T_s$. It implies that $C_z(\tau)$ is sampled at $\tau = n/f_{Nyq} = nT_s/2$ to produce $C_z(n) = J_0^2(\pi n f_d T_s)$. From the previous study, fast fading corresponds to $f_d T_s > 0.423$ and slow fading corresponds to $f_d T_s \ll 0.423$.

Typical channel fluctuations of fast fading and slow fading are shown, respectively, in Figures 3.2 and 3.3, where $f_d T_s$ is set to 0.5 for the former and to 0.06 in the latter. The first method is used for the generation of $\mathbf{y}$. It is observed that the channel with $m = 1.05$ can suffer a much deeper fading than its counterpart with $m = 2.25$.

3.7.2.1 *Estimation of the Nakagami-m parameter*

The power of the Nakagami variable follows the gamma distribution, as shown by

$$f(x; m, \Omega) = \frac{1}{\Gamma(m)} \frac{x^{m-1}}{\Omega^m} \exp(-x/\Omega). \tag{3.90}$$

Given independent samples $x_1, x_2, \cdots, x_n$ of the gamma random variable x, we evaluate their arithmetic and geometric means as

$$a = \frac{1}{n} \sum_{i=1}^{n} x_i, \; g = (x_1 x_2 \cdots x_n)^{1/n}. \tag{3.91}$$

Denote $y = \ln(a/g)$. Then the likelihood estimator of m is determined by the following equation:

$$\ln(m) - \psi(m) = y, \tag{3.92}$$

where $\psi(x)$ is the digamma function defined as

$$\psi(x) = \frac{d \ln \Gamma(x)}{dx}. \tag{3.93}$$

One may use the Newton–Raphson method to iteratively determine the solution of m, but this probably leads to a divergent series. Greenwood and Durand gave a more stable approximation, taking the form

$$\hat{m} = \begin{cases} \frac{0.5000876 + 0.1648852y - 0.0544274y^2}{y}, & 0 < y \le 0.5772 \\\\ \frac{8.898919 + 9.059950y + 0.9775373y^2}{y(17.79728 + 11.968477y + y^2)}, & 0.5772 < y \le 17 \end{cases} \tag{3.94}$$

The maximum errors of the estimators on the first and second lines are 0.0088% and 0.0054%, respectively.

3.7.3   Complex correlated Nakagami channels

Because of its flexibility in fitting empirical data of multipath fading [16], the Nakagami distribution has become popular in the wireless community. The Nakagami distribution is, in essence, the χ distribution. It is named after its investigator, just as $\chi(2)$ is named as Rayleigh fading. Note that both Nakagami and Rayleigh distributions are used to describe the statistical behavior of faded signal magnitudes. In many applications such as performance evaluation of a differential MPSK system, however, we need a model that can also account for both the amplitude and phase behavior. The Rayleigh distribution has such a complex counterpart – the complex Gaussian distribution. It is, therefore, desirable to have a similar counterpart for Nakagami fading. There is no difficulty to generate a single complex Nakagami variable [18], but it is difficult to generate a complex Nakagami process with a correlation matrix of arbitrary Hermitian structure. However, the situation becomes easier if we restrict the correlation to the structure of all real entries such as defined by the Jakes model. With such a structure, the real and imaginary components of the complex Nakagami process can be treated as independent, and each component can be generated by using the technique described in the previous subsection. In particular, we want to generate a complex process $\mathbf{w}$ with fading parameter m and covariance matrix $\mathbf{R}_w$. Then, we can write

$$\mathbf{w} = \mathbf{u} + j\mathbf{v},$$

such that $\mathbf{u}$ and $\mathbf{v}$ are independent, and each is distributed as $\mathcal{NK}(0.5m, \mathbf{R}_w)$. With this setting, the phase process is automatically defined by the two components, although its exact statistical characterization is still difficult to determine. Nevertheless, it is sufficient for use in system simulations. We know that the Jakes model has been derived for homogeneous planar fields with a large number of uniform scatterers. Thus, the Jakes model is usually used alongside Rayleigh fading with justification from the central limit theorem. The mechanism that generates Nakagami fading involves fewer scatterers. Since there is no well-understood correlation structure for Nakagami fading, one may approximately employ the Jakes model for the generation of a complex Nakagami process.

3.7.4   Correlated lognormal shadowing

In channel simulations, we often need to generate a corrected lognormal vector $\mathbf{z} \sim \mathcal{LN}(\mathbf{m}_z, \mathbf{R}_z)$ of mean $\mathbf{m}_z$ and covariance matrix $\mathbf{R}_z$, from a corresponding Gaussian vector $\mathbf{x} \sim \mathcal{N}(\mathbf{m}_x, \mathbf{R}_x)$, such that $\mathbf{z} = \exp(\mathbf{x})$. Given $\mathbf{m}_z$ and $\mathbf{R}_z$, we need to find $\mathbf{m}_x$ and $\mathbf{R}_x$. Let $z(i)$ denote the ith entry of $\mathbf{z}$, and let $R_z(i, j)$ denote the (i, j)th entry of $\mathbf{R}_z$. Similar notation is defined for $\mathbf{x}$ and $\mathbf{R}_x$. Then, it is straightforward to show that [19]

$$\mathbb{E}[z^r(i)] = \exp\left(rm_x(i) + \frac{r^2}{2}R_x(i,i)\right), r = 1, 2, \cdots$$

$$\mathbb{E}[z(i)z(j)] = \exp(m_x(i) + m_x(j) + 0.5R_x(i,i) + 0.5R_x(j,j) + R_x(i,j)). \tag{3.95}$$

In deriving the second assertion, the idea is to represent the exponent of $\mathbb{E}[z(i)z(j)] = e^{x(i)+x(j)}$ as a complete quadratic form such that its expectation equals 1, and the remaining factor is just as shown in the second line. It follows that

$$m_z(i) = \exp(m_x(i) + 0.5R_x(i,i))$$

$$\mathrm{var}[z(i)] = e^{2m_x(i)+R_x(i,i)}(e^{R_x(i,i)} - 1). \tag{3.96}$$

Solving (3.95) for the first two moments of $\mathbf{x}$ leads to

$$m_x(i) = \ln\left[\frac{m_z^2(i)}{\sqrt{\mathbb{E}[z^2(i)]}}\right],$$

$$R_x(i,j) = \begin{cases} \ln[\mathbb{E}[z^2(i)] - \ln[m_z^2(i)], & i = j \\ \ln\mathbb{E}[z(i)z(j)] - \ln[m_z(i)m_z(j)], & i \neq j \end{cases} \tag{3.97}$$

where the noncentral second moment $\mathbb{E}[z(i)z(j)]$ can be calculated by using

$$\mathbb{E}[z(i)z(j)] = R_z(i,j) + m_z(i)m_z(j).$$

Algorithm 3.3 *To generate an $L \times 1$ lognormal vector $\mathbf{z}$ of mean $\mathbf{m}_z = \mathbb{E}[\mathbf{z}]$ and correlation $\mathbf{\Omega} = \mathbb{E}[\mathbf{zz}^T]$, we may employ the following algorithm:*

(a) Determine the mean and covariance of $\mathbf{x} \sim \mathcal{N}(\mathbf{m}_x, \mathbf{R}_x)$ by using

$$m_x(i) = \ln\frac{m_z^2(i)}{\sqrt{\Omega(i,i)}}, \ i = 1, 2, \cdots, L,$$

$$R_x(i,j) = \ln\Omega(i,j) - \ln[m_z(i)m_z(j)], \ i, j = 1, 2, \cdots, L. \tag{3.98}$$

(b) Generate $\mathbf{x}$. Find the square root $\mathbf{Q} = \mathbf{R}_x^{1/2}$ such that $\mathbf{R}_x = \mathbf{QQ}^\dagger$ to obtain $\mathbf{Q}$. Then, invoke the MATLAB routine $randn(L, 1) + \mathbf{Q}^{-1}\mathbf{m}_x$ to generate $\mathbf{e} \sim \mathcal{N}(\mathbf{Q}^{-1}\mathbf{m}_x, \mathbf{I})$, which is then transformed to $\mathbf{x}$ by

$$\mathbf{x} = \mathbf{Qe}. \tag{3.99}$$

If Cholesky decomposition is used, the MATLAB function to generate $\mathbf{Q}$ is $\mathbf{Q} = \mathbf{U}'$, where $\mathbf{U} = chol(\mathbf{R}_x)$. Cholesky decomposition may come across a singular problem. If $\mathbf{R}_x$ is not of full rank, it is better to use eigen decomposition instead.
(c) Generate $\mathbf{z}$ through $\mathbf{z} = \exp(\mathbf{x}) = [e^{x_1}, e^{x_2}, \cdots, e^{x_L}].$

✍ **Example 3.5** _____

In Reference [20], the author assumes that the received signal z is a lognormal variable with a distance-dependent mean and correlation, given by

$$\log\mathbf{z} = \underbrace{K_1 - K_2\log(d)}_{\mu_z} + \mathbf{u}(d), \tag{3.100}$$

where $\mathbf{u} \sim \mathcal{N}(0, \mathbf{R}_u)$ with $R_u(d) = \mathbb{E}[u(d_1)u(d_2)] = \sigma_s^2 \exp(-|d_1 - d_2|/d_0)$. The authors of Ref. [21] also use the lognormal model to fit the L-band (1.3 GHz) and S-band (2.3 GHz) narrowband measurement data: $C(\tau) = \mathbb{E}[\xi(t)\xi(t-\tau)] - \mathbb{E}^2[\xi(t)] = \sigma_\xi^2 e^{-v(\tau)/x}$, where x is the effective correlation distance of shadow fading, and $v = |d_1 - d_2|$ is the velocity.

3.7.5 Fitting a lognormal sum

In cellular mobile communications, it is sometimes of practical interest to approximate a sum of independent random variables by a single variable so as to simplify the analysis of co-channel interference and capacity performance. Two examples are approximation to a lognormal sum and to a chi-square sum. In the previous study of Property 2.4.6 of the chi-square distribution, we have shown how to approximate a chi-square sum by a single chi-square variable. Thus, here we just focus on the approximation to a lognormal sum.

Since a single lognormal variable has a large dynamic range, the sum of a number of, say, m, lognormal variables has an even larger range. It is, therefore, natural to expect that the summation, after a log operation, should have a much smaller dynamic range and approach Gaussianity. This type of reasoning suggests the use of a lognormal approximation. Simulations shown in Figure 3.1 of Ref. [22], however, demonstrate a poor match, especially as the number of lognormal summands increases. The discrepancy is not difficult to explain from statistical theory. Approximating a lognormal sum with a single lognormal variable implies accepting the assertion that the closure property is applicable to the lognormal distribution. Such an assertion, however, is not justified in probability theory. We, therefore, seek a better fitting model from the commonly used Pearson's or Johnson's family of distributions.

Most of distributions used in science and engineering come from the former family, while the latter is often derivable by transforming the original data the logarithm. The Pearson family includes seven types of distributions with PDFs given by Stuart and Ord [23, 24].

$$f(x) = \begin{cases} \frac{1}{B(p,q)} x^{p-1}(1-x)^{q-1}, x \in [0,1] & \text{Type 1} \\[2mm] \frac{1}{a\,B(0.5,m+1)}\left(1 - \frac{x^2}{a^2}\right)^m, x \in [-a,a] & \text{Type 2} \\[2mm] k\left(1 + \frac{x}{a}\right)^p \exp(-px/a), x \in [-a,\infty) & \text{Type 3} \\[2mm] k\left[1 + \left(\frac{x+u}{d}\right)^2\right]^{-m} \exp\{-v\arctan\frac{x+u}{d}\}, -\infty < x < \infty & \text{Type 4} \\[2mm] \frac{\gamma^{p-1}}{\Gamma(p-1)} x^{-p} \exp(-\gamma/x) & \text{Type 5} \\[2mm] \frac{1}{B(p,q)} x^{p-1}(1+x)^{-(p+q)}, x \in [0,\infty) & \text{Type 6} \\[2mm] \frac{1}{a\,B(0.5,m-0.5)}[1 + (x/a)^2]^{-m}, x \in [a,-a] & \text{Type 7.} \end{cases}$$

The normal names for Pearson types 1, 3, 6, and 7 are beta, gamma, beta of the second kind, and Student's-t distribution, respectively. Suppose we want to fit an appropriate model in the Pearson family to a given data sequence. A decision can be made based on the first four moments evaluated from the data. Let μ_1' denote the mean of the lognormal sum data $\{\xi_i\}$, and let μ_k denote the kth moment about μ_1'. The three decision variables are defined as

$$\beta_1 = \mu_3^2/\mu_2^3, \beta_2 = \mu_4/\mu_2^2, \quad \kappa = \frac{\beta_1(\beta_2 + 3)^2}{4(2\beta_2 - 3\beta_1 - 6)(4\beta_2 - 3\beta_1)} \tag{3.101}$$

Table 3.2 *Selection criteria for model type*

Model type	Selection criterion
1	$\kappa < 0$
2	$\kappa = 0, \beta_2 < 3$
3	$2\beta_2 - 3\beta_1 - 6 = 0$, that is $\kappa = \infty$
4	$0 < \kappa < 1$
5	$\kappa = 1$
6	$\kappa > 1$
7	$\kappa = 0, \beta_2 > 3$
Normal	$\kappa = 0, \beta_2 = 3$

by which the decision regions can be partitioned. The resulting decision criteria are tabulated in Table 3.2; they are applicable to general data fitting.

Let us return to the problem of fitting a lognormal sum. Consider the sum $\xi = \xi_1 + \xi_2 + \xi_3 + \xi_4$ of four lognormal variables generated from four independent lognormal constituents with dB spread equal to $\sigma_{\mathrm{dB}} = 6, 8, 10$, and $12\,\mathrm{dB}$, respectively. We do not directly fit ξ, but rather take $z = \ln(\xi)$ and fit z instead. After calculation, we obtain

$$\beta_1 = 0.4344, \beta = 4.0162, \kappa = 0.4968,$$

which, according to the model-selection table, suggests the use of the Pearson type-4 model for fitting z. The corresponding parameters can be determined in terms of the moments of z, yielding

$$u = \frac{a}{2b_2} - \mu_1', m = -\frac{1}{2b_2}, d = \frac{\sqrt{4b_0 b_2 - a^2}}{2|b_2|}, v = \frac{a(1 + 2b_2)}{2b_2^2 d},$$

where

$$b_0 = -[\mu_2(2\mu_2\mu_4 - 3\mu_3^2)]/A,$$
$$b_2 = -[2\mu_2\mu_4 - 3\mu_3^2 - 6\mu_2^3]/A,$$
$$a = -[\mu_3(\mu_4 + 3\mu_2^2)]/A,$$

with A defined by $A = 10\mu_2\mu_4 - 12\mu_3^2 - 18\mu_2^3$.

3.8 Summary

In wireless communication, the propagation of electromagnetic waves over scattering fields is governed by both physical and statistical mechanisms. The physical mechanism enables us to characterize the large-scale average propagation loss behavior by various empirical formulas such as the Okumura and Hata models, and to characterize the local multipath behavior by a tap-delayed line filter with random coefficients. The rms delay spread of a channel filter defines its coherence bandwidth. The coherence bandwidth is a relative concept that can be used to classify physical channels into two categories: flat fading and frequency-selective fading. Two techniques were provided for coherence-bandwidth estimation, one based on an empirical formula and the other based on statistical analysis. The tap coefficients of a channel filter vary with time, due to the relative motion of mobile units. The temporal variation of each tap coefficient is usually characterized by its coherence time, which enables the classification of physical channels into fast-fading and slow-fading. Fast or slow fading is a relative concept caused by Doppler frequency shift versus signaling bandwidth.

Besides their filter attributes, multipath channels exhibit various statistical behaviors dictated by certain statistical mechanisms. Simulating these behaviors is essential for a complete performance evaluation of cellular systems in a laboratory environment. We have described algorithms for the generation of correlated Rayleigh, Rician, lognormal, and Nakagami fading channels.

3.9   Additional reading

For additional reading on channel modeling and characterization, readers are referred to [26–34].

Problems

Problem 3.1   Consider Channel A recommended in the JTC Standards for outdoor urban high-rise and high antenna, with tap parameters given below:

Tap	1	2	3	4	5	6	7	8
Relative Delay (ns)	0	50	250	300	550	800	2050	2675
Average power (dB)	0	−0.4	−6.0	−2.5	−4.5	−1.2	−17.0	−10.0

Determine its coherence bandwidth [3].

Problem 3.2   Employ Algorithm 3.2 to generate a correlated Nakagami process with parameters and correlation of your choice. Then, use the formula (3.94) to numerically estimate its fading parameter and power. Compare your estimation with the true values, and offer your comments.

Problem 3.3   In a Rician channel, denote the total received power as $\Omega = |c|^2 + \sigma^2$ in terms of its mean c and variance σ^2. Represent the PDF given in (3.44) in terms of Ω and the K factor.

Problem 3.4   Derive an algorithm for the generation of a Rician fading process with distribution $\mathcal{CN}(\mathbf{c}, \mathbf{R})$.

Problem 3.5   Denote the chi-variate with m degrees of freedom by $x \sim \chi(m)$, and denote the Nakagami distribution with variance Ω and fading parameter m by $y \sim \mathcal{NK}(m, \Omega)$. Their PDFs are given, respectively, by

$$\chi(m) : f(x) = \frac{x^{m-1}e^{-x^2/2}}{\Gamma(m/2)2^{(m/2)-1}}, \ x > 0$$

$$\mathcal{NK}(m, \Omega) : f(y) = \frac{2y^{2m-1}}{\Gamma(m)}\left(\frac{m}{\Omega}\right)^m \exp\left(-\frac{my^2}{\Omega}\right), \ y > 0.$$

Determine the relationship between x and y.

Problem 3.6   Generate a correlated lognormal shadowing process, which, in dB scale, is denoted as $x(t)$. Assume that $x(t)$ has an exponential type of auto-correlation

$$C_x(k) = \sigma_x^2 a^{|k|},$$

where σ_x^2 denotes the variance of $x(t)$, typically ranging from 3 to 10 dB, a denotes its correlation coefficient, and integer k represents the separation between two of its samples. More specifically, we may write $a = c_d^{vT/d}$, where T is the sampling interval, v is the velocity, and c_d is the correlation between two points separated by distance d [20].

Problem 3.7   In a cellular systems, a desired signal x_0 is usually corrupted by a number of co-channel interference signals, say, $x_i, i = 1, \cdots, M$, whose powers are denoted by $\xi_i = \alpha_i 10^{0.1u_i}$. The factor $\alpha_i = P_i/d_i^n$ represents the average power determined by the transmitted power and transmission distance, and u_i represents random fluctuation due to longnormal shadowing in the dB scale. In other words, $u_i \sim \mathcal{N}(0, \sigma_i^2)$ Gaussian variable with dB spread σ_i. The SIR is then defined by

$$\gamma = \frac{\alpha_0 10^{0.1u_0}}{\sum_{i=1}^{K} \alpha_i 10^{0.1u_i}},$$

where, typically, $K = 6$. For a preset threshold Λ, outage occurs when $\gamma < \Lambda$. Show that the outage probability is upper-bounded by

$$P_{\text{out}} = \Pr\{\gamma < \Lambda\} \leq Q(\Omega/\sigma),$$

where $\sigma^2 = [\sigma_0^2 + (1/K)\sum_{i=1}^{K} \sigma_i^2]$, $\Omega = 10\log_{10}(\alpha\Omega)$, and $\alpha = K(\prod_{i=1}^{K} \alpha_i)^{1/K}/\alpha_0$ [25].

Problem 3.8   In (3.63), assume Rayleigh fading for which all $m_k = 1$. Derive a closed-form solution for the outage probability.

Problem 3.9   The design of cellular reuse pattern is usually based on static SINR. We would like its relation to the expected value of a dynamic model. Determine the mean SINR in a Rayleigh fading environment by assuming K i.i.d. interferers.

Problem 3.10   A student employed the AR(1) process to generate a sequence to simulate a fading channel with Doppler spectrum defined by the Jakes model, and found very poor fitting. Explain why that happened and suggest a method for improvement.

Problem 3.11   Suppose two independent Nakagami variables, $x, y \sim \mathcal{NK}(m/2, \Omega)$, are used to construct a complex Nakagami one, such that $z = x + jy$. Prove that the phase θ of z is distributed as

$$f(\theta) = \frac{\Gamma(m)}{2^m \Gamma^2(m/2)} |\sin(2\theta)|^{m-1}, \quad \text{where} \ \ \theta \in [0, 2\pi).$$

[18].

Problem 3.12   A user in a cellular system receives the intended signal $x_0 \sim \mathcal{CN}(0, \sigma_0^2)$, which is corrupted by three independent co-channel interferers $x_1 \sim \mathcal{CN}(0, \sigma_1^2)$, $x_2 \sim \mathcal{CN}(\mu, \sigma_2^2)$, and $|x_3| \sim \mathcal{NK}(m, \Omega)$. The signal to interference (SIR) is defined by

$$\eta = \frac{|x_0|^2}{|x_1|^2 + |x_2|^2 + |x_3|^2}.$$

(1) Determine the average SIR $\mathbb{E}[\eta]$. (2) Assuming $\sigma_0^2 = 8, \sigma_1^2 = 3, \sigma_2^2 = 2, \mu = 1, m = 2$, and $\Omega = 4$, calculate the value of the average SIR.

Problem 3.13   Given a lognormal vector $\mathbf{z}$ with mean $\mathbf{m}_z = [1.2, 1.25, 1.15, 1.08]^T$ and covariance matrix $\mathbf{R}_z = toeplitz(\mathbf{r})$, where $\mathbf{r} = [1, 0.795, 0.605, 0.375]$. Determine the mean $\mathbf{m}_x$ and covariance matrix $\mathbf{R}_x$ of the corresponding Gaussian vector $\mathbf{x}$.

References

1. B. Sklar, "Rayleigh fading channels in mobile digital communication systems-Part I: characterization," *IEEE Commun. Mag.*, vol. 35, no. 7, pp. 90–100, 1997.
2. Y. Okumura, E. Ohmori, T. Kawano, and K. Fukuda, "Field strength and its variability in VHF and UHF land mobile service," *Rev. Electr. Commun. Lab.*, vol. 16, pp. 825–873, 1968.
3. M.G. Laflin, "Draft final report on RF channel characterization," in *Joint Technical Committee (AIR) Standard Contribution*, Dec. 6, 1993.
4. Q.T. Zhang and S.H. Song, "Exact expression for the coherence bandwidth of Rayleigh fading channels," *IEEE Trans. Commun.*, vol. 55, no. 7, pp. 1296–1299, 2007.
5. K.S. Miller, *Complex Stochastic Processes: An Introduction to Theory and Applications,* Reading, MA: Addison-Wesley, 1974.
6. D.C. Coc and R.R. Leck, "Correlation bandwidth and delay spread multipath propagation statistics for 910-Mhz urban mobile radio path," *IEEE Trans. Commun.*, vol. 23, no. 11, pp. 1271–1280, 1975.
7. T.S. Rappaport, *Wireless Communications: Principles and Practice*, 2nd edn, Upper Saddle River, NJ: Pearson Education, 2002.
8. R.H. Clarke, "A statistical theory of mobile radio reception," *Bell Syst. Tech. J.*, vol. 47, pp. 957–1000, 1968.
9. W.C. Jakes, *Microwave Mobile Communications*, New York: John Wiley & Sons, Inc., 1974.
10. K.S. Gilhousen, I.M. Jacobs, R. Padovani, A.J. Viterbi, L.A. Weaver Jr., and C.E. Wheatley III, "On the capacity of a cellular CDMA system," *IEEE Trans. Veh. Technol.*, vol. 40, no. 2, pp. 303–312, 1991.
11. R.B. Retel and J.H. Reed, "Generation of two equal power correlated Rayleigh fading envelopes," *IEEE Commun. Lett.*, vol. 2, no. 10, pp. 276–278, 1998.
12. B. Natarajan, C.R. Bassar, and V. Chandrasekhar, "Generation of correlated Rayleigh fading envelopes for spread spectrum applications," *IEEE Commun. Lett.*, vol. 4, no. 1, pp. 9–11, 2000.
13. Q.T. Zhang, "A decomposition technique for efficient generation of correlated Nakagami fading channels," *IEEE J. Sel. Areas Commun.*, vol. 18, no. 11, pp. 2385–2392, 2000.
14. K. Zhang, Z. Song, and Y.L. Guan, "Simulation of Nakagami fading channels with arbitrary cross-correlation and fading parameters," *IEEE Trans. Wireless Commun.*, vol. 3, no. 5, 1463–1468, 2004.
15. C.H. Sim, "Generation of Poisson and Gamma random vectors with given marginal and covariance matrix," *J. Stat. Comput. Simul.*, vol. 47, no. 1-2, pp. 1–10, 1993.
16. M. Nakagami, "The m-distribution-A general formula of density distribution of rapid fading," in *Statistical Methods of Radio Propagation*, W.C. Hoffman, Ed. Oxford: Pergamon, 1960, pp. 3–36.
17. G.T. Freitas de Abreu, "On the moment-Determinance and random mixture of Nakagami-m variates," *IEEE Trans. Commun.*, vol. 58, no. 9, pp. 2561–2575, 2010.
18. M.D. Yacoub, G. Fraidenraich, and J.C.S. Santos Filho, "Nakagami-m phase-envelope joint distribution," *Electron. Lett.*, vol. 41, no. 5, 2005.
19. S.J. Press, *Applies Multivariate Analysis*, New York: Holt, Rinehart and Winston, 1972.
20. M. Gadmundson, "Correlation model for shadowing in mobile systems," *Electron. Lett.*, vol. 27, no. 23, pp. 2145–2146, 1991.
21. P. Taaghol and R. Tafazolli, "Correlation model for shadow fading in land-mobile satellite systems," *Electron. Lett.*, vol. 33, no. 15, pp. 1287–1289, 1997.
22. G.L. Stuber, *Principles of Mobile Communications*, 2nd edn, Boston, MA: Kluwer Academic Publishers, 2001, pp. 129–139.
23. A. Stuart and J.K. Ord, *Kendall's Advanced Theory of Statistics*, London: Charles Griffin, 1987, pp. 210–220.
24. M.D. Springer, *The Algebra of Random Variables*, New York: John Wiley & Sons, Inc., 1979, pp. 254–256.
25. F. Berggren and S.B. Slimane, "A simple bound on the outage probability with lognormal interferers," *IEEE commun. Lett.*, vol. 8, no. 5, pp. 271–273, 2004.
26. *LTE-Advanced: 3GPP TS 36.211*, Relase 10, Sep. 2011.
27. N. Beaulieu, "Generation of correlated Rayleigh fading envelopes," *IEEE Commun. Lett.*, vol. 3, no. 6, pp. 172–174, 1999.

28. C.S. Patel, G.L. Stuber, and T.G. Pratt, "Comparative analysis of statistical models for the simulation of Rayleigh faded cellular channels," *IEEE Trans. Commun.*, vol. 53, no. 6, pp. 1107–1026, 2005.

29. D.J. Young and N.C. Beaulieu, "The generation of correlated Rayleigh random variates by inverse Fourier transform," *IEEE Trans. Commun.*, vol. 48, no. 7, pp. 1114–1127, 2000.

30. J. W. Craig, "A new, simple and exact result for calculating the probability of error for two-dimensional signal constellations," *IEEE MILCOM Conference Record*, Boston, MA, 1991, pp. 25.5.1–25.5.5. problems," *Proceedings of the IRE*, vol. 44, 1956, pp. 609–638.

31. D.C. Cox and R.P. Leck, "Correlation bandwidth and delay spread multipath propagation statistics for 910-MHz urban mobile radio channels," *IEEE Trans. Commun.*, vol. 23, no. 11, pp. 1271–1280, 1975.

32. G.L. Turin, F.D. Clapp, T.L. Johnson, S.B. Fine, and D. Lavry, "A statistical model of urban multipath propagation," *IEEE Trans. Veh. Technol.*, vol. 21, pp. 1–9, 1972.

33. D. Giancristofaro, "Correlation model for shadow fading in mobile channels," *IEE Electron. Lett.*, vol. 32, no. 11, pp. 958–959, 1996.

34. E. Biglieri, J. Proakis, and S. Shamai, "Fading channels: information-theoretic and communications aspects," *IEEE Trans. Inf. Theory*, vol. 44, no. 6, pp. 2619–2692, 1998.

4

Digital Modulation

Abstract

Digital modulation, in nature, involves mapping a set of discrete symbols into a set of waveforms for efficient transmission. For wireless communication, a convenient way to implement digital modulation is to use information symbols to modulate the amplitude, phase, or frequency of a sinusoidal carrier. Each set of the modulated signals can be viewed as continuous-time waveforms in the functional space or as signal points in the Euclidean space. This duality in signal representation offers insight into the nature of different digital modulation schemes and provides flexibility in their performance analysis. Besides their geometrical implications, different modulation schemes have their physical significance in terms of spectral efficiency, power efficiency, and reliability in signal detection. In this chapter, we focus on typical modulation schemes such as MPSK, square M-QAM, coherent and noncoherent MFSK, and differential MPSK, with an emphasis on revealing their physical/geometrical nature and demonstrating the ideas, formulation, and methodology used in their performance analysis.

♣

4.1 Introduction

Digital transmission over wireless channels, such as cellular, microwave radio, and satellite links, requires carrier modulation for a number of reasons. First, as a very precious resource, the frequency spectrum is regulated by the governments and international standardization bodies. Different users and different applications are allowed to use only a particular frequency band for communications. The baseband information is, therefore, to translate into that particular frequency band via the technique of digital modulation. The second consideration is the size limit of a transceiver. For radio transmission, a carrier is converted to electromagnetic fields by using an antenna before propagation through space. The size of the antenna highly depends on the carrier frequency, with the dimension of at least half the wavelength λ of the carrier. For a cellular system operating at 900 MHz, $\lambda = 3 \times 10^8/(900 \times 10^6) = 1/3$ m. If the carrier frequency reduces to 30 MHz, the wavelength increases to $\lambda = 10$ m.

Wireless Communications: Principles, Theory and Methodology, First Edition. Keith Q.T. Zhang.
© 2016 John Wiley & Sons, Ltd. Published 2016 by John Wiley & Sons, Ltd.
Companion Website: www.wiley.com/go/zhang7749

A carrier usually takes the form of a sine wave, which has only three parameters to control: that is, amplitude, frequency, and phase. Thus, data symbols can be used to modulate one of these parameters or their combination. In digital modulation, the amplitude, frequency, or phase of a sinusoidal carrier remains unchanged over each symbol period but switching (keying) from one symbol to another according to the pattern of the data symbol sequence. Thus, the four types of basic signaling schemes are amplitude-shift keying (ASK), phase-shift keying (PSK), frequency-shift keying (FSK), and combination of the first two (QAM). Subsequently, M-ary ASK, M-ary PSK, M-ary FSK, and M-ary QAM are abbreviated as MASK, MPSK, MFSK, and M-QAM, respectively.

These signaling schemes have their own advantages and limitations in terms of spectral efficiency, error performance, and implementation complexity. A system designer needs to choose the one that best matches the system and channel requirements. The issues of spectral efficiency and complexity are relatively easy to analyze. The most difficult issue arises from their error performance analysis, even for the simplest case of additive white Gaussian noise (AWGN) channels. The methods and skills used for analysis are flexible, providing a very good training platform for graduate students. Undergraduate students may be more interested in relevant system techniques and knowledge; but they will find the methodology used for formulation interesting.

A powerful tool to study digital modulation and their performance analysis is the concept of signal space. A coherent MPSK or M-QAM signal possesses in-phase and quadrature components, represented in terms of sine and cosine functions. In other words, such a signal has two coordinates that can be more compactly represented as a single complex number. Complex operations are much easier to handle than their two-dimensional vector counterparts, explaining the popularity of the baseband complex representation in digital communication.

4.2   Signals and signal space

In digital communication, information symbols are mapped onto signal waveforms of certain energy, via a process known as *modulation*, to ensure a necessary signal-to-noise ratio (SNR) for reliable transmission over a physical channel. For example, in an M-ary signaling scheme, we need M signal waveforms $\{s_1(t), \cdots, s_M(t)\}$, where $0 \leq t < T_s$, to represent M symbols $\{s_1, \cdots, s_M\}$ each of $\log_2 M$ bits. At the receiver end, we need to recover the original information based on a noisy, and possibly faded, continuous-time received signal waveform under some criteria such as maximum likelihood (ML) or maximum *a posteriori* probability (MAP). Symbol retrieval is, in essence, a statistical decision process, requiring the conversion of the continuous-time received waveform back into a random variable in the Euclidean space to facilitate receiver derivation and performance analysis. Representing all possible signal waveforms in a Euclidean space constitutes what is called *signal constellation*, and the corresponding Euclidean space is called the *signal space*. Transforming signals between the continuous-time and Euclidean spaces reflects the duality of signal representation. Continuous waveforms facilitate physical transmission, whereas a vector representation provides intuitive insight into digital modulation.

A given set of M signal waveforms has its fixed signal constellation, irrespective of the basis functions to be used. The choice of a suitable set of basis functions for signal representation depends on the application. Theoretically, given M arbitrary waveforms $s_k(t)$, one can always construct a complete set of $L \leq M$ orthonormal basis functions $\{\phi_1(t), \cdots, \phi_L(t)\}$ for M signals by using the standard Gram–Schmidt procedure, such that

$$< \phi_i(t), \phi_j(t) > = \begin{cases} 1, & i = j \\ 0, & i \neq j \end{cases}. \tag{4.1}$$

Very often, there is no need to work out these basis functions explicitly; it suffices to know the existence of such basis functions, since our purpose is to derive the optimal receiver structure and basis functions are just a tool to achieve this goal. With these basis functions, all the signals are exactly expressible as their linear combination, as shown by

$$s_k(t) = \sum_{i=1}^{L} a_{ki}\phi_i(t), \tag{4.2}$$

where a_{ki} is the projection $s_k(t)$ on the basis $\phi_i(t)$, given by

$$a_{ki} = \int_0^{T_s} s_k(t)\phi_i(t)\, dt. \tag{4.3}$$

In so doing, the M signals in their functional space are mapped onto M signal points in the L-dimensional signal space given by

$$\mathbf{s}_k = \begin{bmatrix} a_{k1} \\ \vdots \\ a_{kL} \end{bmatrix} \tag{4.4}$$

with $k = 1, \cdots, M$. From (4.2), it is easy to see that

$$\int_0^{T_s} s_k^2(t)\, dt = \sum_{i=1}^{L} |a_{ki}|^2 = \mathbf{s}_k^T \mathbf{s}_k, \tag{4.5}$$

which is a restatement of the Parseval's theorem.

Property 4.2.1 *The mapping in (4.2) has the following properties:*

(a) The signal energy (or equivalently, the length) remains unchanged before and after mapping.
(b) The signal constellation is invariant under the mapping. The use of the Euclidean space, however, makes the signal geometry more visible and intuitive.
(c) Given the properties above, we say the two space are equivalent.

4.3   Optimal MAP and ML receivers

Suppose that a symbol waveform $s_i(t)$, chosen from a signal set $\{s_1(t), \cdots, s_M(t)\}$, is transmitted through an AWGN channel, producing channel output

$$y(t) = s_i(t) + n(t), \quad i = 1, \cdots, M; \ 0 \le t < T_s \tag{4.6}$$

where $n(t)$ is a bandlimited AWGN process with mean zero and two-sided power spectral density $N_0/2$ W/Hz. We need a vector expression for $y(t)$ for ease of design and analysis. Recall that M signal waveforms span a finite signal space of at most M dimensions with its orthonormal basis functions constructed by using, for example, the Gram–Schmidt procedure. Assume that the true signal dimension is $L \le M$, whereby the signal $s_k(t)$ can be represented by an L-vector $\mathbf{s}_k$. However, the noise component usually possesses infinite dimensions and, thus, the same is true for the received signal $y(t)$. Denote the projection of the noise $n(t)$ and the noisy signal $y(t)$ onto the signal space by the L-vectors $\mathbf{n}$ and $\mathbf{y}$, respectively. Thus, in the infinite-dimensional Euclidean space, we can represent the noise vector $\mathbf{n}_\infty$, the signal vector $\mathbf{s}_{k,\infty}$, and the received signal vector $\mathbf{y}_\infty$ as

$$\mathbf{n}_\infty = \begin{bmatrix} \mathbf{n} \\ n_{L+1} \\ \vdots \\ n_\infty \end{bmatrix}, \ \mathbf{s}_{k,\infty} = \begin{bmatrix} \mathbf{s}_k \\ 0 \\ \vdots \\ 0 \end{bmatrix}, \ \mathbf{y}_\infty = \begin{bmatrix} \mathbf{y} \\ y_{L+1} \\ \vdots \\ y_\infty \end{bmatrix} \tag{4.7}$$

whereby the received signal $y(t)$ can be rewritten as

$$\mathbf{y}_\infty = \mathbf{s}_{k,\infty} + \mathbf{n}_\infty = \begin{bmatrix} \mathbf{s}_k + \mathbf{n} \\ n_{L+1} \\ \vdots \\ n_\infty \end{bmatrix}. \tag{4.8}$$

Since only the first L coordinates of $\mathbf{y}_\infty$ contain signal components, it is sufficient to use $\mathbf{y}$ for signal detection. This observation is known as the *theorem of irrelevance* in the early literature. The above vector expression will be used for the subsequent derivation of the optimum receivers.

Let $p(s_i)$ denote the *priori* occurrence probability of waveform $s_i(t)$, and suppose that MAP is used as the criterion for receiver optimization. Given a received signal $y(t)$, the transmitted symbol is estimated as

$$\hat{s}_\ell = \arg \max_i \{p(s_i) \; \Pr\{y(t)|s_i\}\}$$

$$= \arg \max_i \left\{ p(s_i) \exp\left(-\frac{1}{N_0} \|\mathbf{y}_\infty - \mathbf{s}_{i,\infty}\|^2 \right) \right\}$$

$$= \arg \min_i \{\|\mathbf{y}_\infty - \mathbf{s}_{i,\infty}\|^2 - N_0 \ln p(s_i)\} \tag{4.9}$$

$$= \arg \min_i \{\|\mathbf{y} - \mathbf{s}_i\|^2 - N_0 \ln p(s_i)\}\} \tag{4.10}$$

$$= \arg \max_i \{\Re\{\mathbf{y}^\dagger \mathbf{s}_i\} + \alpha_i\}, \tag{4.11}$$

where

$$\alpha_i = \frac{N_0}{2} \ln p(s_i) - \frac{1}{2} E_i,$$

$$E_i = \|s_i(t)\|^2 = \|\mathbf{s}_i\|^2 = \int_0^{T_s} s_i^2(t) dt. \tag{4.12}$$

An MAP receiver to implement the coherent detection of M arbitrary waveforms is sketched in Figure 4.1.

It is clear that the optimum MAP receiver has a correlator structure comparing the resemblance between the received signal $y(t)$ and all possible waveforms. The receiver selects the most resembling waveform as the estimated symbol. The optimum MAP receiver is a *coherent* receiver since it assumes perfect knowledge

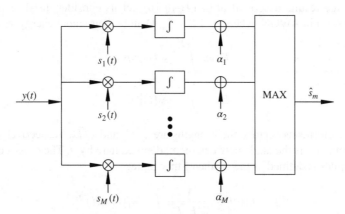

Figure 4.1   *MAP receiver for coherent detection of M arbitrary signal waveforms*

of waveform vectors at the local references for correlation. From (4.9)–(4.10), it is also clear that the extra dimensions beyond the signal space have no influence on symbol estimation, implying that signal detection conducted on the signal space is the same as that conducted on the infinite-dimensional space. The expression on the infinite space, however, can easily help transfer the vector operation to the operation on continuous-time functions. In particular, applying Parseval's theorem to (4.9) enables us to rewrite the decision variable and decision rule alternatively as

$$
\hat{s}_\ell = \arg\ \min_i \left\{ \int_0^{T_s} [y(t) - s_i(t)]^2 dt - N_0 \ln p(s_i) \right\}
$$

$$
= \arg\ \max_i \left\{ \int_0^{T_s} y(t) s_i(t) dt + \alpha_i \right\}. \tag{4.13}
$$

The vector form (4.10) and its continuous-time counterpart (4.13) are equivalent.

In the subsequent study, we focus on the case of equiprobable symbols as commonly encountered in practice; that is, $p(s_1) = p(s_2) = \cdots = p(s_M)$. In this case, the symbol decision rule reduces to

$$
\hat{s}_\ell = \arg\ \min_i \{ \|\mathbf{y} - \mathbf{s}_i\|^2 \}\} \tag{4.14}
$$

$$
= \arg\ \max_i \left\{ \Re\{\mathbf{y}^\dagger \mathbf{s}_i\} - \frac{1}{2} E_i \right\} \tag{4.15}
$$

$$
= \arg\ \max_i \left\{ \int_0^{T_s} y^\dagger(t) s_i(t) dt - \frac{1}{2} E_i \right\}. \tag{4.16}
$$

From these expressions, we can make the following observations:

- Symbol decision can be made based either on the signal space or on the original waveforms; they are equivalent.
- Performance analysis can also be done on the two domains, as subsequently illustrated in more detail.
- The advantage of the signal-space method is its geometrical intuition and flexibility in signal processing.

4.4   Detection of two arbitrary waveforms

As a special case of the results described in (4.14)–(4.16), let us consider the detection of two arbitrary waveforms $s_1(t)$ and $s_2(t)$ in AWGN, where $0 \leq t < T_b$. Denote their signal energy, respectively, by

$$
E_1 = \int_0^{T_b} [s_1(t)]^2 dt
$$

$$
E_2 = \int_0^{T_b} [s_2(t)]^2 dt. \tag{4.17}
$$

If we treat the two waveforms as vectors, their lengths are $\sqrt{E_1}$ and $\sqrt{E_2}$, respectively. Besides the length, another important parameter is the angle between them, denoted here by θ. The cross-correlation coefficient ρ between the two signals is defined as the cosine of this angle:

$$
\rho = \cos\theta = \frac{1}{\sqrt{E_1 E_2}} \int_0^{T_b} s_1(t) s_2(t) dt. \tag{4.18}
$$

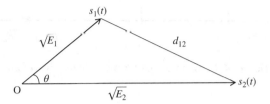

Figure 4.2 *Geometry of two arbitrary signals*

With these parameters, the geometry is as depicted in Figure 4.2. In accordance with the cosine rule, we can find the squared distance between signals $s_1(t)$ and $s_2(t)$ to be

$$d_{12}^2 = \int_0^{T_s} |s_1(t) - s_2(t)|^2 dt = E_1 + E_2 - 2\rho\sqrt{E_1 E_2}, \tag{4.19}$$

which, as will be shown subsequently, is important to the detection performance. For the special case of two equal-power signals with $E_1 = E_2 = E$, the squared distance reduces to $d_{12}^2 = 2E(1 - \rho)$.

As mentioned in the preceding section, we can work equivalently on the signal space or with the continuous-time waveforms. We choose the latter for illustration. The decision rule in (4.16) can be simplified as an inequality, as shown by

$$\underbrace{\int_0^{T_b} y(t)[s_1(t) - s_2(t)]dt}_{\xi} \begin{cases} > \Lambda, & s_1 \\ < \Lambda, & s_2 \end{cases}, \tag{4.20}$$

where the threshold is given by

$$\Lambda = \frac{1}{2}(E_1 - E_2). \tag{4.21}$$

Since this correlation-receiver structure is derived from the minimum distance criterion, it is important to note that the use of the decision variable ξ and the threshold Λ in (4.20) *automatically* implies that the decision boundary is simply the bisector of the line segment that links the two signal points. This advantage will be exploited later in the performance analysis of MPSK signals. Recall that $y(t)$ is of infinite dimensions; but only its projection onto the line linking $s_1(t)$ and $s_2(t)$ carries useful information for detection. This is the geometrical significance of ξ defined in (4.20).

We are interested in the average error probability P_b. Since the symbols are equally likely, we have $p(s_1) = p(s_2) = 1/2$. Thus,

$$P_b = \frac{1}{2}\Pr\{\xi < \Lambda | s_1\} + \frac{1}{2}\Pr\{\xi > \Lambda | s_2\}. \tag{4.22}$$

To find $\Pr\{\xi > \Lambda | s_2\}$, we need to find the PDF of ξ conditioned on the transmission of symbol s_2, for which

$$y(t) = s_2(t) + n(t). \tag{4.23}$$

We can easily determine the corresponding mean and variance of ξ as

$$\mathbb{E}[\xi | s_2] = \rho\sqrt{E_1 E_2} - E_2$$

$$\text{var}[\xi | s_2] = \mathbb{E}\left[\left(\int_0^{T_b} n(t)[s_1(t) - s_2(t)]dt\right)^2\right]$$

$$= \frac{N_0 d_{12}^2}{2}. \tag{4.24}$$

Figure 4.3   *Signal space for two arbitrary waveforms*

In obtaining the variance of ξ on the last line, we represent the squared integral as the product of two integrals with different integration variables, for example, t and τ, and then use the property that the correlation of white noise is a delta function to write $\mathbb{E}[n(t)n(\tau)] = \frac{N_0}{2}\delta(t-\tau)$, whereby the result can be simplified. It follows that, when s_2 is sent,

$$\xi \sim \mathcal{N}\left(\rho\sqrt{E_1 E_2} - E_2, \frac{N_0 d_{12}^2}{2}\right). \tag{4.25}$$

Using this PDF, we can express $\Pr\{\xi > \Lambda | s_2\}$ in the form of a Gaussian integral, which, after simplification, gives

$$\Pr\{\xi > \Lambda | s_2\} = Q\left(\sqrt{\frac{d_{12}^2}{2N_0}}\right). \tag{4.26}$$

In a similar manner, we can determine another conditional error probability, $\Pr\{\xi < \Lambda | s_1\}$, leading to the same expression as the above. Hence, we obtain the average error probability

$$P_b = Q\left(\sqrt{\frac{d_{12}^2}{2N_0}}\right). \tag{4.27}$$

In fact, the geometry of ξ is also very clear, as depicted in Figure 4.3; it has the mean location under s_1 and s_2 at $E_1 - \rho\sqrt{E_1 E_2}$ and at $\rho\sqrt{E_1 E_2} - E_2$, respectively. The threshold is just the midpoint of the line that links the two points, equal to $\Lambda = \frac{1}{2}(E_1 - E_2)$. Using the midpoint between the two signals as the threshold is the strategy of the ML receiver with two equally likely symbols.

Property 4.4.1   *The optimum coherent ML receiver for two arbitrary waveforms of equal probabilities is defined in (4.20). Its error probability depends only on the signals through their distance, given by*

$$P_b = Q\left(\sqrt{\frac{d_{12}^2}{2N_0}}\right) \tag{4.28}$$

with d_{12}^2 given by

$$d_{12}^2 = \int_0^{T_s} |s_1(t) - s_2(t)|^2 dt = (\mathbf{s}_1 - \mathbf{s}_2)^\dagger(\mathbf{s}_1 - \mathbf{s}_2) = E_1 + E_2 - 2\rho\sqrt{E_1 E_2}. \tag{4.29}$$

This is a general formula for pairwise error probability applicable to arbitrary waveforms, no matter whether they are baseband or band-pass signals.

Some remarks are necessary here. First, in the above formula, the local references assume the exact knowledge of the signal waveforms arriving at the receiver, including random phase and time delay. Second, the squared distance d_{12}^2 can be evaluated using any one of the expressions shown above; which one is a better choice depends on a particular application.

✍ **Example 4.1** _____

The following waveforms are used in a coherent binary system to transmit bit 1 and bit 0, respectively:

$$s_1(t) = \begin{cases} A, & 0 \le t < T_b \\ 0, & \text{elsewhere} \end{cases},$$

$$s_2(t) = \begin{cases} A, & 0 \le t < 3T_b/4 \\ -A, & 3T_b/4 \le t < T_b. \\ 0, & \text{elsewhere} \end{cases}$$

Assuming that $A^2 T_b = 2$ μW, and the single-sided noise density is equal to $N_0 = 0.2 \times 10^{-6}$ W/Hz, sketch the geometry of the two signals and determine the bit error probability.

Solution

Determine the energy of the two signals and their cross product as

$$E_1 = \int_0^{T_b} s_1^2(t)dt = A^2 T_b, \quad E_2 = \int_0^{T_b} s_2^2(t)dt = A^2 T_b$$

$$E_{12} = \int_0^{T_b} s_1(t)s_2(t)dt = \frac{3A^2 T_b}{4} - \frac{A^2 T_b}{4} = \frac{A^2 T_b}{2}.$$

Thus, we obtain

$$d_{12}^2 = E_1 + E_2 - 2E_{12} = A^2 T_b$$

$$\rho = \cos\theta = \frac{E_{12}}{\sqrt{E_1 E_2}} = \frac{1}{2} \rightarrow \theta = 60°.$$

Hence, by the same token as we obtained in Figure 4.2, we can plot the geometry of these two signals and obtain the bit error rate (BER) as

$$P_b = Q\left(\sqrt{\frac{d_{12}^2}{2N_0}}\right) = Q(\sqrt{5}) = 0.0127.$$

○

〰✍

4.5 MPSK

In the previous two sections, we have derived the ML and MAP frameworks for arbitrary signals without the need to distinguish between baseband and band-pass signals. Indeed, their mathematical treatment is exactly the same, except for a possible difference in the choice of the basis functions. However, baseband and band-pass signals have different physical implications in many important issues such as frequency bandwidth requirements and propagation characteristics.

4.5.1 BPSK

In this simplest case of binary PSK (BPSK), each symbol contains one bit by mapping, for example, bit 1 to phase zero and bit 0 to phase π. Thus, the BPSK signal over one symbol interval $0 \le t < T_b$ can be written as

$$s(t) = \begin{cases} s_1(t) = A\cos(\omega_c t + 0) = A\cos\omega_c t, & \text{bit 1} \\ s_2(t) = A\cos(\omega_c t + \pi) = -A\cos\omega_c t, & \text{bit 0} \end{cases}. \tag{4.30}$$

For BPSK, both signals are defined by the same cosine function with the only difference being in its sign. Thus, if we define

$$\phi(t) = \sqrt{\frac{2}{T_b}} \cos \omega_c t, \ 0 \le t < T_b \tag{4.31}$$

to meet the condition of unit energy, then we can rewrite

$$s(t) = \pm \sqrt{E_b} \phi(t), \tag{4.32}$$

where $E_b = A^2 T_b/2$ is the symbol energy (also the bit energy for a binary signaling scheme). This expression indicates that, if a coherent local reference $\phi(t)$ is used for correlation, the two signal points are located at $s_1 = \sqrt{E_b}$ and $s_2 = -\sqrt{E_b}$ of a one-dimensional signal space, similar to that illustrated in Figure 4.3. In fact, the same geometry can be obtained by directly determining the lengths of $s_1(t)$ and $s_2(t)$ and the angle between them. To determine the bandwidth requirement, we assume rectangular shaping for ease of illustration. The null-to-null baseband bandwidth is $1/T_b$, which, after double sideband carrier modulation, leads to the bandwidth requirement of BPSK:

$$B = \frac{2}{T_b} = 2R_b, \tag{4.33}$$

numerically twice the bit rate R_b.

After transmitting over an AWGN channel, the received signal is given by

$$y(t) = \pm \sqrt{E_b} \ \phi(t) + n(t), \tag{4.34}$$

which takes the positive sign for bit 1 and negative sign for bit 0. The two signals are of equal energy; the threshold reduces to $\Lambda = (E_1 - E_2)/2 = 0$. Following the general form of the ML detection, and noting that scaling $[s_1(t) - s_2(t)] = 2\sqrt{E_b} \ \phi(t)$ to $\phi(t)$ does not alter the decision rule, we obtain

$$\xi = \int_0^{T_b} \phi(t)y(t)dt \begin{cases} > 0, & \text{bit 1} \\ < 0, & \text{bit 0'} \end{cases} \tag{4.35}$$

which defines a correlation receiver. From the signal space, we observe that $d_{12} = 2\sqrt{E_b}$, and, thus, according to the results in the preceding section, we have

$$P_b = Q\left(\sqrt{\frac{2E_b}{N_0}}\right). \tag{4.36}$$

In the above, d_{12} is *indirectly* obtained from the signal space. In fact, if we want to go around the procedure of finding the signal space and directly work on the signal waveforms, it is quite simple to obtain

$$d_{12}^2 = \int_0^{T_b} |s_1(t) - s_2(t)|^2 dt = 4A^2 \int_0^{T_b} \cos^2\omega_c t \ dt = 4E_b. \tag{4.37}$$

✎ **Example 4.2** _____

Suppose that in a BPSK system, a phase error θ_0 is introduced to the received signal such that

$$x(t) = \pm A \cos(\omega_c t + \theta_0) + n(t) \tag{4.38}$$

during transmission. Because of the absence of phase-error information, the local reference used at the receiver takes the form

$$\phi(t) = \sqrt{2/T_s} \cos \omega_c t. \tag{4.39}$$

Determine the BER if $\theta_0 = 20°$.

Solution

This is not a coherent receiver in the strict sense and, thus, we cannot directly employ the formula (4.36). We, therefore, need to calculate the projection of the signal onto the local reference, yielding

$$\int_0^{T_b} A\cos(\omega_c t + \theta_0)\phi(t)dt = \sqrt{E_b}\cos\theta_0.$$

Without the knowledge of the random phase rotation θ_0, we have no choice but to use the unrotated reference for reception. Thus, the distance between the two signal points reduces to $d_{12} = 2\sqrt{E_b}\cos\theta_0$. When a normalized reference is used, the output noise variance remains unchanged (equal to $N_0/2$) and, thus, there is no need for recalculation. It follows that the SNR equals $d_{12}^2/(2N_0) = 2E_b\cos^2\theta_0/N_0$, whereby the BER

$$P_e = Q\left(\sqrt{\frac{2E_b\cos^2\theta_0}{N_0}}\right).$$

4.5.2   QPSK

In quadriphase-shift keying (QPSK), each symbol represents two bits. Thus, its bit energy E_b is related to the symbol energy by $E_b = E_s/2$. Denote the kth symbol by $s_k = (a_k, b_k)$, with both bits taking values of 1 or -1. QPSK can be regarded as a multiplexing scheme where two binary sequences are transmitted, in parallel, over the mutually orthogonal cosine (I) and sine (Q) subchannels. As such, we can write

$$s(t) = \frac{A}{\sqrt{2}}(a_k\cos\omega_c t - b_k\sin\omega_c t),\ 0 \le t < T_s, \tag{4.40}$$

where the symbol interval T_s is chosen to be a multiple of the period of the carrier $T_c = 1/f_c$, and the amplitude has been normalized so that the symbol energy equals $E_s = A^2 T_s/2$. This expression suggests an easy way of generating a QPSK modulated signal with symbol energy E_s, as illustrated in Figure 4.4(a).

Noting that $a_k^2 + b_k^2 = 2$, we can define

$$\cos\varphi_m = \frac{a_k}{\sqrt{2}}$$

$$\sin\varphi_m = \frac{b_k}{\sqrt{2}}, \tag{4.41}$$

and rewrite $s(t)$ in the form of phase modulation to obtain

$$s(t) = A(\cos\varphi_m\cos\omega_c t - \sin\varphi_m\sin\omega_c t)$$
$$= A\cos(\omega_c t + \varphi_m). \tag{4.42}$$

The dibits-to-angle mapping defined in (4.41) can be tabulated in Table 4.1. It can be seen that any two adjacent symbols differ only by one bit. This type of mapping is called *Gray-coded mapping*. The reader can examine that, if we replace $-\sin\omega_c t$ with $\sin\omega_c t$ in the QPSK generator equation (4.40), the result is still a Gray-code mapping but with a different mapping table. In fact, by defining the symbol energy $E_s = A^2 T_s/2$ and the bit energy $E_b = E_s/2$, we can rewrite (4.40) and (4.42) as

$$s(t) = \sqrt{E_b}\ [a_k\phi_1(t) + b_k\phi_2(t)] \tag{4.43}$$
$$= \sqrt{E_s}\ [\cos\varphi_m\phi_1(t) + \sin\varphi_m\phi_2(t)] \tag{4.44}$$

Table 4.1 *Mapping table*

Input dibit	Phase
11	$\pi/4$
01	$3\pi/4$
00	$5\pi/4$
10	$7\pi/4$

with the basis functions defined by

$$\phi_1(t) = \sqrt{\frac{2}{T_s}} \, \cos\omega_c t$$

$$\phi_2(t) = -\sqrt{\frac{2}{T_s}} \, \sin\omega_c t, 0 \leq \ t < T_s. \tag{4.45}$$

The important thing is that, if we use $\phi_1(t)$ and $\phi_2(t)$ for QPSK generation, the decision rule remains the same as long as we use the same set of ϕ_1 and ϕ_2 for demodulation at the receiver. In analogy to the representation of a vector $\vec{v} = 2\vec{i} + 3\vec{j}$ as a point $(2,3)$ in the two-dimensional space, the basis functions $\phi_1(t)$ and $\phi_2(t)$ possess similar properties as $\vec{i}$ and $\vec{j}$. Thus, we can write the signal points of QPSK as

$$\mathbf{s}_m = (\sqrt{E_s} \, \cos\varphi_m, \sqrt{E_s} \, s\sin\varphi_m) \tag{4.46}$$

in the space spanned by $\phi_1(t)$ and $\phi_2(t)$. It is more convenient to represent the coordinates in the polar form as

$$s_m = \sqrt{E_s} \, (\cos\varphi_m + j\sin\varphi_m) = \sqrt{E_s}e^{j\varphi_m}. \tag{4.47}$$

The signal constellation is depicted in Figure 4.4(b).

After transmission over an AWGN channel, the received noisy signal is expressible as

$$y(t) = \sqrt{E_b} \, (a_k\phi_1(t) + b_k\phi_2(t)) + n(t). \tag{4.48}$$

Since the I and Q channels are orthogonal, the binary decision on $\{a_k\}$ and $\{b_k\}$ can be done separately. The decision variables for the I and Q channels are given by

$$u = \langle y(t), \phi_1(t) \rangle = a_k\sqrt{E_b} + n_1, \text{ and } v = \langle y(t), \phi_1(t) \rangle = b_k\sqrt{E_b} + n_2, \tag{4.49}$$

respectively, where $n_i \sim \mathcal{N}(0, N_0/2)$. The two bits are estimated as

$$\hat{a}_k = \text{sgn}(u)$$

$$\hat{b}_k = \text{sgn}(v), \tag{4.50}$$

where $\text{sgn}(c)$ is the sign operator taking the value of 1 if $c > 0$ and -1 if $c < 0$. The angle can be estimated as

$$\hat{\varphi}_m = \text{sgn}(v)[1 - \text{sgn}(u)]\frac{\pi}{2} + \arctan\left(\frac{v}{u}\right). \tag{4.51}$$

The corresponding receiver is shown in Figure 4.5.

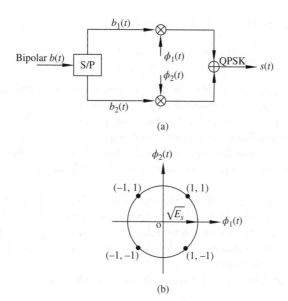

Figure 4.4   QPSK signals. (a) Generator. (b) Constellation

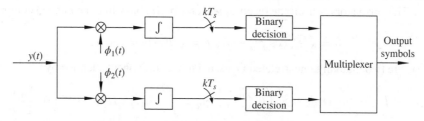

Figure 4.5   Coherent detection of QPSK signals

A correct symbol decision requires a correct decision on both bits a_k and b_k. It follows that the probability of a correct symbol decision is given by

$$P_c = [1 - Q(\sqrt{E_s/N_0})]^2. \tag{4.52}$$

Hence, the symbol error probability equals

$$P_e = 1 - P_c = 2Q(\sqrt{E_s/N_0}) - Q^2(\sqrt{E_s/N_0}). \tag{4.53}$$

4.5.3 MPSK

In the MPSK signaling scheme, the symbol s_m is mapped into the phase $\varphi_m = 2\pi(m-1)/M$ of a carrier to produce

$$s_m(t) = A\cos(\omega_c t + \varphi_m), 0 \le t < T_s, \tag{4.54}$$

where we focus on one symbol period for illustrative convenience. Each symbol contains $\log_2 M$ bits. To see its signal constellation, rewrite $s_m(t)$ in terms of the basis functions defined in (4.45), as

$$s_m(t) = \sqrt{E_s}\cos\varphi_m\phi_1(t) + \sqrt{E_s}\sin\varphi_m\phi_2(t). \tag{4.55}$$

More compactly, we can represent $s_m(t)$ as a complex number:

$$s_m(t) \overset{map}{\to} \sqrt{E_s} \begin{bmatrix} \cos \varphi_m \\ \sin \varphi_m \end{bmatrix} \overset{map}{\to} \sqrt{E_s} e^{j\varphi}. \qquad (4.56)$$

The mapping here has two implications. First, it maps continuous-time waveforms into signal points in the two-dimensional Euclidean space, so that we can clearly visualize the geometry of the MPSK signals; that is, all the MPSK signal points are symmetrically distributed over a circle of radius $\sqrt{E_s}$. Second, it converts a band-pass signal into a baseband form. Since a pair of two real coordinates is isomorphic to a complex number, we can further write them as a complex number, which can be regarded as the signal representation in the polar coordinate system. The three representations are equivalent and useful in the analysis and understanding of MPSK signals.

At this point, it is interesting to compare MPSK with M-ary ASK (or M-ary PAM) since they have something similar in philosophy. The M-ASK has its signal points uniformly located on a straight line, whereas the signal points of MPSK are uniformly located on a circle of radius equal to $\sqrt{E_s}$. The difference is that, unlike M-ASK with parallel decision boundaries, the decision boundaries for MPSK have intersection, making it more difficult to analyze its performance.

An MPSK signal, when transmitted over AWGN channels, produces an input at the receiver given by

$$y(t) = s_m(t) + n(t). \qquad (4.57)$$

An MPSK receiver usually consists of two correlators with local reference signals $\phi_1(t)$ and $\phi_2(t)$ or their scaled versions. The correlators convert the received waveform $y(t)$ back into the I and Q components, u and v, given by

$$u = \sqrt{E_s} \cos \varphi_m + n_1, \ v = \sqrt{E_s} \cos \varphi_m + n_2, \qquad (4.58)$$

where n_1 and n_2 are two mutually uncorrelated Gaussian noise components defined by

$$n_1 = \int_0^{T_s} n(t)\phi_1(t)dt \sim \mathcal{N}\left(0, \frac{N_0}{2}\right), \ n_2 = \int_0^{T_s} n(t)\phi_2(t)dt \sim \mathcal{N}\left(0, \frac{N_0}{2}\right). \qquad (4.59)$$

The I and Q components u and v are then combined as a single complex number

$$z = u + jv = \sqrt{E_s} \, e^{j\varphi_m} + \underbrace{n_1 + jn_2}_{n} \qquad (4.60)$$

for which, n is complex AWGN with $n \sim \mathcal{CN}(0, N_0)$.

Before proceeding further, it is worth spending sometime to clarify the definition of SNR used in digital modulation. First, we need to distinguish n and $n_i, i = 1, 2$ from $n(t)$. Note that $n(t)$ is an AWGN process of power spectral density $N_0/2$ measured in Watt/Hz. Its passage through the correlators $\phi_i(t), i = 1, 2$ produces random noise samples n_1 and n_2, each of variance $N_0/2$ measured in Watt. Second, we must find out whether the noise come from one branch or two branches. For MPSK, MDPSK and MFSK modulation, the received information is carried by both the I/Q channels, which are required to combine to form a single complex output. The energy of the combined output remains to be E_s while the combined noise sample n has the variance of $\sigma_n^2 = N_0$. As such, the SNR at the correlators output is equal to E_s/N_0. However, for coherent BPSK signaling, the signal energy exists only in a single branch, I or Q channel, and hence, the noise variance is $\sigma_n^2 = N_0/2$ whereby the SNR becomes $2E_s/N_0$. Nevertheless, we basically adopt the tradition of defining the SNR as $\gamma = E_s/N_0$ unless otherwise stated. This remark is also applicable to M-QAM except that E_s is replaced by the average symbol energy.

Let us see how to determine the symbol error performance P_e of MPSK in AWGN. Without loss of generality, assume that the symbol of $\varphi_m = 0$ is sent, with its correct decision zone depicted in Figure 4.6.

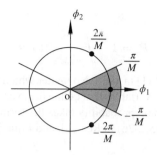

Figure 4.6   *Correct decision region for signal point $\varphi_m = 0$*

Correspondingly, the received noisy signal is $z = \sqrt{E_s} + n$. There are many ways to formulate the event of correction decision. Some formulations are listed below:

$$-\frac{\pi}{M} < \tan^{-1}\left(\frac{n_2}{n_1 + \sqrt{E_s}}\right) < \frac{\pi}{M}$$

$$-\tan(\pi/M) < \frac{n_2}{n_1 + \sqrt{E_s}} < \tan(\pi/M)$$

$$(\Im(z\, e^{j\pi/M}) > 0) \cap (\Im(z\, e^{-j\pi/M}) < 0). \tag{4.61}$$

Different formulations can lead to different expressions for the performance formulas, as illustrated below:

4.5.3.1 Approach 1

This is a straightforward method. The idea is to find the PDF of the phase angle $\tan^{-1}(v/u)$ estimated from samples with coordinates

$$u = \sqrt{E_s} + n_1$$

$$v = n_2. \tag{4.62}$$

They are jointly Gaussian-distributed, with the PDF given by

$$f(u, v) = \frac{1}{\pi N_0} \exp\left[-\frac{(u - \sqrt{E_s})^2 + v^2}{N_0}\right]. \tag{4.63}$$

We convert it to the PDF of the variables defined in polar coordinates by the change of variables

$$r = \sqrt{u^2 + v^2}\,,$$

$$\theta = \frac{v}{u}, \tag{4.64}$$

yielding

$$f(r, \theta) = \frac{r}{\pi N_0} \exp\left(-\frac{r^2 + E_s - 2r\sqrt{E_s}\cos\theta}{N_0}\right), \tag{4.65}$$

which, when averaged over r, leads to the PDF of the phase angle expressible in terms of SNR $\gamma = E_s/N_0$:

$$f(\theta) = \frac{1}{2\pi} e^{-\gamma\sin^2\theta} \int_0^\infty r e^{-(r - \sqrt{4\gamma}\cos\theta)^2}\, dr. \tag{4.66}$$

The symbol error rate is then given by

$$P_e = 1 - \int_{-\pi/M}^{\pi/M} f(\theta)d\theta. \tag{4.67}$$

This result was first obtained by Bennett [1] in 1956. No matter which formulation is used, we can easily find the joint distribution of the two Gaussian variables that define the event of a correct decision. With this type of formulations, the resulting correct decision probability takes the form of a double integral; see for example, Equation (4.131) in Ref. [2], Equation (5-2-26) in Ref. [3], and Equation (6.15) in Ref. [4].

4.5.3.2   Approach 2

Pawula et al. [5] pursue an idea similar to Approach 1, but starting from two random vectors corrupted by AWGN to derive the distribution function of the angle between them. A number of unconventional techniques are used in their derivation. First, the joint CHF, rather than the joint PDF, of the two random vectors is determined. Meanwhile, the expectation of high- frequency terms involved therein vanishes based on some physical arguments. Second, the joint CHF is then converted back to the joint CDF through a special transformation. Third, to obtain the PDF for MPSK, they need to pass γ_2 to infinity and set $\Delta\phi = 0$. A number of skills are required in a limiting treatment, including the location of a peak in the integrand and the Taylor expansion of the integrand in the vicinity of the peak, followed by the change of variables. Finally, they obtain an elegant formula

$$P_e(M) = \frac{1}{\pi} \int_0^{\pi-\pi/M} \exp\left(-\frac{\gamma\sin^2(\pi/M)}{\sin^2\theta}\right) d\theta, \tag{4.68}$$

which is identical to the result of Weinstein [6].

4.5.3.3   Approach 3

Assume, again, that symbol s_0 is transmitted. Its immediate adjacent points are $s_i, i = \pm 1$. Let

$$\ell_i = \int_0^{T_s} [s_0(t) - s_i(t)][s_0(t) + n(t)]dt, \; i = 1, -1, \tag{4.69}$$

denote the half-plane decision variables between s_0 and $s_{\pm 1}$. Their partition boundaries are $\varphi = \pm\pi/M$, respectively. Hence, the correct decision region of s_0 is expressible as $P_c = \Pr\{(\ell_1 > 0) \cap (\ell_{-1} > 0)\}$.

To evaluate P_c, we need an alternative expression for the correction-decision event. By direct calculation, it is easy to examine that

$$\mathbb{E}[\ell_i] = E_s(1 - \rho); \; \text{var}[\ell_i] = (1 - \rho)N_0 E_s.$$
$$\mathbb{E}[(\ell_1 - \mathbb{E}[\ell_1])(\ell_{-1} - \mathbb{E}[\ell_{-1}])] = \rho(\rho - 1)N_0 E_s \tag{4.70}$$

with

$$\rho = \cos(2\pi/M) \tag{4.71}$$

representing the cross-correlation between two adjacent symbols. In the calculation of variance and covariance shown above, we have used the property of white noise, namely $\mathbb{E}[n(t_1)n(t_2)] = \frac{N_0}{2}\delta(t_1 - t_2)$, and the property of the delta function. For notational simplicity, define

$$\xi_i = \ell_i/\sqrt{(1 - \rho)N_0 E_s}, i = 1, -1,$$
$$\alpha = \sqrt{(1 - \rho)E_s/N_0}. \tag{4.72}$$

Then, we have

$$\begin{pmatrix} \xi_1 \\ \xi_{-1} \end{pmatrix} \sim \mathcal{N}\left(\alpha \begin{bmatrix} 1 \\ 1 \end{bmatrix}, \begin{pmatrix} 1 & -\rho \\ -\rho & 1 \end{pmatrix}\right). \tag{4.73}$$

Note that the covariance matrix of ξ_1 and ξ_{-1} is very special with eigenvalues $(1-\rho)$ and $(1+\rho)$, and the corresponding eigenvectors given by $[1;1]/\sqrt{2}$ and $[1;-1]/\sqrt{2}$, respectively. We can define a transform by using their eigenvectors, resulting in a pair of uncorrelated variables

$$\begin{pmatrix} z_1 \\ z_2 \end{pmatrix} = \frac{1}{\sqrt{2}} \begin{pmatrix} 1 & 1 \\ 1 & -1 \end{pmatrix} \begin{pmatrix} \xi_1 \\ \xi_{-1} \end{pmatrix} \sim \mathcal{N}\left(\sqrt{2\alpha} \begin{bmatrix} 1 \\ 0 \end{bmatrix}, \begin{bmatrix} 1-\rho & 0 \\ 0 & 1+\rho \end{bmatrix}\right). \tag{4.74}$$

From the transformation, we have $z_1 + z_2 = \sqrt{2}\,\xi_1$ and $z_1 - z_2 = \sqrt{2}\,\xi_{-1}$, from which the correct decision zone $(\xi_1 > 0) \cap (\xi_{-1} > 0)$ can alternatively be represented as $(-z_1 < z_2 < z_1) \cap (0 < z_1 < \infty)$. As such, we have

$$P_c = \Pr\{(-z_1 < z_{-1} < z_1) \cap (0 < z_1 < \infty)\}$$

$$= \int_0^\infty \frac{1}{\sqrt{2\pi(1-\rho)}} e^{-\frac{(z_1-\sqrt{2\alpha})^2}{2(1-\rho)}} \, dz_1 \int_{-z_1}^{z_1} \frac{1}{\sqrt{2\pi(1+\rho)}} e^{-\frac{z_2^2}{2(1+\rho)}} \, dz_2. \tag{4.75}$$

The second integral can be represented in terms of the Q function. Defining

$$\gamma = \frac{2E_s}{N_0}, \tag{4.76}$$

changing the variable, simplifying, and noting that $\sqrt{2\alpha/(1-\rho)} = \sqrt{\gamma}$ and $\sqrt{(1-\rho)/(1+\rho)} = \tan(\pi/M)$, we obtain

$$P_c = \int_{-\sqrt{\gamma}}^\infty \frac{1}{\sqrt{2\pi}} e^{-\frac{x^2}{2}} \left\{ 1 - 2Q\left[(x+\sqrt{\gamma})\tan\frac{\pi}{M}\right] \right\} dx. \tag{4.77}$$

Hence, the symbol error probability $P_e = 1 - P_c$ is expressible as

$$P_e = Q(\sqrt{\gamma}) + 2 \int_{-\sqrt{\gamma}}^\infty \frac{1}{\sqrt{2\pi}} e^{-\frac{x^2}{2}} Q\left[(x+\sqrt{\gamma})\tan\frac{\pi}{M}\right] dx. \tag{4.78}$$

Numerical comparison between the above formula and (4.68) of Pawula is shown in Figure 4.7. The two formulas provide the same results, showing that the formulation for performance evaluation is not unique but rather is very flexible.

4.6 *M*-ary QAM

We next turn to M-QAM. A signal generator for square M-QAM is sketched in Figure 4.8, and a receiver is shown in Figure 4.9. Assuming $M = L^2$, the M-QAM signal can be written as

$$s_{ij}(t) = a_i \cos\omega_c t + b_j \sin\omega_c t, \ 0 \le t < T_s, \tag{4.79}$$

where

$$a_i = (i - 0.5)d; \ b_j = (j - 0.5)d, \ i = \pm1, \pm2\cdots, \pm L/2; \tag{4.80}$$

and $d > 0$ is a constant to be determined such that the average symbol energy equals E_s. The signal constellation is shown in Figure 4.10.

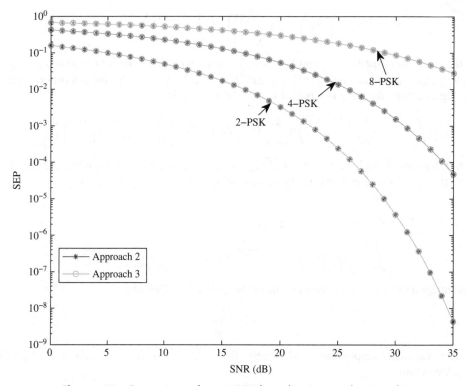

Figure 4.7 *Comparison of two MPSK formulas. Approaches 2 and 3*

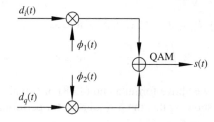

Figure 4.8 *Generation of M-QAM signals*

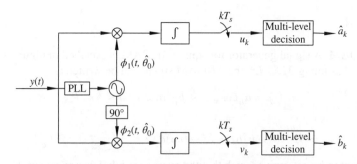

Figure 4.9 *Coherent receiver for M-QAM signals*

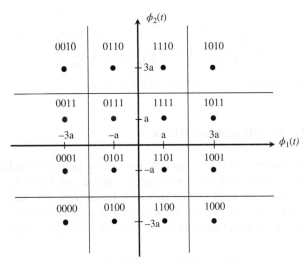

Figure 4.10 *Constellation of square 16-QAM signals*

Denote $E_0 = d^2 T_s$ and rewrite

$$s_{ij}(t) = \underbrace{\left(i - \frac{1}{2}\right)\sqrt{E_0}}_{\alpha_i} \underbrace{\sqrt{2/T_s}\,\cos\omega_c t}_{\phi_1(t)} + \underbrace{\left(j - \frac{1}{2}\right)\sqrt{E_0}}_{\beta_j}\underbrace{\sqrt{2/T_s}\,\sin\omega_c t}_{\phi_2(t)}. \tag{4.81}$$

Then, (α_i, β_j) represents the coordinates of the signal point s_{ij} in the signal space. A decision can be made on the I and Q channels separately. The distance between two nearest neighbor points is

$$d_{12} = \sqrt{E_0}\,. \tag{4.82}$$

We have four signal points at the corners, $(L-2)^2$ interior points, and $4(L-2)$ exterior points at edges. Each point in the same category has the same correct decision probability and, thus, only their representative correct decision events are listed, as follows:

$$\text{Corner points}: \ (n_1 > -0.5d_{12}) \cap (n_2 > -0.5d_{12})$$

$$\text{Interior points}: \ (-0.5d_{12} < n_1 < 0.5d_{12}) \cap (-0.5d_{12} < n_2 < 0.5d_{12})$$

$$\text{Edge points}: \ (n_1 > -0.5d_{12}) \cap (-0.5d_{12} < n_2 < 0.5d_{12}).$$

It follows that the average correct symbol decision probability is

$$P_c = \frac{4}{L^2}\left[1 - Q\left(\sqrt{\frac{E_0}{2N_0}}\right)\right]^2 + \frac{(L-2)^2}{L^2}\left[1 - 2Q\left(\sqrt{\frac{E_0}{2N_0}}\right)\right]^2$$

$$+\frac{4(L-2)}{L^2}\left[1 - Q\left(\sqrt{\frac{E_0}{2N_0}}\right)\right]\left[1 - 2Q\left(\sqrt{\frac{E_0}{2N_0}}\right)\right]. \tag{4.83}$$

Finally, we need to represent E_0 in terms of the average symbol energy E_s. The latter is given by

$$E_s = \frac{1}{(L/2)^2}\sum_{i=1}^{L/2}\sum_{j=1}^{L/2}\left[\left(\frac{2i-1}{2}\right)^2 + \left(\frac{2j-1}{2}\right)^2\right]E_0$$

$$= \frac{1}{6}(M-1)E_0 \tag{4.84}$$

whereby we can further write

$$Q\left(\sqrt{\frac{E_0}{2N_0}}\right) = Q\left(\sqrt{\frac{3E_s}{(M-1)N_0}}\right). \tag{4.85}$$

The average symbol error probability can be calculated by $P_e = 1 - P_c$.

4.7   Noncoherent scheme–differential MPSK

In differential MPSK, information symbols are not directly mapped onto the absolute phase of a carrier. But rather, they are carried by the phase difference between two successive symbol periods, so that the phase ambiguity introduced by a fading channel can be removed. In other words, the previous signal sample is used as a reference for the demodulation of the current symbol to avoid the need for channel gain estimation. The resulting receiver is a simple product demodulator. Thus, two consecutive received samples, in baseband form, can be written as

$$x(k) = \sqrt{\gamma}e^{j\theta_k} + n(k)$$
$$x(k-1) = \sqrt{\gamma}e^{j\theta_{k-1}} + n(k-1), \tag{4.86}$$

where the noise components are assumed to be normalized to have unit variance. Unlike coherent MPSK, symbol information is mapped onto the phase difference

$$\varphi_k = \theta_k - \theta_{k-1} \tag{4.87}$$

according to a preset mapping table. Such a modulation scheme is known as differential PSK (DPSK). Upon the receipt of $x(k)$ and $x(k-1)$, symbol estimation is based on the product $x(k)x^*(k-1)$ by using, for example, the following criterion:

$$\hat{\varphi}_k = \arg\max_i \Re\{x(k)x^*(k-1)e^{-j\varphi_i}\}, \tag{4.88}$$

which is of the same form as that for MPSK except that $x(k)$ is replaced by a product. In performance analysis, we can assume that symbol $\varphi_k = 0$ is sent and, in a manner similar to MPSK, represents the correct decision event as

$$(\Im(x(k)x^*(k-1)e^{j\pi/M}) > 0) \cap (\Im(x(k)x^*(k-1)e^{-j\pi/M}) < 0). \tag{4.89}$$

However, its error performance is generally not easy to analyze. The difficulty arises from the fact that the correct or erroneous decision event for general *M*-ary DPSK (MDPSK) is defined by two quadratic-form random variables. Finding the joint PDF of these random variables is usually not a simple matter.

4.7.1   Differential BPSK

Let us begin with the simplest case of binary DPSK for which the symbol mapping is defined by

$$\varphi_k = \theta_k - \theta_{k-1} = \begin{cases} 0, & b_k = 1 \\ \pi, & b_k = 0 \end{cases}. \tag{4.90}$$

The decision variable and decision rule are then given by

$$D_k = 2\Re\{x(k)x^*(k-1)\} \begin{cases} \geq 0, & \hat{b}_k = 1 \\ < 0, & \hat{b}_k = 0 \end{cases}. \tag{4.91}$$

To find the error performance, it is more convenient to rewrite the decision variable in quadratic form. To this end, define $\mathbf{x}_k = [x(k), x(k-1)]^T$, $\mathbf{e} = [1, 1]^T$, and

$$\boldsymbol{\Omega} = \begin{pmatrix} 0 & 1 \\ 1 & 0 \end{pmatrix} \tag{4.92}$$

which allows us to rewrite $D_k = \mathbf{x}_k^\dagger \boldsymbol{\Omega} \mathbf{x}_k$ in quadratic form. Without loss of generality, assume that bit one is sent, i.e., $b_k = 1$, whereby $\mathbf{x}_k \sim \mathcal{CN}(\sqrt{\gamma}e^{j\theta_{k-1}}\mathbf{e}, \mathbf{I})$, and we obtain the CHF of D_k as

$$\Phi_{D_k}(s) = \det(\mathbf{I} - s\boldsymbol{\Omega})^{-1} \exp[s\gamma \mathbf{e}^\dagger (\boldsymbol{\Omega}^{-1} - s\mathbf{I})^{-1}\mathbf{e}]$$

$$= \frac{1}{(1+s)(1-s)} \exp\left(\frac{2s\gamma}{1-s}\right). \tag{4.93}$$

Note that D_k is a quadratic form in a nonzero-mean Gaussian vector, and usually has two essential singular points in its CHF. Fortunately, the special structure of $\boldsymbol{\Omega}$, along with the diagonal covariance matrix of $\mathbf{x}_k$, makes one such singular point removable. By using the residue theorem, it follows that

$$P_e = -\text{Res}\left\{\frac{\Phi_{D_k}(s)}{s}, \text{ poles in } \Re(s) < 0\right\}$$

$$= -\frac{1}{s(1-s)} \exp\left(\frac{2s\gamma}{1-s}\right) \Big|_{s=-1}$$

$$= \frac{1}{2}e^{-\gamma}. \tag{4.94}$$

4.7.2 Differential MPSK

The most elegant result is derived by Pawula et al. [5] on the basis of the distribution of the phase angle between two noisy random vectors. The symbol error probability takes the form of a Gaussian integral, as shown by

$$P_e(\gamma) = \frac{1}{\pi} \int_0^{\pi - \pi/M} \exp\left(-\frac{\gamma \sin^2(\pi/M)}{1 + \cos(\pi/M)\cos\theta}\right) d\theta. \tag{4.95}$$

4.7.3 Connection to MPSK

As mentioned above, the receiver for MDPSK signals follows a similar idea as its counterpart for MPSK except for the use of the previous sample as a local reference for demodulating the current symbol. A natural question is: what is the performance degradation caused by a noisy sample reference, as compared to a coherent MPSK receiver? Recall that the differential detection is based on the sample product

$$x(k)x^*(k-1) = \gamma e^{j(\theta_k - \theta_{k-1})} + \sqrt{\gamma}(n_k + n_{k-1}^*) + n_k n*_{k-1}. \tag{4.96}$$

The source of difficulty is the nonlinear noise term, which is negligible for large SNRs. Thus, for large SNRs, the decision variable reduces to

$$x(k)x^*(k-1) = \gamma e^{j(\theta_k - \theta_{k-1})} + \sqrt{\gamma}(n_k + n_{k-1}^*) \tag{4.97}$$

having the same form as the MPSK receiver except that the noise variance is doubled. This implies that the error performance of the MDPSK receiver is 3 dB inferior to its coherent MPSK counterpart. When expressed in terms of Pawula's formula, the error probability of MDPSK for large SNRs can be calculated by using (4.68) except replacing γ with $\gamma/2$.

4.8   MFSK

In MFSK, each symbol is mapped into a carrier of a different frequency, as shown by

$$s_i(t) = A \cos \omega_i t, \ 0 \leq t < T_s, \tag{4.98}$$

for $i = 0, 1, \cdots, M - 1$. Two kinds receivers, namely coherent and noncoherent, can be used for the detection of MPSK signals. Coherent MFSK requires a smaller separation between two adjacent carrier frequencies at the cost of a coherent receiver; noncoherent MFSK requires a wider frequency separation spacing but a much simpler receiver.

4.8.1   BFSK with coherent detection

Coherent detection is seldom employed in MFSK systems because of its inferior performance yet wider bandwidth requirements as compared to its MPSK counterpart. This is easy to explain if we consider a coherent binary FSK (BFSK) system for which bits $b_k = 1$ and $b_k = 0$ are transmitted by using $\phi_1(t) = \sqrt{2/T_b} \cos \omega_1 t$ and $\phi_2(t) = \sqrt{2/T_b} \cos \omega_2 t$, respectively. A coherent receiver consists of two parallel correlators with one matched to $\phi_1(t)$ and the other matched to $\phi_2(t)$. A decision is made by comparing the outputs of the two correlators. The correlators perform the following operations:

$$x_1 = \int_0^{T_s} x(t)\phi_1(t) \ dt,$$

$$x_2 = \int_0^{T_s} x(t)\phi_2(t) \ dt. \tag{4.99}$$

The decision rule is

$$\ell \overset{\Delta}{=} x_1 - x_2 \begin{cases} > 0, & \hat{b}_k = 1 \\ < 0, & \hat{b}_k = 0 \end{cases}. \tag{4.100}$$

For coherent detection of two arbitrary waveforms, the general formula (4.28) is applicable. The distance between the two FSK signals is given by

$$d_{12}^2 = A \int_0^{T_b} [\cos \omega_1 t - \cos \omega_2 t]^2 dt$$

$$= 2E_b - 2E_{12}, \tag{4.101}$$

where the correlation

$$E_{12} = A^2 \int_0^{T_b} \cos \omega_1 t \cos \omega_2 t \ dt = \frac{A^2}{2} \int_0^{T_b} \cos(\omega_1 - \omega_2)t \ dt, \tag{4.102}$$

which vanishes for orthogonal BFSK and is nonzero for nonorthogonal BFSK signals. Clearly, for BFSK even with orthogonal carriers, the signal separation is simply equal to $d_{12} = \sqrt{2E_b}$ as illustrated in Figure 4.11. It is much less than that of the BPSK counterpart of the same symbol energy, which is given by $d_{12} = 2\sqrt{E_b}$. Thus, for orthogonal BSFK with coherent detection, the BER is given by

$$P_b = Q(\sqrt{E_b/N_0}), \tag{4.103}$$

which is 3 dB less than that of its coherent BPSK counterpart. Because of its inferior performance and demanding bandwidth requirement, as subsequently addressed, coherent MFSK is seldom employed in practice.

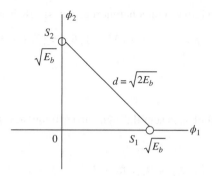

Figure 4.11 *Distance between two BFSK signals*

To determine the minimum frequency separation between two adjacent signals of coherent MFSK, we need to determine the cross-correlation between $s_1(t)$ and $s_2(t)$. By definition, we have

$$\rho = \frac{\int_0^{T_s} s_1(t)s_2(t)dt}{\sqrt{E_1 E_2}}$$

$$= \frac{2}{T_s} \int_0^{T_s} \cos\omega_1 t \cos\omega_2 t dt$$

$$= \underbrace{\frac{1}{T_s} \int_0^{T_s} \cos 2\pi(f_1 + f_2)t dt}_{=0: \, ?} + \frac{1}{T_s} \int_0^{T_s} \cos 2\pi(f_1 - f_2)t dt \qquad (4.104)$$

$$= \frac{1}{T_s} \frac{\sin 2\pi(f_1 - f_2)t}{2\pi(f_1 - f_2)} \Big|_{t=0}^{T_s}$$

$$= \frac{1}{T_s} \frac{\sin 2\pi(f_1 - f_2)T_s}{2\pi(f_1 - f_2)}. \qquad (4.105)$$

The minimum separation requirement is fulfilled when the cross-correlation $\rho = 0$, leading to

$$2\pi|f_1 - f_2|T_s = \pi \rightarrow \Delta f = \frac{1}{2T_s} = \frac{R_s}{2}. \qquad (4.106)$$

This is the requirement of orthogonality between carriers and is a necessary condition for coherent detection of MFSK signals. The minimum separation between two BFSK signals is shown in Figure 4.12. In general, the minimum bandwidth of coherent MFSK is equal to

$$B = M \times (2/T_s) + (M - 1) \times (R_s/2) = (M + 3)R_s/2. \qquad (4.107)$$

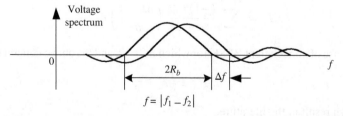

Figure 4.12 *Illustrating the minimum bandwidth of orthogonal BFSK*

For comparison, the system bandwidth of noncoherent receiver MFSK is given by

$$B = M \times (2/T_s) + (M-1)R_s = (3M-1)R_s. \tag{4.108}$$

4.9   Noncoherent MFSK

Suppose that the information symbol s_0 is sent through carrier frequency f_0 so that we have

$$x_0 = \sqrt{\gamma} + n_0,$$
$$x_m = n_m, \ m = 1, \cdots, M-1, \tag{4.109}$$

where we assume, without loss of generality, that the AWGN has been scaled to have $n_m \sim \mathcal{CN}(0,1)$ for $m = 0, \cdots, M-1$, so that γ represents the SNR on the channel f_0. The decision variables are the output energy from the M band-pass filters, given by $\xi_m = |x_m|^2$, with which a correct decision is made if

$$\xi_m < \xi_0, \tag{4.110}$$

for all $m = 1, \cdots, M-1$. By denoting $z = \xi_0$ and conditioned on z, all the events $\xi_m < z$ are independent, thus allowing us to write the conditional correct decision probability as

$$P_c(z) = \Pr\{\cap_{m=1}^{M-1}(\xi_m < z|z)\}$$
$$= [\Pr\{\xi_m < z|z\}]^{M-1}. \tag{4.111}$$

Note that $\xi_m \sim \frac{1}{2}\chi^2(2)$ for $1 \leq m \leq M-1$, whose PDFs are explicitly expressible as $f_{\xi_m}(u) = e^{-u}$. It is thus easy to obtain

$$\Pr\{\xi_m < z|z\} = \int_0^z e^{-u} \, du = 1 - e^{-z}, \tag{4.112}$$

whereby

$$P_c(z) = (1 - e^{-z})^{M-1}$$
$$= 1 + \sum_{m=1}^{M-1} (-1)^m \binom{M-1}{m} e^{-mz}. \tag{4.113}$$

Thus, averaging $P_c(z)$ over z is equivalent to finding the CHF of z, that is, $\mathbb{E}_z[e^{mz}]$, with the transform variable $-m$. Recall that $z = |\sqrt{\gamma} + n_0|^2 \sim \frac{1}{2}\chi^2(2, 2\gamma)$, with the CHF given by

$$\mathbb{E}[e^{-mz}] = (1+m)^{-1} \exp\left(\frac{-m\gamma}{1+m}\right). \tag{4.114}$$

It, when combined with (4.113) and the fact that $P_e = 1 - \mathbb{E}_z[P_c(z)]$, gives the symbol error probability

$$P_e = \sum_{m=1}^{M-1} \frac{(-1)^{m+1}}{m+1} \binom{M-1}{m} e^{-\frac{m\gamma}{m+1}}. \tag{4.115}$$

Consider the special case of noncoherent BFSK for which $M = 2$, and the above general expression reduces to

$$P_e = \frac{1}{2} e^{-\gamma/2}, \tag{4.116}$$

which is a well-known result in the literature.

4.10   Bit error probability versus symbol error probability

It is of practical importance to compare various modulation schemes on a common measure, for example, their bit error probability. The conversion of the symbol error probability P_s to its bit error counterpart P_b is generally difficult. The situation can be made easier if we just consider the case of modulation schemes operating in large SNRs and with Gray-code mapping in the sense that any adjacent signal points differ only by one bit. Large SNRs imply that erroneous events in MPSK and MASK, if they occur, are dominated by mistaking a decision from the true signal point to its immediately adjacent decision regions. For an M-ary modulation scheme, each symbol contains $m = \log_2 M$ bits. Thus, for an m-bit Gray-mapped symbol operating in high SNRs, the receiver makes a symbol error (dominated by one bit) with probability P_s and makes no error with probability $(1 - P_s)$. As such, each m-bit symbol experiences $1 \times P_s + 0 \times (1 - P_s) = P_s$ bits on average, implying that the BER is given by

$$P_b = \frac{P_s}{\log_2 M}. \tag{4.117}$$

4.10.1   Orthogonal MFSK

For orthogonal MFSK, an erroneous symbol decision can happen by mistaking it for any other $M - 1$ symbols with equal probability $P_s/(M - 1)$. To determine the average number of bits in error, note that among these $M - 1$ possible erroneous symbols, the number of symbols with k bits in error is same as that of choosing k objects from a total of m. With this weighting, the average number of bits in error is thus equal to

$$\ell = \sum_{k=1}^{m} k \binom{m}{k} \frac{P_s}{M - 1} = \frac{P_s}{M - 1} \underbrace{\sum_{k=1}^{m} k \binom{m}{k}}_{A}. \tag{4.118}$$

To determine A, we note that it is related to

$$B = \sum_{k=1}^{m} \binom{m}{k} x^k = (1 + x)^m - 1$$

by

$$A = x \frac{dB}{dx}\Big|_{x=1} = mx(1 + x)^{m-1}\Big|_{x=1} = m \cdot 2^{m-1}. \tag{4.119}$$

Inserting A into (4.118) and dividing both sides by m, we obtain the average BER for orthogonal MFSK as

$$P_b = \frac{M P_s}{2(M - 1)}. \tag{4.120}$$

4.10.2   Square M-QAM

Consider a square M-ary QAM with *Gray mapping*. Each QAM symbol contains $m = \log_2 M$ bits. For such a system, the probability of error occurring at bit k can be shown to be [7]

$$P_b(k) = \frac{1}{\sqrt{M}} \sum_{i=0}^{\kappa} \left\{ 2(-1)^{a_k} \left(2^{k-1} - \lfloor a_k + 0.5 \rfloor \right) Q \left((2i + 1) \sqrt{\frac{3m\gamma_b}{M - 1}} \right) \right\}, \tag{4.121}$$

for $k = 1, \cdots, m/2$, where $\gamma_b = E_b/N_0$ is the bit energy to noise ratio, $\lfloor z \rfloor$ denotes the integral part of a nonnegative number z, and

$$\kappa = (1 - 2^{-k})\sqrt{M} - 1,$$

$$a_k = \left\lfloor \frac{i2^{k-1}}{\sqrt{M}} \right\rfloor$$

The average BER is thus equal to

$$P_b = \frac{2}{m} \sum_{k=1}^{m/2} P_b(k). \tag{4.122}$$

In particular, for 16-QAM, it is easy to check that

$$P_b = \frac{1}{4} \left[3Q\left(\sqrt{\frac{4\gamma_b}{5}}\right) + 2Q\left(3\sqrt{\frac{4\gamma_b}{5}}\right) - Q\left(\sqrt{\frac{4\gamma_b}{5}}\right) \right]. \tag{4.123}$$

Note that here $4\gamma_b$ can be alternatively written as the symbol energy-to-noise ratio γ_s.

4.10.3   Gray-mapped MPSK

The BER of Gray-mapped MPSK depends on the rule chosen for Gray mapping. The cyclic Gray code widely adopted in current use was originated by Frank Gray [8]. By cyclic Gray code, we mean that all the codewords are organized in such a way that each adjacent pair differs only by one bit, including the first and the last codeword in the list.

The rule for generating cyclic Gray codes is simple, and is executable in a recursive manner. Denote all the codewords for MPSK by the rows of an $M \times M$ matrix $\mathbf{V}_M$. Then, the matrix $\mathbf{V}_{m+1}$ can be generated from $\mathbf{V}_m$ by using the following recursion:

$$\mathbf{V}_m \rightarrow \underbrace{\begin{bmatrix} \mathbf{0}_{2^m \times 1} & \mathbf{V}_m \\ \mathbf{1}_{2^m \times 1} & \mathbf{Q}_m\mathbf{V}_m \end{bmatrix}}_{\mathbf{V}_{m+1}}, \tag{4.124}$$

where $\mathbf{0}_{2^m \times 1}$ denotes the all-zero 2^m-dimensional column vector, $\mathbf{1}_{2^m \times 1}$ denotes the all-one 2^m-dimensional column vector, and $\mathbf{Q}_m$ denotes the $2^m \times 2^m$ opposite diagonal matrix. For example,

$$\mathbf{Q}_1 = \begin{pmatrix} 0 & 1 \\ 1 & 0 \end{pmatrix}. \tag{4.125}$$

It is easy to generate $\mathbf{Q}_m$ in MATLAB through $Q = \text{fliplr}(\text{eye}(M))$, where $\text{M} = 2^m$. The iteration is initialized with $\mathbf{V}_1 = [0; 1]$. The first two iterations evolve as follows:

$$\underbrace{\begin{bmatrix} 0 \\ 1 \end{bmatrix}}_{\mathbf{V}_1} \rightarrow \underbrace{\begin{bmatrix} 0 & 0 \\ 0 & 1 \\ 1 & 1 \\ 1 & 0 \end{bmatrix}}_{\mathbf{V}_2} \rightarrow \underbrace{\begin{bmatrix} 0 & 0 & 0 \\ 0 & 0 & 1 \\ 0 & 1 & 1 \\ 0 & 1 & 0 \\ 1 & 1 & 0 \\ 1 & 1 & 1 \\ 1 & 0 & 1 \\ 1 & 0 & 0 \end{bmatrix}}_{\mathbf{V}_3},$$

where the codewords are read row-wise.

The resulting $M = 2^m$ Gray cyclic codewords are then evenly mapped to a circle with radius $\sqrt{E_s}$ and center at the origin of the complex plane. The spacing between any two adjacent symbols (i.e., signal points) is equal to

$$\phi = \frac{2\pi}{M}.$$

As a convention, the all-zero codeword is placed at the point with coordinates $(\sqrt{E_s}, 0)$. To determine the BER for coherent MPSK on an AWGN channel, the key is to find how many bits are in error on average. Suppose signal ℓ is sent, which is located at $e^{j\ell\phi}$, and the received signal is

$$x = \sqrt{E_s}e^{j\ell\phi} + n,$$

where $n \sim \mathcal{CN}(0, N_0)$. The correction decision region for symbol ℓ is

$$-\frac{\phi}{2} \leq \arg(x) - \ell\phi < \frac{\phi}{2}.$$

An erroneous decision can be made in favor of symbol k if the phase angle $\arg(x)$ falls into the decision region for k for any $k \neq \ell$. Each erroneous decision implies a particular number of bits in error, very much depending upon the codeword pattern. The average number of bits in error per codeword is given by

$$\overline{n}_b = \sum_{\ell=0}^{M-1} \sum_{q=0,\neq\ell}^{M-1} \underbrace{\|\mathbf{s}_\ell - \mathbf{s}_q\|}_{n_b(q,\ell)} \Pr\{\mathbf{s}_q|\mathbf{s}_\ell\}\underbrace{\Pr\{\mathbf{s}_\ell\}}_{=1/M}. \tag{4.126}$$

The transitional probability $\Pr\{\mathbf{s}_q|\mathbf{s}_\ell\}$ represents the symbol $\mathbf{s}_\ell$ is sent but a decision is made in favor of $\mathbf{s}_q$ where each symbol is represented as a vector of entries 0 and 1. The number of bits in error associated with such an erroneous decision is equal to the norm of $\|\mathbf{s}_\ell - \mathbf{s}_q\|$. Since all the entries of $(\mathbf{s}_\ell - \mathbf{s}_q)$ are zero or 1, $n_b(q, \ell)$ is numerically the same as the squared norm. Because of the geometric symmetry of the MPSK signal constellation, the transitional probability $\Pr\{\mathbf{s}_q|\mathbf{s}_\ell\}$ only depends on the angle $|q - \ell|\phi$ that separates the two symbols. Thus, we can write it as a function of $(q - \ell)$, as shown by

$$\Pr\{\mathbf{s}_q|\mathbf{s}_\ell\} = P(q - \ell). \tag{4.127}$$

The situation for $n_b(q, \ell)$ is different. Although the Gray codes can guarantee any two *adjacent* symbols to differ only in one bit, there is *no* guarantee that any two symbols separated by angle $k\phi$ for $k > 1$ have the same number of bits in difference, implying that we cannot write $n_b(q, \ell) = n_b(q - \ell)$. On the basis of these arguments, we rewrite (4.126) as

$$\overline{n}_b = \sum_{k=1}^{M-1} P(k) \underbrace{\left[\frac{1}{M} \sum_{\ell=0}^{M-1} \|\mathbf{s}_\ell - \mathbf{s}_{\ell+k}\|^2\right]}_{\overline{n}_b(k)}, \tag{4.128}$$

where $\overline{n}_b(k)$ signifies the average number of bits in error for the set of symbol pairs separated by angle $k\phi$, and the index $\ell + k$ is modulo-M addition. When using MATLAB, the index modulo-M operation can be easily avoided if we form an augmented matrix $\mathbf{V} = [\mathbf{V}_m; \mathbf{V}_m]$ by stacking the codeword matrix $\mathbf{V}_m$ for the MPSK one over another.

Through a simple calculation using MATLAB, we can find $\overline{n}_b(k)$ for MPSK of different sizes, as tabulated below. Only half of $\overline{n}_b$ is listed for saving space. The remaining part is symmetric about the last number in the list. For example, the complete list for 8-PSK should be $[1, 2, 2, 2, 2.5, 3, 2.5, 2, 2.5, 3, 2.5, 2, 2, 2, 1]$.

Table 4.2   *Average number of error bits $\overline{n}_b(k)$ versus k for commonly used MPSK*

Schemes	$\overline{n}_b(k)$							
	$k = 1$	2	3	4	5	6	7	8
BPSK	1							
QPSK	1	2						
8-PSK	1	2	2	2				
16-PSK	1	2	2	2	2.5	3	2.5	2
32-PSK ($\overline{n}_b(1:8)$	1	2	2	2	2.5	3	2.5	2
($\overline{n}_b(9:16)$	2.75	3.5	3.25	3	3.25	3.5	2.75	2

From Table 4.2, we can see that nonintegral numbers occur in the list of 16-PSK and 32-PSK. They indicate that the symbol pairs with these angle separations must have unequal distance. This interesting finding is originally due to Lassing et al. [9]. It clarifies and refines the early seminal work of P.J. Lee [10].

It remains to calculate $P(k)$. To this end, we may invoke the result of [10]

$$P(k) = \int_0^\infty \frac{1}{\sqrt{2\pi}} e^{\xi^2/2} \left[Q\left(\sqrt{\gamma_s} + \xi \tan\left[\frac{(M - 4k - 2)\phi}{4} \right] \right) \right.$$
$$\left. - Q\left(\sqrt{\gamma_s} + \xi \tan\left[\frac{(M - 4k + 2)\phi}{4} \right] \right) \right] d\xi, \qquad (4.129)$$

where $\gamma_s = 2E_s/N_0$ and $k = 1, 2 \cdots, M/2 - 1$. Inserting $\overline{n}_b(k)$ and $P(k)$ into (4.126) and dividing both sides with $m = \log_2 M$ lead to the BER of MPSK with Gray mapping

$$P_b = \frac{\overline{n}_b}{m}. \qquad (4.130)$$

4.11   Spectral efficiency

For M-ary systems, the bit rate is equal to $R_b = R_s \log_2 M$. To transmit at such a bit rate, an MPSK system needs a null-to-null bandwidth of $2R_s$, implying that the spectral efficiency

$$\eta = \frac{R_s \log_2 M}{2R_s} = \frac{1}{2} \log_2 M. \qquad (4.131)$$

The question is, using modulation alone can we approach the theoretical bound? To answer this question, let us consider M-ary orthogonal modulation with coherent demodulation. Assuming that $s_k(t) \in \{s_0(t), \cdots, s_{M-1}(t)\}$ is transmitted, a filter bank of M coherent receivers is employed to detect the transmitted signal, with each receiver tuned to an orthogonal waveform. The output signals are, respectively, given by

$$x_i = \sqrt{\gamma} \, \delta(i - k) + n_i, \qquad (4.132)$$

where $\gamma = 2E_s/N_0$, $\delta(i)$ is the Dirac delta function, and $n_i \sim \mathcal{N}(0, 1)$ are i.i.d. AWGN. As such, the error probability, conditioned on $\xi = x_k = \sqrt{\gamma} + n_k$, is then given by

$$P_M = 1 - [1 - Q(\sqrt{\xi})]^{M-1}. \tag{4.133}$$

The average error performance $\overline{P}_M$ can be obtained by averaging P_M over the distribution of ξ. As $M \to \infty$, we would like to compare the bit rate R with the channel capacity C_∞. To this end, we relate γ to C_∞ of the AWGN channel with bandwidth B and single-sided power spectral density N_0 through the capacity formula, as shown by

$$C_\infty = \lim_{B \to \infty} \log_2 \left(1 + \frac{P_{av}/B}{N_0} \right) = P_{av}/(N_0 \ln 2). \tag{4.134}$$

whereby we can write

$$\gamma = 2E_s/N_0 = 2P_{av}T/N_0 = 2C_\infty \, T \ln 2. \tag{4.135}$$

It follows that $\xi = \sqrt{\gamma} + n_k \sim \mathcal{N}(\sqrt{\gamma}, 1)$. We further relate the bit rate R to M by $M = 2^{RT}$, so that we can rewrite

$$\overline{P}_M = \int_{-\infty}^{\infty} [1 - (1 - Q(\sqrt{\xi}))]^{RT-1} f(\xi) d\xi$$

$$= \int_{-\infty}^{\infty} [1 - (1 - Q(\sqrt{\xi}))]^{RT-1} \frac{1}{\sqrt{2\pi}} e^{-(\xi - \sqrt{\gamma})^2/2} d\xi. \tag{4.136}$$

The idea is to study the behavior of $\overline{P}_M$ as $M \to \infty$ (or equivalently, as $T \to \infty$) by following the technique originated by J.G. Proakis [3]. Partition the integral region at $a = \sqrt{2 \ln M}$, and upper-bound the $\overline{P}_M$ as

$$\overline{P}_M < \int_{-\infty}^{a} \frac{1}{\sqrt{\pi}} e^{-(\xi - \sqrt{\gamma})^2/2} d\xi + \frac{M}{\sqrt{2\pi}} \int_{a}^{\infty} e^{-\xi^2/2} e^{-(\xi - \sqrt{\gamma})^2/2} d\xi. \tag{4.137}$$

In the first term, we have used the fact $P_M < 1$ to replace P_M with unity; in the second term, we have used $1 - [1 - Q(\xi)]^{M-1} \leq (M-1)Q(\xi) < MQ(\xi)$ and $Q(x) < e^{-x^2/2}$ for $x > 0$.

We bound the first and the second integrals on the right by a Gaussian function and a Gaussian integral respectively. It turns out that

$$\overline{P}_M < \begin{cases} 2 \times 2^{-T[\frac{1}{2}C_\infty - R]}, & 0 \leq R \leq \frac{1}{4}C_\infty \\ 2 \times 2^{-T(\sqrt{C_\infty} - \sqrt{R})^2}, & \leq \frac{1}{4}C_\infty \leq R \leq C_\infty \end{cases}. \tag{4.138}$$

Thus, we conclude that as $T \to \infty$, $\overline{P}_M$ approaches zero as long as $R < C_\infty$, implying that the M-ary orthogonal scheme reaches the channel capacity at the location of near-zero efficiency. It should be pointed out that in general cases, modulation alone cannot reach the capacity.

4.12 Summary of symbol error probability for various schemes

The probability of symbol error for various modulation schemes on an AWGN channel is summarized in Table 4.3. See also Ref. [11]. The Q-function and squared Q-function in Table 4.3 can be alternatively

Table 4.3   *Summary of formulas for symbol error rate (SER)*

Scheme	SER $P_s(\gamma)$ versus $\gamma = E_s/N_0$
BPSK	$Q(\sqrt{2\gamma})$
QPSK	$2Q(\sqrt{\gamma}) - Q^2(\sqrt{\gamma})$
MPSK	$\dfrac{1}{\pi} \displaystyle\int_0^{\pi - \pi/M} \exp\left(-\dfrac{\gamma\sin^2(\pi/M)}{\sin^2\theta}\right)\, d\theta$, or alternative formula given in (4.78)
Square M-QAM	$aQ(\sqrt{b\gamma}) - (a/2)^2 Q^2(\sqrt{b\gamma})$ where $a = 4(1 - \sqrt{1/M})$, $b = \frac{3}{M-1}$
Binary DPSK	$0.5\exp(-\gamma)$
Gray-coded $\frac{\pi}{4}$-DQPSK [11]	$\dfrac{1}{2\pi} \displaystyle\int_0^{\pi} \exp\left(\dfrac{-2\gamma}{2 - \sqrt{2}\,\cos\theta}\right)\, d\theta$
MDPSK [12]	$\dfrac{1}{\pi} \displaystyle\int_0^{\pi - \pi/M} \exp\left(-\dfrac{\gamma\sin^2(\pi/M)}{1 + \cos(\pi/M)\cos\theta}\right)\, d\theta$
MSK	$2Q(\sqrt{\gamma}) - Q^2(\sqrt{\gamma})$
CBFSK	$Q(\sqrt{\gamma})$
NBFSK	$0.5\exp(-0.5\gamma)$
NMFSK	Formula in (4.115)

Table 4.4   *Bit error rate P_b for various schemes, where P_s denotes SER*

Schemes	Formulas for BER	
	Exact P_b	Approximation
MFSK	$\dfrac{MP_s}{2(M-1)}$ (see (4.120))	
Square M-QAM with Gray mapping	Eqs. (4.121)–(–4.122)	$\dfrac{4(\sqrt{M} - 1)}{\sqrt{M}\log_2 M} Q\left(\sqrt{\dfrac{3\gamma_b \log_2 M}{M - 1}}\right)$ for large γ_b
Gray-mapped MPSK	Eqs. (4.128)–(4.130)	$\dfrac{P_s}{\log_2 M}$

represented as exponential integrals, by virtue of (2.69). The resulting expressions are very useful for the performance analysis of diversity systems to be addressed in Chapter 7.

4.13   Additional reading

For supplementary materials on digital modulation, we provide references [13]–[20] for additional reading.

Problems

Problem 4.1   In the derivation of the error performance of MPSK using approach 3, we formulate the correct decision event directly based on the received waveforms. Show that the same performance result can be obtained by formulating the correct event based on the baseband received signals, as given in third line of (4.61).

Problem 4.2 In a coherent QPSK system, the mapping table is

$$11 \rightarrow -\pi/4, \quad 10 \rightarrow \pi/4$$
$$01 \rightarrow 5\pi/4, \quad 00 \rightarrow 3\pi/4.$$

Design a transmitter to implement the above mapping, and a receiver to implement binary decisions, separately, on the in-phase and quadratic branches.

Problem 4.3 In a BFSK system, suppose the carriers for bit 0 and bit 1 are f_1 and f_2, respectively. Further assume that f_1 and f_2 are related to the bit interval T_b by $f_1 = 1/T_b$ and $f_2 = 3/T_b$. If the transmitted bit sequence is 1001011011, sketch and label the corresponding waveform.

Problem 4.4 Sketch a receiver block diagram for coherent detection of BFSK signals.

Problem 4.5 Let $p(t; \tau)$ denote the rectangular pulse of unit magnitude over $0 \le t < \tau$ such that

$$p(t; \tau) = \begin{cases} 1 & 0 \le t < \tau \\ 0 & \text{elsewhere.} \end{cases}$$

Four waveforms are used to represent four 2-bit symbols.

$$s_1(t) = p(t; 2\tau) - p(t - 2\tau; \tau)$$
$$s_2(t) = p(t|\tau) - p(t - \tau|2\tau)$$
$$s_3(t) = -p(t|\tau) + p(t - \tau|2\tau)$$
$$s_4(t) = -(t|2\tau) + p(t - 2\tau|\tau).$$

(1) Find the angles between any two signals. (2) Sketch the signal points in the signal constellation.

Problem 4.6 A coherent receiver is used to detect a QPSK signal in AWGN

$$s(t) = A \cos(\omega_c t + \phi_m) + n(t)$$

where the information symbol $\phi_m \in \{\pm \pi/4, \pm 3\pi/4\}$, and $n(t)$ is the zero-mean AWGN with two-sided spectral density $N_0/2$ W/Hz. Assume that the two reference local signals in the in-phase and quadratic branches, for some reason, carry a phase error θ_0, as shown by $\psi_1(t) = \cos(\omega_c t + \theta_0)$ and $\psi_2(t) = \sin(\omega_c t + \theta_0)$. (1) Determine the probability of symbol error P_e. (2) Determine P_e if $\theta_0 = 12°$, $A = 0.4$ V, $N_0/2 = 10^{-5}$ W/Hz, and the symbol interval $T_s = 100$ μs.

Problem 4.7 A binary modulation system transmits the two symbols by using the carriers

$$s_1(t) = \cos 2\pi \times 10^5 t$$
$$s_2(t) = \cos 2\pi (10^5 + 20{,}000)t$$

over the symbol interval $0 < t < 20$ μs. (1) Plot the signal constellation. (2) Determine the error performance if a coherent receiver is used to operate in AWGN with single-sided power spectral density $N_0 = 10^{-6}$ W/Hz..

Problem 4.8 Consider a binary communication system in which bits 1 and 0 are represented by

$$s(t) = \begin{cases} 1 & \text{if } b = 1 \\ \alpha & \text{if } b = 0' \end{cases}$$

over $0 \le t < T_b, -1 \le \alpha \le 1$. The system operates on an AWGN channel.

(a) Sketch and label the optimal receiver structure.
(b) Find the optimal value of α in the above range that maximizes the bit error probability.
(c) If $s(t)$ is given, instead, by

$$s(t) = \begin{cases} \cos(2\pi \, f_c t) & \text{if } b = 1 \\ \sin(2\pi \, f_c t + \beta \pi) & \text{if } b = 0 \text{'} \end{cases}$$

over $0 \leq t < T_b, -1 \leq \beta \leq 1$, what is the optimal β to maximize the bit error probability?

Problem 4.9 Prove that the error probability of a coherence communication system operating on an AWGN channel falls off exponentially with SNR.

Problem 4.10 Using the definition of null-to-null bandwidth, determine the spectral efficiency of (1) 16-PSK, (2) 256-QAM, (3) coherent 16-FSK, and (4) non-coherent 16-FSK.

Problem 4.11 Coherent detection of BPSK signals with a phase error θ_0 in the reference signal is shown in Example 4.2.

(a) Suppose that without phase error, the BER of the system is found to be 10^{-5}. What is the maximum tolerable phase error in order to maintain the error performance not less than 10^{-4}?
(b) If the phase error is uniformly distributed over the range of $\theta_0 \in [-\epsilon, \epsilon)$ radians with $0 < \epsilon < \pi$. Determine the average probability of bit error.

Problem 4.12 Design a modulation scheme for transmission over a bandpass AWGN channel of bandwidth 4000 Hz and the signal to noise single-sided spectral density ratio $E_s/N_0 = 10$ dB to achieve the BER not less than $P_b = 10^{-3}$.

Problem 4.13 A coherent BPSK system operates in AWGN with phase error $\theta_0 = 10°$. Suppose that its bit error performance $P_b = 10^{-3}$. Determine P_b if the phase error is removed.

Problem 4.14 Consider a coherent binary communication system with two transmitted waveforms $s_1(t)$ of symbol energy E_1 and $s_2(t)$ of symbol energy E_2 over an AWGN channel. Assume that local reference signals with phase error, that is, $s_1(t)e^{j\theta}$ and $s_2(t)e^{j\theta}$, are used. Determine the bit error performance.

Problem 4.15 For a coherent MPSK system in AWGN with large SNR ρ_s, its error performance is dominated by erroneous decisions in favor of two adjacent signal points. Use this fact to show that, for large ρ_s, the probability of symbol error can be approximated by its lower bound

$$P_e \approx 2Q[\sqrt{2\rho_s} \, \sin(\pi/M)].$$

Problem 4.16 Suppose 16-PSK and 16-square QAM are two candidate schemes in a system design. Let d_1 and d_2 denote the distance between two nearest neighbor signal points in the two schemes, respectively. If the two schemes have the same average signal power,

(a) find the relationship between d_1 and d_2;
(b) compare the average probability of symbol error of the two schemes in AWGN.

Problem 4.17 A channel of bandwidth 0.5 MHz is supposed to support a data rate of 1.2 Mb/s. Consider the following schemes for transmission: (1) MPSK, (2) M-QAM, (3) noncoherent MFSK, and (4) MDPSK. Determine the minimum required values of M for these schemes.

Problem 4.18 In Problem 4.17, with the values of M you have chosen for transmission over an AWGN channel, determine the corresponding BER for SNR $\rho = 15$ dB.

Problem 4.19 Consider a 16-ary differential amplitude and phase shift keying (16-DAPSK) in which each word $w = [b_0 \, b_1 \, b_2 \, b_3]$ consists of four bits taking values of 0 or 1. Bit b_0 is differently modulated into the amplitude, and the remaining three bits are differentially modulated into the phase difference, so that the DAPSK modulated signal at time instant i is given by $s_i = a_i e^{j(\phi_{i-1} + \Delta\psi_i)}$. The amplitude a_i can take one of two possible levels, namely A_L or A_H where $A_H > A_L$. The two levels A_H and A_L are virtually mirror symmetric about their geometric mean $A = \sqrt{A_L A_H}$. Thus, by denoting $\alpha = \sqrt{A_H / A_L}$, we can write $A_H = \alpha A$ and $A_L = A/\alpha$. The current amplitude a_i is updated from the previous one a_{i-1} under the control of bit b_0, as shown by

$$a_i = a_{i-1} \left(\frac{A}{a_{i-1}} \right)^{2b_0}.$$

The phase is updated by

$$\phi_i = \phi_{i-1} + \Delta\phi_i,$$

where the mapping between the remaining three bits and $\Delta\phi_i$ is defined by the following table:

$\Delta\phi_i$	0	$\dfrac{\pi}{4}$	$\dfrac{\pi}{2}$	$\dfrac{3\pi}{4}$	π	$\dfrac{5\pi}{4}$	$\dfrac{3\pi}{2}$	$\dfrac{7\pi}{4}$
$[b_1 b_2 b_3]$	000	001	011	010	110	111	101	100

Given the noisy received signal $x_i = s_i + n_i$, where $n_i \sim \mathcal{CN}(0, \sigma_n^2)$, determine two decision variables for amplitude and phase, respectively, and specify the corresponding decision thresholds.

Problem 4.20 For the square M-ary QAM with Gray mapping, show that if only the two dominating terms are kept, the BER can be approximated as

$$P_b = \alpha Q \left(\sqrt{\frac{3m\gamma_b}{M-1}} \right) + \beta Q \left(3\sqrt{\frac{3m\gamma_b}{M-1}} \right),$$

where $\gamma_b = E_b / N_0$ is the bit energy-to-noise ratio, $m = \log_2 M$, and

$$\alpha = \frac{4(\sqrt{M} - 1)}{m\sqrt{M}}, \ \beta = \frac{4(\sqrt{M} - 2)}{m\sqrt{M}}$$

[7].

Problem 4.21 Write a MATLAB program to generate all the codewords for Gray-mapped 64-PSK, and give a full list of $\overline{n}_b(k)$, the average number of Hamming distance between all the symbol pairs separated by phase angle $k\pi/32$.

Problem 4.22 As shown in the text, the computational burden for determining the BER for Gray-coded coherent MPSK signaling over an AWGN channel comes from $P(k)$, the probability that the received signal x for the current symbol mistakenly falls into the decision sectorial region for another symbol $k2\pi/M$ apart. There are many ways to determine $P(k)$. By assuming that the all-zero codeword is sent, one way is to formulate the error event as $-\phi/2 \le \arg\{xe^{-jk\phi}\} < \phi/2$ where $\phi = 2\pi/M$. Based on this formulation, show that $P(k)$ can be represented as

$$\int_0^\infty \frac{1}{\sqrt{2\pi}} e^{-\frac{(\xi - \gamma_s \cos k\phi)^2}{2}} d\xi \left[\int_{-\xi \tan\frac{\phi}{2} + \gamma_s \sin k\phi}^{\xi \tan\frac{\phi}{2} + \gamma_s \sin k\phi} \frac{1}{\sqrt{2\pi}} e^{-z^2/2} dz \right],$$

where $\gamma_s = 2E_s / N_0$.

References

1. W.R. Bennett, "Methods of solving noise problems," *Proc. IRE*, vol. 44, pp. 609–638, 1956.
2. M.K. Simon, W.C. Lindsey, and S.M. Hinedi, *Digital Communication Techniques*, Englewood Cliffs, NJ: PTR Prentice Hall, 1995.
3. J.G. Proakis, *Digital Communications*, 3rd ed., New York: McGraw Hill, 1995.
4. A. Goldsmith, *Wireless Communications*, Cambridge University Press, 2005.
5. R.F. Pawula, S.O. Rice, and J.H. Roberts, "Distribution of the phase angle between two vectors perturbed by Gaussian noise," *IEEE Trans. Commun.*, vol. 30, pp. 1828–1841, 1982.
6. F.S. Weinstein, "A table of the cumulative probability distribution of the phase of a sine wave in narrow-band normal noise," *IEEE Trans. Inf. Theory*, vol. IT-23, pp. 640–643, 1977.
7. K. Cho and D. Yoon, "On the general BER expression of one- and two-dimensional amplitude modulations," *IEEE Trans. Commun.*, vol. 50, no. 7, pp. 1074–1080, 2002.
8. F. Gray, "Pulse Code Communications," U.S. Patent 2 632 058, Mar. 17, 1953.
9. J. Lassing, E.G. Strom, E. Agrell, and T. Ottosson, "Computation of the exact bit-error rate of coherent M-ary PSK with Gray code bit mapping," *IEEE Trans. Commun.*, vol. 51, no. 11, 1758–1760, 2003.
10. P.J. Lee, "Computation of the bit-error rate of coherent M-ary PSK with Gray code bit mapping," *IEEE Trans. Commun.*, vol. 34, pp. 488–491, 1986.
11. A. Annamalai, C. Tellumbura, and V.K. Bhargava, "Equal-gain diversity receiver performance in wireless channels," *IEEE Trans. Commun.*, vol. 48, no. 10, pp. 1732–1745, 2000.
12. R.F. Pawula, "Generic error probabilities," *IEEE Commun. Lett.*, vol. 2, no. 10, pp. 271–272, 1998.
13. M.K. Simon and M.S. Alouini, *Digital Communication Over Fading Channels: A Unified Approach to Performance Analysis*, New York: John Wiley & Sons, Inc., 2000.
14. R.E. Ziemer and R.L. Peterson, *Introduction to Digital Communication*, 2nd edn, Upper-Saddle River, NJ: Prentice Hall, 2001.
15. M. Chiani, D. Dardari, and M.K. Simon, "New exponential bounds and approximations for the computation of error probability in fading channels," *IEEE Trans. Wireless Commun.*, vol. 2, no. 4, pp. 840–845, 2003.
16. J.W. Craig, "A new, simple and exact result for calculating the probability of error for two-dimensional signal constellations," *IEEE MILCOM'91*, MILCOM Conference Record, vol. 2, pp. 571–575, McLean, VA, Nov. 4-7, 1991.
17. F.S. Weinstein, "Simplified relationships for the probability distribution of phase of a sine wave in narrow-band normal noise," *IEEE Trans. Inf. Theory*, vol. IT-20, pp. 658–661, 1974.
18. R.F. Pawula, "Generic error probabilities," *IEEE Trans. Commun.*, vol. 47, no. 5, pp. 697–702, 1999.
19. M.K. Simon and D. Divsalar, "Some new twists to problems involving the Gaussian probability integral," *IEEE Trans. Commun.*, vol. 46, pp. 200–210, 1998.
20. J. Park and S. Park, "Approximation for the two-dimensional Gaussian Q-function and its approximations," *ETRI J.*, vol. 32, no. 1, pp. 145–147, 2010.

5

Minimum Shift Keying

Abstract

Given its important role played in 2G cellular systems, minimum shift keying (MSK) is worth a dedicated chapter for its discussion to serve as an in-depth record. The reader can skip this chapter without losing reading continuity. The drawback of QPSK is its phase discontinuity such that the phase shift can jump by as much as 180° from one symbol to the next, causing spectrum spillover to the adjacent channels. In MSK, a different philosophy is employed. The information-modulated phase is no longer a constant over a symbol duration. But rather, it varies with time as a straight line. In doing so, two degrees of freedom are used to control the phase trajectory: the slope of the straight line, and its intercept on the y-axis. The two parameters are chosen so that the variation of the phase is a continuous function of time. Clearly, two consecutive symbols need to jointly control the two phase parameters. That is the characteristics of the MSK.

♣

5.1 Introduction

Digital modulation/demodulation is a key technology to cellular mobile communication. A cellular system is a communication system that supports a large number of voice channels. Any out-of-band radiation (leakage) will produce interference with adjacent channels, thereby degrading the system performance. To ensure the necessary transmission quality, the signal to adjacent interference should be greater that 20 dB. Furthermore, as a mobile unit moves, the depth of fading can range between 40 and 60 dB. As such, the leakage of a modulated signal to adjacent channels must be maintained at a level from −60 to −80 dB.

In TDMA cellular such as IS-54 and IS-136, the RF channel bandwidth is 30 kHz while the data rate generated by digital voice is 16 kb/s. Thus, it requires that the frequency efficiency be better than 1 bit/s/Hz (baseband).

The modulation scheme must be power-efficient because a mobile unit cannot carry a big battery. Accordingly, QPSK and QAM were not used in the second-generation (2G) mobile communication, since the former was frequency-inefficient and the latter was power- inefficient. The two schemes chosen for the 2G TDMA cellular systems were MSK and $\frac{\pi}{4}$-QPSK. We only address MSK in this chapter.

Before proceeding, let us briefly review the QPSK signaling scheme, in which each two-bit symbol is mapped onto a phase angle, as defined by

$$s_k(t) = A\cos(\omega_c t + \underbrace{(2k-1)\pi/4}_{\theta_k}), \quad 0 \le t < T_s.$$

Wireless Communications: Principles, Theory and Methodology, First Edition. Keith Q.T. Zhang.
© 2016 John Wiley & Sons, Ltd. Published 2016 by John Wiley & Sons, Ltd.
Companion Website: www.wiley.com/go/zhang7749

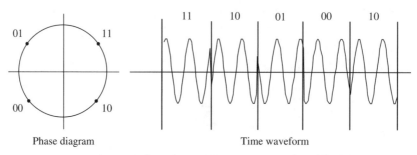

Figure 5.1   *Discontinuity of QPSK waveform*

Suppose we use the following Gray-coded mapping:

Symbol	11	01	00	10
Phase θ_k	$\frac{\pi}{4}$	$\frac{3\pi}{4}$	$\frac{5\pi}{4}$	$\frac{7\pi}{4}$

The output QPSK waveform corresponding to the message 1110010010 is depicted in Figure 5.1.

The waveform discontinuity is clearly indicated, which is due to a phase jump. Phase transition from one symbol to the next can be as large as 180°, which occurs, for instance, from symbol 10 to 01. A large, sudden change in phase causes a large instantaneous frequency $f_i(t)$, given that the latter is related to the instantaneous phase by the rule of differentiation:

$$f_i(t) = \frac{d\theta(t)}{dt}. \tag{5.1}$$

It is also observed that the envelope is not strictly constant, implying that linear amplifiers should be used.

5.2 MSK

The idea for bandwidth conservation is to maintain the phase continuity during the transition from one symbol to the next. Minimum-shift keying (MSK) is one such technique.

MSK modulation can be introduced in different ways. Since phase continuity is the major consideration, we start from the instantaneous phase and represent the MSK modulated signal over the bit interval $(n-1)T_b \le t < nT_b$ as

$$s_n(t) = A \cos[\omega_c t + \theta_n(t, \alpha)], \tag{5.2}$$

where A and ω_c denote the amplitude and carrier frequency in radians/second, respectively. There are different ways in which a bipolar message sequence $\alpha = [\alpha_0, \alpha_1, \cdots, \alpha_n]$ is modulated into the instantaneous phase $\theta(t, \alpha)$. In this chapter, we focus on two approaches. In the first approach, the message sequence is converted to a bipolar sequence with $\alpha_n \in \{-1, 1\}$ before modulating the instantaneous phase to produce

$$\theta_n(t, \alpha) = \alpha_n \left(\frac{\pi t}{2T_b} \right) + \underbrace{\phi_n(\alpha_n, \alpha_{n-1})}_{\pi u_n}, \quad nT_b \le t < (n+1)T_b. \tag{5.3}$$

This method was introduced by de Buda [1].

The second approach was introduced in Refs [2, 3], where a unipolar information bit sequence $\alpha_n \in \{0, 1\}$ is encoded into the phase differently, as shown by

$$\theta_n(t, \alpha) = \alpha_n \pi \left(\frac{t}{T_b} - n \right) + \underbrace{\phi_n(\alpha_0, \cdots, \alpha_n)}_{\pi v_n}, \quad nT_b < t < (n+1)T_b. \tag{5.4}$$

In both techniques, we need to choose the initial phase ϕ_n such that the instantaneous phase transits continuously from one symbol period to the next. These two different symbol-to-phase mappings lead to different phase behavior and, thus, different transceiver structures, which will be treated separately. We mainly introduce the first approach, and postpone a brief description of the second approach to Section 5.8.

5.3 de Buda's approach

Let us proceed with the first approach due to de Buda. It is observed from (5.3) that the instantaneous phase $\theta(t, \alpha)$ consists of two terms: a piecewise linear term in time, and an initial phase-state term ϕ_n, which depends on both the current and previous symbols. Geometrically, $\theta(t, \alpha)$ is a piecewise straight line, with ϕ_n representing its intercept with the y-axis. The slope of the straight-line segment depends on the sign of the current symbol. The slope is positive if $\alpha_n > 0$ and negative otherwise. The linear segment creates a phase increment of $\frac{\pi}{2}$ over a symbol interval T_b.

5.3.1 The basic idea and key equations

We are interested in how MSK makes the phase $\theta(t, \alpha)$ *smoothly* transit from one symbol to the next. Over the period $nT_b \leq t < (n+1)T_b$, the phase is expressible as

$$\theta_n(t, \alpha) = \alpha_n \left(\frac{\pi t}{2T_b} \right) + \phi_n, \tag{5.5}$$

whereas over $(n-1)T_b \leq t < nT_b$, the phase is given by

$$\theta_n(t, \alpha) = \alpha_{n-1} \left(\frac{\pi t}{2T_b} \right) + \phi_{n-1}. \tag{5.6}$$

To maintain phase continuity at symbol transition at $t = nT_b$, the two phases shown above must be equal; namely

$$\alpha_n \left(\frac{\pi t}{2T_b} \right) + \phi_n|_{t=nT_b} = \alpha_{n-1} \left(\frac{\pi t}{2T_b} \right) + \phi_{n-1}|_{t=nT_b}, \tag{5.7}$$

which is simplified to

$$\phi_n = \phi_{n-1} + \frac{n\pi}{2}(\alpha_{n-1} - \alpha_n), \ mod \ 2\pi, \tag{5.8}$$

with initial setting $\phi_0 = 0$ (must). This is the key equation for MSK, indicating how the information symbol is used for encoding the phase sequence. The phase trellis of an MSK signal based on this equation is shown in Figure 5.2. Sometimes, it is more convenient to normalize ϕ_n by π, such that $u_n = \phi_n/\pi$ whereby the above recursive equation is expressible as

$$u_n = u_{n-1} + \frac{n}{2}(\alpha_{n-1} - \alpha_n) \tag{5.9}$$

for which $u_0 = 0$ (must). The above equations lead to two observations. First, the MSK is, in essence, a differential coding scheme whereby the difference of two consecutive bits is differentially encoded into two

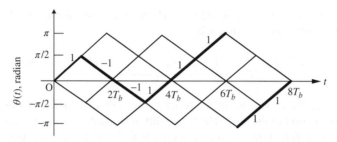

Figure 5.2   *Illustrating the phase transition of an MSK signal*

consecutive phases. Second, the instantaneous phase evolves form one symbol to the next, according to an order-1 Markov process.

Equation (5.8) offers a foundation to fully understand MSK, and different understandings lead to different transceiver structures.

5.4   Properties of MSK signals

We usually start from $\phi_0 = 0$. After a simple recursive calculation of (5.8) with modulo-2 addition, it is easy to assert

$$\phi_n \in \{0, \pi\}. \tag{5.10}$$

Taking the cosine of both sides of (5.8), denoting $\xi_n = (\alpha_{n-1} - \alpha_n)/2$, and using the fact that $\xi_n \in \{0, 1, -1\}$, we have

$$\cos \phi_n = \cos(\phi_{n-1} + n\xi_n \pi) = (-1)^{n\xi_n} \cos \phi_{n-1}, \tag{5.11}$$

where we have used the rule $\cos(x + m\pi) = (-1)^m \cos x$ for an arbitrary integral m. It follows that

$$\cos \phi_n = \begin{cases} \cos \phi_{n-1}, & \text{if } n = 2k \text{ or } \xi_n = 0 \\ -\cos \phi_{n-1}, & \text{iff } n = 2k + 1 \text{ and } \xi_n \neq 0 \end{cases}. \tag{5.12}$$

Here, the symbol iff means if and only if. This equation clearly indicates that the variational behavior of the coded phase is much easier to capture by the cosine function of ϕ_n than the phase itself. Accordingly, we define

$$a_n = \cos \phi_n, \ b_n = \alpha_n \cos \phi_n = \alpha_n a_n, \tag{5.13}$$

which can be considered a coded version of the message sequence which, as shown in the next section, is actually used to modulate the in-phase and quadrature components of the carrier. With this notation, (5.12) is further expressible as

$$a_n = \begin{cases} a_{n-1}, & \text{if } n = 2k \text{ or } \alpha_n = \alpha_{n-1} \\ -a_{n-1}, & \text{if } (n = 2k + 1) \cap (\alpha_n = -\alpha_{n-1}) \end{cases}. \tag{5.14}$$

The physical significance of this expression, when stated explicitly, is as follows.

Property 5.4.1   *In the MSK system described by (5.3), the relationship $a_{2k} = a_{2k-1}$ always holds, regardless of the values of α_{2k} and α_{2k-1}. In other words, the symbol a_n remains unchanged over the two consecutive intervals $(2k - 1)T_b \leq t < (2k + 1)T_b$, centered at $t = 2kT_b$. The only possible locations at which a_n changes its sign are $t = (2k + 1)T_b$, and this happens only when $\alpha_{2k} = -\alpha_{2k-1}$.*

Therefore, in estimating a_{2k} from a received MSK signal, one should accumulate useful information over $2T_b$ duration from $t = (2k - 1)T_b$ to $t = 2kT_b$.

To find the variational rule for b_n, we can concisely represent the second line of (5.12) by multiplying the equation there on with the required condition $\alpha_n = -\alpha_{n-1}$ to form a single equation. The result is $b_{2k+1} = b_{2k}$. From the first line, we have $b_{2k} = b_{2k-1}$ if $\alpha_{2k} = \alpha_{2k-1}$, and $b_{2k} = -b_{2k-1}$ otherwise. In summary, we conclude

$$b_n = \begin{cases} b_{n-1}, & \text{if } n = 2k + 1 \text{ or } \alpha_n = \alpha_{n-1} \\ -b_{n-1}, & \text{iff } (n = 2k) \cap (\alpha_n = -\alpha_{n-1}) \end{cases}. \tag{5.15}$$

Property 5.4.2 *In the MSK system described by (5.3), the relationship $b_{2k} = b_{2k+1}$ always holds, irrespective of the values of α_{2k+1} and α_{2k}. The only possible locations for b_n to change its sign are $t = 2kT_b$, and this happens only when $\alpha_{2k+1} = -\alpha_{2k}$.*

Therefore, in estimating b_{2k} from a received MSK signal, one should accumulate useful information over $2T_b$ duration from $t = 2kT_b$ to $t = (2k + 2)T_b$.

The above two properties can be more intuitively illustrated in Figure 5.3. At this point, it is interesting to note that a_n has the same sign-changing behavior as $\cos\frac{\pi t}{2T_b}$, while b_n has the sign-changing rule as $\sin\frac{\pi t}{2T_b}$.

Property 5.4.3 *Symbol a_{2k} has its sign unchanged over $[(2k - 1)T_b, (2k + 1)T_b]$, having exactly the same sign change rule as $\cos\frac{\pi t}{2T_b}$, whereas symbol b_{2k} keeps its sign unchanged over $[(2kT_b, 2(k + 1)T_b]$, having exactly the same sign change rule as $\sin\frac{\pi t}{2T_b}$.*

In summary, a_n and b_n assume values of -1 or 1. The symbol a_n *can* change its sign only at the beginning of the $n = (2k - 1)$th bit interval, corresponding to the zero-crossings of $\cos\frac{\pi t}{2T_b}$. The symbol b_n *can* change its sign only at the beginning of the $n = 2k^{th}$ bit interval, corresponding to the zero-crossings of $\sin\frac{\pi t}{2T_b}$.

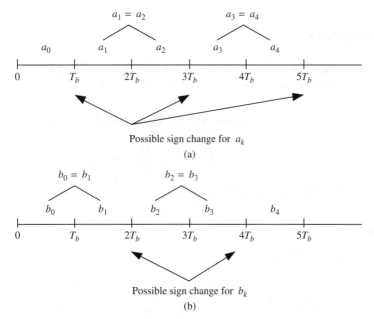

Figure 5.3 *Variational rules for a_n (a) and b_n (b)*

Table 5.1   *Example illustrating the generation of coded MSK sequences*

n	α_n	ϕ_n	$a_n = \cos \phi_n$	$b_n = \alpha_n a_n$	Time interval
0	1	0	1	1	$t \in (0, T_b)$
1	−1	π	−1	1	$(T_b, 2T_b)$
2	−1	π	−1	1	$(2T_b, 3T_b)$
3	1	0	1	1	$(3T_b, 4T_b)$
4	1	0	1	1	$(4T_b, 5T_b)$
5	1	0	1	1	$(5T_b, 6T_b)$
6	−1	0	1	−1	$(6T_b, 7T_b)$
7	1	π	−1	−1	$(7T_b, 8T_b)$
8	−1	π	−1	1	$(8T_b, 9T_b)$

✍ Example 5.1

The input sequence $\{\alpha_n, n = 1, \cdots, 8\} = \{-1 \ -1 \ 1 \ 1 \ 1 \ -1 \ 1 \ -1\}$ is applied to the MSK modulator. Determine the corresponding coded sequence $\{a_n, b_n\}$ by setting $\alpha_0 = 1$.

Solution

To begin with, we need to set the initial conditions: $\phi_0 = 0$ (must) and $\alpha_0 = 1$ or 0. Here, we set $\alpha_0 = 1$. The basic equations we use are $a_n = \cos \phi_n$ and $b_n = \alpha_n a_n$, and $\phi_n = \phi_{n-1} + \frac{n\pi}{2}(\alpha_{n-1} - \alpha_n)$, where addition is modulo-2π. The results are summarized in Table 5.1. As expected, a_n can change its sign only at $(2k - 1)T_b$ and b_n can change its sign only at $2kT_b$.   ◯

〜〜〜✍

5.5   Understanding MSK

We can get insight into MSK modulation from different perspectives. Given that $\phi_n = 0$ or π, we have $\sin \phi_n = 0$. By using the rule for expanding $\cos(x + y)$, we can obtain

$$s_n(t) = A \cos \left(\omega_c t + \alpha_n \frac{\pi t}{2T_b} + \phi_n \right)$$

$$= A \cos \left(\omega_c t + \alpha_n \frac{\pi t}{2T_b} \right) \cos \phi_n, \tag{5.16}$$

which allows understanding MSK in two different ways.

5.5.1   MSK as FSK

For the first, we further write (5.16) as

$$s_n(t) = \begin{cases} A a_n \cos \left(\omega_c t + \frac{\pi t}{2T_b} \right), & if \ \alpha_n = 1 \\[2mm] A a_n \cos \left(\omega_c t - \frac{\pi t}{2T_b} \right), & if \ \alpha_n = -1 \end{cases}. \tag{5.17}$$

Thus, we can conclude that MSK is a continuous-phase FSK with frequency separation between the two tones given by

$$\Delta f = \frac{1}{2T_b} = \frac{1}{2} R_b, \tag{5.18}$$

which is equal to the minimum frequency separation for coherent FSK. As such, MSK is called fast FSK. The modulation index, defined as the frequency separation normalized by R_b, is equal to

$$\frac{\Delta f}{R_b} = \frac{1}{2}. \tag{5.19}$$

5.5.2   MSK as offset PSK

To proceed for an alternative understanding, let $p(t - nT_b|\tau)$ denote the unit-amplitude rectangular pulse with width τ starting from $t = nT_b$. Thus, by denoting

$$u(t) = \cos\left(\frac{\pi t}{2T_b}\right) \cos(\omega_c t),$$

$$v(t) = \sin\left(\frac{\pi t}{2T_b}\right) \sin(\omega_c t), \tag{5.20}$$

we can rewrite (5.16) in quadrature form to yield

$$s_{2k}(t) = A[a_{2k+1}\, p(t - (2k+1)T_b|2T_b)u(t) - b_{2k}\, p(t - 2kT_b|2T_b)v(t)]. \tag{5.21}$$

When modulated with a message sequence, the MSK signal over the entire time axis is given by

$$s(t) = \sum_{k=-\infty}^{\infty} A[a_{2k+1}\, p(t - (2k+1)T_b|2T_b)u(t) - b_{2k}\, p(t - 2kT_b|2T_b)v(t)], \tag{5.22}$$

which differs from a standard QPSK signal in two aspects. The first is their shaping; the MSK employs a sinusoidal shaping waveform, while the QPSK uses a rectangle pulse. The second difference is that the shaping pulses used in the I and Q channels of MSK are offset with respect to each other.

5.6   Signal space

Since two offset shaping pulses are used in the I and Q channels of MSK and since each coded symbol is spilled over two consecutive intervals, the way to determine the signal space of MSK will differ from its QPSK counterpart. Clearly, we cannot simply project $s(t)$ onto the "basis function" $\cos\left(\frac{\pi t}{2T_b}\right) \cos(\omega_c t)$ over $(2k-1)T_b \le t < (2k+1)T_b$ to obtain the in-phase component, or project it onto $\sin\left(\frac{\pi t}{2T_b}\right) \sin(\omega_c t)$ over $2kT_b \le t < (2k+1)T_b$ to obtain the quadrature component. The reason is that the two basis functions have different supports, in the sense that they are defined over *different* intervals.

However, the difficulty can be circumvented by defining the basis functions over an interval of width T_b, as follows:

$$\varphi_1(t) = \sqrt{4/T_b}\, p(t - nT_b|T_b) \cos\left(\frac{\pi t}{2T_b}\right) \cos(\omega_c t),$$

$$\varphi_2(t) = \sqrt{4/T_b}\, p(t - nT_b|T_b) \sin\left(\frac{\pi t}{2T_b}\right) \sin(\omega_c t). \tag{5.23}$$

It is easy to see the orthogonality of the two set of functions by determining their inner product for $i, j = 1, 2$,

$$\langle \varphi_i(t), \varphi_j(t) \rangle = \delta(i - j), \tag{5.24}$$

where $\delta(k)$ is the Dirac delta function. It is this orthogonality that enables independent symbol decisions on the I and Q channels.

We proceed to determine the projection of the MSK signal $s(t)$ onto the basis functions, namely the inner products $\langle s(t), \varphi_i(t) \rangle$ for $i = 1, 2$. By calculation, we obtain

$$I_n = \langle s(t), \varphi_1(t) \rangle = \begin{cases} a_{2k-1}\sqrt{E_b/2}, & n = 2k \\ a_{2k+1}\sqrt{E_b/2}, & n = 2k + 1 \end{cases},$$

$$Q_n = \langle s(t), \varphi_2(t) \rangle = \begin{cases} b_{2k}\sqrt{E_b/2}, & n = 2k \\ b_{2k+2}\sqrt{E_b/2}, & n = 2k + 1 \end{cases}. \tag{5.25}$$

Recall that $a_{2k} = a_{2k-1}$ and $b_{2k} = b_{2k+1}$. Thus, the coordinates of $s(t)$ over $(n-1)T_b \leq t < nT_b$ are expressible as

$$(I_n, Q_n) = (a_n\sqrt{E_b/2}, b_n\sqrt{E_b/2}), \tag{5.26}$$

which uncovers the mapping from $s(t)$ to its signal space:

$$s(t) \leftrightarrow (a_n\sqrt{E_b/2}, b_n\sqrt{E_b/2}). \tag{5.27}$$

This understanding can be used to construct a receiver by using matched filters $\varphi_1(t)$ and $\varphi_2(t)$ for the I and Q channels, respectively. We sample the I channel output at every T_b seconds, and combine samples at $t = (2k-1)T_b$ and $t = 2kT_b$ for the detection of symbol $a_{2k} = a_{2k-1}$. The combined SNR is equal to $\gamma = E_b/\sigma_n^2$. A similar procedure is applicable for the detection of $b_{2k} = b_{2k+1}$.

Information combining over two consecutive intervals for as and bs can be implemented in a single step by using the local references defined over a $2T_b$ interval as

$$\phi_1(t) = \sqrt{2/T_b}\, p(t - (2k+1)T_b|2T_b)u(t),$$

$$\phi_2(t) = \sqrt{2/T_b}\, p(t - 2kT_b|2T_b)v(t), \tag{5.28}$$

for I and Q channels, respectively. The constellation of MSK signals is shown in Figure 5.4. It is straightforward to check that

$$\langle s(t), \phi_1(t) \rangle = a_{2k+1}\sqrt{E_b},$$

$$\langle s(t), \phi_2(t) \rangle = b_{2k}\sqrt{E_b}, \tag{5.29}$$

which is left to the reader as an exercise.

In MSK scheme, each coded symbol a_k or b_k is transmitted twice over two consecutive intervals. Thus, we can combine them coherently in much the same way as equal gain combining.

5.7   MSK power spectrum

As previously described, higher spectral efficiency is a major motivation for introducing MSK, so it is interesting to have a close look at its power spectrum. The method to determine the MSK power spectrum is to

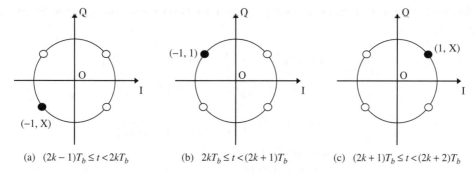

(a) $(2k-1)T_b \leq t < 2kT_b$ (b) $2kT_b \leq t < (2k+1)T_b$ (c) $(2k+1)T_b \leq t < (2k+2)T_b$

Figure 5.4 *MSK signal space*

represent the MSK signal as the output of a linear filter in response to a statistically independent input so that the Wiener–Khinchint theorem is applicable. To this end, rewrite (5.22) as the real part of a complex signal

$$s(t) = \Re \left\{ \sum_{k=-\infty}^{\infty} \left[a_{2k+1} p(t - (2k+1)T_b | 2T_b) \cos \left(\frac{\pi t}{2T_b} \right) - jb_{2k}\, p(t - 2kT_b | 2T_b) \right. \right.$$

$$\left. \left. \times \sin \left(\frac{\pi t}{2T_b} \right) \right] \exp \left(j\omega_c t \right) \right\}.$$ (5.30)

Since modulation has no influence on the power spectrum shape except for a translation to a new frequency location, we can just focus on the complex envelope of $s(t)$, denoted here by $\tilde{s}(t)$, and appropriately represent the sine and cosine of $\pi t/(2T_b)$ therein in the form of a convolution. The result is

$$\tilde{s}(t) = \sum_{k=-\infty}^{\infty} \left[a_{2k} p(t - 2kT_b) \cos \left(\frac{\pi t}{2T_b} \right) - jb_{2k}\, p(t - (2k+1)T_b) \sin \left(\frac{\pi t}{2T_b} \right) \right]$$

$$= \sum_{k=-\infty}^{\infty} \left[a_{2k}(-1)^k p(t - 2kT_b) \cos \left(\frac{\pi(t - 2kT_b)}{2T_b} \right) + jb_{2k}(-1)^{k+1} \right.$$

$$\left. \times p(t - (2k+1)T_b) \sin \left(\frac{\pi(t - (2k+1)T_b)}{2T_b} \right) \right].$$ (5.31)

As such, the real part can be considered to have been generated by applying a delta sequence $e_r(t) = \sum_{k=-\infty}^{\infty} (-j)^{2k} a_{2k} \delta(t - 2kT_b)$ to the filter with impulse response $h_r(t) = p(t) \cos \left(\frac{\pi t}{2T_b} \right)$. Note that the sequence $\{a_{2k}\}$ is a white sequence with unit variance, implying that the power spectrum density of $e_r(t)$ is equal to

$$S_e(f) = \frac{1}{2T_b}.$$ (5.32)

It follows that the power spectrum of the real part of $\tilde{s}(t)$ is given by

$$S_r(f) = S_e(f)|H_r(f)|^2 = \frac{|H_r(f)|^2}{2T_b},$$ (5.33)

where $H_r(f)$ denotes the Fourier transfer of $h_r(t)$. To determine $H_r(f)$, we note that the rectangular pulse $p(t)$ has the Fourier transform

$$p(t) = \mathrm{rect} \left(\frac{t}{2T_b} \right) \; FT{\rightarrow} 2T_b \, \mathrm{sinc}\, (2fT_b) = \frac{\sin 2\pi fT_b}{\pi f},$$ (5.34)

which, when used along with of the modulation property of the Fourier transform, enables the determination of $H_r(f)$ as

$$
\begin{aligned}
H_r(f) &= \frac{\sin 2\pi T_b \left(f + \frac{1}{4T_b} \right)}{2\pi \left(f + \frac{1}{4T_b} \right)} + \frac{\sin 2\pi T_b \left(f - \frac{1}{4T_b} \right)}{2\pi \left(f - \frac{1}{4T_b} \right)} \\
&= \frac{2T_b}{\pi} \cos 2\pi f T_b \left[\frac{1}{1 + 4fT_b} + \frac{1}{1 - 4fT_b} \right] \\
&= \frac{4T_b}{\pi} \frac{\cos 2\pi f T_b}{1 - 16 f^2 T_b^2}.
\end{aligned}
\tag{5.35}
$$

Inserting $H_r(f)$ into (5.33) leads to

$$
S_r(f) = \frac{32 T_b}{\pi^2} \left[\frac{\cos 2\pi f T_b}{1 - 16 f^2 T_b^2} \right]^2.
\tag{5.36}
$$

By the same token, the power spectrum of the imaginary part of $\tilde{s}(t)$ is determined, ending up with the same result as $S_r(f)$. Given that the random sequences in the real and imaginary parts are independent, we can add the two spectra to get the power spectrum of MSK signals to yield

$$
S_{\mathrm{MSK}}(f) = \frac{64 T_b}{\pi^2} \left[\frac{\cos 2\pi f T_b}{1 - 16 f^2 T_b^2} \right]^2,
\tag{5.37}
$$

which is shown in Figure 5.5 for comparison of the bandwidth efficiency between QPSK and MSK.

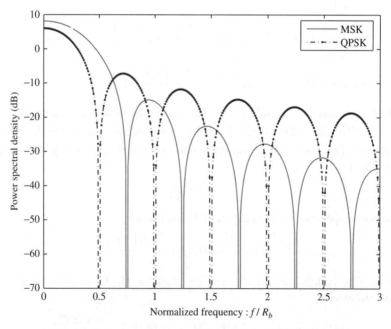

Figure 5.5 *Comparison of power spectra of MSK and QPSK*

5.8 Alternative scheme–differential encoder

The MSK coding scheme we have studied so far is based on the following procedure to produce $\{a_n, b_n\}$:

$$\{\alpha_n\} \rightarrow \{\phi_n\} \rightarrow \{a_n, b_n\}. \tag{5.38}$$

In fact, the same $\{a_n, b_n\}$ sequence can be alternatively generated by a using more efficient recursive equation

$$\beta_n = \alpha_n \beta_{n-1}. \tag{5.39}$$

To initialize the recursion, we set $\beta_{-1} = 1$. The idea is illustrated below.

$$(\beta_{-1} = 1) \rightarrow \{\alpha_n\} \overset{(5.39)}{\rightarrow} \{\beta_n\} \rightarrow \begin{cases} \beta_{2k+1} \rightarrow a_{2k+1} \\ \beta_{2k} \rightarrow b_{2k} \end{cases}. \tag{5.40}$$

Once the differentially encoded version $\{\beta_n\}$ of the original sequence $\{\alpha_n\}$ is generated, its odd and even subsequences are assigned to a_{2k-1} and b_{2k}, respectively. More specifically, the algorithm is stated as follows:

Algorithm 5.1 *Given input sequence* $\alpha_0, \alpha_1, \cdots$, *we can generate a new sequence* $\{\beta_n\}$ *using the recursion*

$$\beta_n = \alpha_n \beta_{n-1} \quad \text{initialized with} \quad \beta_{-1} = 1. \tag{5.41}$$

Then, the $\{a_n, b_n\}$ *sequence can be obtained by splitting* $\{\beta_n\}$ *as follows:*

$$[a_1, a_2, a_3, a_4, \cdots] = [\beta_1, \beta_1, \beta_3, \beta_3 \cdots],$$
$$[b_1, b_2, b_3, b_4, \cdots] = [\beta_0, \beta_0, \beta_2, \beta_2 \cdots]. \tag{5.42}$$

To prove the validity of this algorithm, we multiply both sides of (5.39) with β_{n-1} to yield

$$\alpha_n = \beta_n \, \beta_{n-1}. \tag{5.43}$$

Here, we have used the fact that $\beta_{n-1}^2 = 1$. Let us start with the initial conditions $\beta_{-1} = 1$ and $\phi_0 = 0$. The phase over the interval $T_b \leq t \leq 2T_b$ is then given by

$$\phi_1 = \phi_0 + (\alpha_0 - \alpha_1)\frac{\pi}{2} = \beta_0(\beta_{-1} - \beta_1)\frac{\pi}{2},$$

which, using some trigonometric identities and $\beta_{-1} = 1$, enables us to write

$$\cos \phi_1 = \cos \frac{\beta_0(\beta_{-1} - \beta_1)\pi}{2}$$
$$= \sin(\beta_1 \pi/2)$$
$$= \beta_1.$$

In other words, $a_1 = \beta_1$. Multiplying the above expression with α_1 and noting that $\alpha_1 = \beta_0 \beta_1$, we obtain

$$\alpha_1 \cos \phi_1 = (\beta_1 \, \beta_0) \, \beta_1 = \beta_0$$
$$b_1 = \beta_0.$$

Consider next the phase defined over the interval $2T_b \leq t \leq 3T_b$:

$$\phi_2 = \phi_1 + (\alpha_1 - \alpha_2)\pi,$$

where the second term is equal to zero or $\pm 2\pi$. Hence, we have $\cos \phi_2 = \cos \phi_1$, or

$$a_2 = a_1.$$

Proceed to obtain

$$\cos \phi_3 = \cos(\phi_2 + [a_2 - a_3]3\pi/2)$$

$$= \cos \phi_2 \cos([a_2 - a_3]3\pi/2) - \sin \phi_2 \underbrace{\sin([a_2 - a_3]3\pi/2)}_{0}$$

$$= \cos \phi_2 \sin(a_2 3\pi/2) \sin(a_3 3\pi/2)$$

$$= a_2 a_3 \cos \phi_2,$$

which, along with $\alpha_n = \beta_n \beta_{n-1}$ and $\cos \phi_2 = a_2 = a_1 = \beta_1$, gives

$$a_3 = \beta_3.$$

Multiplying both sides of the above with $\alpha_3 = \beta_3 \beta_2$ enables the assertion $b_3 = \beta_2$.
In general, over $(2k + 1)T_b \leq t \leq (2k + 3)T_b$, we obtain

$$\cos \phi_{2k+1} = \cos \phi_{2k+2} = \beta_{2k+1}, \quad k = 0, 1, \cdots,$$

and over $2kT_b \leq t \leq (2k + 2)T_b$,

$$\alpha_{2k} \cos \phi_{2k} = \alpha_{2k+1} \cos \phi_{2k+1} = \beta_{2k}, \quad k = 0, 1, \cdots,$$

Alternatively, they can be written as

$$\beta_{2k-1} = a_{2k-1} = a_{2k},$$

$$\beta_{2k} = b_{2k} = b_{2k+1}.$$

Accordingly, the MSK signal can also be expressed as

$$s(t) = \cos\left(\omega_c t + \beta_n \beta_{n-1} \frac{\pi t}{2T_b} + \phi_n\right)$$

$$= \cos\left(\omega_c t + \alpha_n \frac{\pi t}{2T_b} + \phi_n\right). \tag{5.44}$$

✍ **Example 5.2**

Consider the same input sequence as discussed in Example 5.1, but using the differential encoder (5.39) to generate the coded output sequence.

Solution

The result is shown in the following table.

$n - 1$	01	23	45	67	8
α_n	$1 - 1$	$-1 1$	$1 1$	$-1 1$	-1
$\beta_n 1$	$1 - 1$	$1 1$	$1 1$	$-1 - 1$	1

The resulting $\{\beta_n\}$ is split into odd and even subsequences, and each of their entries is repeated to account for the property of a_n and b_n. The results are shown in order.

$$\{\beta_{2k-1}\} = [\underbrace{-1\ -1}_{\beta_1}\ \underbrace{1\ 1}_{\beta_3}\ 1\ 1\ -1\ -1] = \{a_{2k-1}\},$$

$$\{\beta_{2k}\} = [\underbrace{1\ 1}_{\beta_0}\ \underbrace{1\ 1}_{\beta_2}\ 1\ 1\ -1\ -1\ 1\ 1] = \{b_{2k}\},$$

which is exactly the same as what we obtained in the previous Table 5.1.

5.9 Transceivers for MSK signals

Refer to (5.22). A key step to synthesize an MSK signal is to generate $u(t)$ and $v(t)$ as defined in (5.20). A block diagram for their generation is shown in Figure 5.6. The product $\cos\frac{\pi t}{2T_s}$, after bandpass filtering, produces $u(t)$ and $v(t)$ at the upper and lower branches, respectively.

Different insights into the MSK signal lead to different receiver structures. Understanding MSK as an FSK signal, as represented in (5.17), enables the use of a frequency discriminator for signal detection. On the other hand, the understanding of MSK as an offset QPSK signal, as shown in (5.22), forms a basis for coherent detection. In what follows, we focus on coherent detection. A matched-filter-based receiver can be constructed by slightly modifying the basis functions $\phi_i(t)$ given in (5.28), as sketched in Figure 5.7. The received signal over the AWGN channel is a noisy version of (5.22) given by

$$y(t) = A \sum_{k=-\infty}^{\infty} a_{2k+1}\, p(t-(2k+1)T_b)u(t) - b_{2k}\, p(t-2kT_b)v(t) + n(t), \tag{5.45}$$

where $n(t)$ is a zero-mean white Gaussian process with two-sided power spectrum density $N_0/2$ W/Hz.

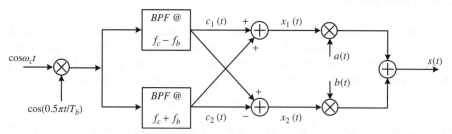

Figure 5.6 *Generation of MSK signals where $f_b = 1/(4T_b)$*

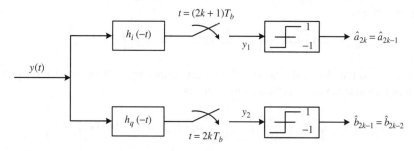

Figure 5.7 *Coherent MSK receiver, where the matched filters for the I and Q channels are defined by $h_i(t) = \cos(0.5\pi/T_b)\cos\omega_c t$ and $h_q(t) = \sin(0.5\pi t/T_b)\sin\omega_c t$, respectively*

The two branches of the receiver correlate $y(t)$ with locally generated reference signals $\phi_1(t)$ and $\phi_2(t)$, respectively, to accumulate symbol information about a_{2k} and b_{2k} over two consecutive bit intervals, producing the corresponding decision variables

$$y_1 = \int_{(2k-1)T_b}^{(2k+1)T_b} y(t)\phi_1(t)dt = a_{2k}\sqrt{E_b} + n_1,$$

$$y_2 = \int_{2kT_b}^{(2k+2)T_b} y(t)\phi_2(t)dt = b_{2k}\sqrt{E_b} + n_2. \tag{5.46}$$

Here, n_1 and n_2 are the projection of $n(t)$ onto the two orthonormal basis functions. Hence they are independent, with each of distribution $\mathcal{N}(0, N_0/2)$. The two branches are independent, justifying the use of a separate binary decision strategy for each branch. The error performance for as and bs is thus given by

$$P_{e,a} = \Pr\{a_{2k} \text{ in error}\} = Q(\sqrt{2\gamma}),$$

$$P_{e,b} = \Pr\{b_{2k} \text{ in error}\} = Q(\sqrt{2\gamma}).$$

where $\gamma = E_b/N_0$. A correct decision of the original message symbol $\alpha_{2k} = a_{2k}b_{2k}$ requires correct decisions on both branches. As such, the error probability of α_{2k} is equal to

$$P_{e,\alpha} = 1 - [1 - Q(\sqrt{2\gamma})]^2$$

$$= 2Q(\sqrt{2\gamma}) - Q^2(\sqrt{2\gamma}). \tag{5.47}$$

For a large SNR, the square term is negligible. Clearly, the error probability for coded symbols is $Q(\sqrt{2\gamma})$. It is doubled, however, for original symbols.

5.10 Gaussian-shaped MSK

The standard MSK signal described above, however, fails to meet the spectrum requirements by ETSI/TC GSM. The main problem lies in out-band energy radiation. To overcome this drawback, a premodulation filter is needed to compress the MSK spectrum. A commonly used premodulation filter possesses a Gaussian-shaped frequency transfer function, whose impulse response takes the form

$$h(t) = B_b \exp\left(-2\pi^2 \alpha B_b^2 t^2\right), \tag{5.48}$$

where α is an appropriately designed parameter, so that B_b represents the 3-dB bandwidth of the Gaussian filter. A symbol pulse $p(t) = \text{rect}\left(\frac{t}{T_b}\right)$ of duration T_b, when passing through the filter, is converted into the desirable output waveform

$$g(t) = Q\left[2\pi\beta B_b\left(t - \frac{T_b}{2}\right)\right] - Q\left[2\pi\beta B_b\left(t + \frac{T_b}{2}\right)\right], \tag{5.49}$$

where $Q(\cdot)$ denotes the Q function and β is a coefficient determined by α. With this shaping pulse, the GMSK signal driven by input sequence $\{a_n\}$ is then expressible as

$$s(t) = \cos\left\{w_c t + \frac{\pi}{2T_b}\int_{-\infty}^{t}\left[\sum_n a_n g(\tau - nT_b - 0.5T_b)\right]d\tau\right\}. \tag{5.50}$$

The falloff of the Gaussian-shaped MSK power spectrum depends on the product $B_b T_b$. The smaller the product, the faster the falloff. In GSM, the product $B_b T_b = 0.3$ is used.

5.11 Massey's approach to MSK

We now turn to Massey's approach to implementing MSK phase modulation, as described in (5.4). In the previous discussions, the message symbols $\{\alpha_n\}$ were directly used to modulate the phase. Massey's scheme, on the contrary, converts the information sequence $\{\alpha_n\}$ into a new sequence $\{v_n\}$, and employs the latter to modulate the phase.

5.11.1 Modulation

To derive the required v_n sequence, we start from the phase

$$\theta_n(t) = \pi(v_n + \alpha_n(t - nT_b)/T_b), \tag{5.51}$$

where v_n should be chosen to keep the phase continuity of $\theta_n(t)|_{t=nT_b} = \theta_{n-1}(t)_{t=nT_n}$. As such, we have

$$\pi v_n + \pi \frac{t - nT_b}{T_b}\alpha_n|_{t=nT_b} = \pi v_{n-1} + \pi \frac{t - (n-1)T_b}{T_b}\alpha_{n-1}|_{t=nT_b}, \tag{5.52}$$

leading to a recursive coding formula

$$v_n = v_{n-1} + \alpha_{n-1}, \tag{5.53}$$

where addition is a modulo-2 operation. It is clear that information symbols are differentially encoded into a new sequence $\{v_n\}$ for transmission. Since $\alpha \in \{0, 1\}$ and we usually set $v_0 = 0$, we have $v_n \in \{0, 1\}$. Thus, by using the relationship $\cos(A + n\pi) = (-1)^n \cos A$, we can rewrite (5.2) as

$$s(t) = \cos(\omega_0 t + \pi v_n + \alpha_n \pi(t/T_b - n))$$
$$= (-1)^{v_n + n\alpha_n} \cos(\omega_0 t + \alpha_n \pi t/T_b). \tag{5.54}$$

Clearly, the message information is modulated into both the carrier frequency and the phase. Defining

$$f_1 = f_0 + \frac{1}{2T_b}, \tag{5.55}$$

we can further write

$$s(t) = \begin{cases} (-1)^{v_n} \cos \omega_0 t, & \alpha_n = 0 \\ (-1)^{v_n + n} \cos \omega_1 t, & \alpha_n = 1 \end{cases}. \tag{5.56}$$

It is more intuitive to represent this equation in the form of a trellis diagram, as shown in Figure 5.8. The branch label "$1/(-f_1)$" means that with input $\alpha = 1$, the modulation will produce an output of amplitude -1 and frequency f_1. This trellis diagram offers a powerful means for symbol detection directly based on v_n.

5.11.2 Receiver structures and error performance

Let us observe Figure 5.8 to find the rules for detection. First, consider an even number of $n = 2k$, say $n = 4$. There are totally two parallelograms and two triangles centered at $n = 4$. Denote

$$g_{1i} = \int_{(2k-1)T_b}^{2kT_b} r(t) \cos \omega_i t \, dt,$$

$$g_{2i} = \int_{2kT_b}^{(2k+1)T_b} r(t) \cos \omega_i t \, dt, \tag{5.57}$$

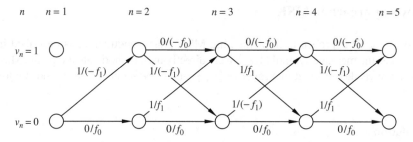

Figure 5.8   *Trellis for Massey's MSK. The branch label $1/-f_1$ means that input $\alpha = 1$ produces $s(t)$ of amplitude -1 and frequency f_1*

for $i = 0, 1$. Suppose that the current state v_{2k} evolves from the previous state $v_{2k-1} = 1$ and re-merges at $v_{2k+1} = 1$; namely

$$\text{path}: \quad v_{2k-1} = 1 \rightarrow v_{2k} = 1 \rightarrow v_{2k+1} = 1$$

$$\text{versus path}: \quad v_{2k-1} = 1 \rightarrow v_{2k} = 0 \rightarrow v_{2k+1} = 1.$$

A decision is made in favor of $v_{2k} = 1$ if the first path is more likely, and $v_{2k} = 0$ otherwise. To this end, compare the similarity of the received signal $r(t)$ with each of the two paths, yielding the decision variable

$$-(g_{10} + g_{20}) > g_{11} + g_{21}, \tag{5.58}$$

whereas those based on the two triangles are given by

$$-(g_{10} + g_{20}) > g_{11} + g_{21}$$
$$-(g_{10} + g_{21}) > g_{11} + g_{20}. \tag{5.59}$$

All the decision rules lead to the same rule as $-(g_{11} + g_{21}) > g_{10} + g_{20}$; namely, $v_{2k} = 1$ if

$$-\int_{(2k-1)T_b}^{(2k+1)T_b} r(t) \cos \omega_1 t \, dt > \int_{(2k-1)T_b}^{(2k+1)T_b} r(t) \cos \omega_0 t \, dt, \tag{5.60}$$

or

$$-\int_{(2k-1)T_b}^{(2k+1)T_b} r(t)(\cos \omega_1 t + \cos \omega_0 t) \, dt > 0. \tag{5.61}$$

We next consider the decision rules for v_{2k+1}. Again, centered at $n = 2k + 1$, we have four geometries for which the decision rules are given, respectively, by

$$-g_{10} + g_{21} > -g_{11} + g_{20}$$
$$-g_{10} - g_{20} > -g_{11} - g_{21}$$
$$g_{11} - g_{20} > g_{10} - g_{21}$$
$$g_{11} + g_{21} > g_{10} + g_{20}. \tag{5.62}$$

All these rules are equivalent and identical to

$$v_{2k+1} = \begin{cases} 1, & g_{11} + g_{21} > g_{10} + g_{20} \\ 0, & g_{11} + g_{21} < g_{10} + g_{20}. \end{cases} \tag{5.63}$$

We combine the two cases into one expression, yielding

$$v_n = \begin{cases} 1, & \xi_n > 0 \\ 0, & \xi_n < 0 \end{cases},$$ (5.64)

where the decision variable is defined by

$$\xi_n = \int_{(n-1)T_b}^{(n+1)T_b} r(t)[(-1)^{n+1} \cos \omega_1 t - \cos \omega_0 t] dt.$$

This general decision rule is independent of the previous value of v_{n-1}.

5.12   Summary

MSK has a constant envelope, thus allowing the use of a nonlinear class C amplifier. Its advantages include high battery efficiency, good channel performance, and self-synchronization capability. In MSK, the requirement of continuous phase evolution is equivalent to encoding an input message sequence into two parallel subsequences for modulating the in-phase and quadrature components of the carrier. Alternatively, an input sequence can be coded with a differential encoder. The MSK can be understood either as fast FSK or as off-set QPSK, leading to two different types of receiver structures. Readers are referred to [5–10] for additional reading.

Problems

Problem 5.1   Suppose the message sequence to MSK is $\{\alpha_n, n = 1, \cdots, 8\} = [1 \ -1 \ -1 \ 1 \ - \ 1 \ 1 \ 1 \ - \ 1]$. Determine its coded output sequence $\{a_n\}$ and $\{b_n\}$ for two initial settings: (a) $\alpha_0 = 1$ and (b) $\alpha_0 = -1$.

Problem 5.2   Show that the transmitter shown in Figure 5.6 can indeed generate the MSK signal, and label the signs of the inputs to the upper and lower branch adders.

Problem 5.3   With the MSK signal $s(t)$ defined in (5.22) and local reference signals, $\phi_1(t)$ and $\phi_2(t)$, defined in (5.28), determine the projection of $s(t)$ onto the two references.

Problem 5.4   Define

$$c_1(t) = \frac{1}{2} \left[\cos 2\pi \left(f_c + \frac{1}{4T_b} \right) t + \frac{1}{2} \cos 2\pi \left(f_c - \frac{1}{4T_b} \right) t \right],$$

$$c_2(t) = \frac{1}{2} \left[\cos 2\pi \left(f_c + \frac{1}{4T_b} \right) t - \frac{1}{2} \cos 2\pi \left(f_c - \frac{1}{4T_b} \right) t \right].$$ (5.65)

Synthesize an MSK signal by using the functions $u(t)$ and $v(t)$ defined by

$$u(t) = c_1(t) + c_2(t),$$
$$v(t) = c_1(t) - c_2(t).$$ (5.66)

Problem 5.5   Design a frequency discriminator for the detection of MSK signals and derive its error performance in an AWGN channel.

Problem 5.6   Show that the bit error rate of differentially decoded $\frac{\pi}{4}$-DQPSK modulation in an i.i.d. AWGN with two-sided power spectrum $N_0/2$ is given by

$$P_b = \frac{1}{2}[1 - Q(\sqrt{a\rho}, \sqrt{b\rho}) + Q(\sqrt{b\rho}, \sqrt{a\rho})],$$

where $a = 2 + \sqrt{2}$ and $b = 2 - \sqrt{2}$, $Q(\alpha, \beta)$ denotes the Marcum's Q function, and $\rho = E_b/N_0$ denotes the SNR. Further, show that the above expression can be approximated by Miller and Lee [4]

$$P_b = Q(\sqrt{1.1716\rho}\,).$$

Problem 5.7   Show that the average bit error rate of differentially decoded $\frac{\pi}{4}$-DQPSK modulation in a Rayleigh fading channel is given by

$$P_b = \frac{1}{2}\left[1 - \frac{\overline{\gamma}}{\sqrt{\frac{1}{2} + 2\overline{\gamma} + \overline{\gamma}^2}}\right],$$

where $\overline{\gamma}$ denotes the average SNR at the receiver [4].

References

1. R. deBuda, "Coherent demodulation of frequency shift keying with low deviation ratio," *IEEE Trans. Commun.*, vol. 20, pp. 429–435, 1972.
2. J.L. Massey, "A generalized formulation of minimum shift keying modulation," *IEEE International Conference on Communications, ICC'1980, Proceedings*, pp. 26.5.1–26.5.4, Seattle, WA, June 1980.
3. B. Rimoldi, "*Five views of differential MSK: a unified approach*," in *Communications and Cryptography: Two Sides of One Tapestry*, E.E. Blahut, D.J. Costello Jr., et al., Boston, MA: Kluwer Academic Publishers, 1994, pp. 333–342.
4. L.E. Miller and J.S. Lee, "BER expressions for differentially decoded $\pi/4$ DQPSK modulation," *IEEE Trans. Commun.*, vol. 46, no. 1, pp. 71–81, 1998.
5. S. Pasupathy "Minimum shift key: a spectrally efficient modulation," *IEEE Commun. Mag.*, vol. 17, pp. 14–22, 1976.
6. A.F. Molisch, J. Fuhl, and P. Proksch, "Error floor of MSK modulation in a mobile-radio channel with two independently fading paths," *IEEE Trans. Veh. Technol.*, vol. 45, no. 2, pp. 303–309, 1996.
7. I. Korn, "GMSK with limited discriminator detection in satellite mobile channel," *IEEE Trans. Commun.*, vol. 39, no. 1, pp. 94–101, 1991.
8. I. Korn, "GMSK with frequency selective Rayleigh fading and cochannel interference," *IEEE J. Sel. Areas Commun.*, vol. 10, 506–515, 1992.
9. I. Korn and J.P. Fonseka, "GMSK with limiter-discriminator detector with and without selection combining," *IEEE Trans. Commun.*, vol. 51, no. 8, pp. 1271–1273, 2003.
10. S.M. Elnoubi, "Analysis of GMSK with discriminator detection in land mobile radio channels," *IEEE Trans. Veh. Technol.*, vol. 35, 71–76, 1986.

6

Channel Coding

Abstract

A communication system designer has a limited resource of energy at hand; but where he should invest this resource? Part of the energy must be invested in digital modulation in order to implement effective transmission over a physical channel. The transmission reliability on AWGN channels improves with the symbol energy invested for signaling, in the form of the Q function. However, the reliability return is even more effective if part of the energy is invested on code structures, with the resulting error probability dropping much faster than that defined by the Q function. Code structures, once added to the information sequence, no longer solely belong to the encoder or the decoder. But rather, these structures should be regarded as the feature of the entire received signal and should be exploited for information-symbols retrieval. Channel coding is implemented through various algebraic structures, which can be in the form of parity-check matrix as used in linear block codes, Galois field structures as used in cyclic codes, filter structures as used in convolutional codes, or constellation structures as used in trellis coded modulation. Each code structure is well justified in its own right. Then, what is the best structure? How to share the energy between modulation and coding, and how to find a good structure for coding and decoding are two challenges to the designer.

♣

6.1 Introduction and philosophical discussion

Coding and digital modulation are two aspects of the same entity, as pointed out by G. Ungerboeck. In communication, it is often the case that only a finite amount of energy and frequency bandwidth is available. A natural question is, therefore, how to allocate the limited resource between coding and digital modulation for a better error performance.

Digital modulation is a fundamental means to provide the necessary data rate for efficient transmission over a physical channel. However, its use alone cannot guarantee reliable communication over an impaired channel, leaving a big gap from the theoretical bound predicted by Shannon. In fact, the achievable error performance by investing all the symbol energy E_s on modulation is dictated by the Q function, taking the form of $Q(\sqrt{2E_s/N_0})$ for BPSK signaling in AWGN with two-sided power spectral density $N_0/2$. A better strategy is to invest part of the symbol energy, say $(n-k)E_s/n$, in coding, to trade for a faster dropping

Wireless Communications: Principles, Theory and Methodology, First Edition. Keith Q.T. Zhang.
© 2016 John Wiley & Sons, Ltd. Published 2016 by John Wiley & Sons, Ltd.
Companion Website: www.wiley.com/go/zhang7749

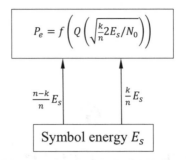

$$P_e = f\left(Q\left(\sqrt{\tfrac{k}{n}2E_s/N_0}\right)\right)$$

$$\frac{n-k}{n}E_s \qquad \frac{k}{n}E_s$$

Symbol energy E_s

Figure 6.1 *Illustrating energy allocation between coding and modulation, where a k-bit message is encoded into an n-bit codeword.*

slope defined by a composite function $f(Q(\cdot))$, as illustrated in Figure 6.1. The particular form of $f(Q(\cdot))$ depends on the code to be used.

The idea of coding could be found everywhere in daily life, probably well before its use in communication engineering. For example, when you chat with your friend in a noisy environment and fail to catch the meaning of his words, you may say "pardon" to request him to repeat the same message. This is exactly the principle of automatic repeat request (ARQ) widely adopted in modern communications. On the other hand, when an author introduces a new concept in his book or paper, he has a one-way communication link with his reader. For correctly conveying information, he first gives a rigorous definition followed by an explanation often headed by "stated in another way." The second part represents a form of redundancy to prevent possible ambiguity in his statement. Coding for reliable communications without a feedback channel follows a similar thought by adding appropriate redundancy to an original message, for example, in the form of parity check. Resulting codes are collectedly called feed-forward error correction (FEC) codes.

FEC requires only a one-way link between the transmitter and the receiver. The receiver performs no other processing than detection and decoding, regardless of whether the decoding of the received codeword is successful. On the contrary, ARQ represents a different philosophy in tackling the error-control problem, in the sense that it employs redundancy merely for the purpose of error detection.

Upon the detection of an error in a transmitted codeword, the receiver requests a repeat transmission of the corrupted codeword, thus requiring a feedback channel from the receiver to the transmitter. Traditionally, FEC codes can be further classified into linear block codes and convolutional codes.

The history of coding can be dated back to late 1940s. In his classical paper published in 1948 [1], C.E. Shannon laid a theoretic foundation for modern communications, conceptually showing how to approach the channel capacity by using random codes of infinite length. R.W. Hamming, a contemporary scientist of Shannon, endeavored along a different path, targeting to implement reliable communications through feasible FEC codes. He extended the idea of coding with even-parity check to a parity-check matrix, resulting in the Hamming code of one-bit correction capability in 1950 [2]. Afterward, several powerful codes were proposed, including the BCH codes in 1959 [3, 4] and the Reed–Solomon codes in 1960. The former have multi-bit error-correction capability and can be decoded by using the Berlekamp–Massey technique, while the latter, a subset of BCH codes, reach the highest minimum Hamming distance. Reed–Solomon [5] codes have been used in deep space communications and compact disks. Trellis codes, invented by Urgerboeck in 1978, are also of multi-bit correction capability, albeit their high decoding complexity. The aforementioned codes are all of algebraic structures, defined either by parity-check matrices or in finite fields. In 1955, Elias [6] employed finite impulse response (FIR) filters to construct his convolutional codes. The inherent Markovian properties of convolutional codes are exploited by Viterbi to derive an efficient algorithm in 1967 for convolutional decoding. Codes such as Hamming, BCH, Reed–Solomon have very strict algebra whereby decoding algorithms of lower complexity can be devised. Indeed, as asserted by Will Durant, "Every science begins as philosophy and ends as art." Each commonly used code is a masterpiece, born out of a profound philosophy. Several celebrated codes and their algebraic origins are tabulated as follows.

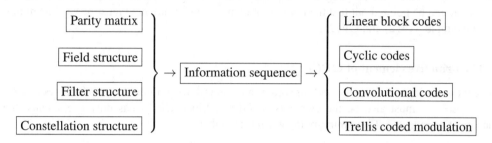

This chapter is focused on FEC codes. The *central idea* of FEC coding is to embed appropriate algebraic structures into an information sequence whereby the receiver is capable of detecting errors introduced during transmission and making a correction. These algebraic structures usually have their foundation in abstract algebra. For example, Hamming codes are rooted in finite Galois fields; cyclic codes are rooted in polynomial rings; and long block codes find their structures in algebraic geometry. A commonly used method is first to identify a good algebraic structure and then to search for good codes. This construction method is pioneered by Hamming. For an up-to-date and most comprehensive treatment of error-control coding, the reader is referred to the seminal book by Lin and Costello [7].

Introducing redundancy in codewords for reliable transmission implies possible bandwidth expansion and time delay in decoding. Clearly, there exists a tradeoff between the error performance, spectral efficiency, power efficiency, and time delay in decoding. A good code should compromise these requirements.

6.2 Preliminary of Galois fields

Before proceeding, let us briefly introduce some basic concepts of Galois fields.

6.2.1 Fields

A field is a set F, which is closed under binary *addition* $(+)$ and *multiplication* $(\times)$, such that the associative, commutative, and distributive laws are applicable. Specifically, given arbitrary $a,\ b \in F$, we always have

$$a + (b + c) = (a + b) + c$$
$$a + b = b + a$$
$$a \times (b \times c) = (a \times b) \times c$$
$$a \times b = b \times a$$
$$a \times (b + c) = a \times b + a \times c. \tag{6.1}$$

Moreover, there exist

- an element 0 and the additive inverse $-a \in F$ such that $a + 0 = a$, $(-a) + a = 0$;
- an identity element $1 \in F$ such that $1 \times a = a$; and
- the multiplicative inverse $a^{-1} \in F$ for each $a \neq 0$ such that $a \times a^{-1} = 1$.

6.2.2   Galois fields

A field that contains only a finite number of elements is called a finite field or *Galois field*. The simplest example is GF(2), which is formed by the set $\{0, 1\}$ under the operations of modulo-2 addition and multiplication. It should be pointed out that the number of elements in a finite field must be a prime number or its power, for example, GF(2) and GF(2^4).

6.2.3   The primitive element of GF(q)

The concept of primitive element and its order are central to a finite Galois field. For $\alpha \in$ GF(q), $\alpha^0, \alpha^1, \cdots, \alpha^n, \cdots$ must also be the elements of GF(q). Since GF(q) has only q elements, the above sequence must be repeated. The minimum integer n that makes

$$\alpha^n = 1 \tag{6.2}$$

is called the *order* of α. An element $\alpha \in F(q)$ is said to *primitive* if $(q-1)$ is the minimum positive integer for which

$$\alpha^{q-1} = 1. \tag{6.3}$$

Property 6.2.1 *A finite Galois field possesses the following properties:*

(a) Each finite field has at least one primitive element.
(b) All the nonzero elements in a finite field can be represented as powers of a primitive element.
(c) If the order of $\alpha \in$ GF(q) is n, then n divides $(q-1)$.
(d) For any nonzero element $a \in GF(q)$, it always holds that $a^{q-1} = 1$.

✐ **Example 6.1** _____

Consider GF(7). It is easy to examine that $3^{7-1}|_{\text{mod } 7} = 1$ and $3^k|_{\text{mod } 7} \neq 1$ for $1 \leq k < (q-1)$. Hence, 3 is a primitive element. The order of 4 is 3, which is a factor of 6.   ◯

~~~~✐

✐ **Example 6.2** _____

Check whether 2 is a primitive element in GF(11).

Solution

$2^1 = 1$, $2^2 = 4$, $2^3 = 8$, $2^4 = 5$, $2^5 = 10$, $2^6 = 9$,
$2^7 = 7$, $2^8 = 3$, $2^9 = 6$, $2^{10} = 1$. Hence, 2 is a primitive element.   ◯

~~~~✐

6.2.4   Construction of GF(q)

First consider the simple case of GF(q) with a prime number q. In this case, GF(q) is defined under modulo-q addition and multiplication. One such example is GF(7) = $\{0, 1, \cdots, 6\}$, with its addition and multiplication tabulated follows:

+	0	1	2	3	4	5	6
0	0	1	2	3	4	5	6
1	1	2	3	4	5	6	0
2	2	3	4	5	6	0	1
3	3	4	5	6	0	1	2
4	4	5	6	0	1	2	3
5	5	6	0	1	2	3	4
6	6	0	1	2	3	4	5

×	0	1	2	3	4	5	6
0	0	0	0	0	0	0	0
1	0	1	2	3	4	5	6
2	0	2	4	6	1	3	5
3	0	3	6	2	5	1	4
4	0	4	1	5	2	6	3
5	0	5	3	1	6	4	2
6	0	6	5	4	3	2	1

All the nonzero elements of $GF(7)$ are expressible as the powers of the primitive element 3.

6.2.4.1 *Construction of* $\mathbf{GF}(q^m)$

Next, consider a more complex case of $GF(q^m)$ with a prime q and integer $m > 0$. There are q^m elements in $GF(q^m)$, which are the roots of the polynomial

$$x^{q^m - 1} + 1 = 0. \tag{6.4}$$

$GF(q^m)$ with $m > 1$ cannot be simply constructed by mod-q^m addition and multiplication. Rather, it is constructed by using an mth-degree *primitive* polynomial of $x^{q^m - 1} + 1$, denoted here as $p(x)$. Let α be a root of $p(x)$. Then, it is a primitive element and its powers generate all nonzero elements of $GF(q^m)$. For the definition and details of primitive polynomials, we have more to say in Chapter 12. In most applications, q is set to 2.

✍ Example 6.3

Consider the construction of $GF(2^3)$. We first factorize $(x^7 + 1)$ over $GF(2)$, yielding

$$x^7 + 1 = (x^3 + x + 1)(x^3 + x^2 + 1)(x + 1).$$

The first two factors are both *primitive* polynomials of degree $m = 3$, and any of them can be used for constructing $GF(2^3)$.

Suppose that the polynomial $\alpha^3 + \alpha + 1 = 0$ is used for generating $GF(2^3)$. Note that $GF(2^3)$ can adopt either an exponential or a polynomial representation. A simple way is starting from the exponential representation, with the resulting $GF(2^3)$ tabulated below.

No.	Exponential	Polynomial
1	0	0
2	1	1
3	α^1	α
4	α^2	α^2
5	α^3	$\alpha + 1$
6	α^4	$\alpha^2 + \alpha$
7	α^5	$\alpha^2 + \alpha + 1$
8	α^6	$\alpha^2 + 1$

The conversion from an exponential representation into a polynomial one can be easily done by using a long division and noting that $\alpha^3 + \alpha + 1 = 0$. For example,

$$\alpha^6\big|_{\mathrm{mod}(\alpha^3+\alpha+1)} = \alpha^2 + 1.$$

Operations with exponential representation are also easy to perform, for example,

$$\alpha^3 + \alpha^6 = \alpha^2 + \alpha + 1 + 1 = \alpha^2 + \alpha = \alpha^4$$

$$\alpha^3 \cdot \alpha^5 = \alpha^8 = \alpha, \; (\because \alpha^7 = 1).$$

Note that all the coefficient addition and multiplication are modulo-2.

The addition with polynomial representation is very simple. However, its multiplication counterpart, though straightforward, is a bit complicated. For illustration, consider the product of the two elements

$$c = (\alpha^2 + 1)(\alpha^2 + \alpha + 1)$$
$$= \alpha^4 + \alpha^3 + \alpha + 1$$
$$= \alpha^2 + \alpha. \tag{6.5}$$

The last line is obtained after mod-$(\alpha^3 + \alpha + 1)$ operation. From the above table, we see that a polynomial representation makes it easy to represent each element as a binary sequence. It is also clear from (6.5) that mod-2^3 addition and multiplication are not applicable to these binary sequences.

6.3   Linear block codes

In an $(n,\, k)$ linear code, a block of k message bits, denoted by a row vector $\mathbf{m}$, is mapped into an n-bit codeword, denoted by the row vector $\mathbf{u}$, through a $m \times n$ code generator matrix $\mathbf{G}$, so that

$$\mathbf{u} = \mathbf{mG}. \tag{6.6}$$

In doing so, certain algebraic structure is embedded into the codeword in the form of the generator matrix. The code rate is defined as $r = k/n$, while $(n-k)/n$ measures the percentage of redundancy in the code. The coded bit rate R_c is related to the message bit rate R_b by

$$R_c = (n/k)R_b. \tag{6.7}$$

The code is said to be linear if any linear combination of the codewords is still a codeword. A linear code includes the all-zero codeword. A linear block code is systematic if its codewords take the form

$(c_1, \cdots, c_k, m_1, \cdots, m_k)$, with one part representing the message and the other representing the check bits. The codes so constructedconstitute an important subclass of linear block codes, often known as *Hamming codes* to honor their inventor.

Consider the simple even-parity $(4, 3)$ Hamming code where the message vector $\mathbf{m} = (m_1, m_2, m_3)$ is prefixed with a parity bit $m_0 = m_1 + m_2 + m_3$ to form a codeword

$$\mathbf{u} = (m_0, m_1, m_2, m_3) \text{ or in matrix form, } \mathbf{u} = \mathbf{mG}, \tag{6.8}$$

with the generator matrix defined by

$$\mathbf{G} = \begin{pmatrix} 1 & 1 & 0 & 0 \\ 1 & 0 & 1 & 0 \\ 1 & 0 & 0 & 1 \end{pmatrix}. \tag{6.9}$$

The above generator matrix consists of two submatrices. The first submatrix is an all-1 vector used to form the parity bit, whereas the second is the identity matrix used to reproduce the message vector. Using the redundancy of a single parity bit endows the code with only a very limited error-detection capability. For a more powerful linear block code, we need to replace the parity-check vector with a more powerful parity-check matrix. As such, a general linear block code can be represented as

$$\mathbf{u} = \mathbf{m}\underbrace{[\mathbf{P}_{k, n-k}, \mathbf{I}_k]}_{\mathbf{G}} = [\underbrace{\mathbf{mP}_{k, n-k}, \mathbf{m}}_{\mathbf{c}}], \tag{6.10}$$

which maps a 1-by-k message vector $\mathbf{m}$ into a 1-by-n code vector $\mathbf{u}$. Here, we use subscripts to explicitly indicate the dimension of the parity check and the identity matrices for clarity. Their indexes will be dropped wherever no ambiguity is introduced. Physically, we add a check vector $\mathbf{c}$ to the message $\mathbf{m}$, introducing an $(n - k)$-bit redundancy. From the geometric point of view, we construct a k-dimensional codeword subspace in an n-space, which is dictated by $\mathbf{P}$. Denoting the row vectors of $\mathbf{G}$ by $\mathbf{g}_i$, each code vector is expressible as a linear combination, as shown by

$$\mathbf{u} = \sum_{i=1}^{k} m_i \mathbf{g}_i. \tag{6.11}$$

Note, here, that $\mathbf{g}$'s are usually not a set of orthogonal bases, yet they can be used to span a codeword subspace. Given $\mathbf{G}$, there must be an orthogonal complementary subspace, spanned by $\mathbf{H}$, which carries useful features for decoding and error correction.

For binary linear block codes, the determination of $\mathbf{H}$ is quite simple. Since $\mathbf{u} = [\mathbf{c}, \mathbf{m}]$ and $\mathbf{c} = \mathbf{mP}$, addition over $\mathrm{GF}(2)$ enables us to write

$$\mathbf{c} + \mathbf{mP} = \mathbf{0} \rightarrow \mathbf{u}\underbrace{\begin{pmatrix} \mathbf{I}_{n-k} \\ \mathbf{P}_{k, n-k} \end{pmatrix}}_{\mathbf{H}^T} = \mathbf{0}, \tag{6.12}$$

where the superscript T denotes transposition. The matrix $\mathbf{H}$ shown above defines the complementary subspace. We can further write the last equation as

$$\mathbf{mGH}^T = \mathbf{0} \rightarrow \mathbf{GH}^T = \mathbf{0}, \tag{6.13}$$

showing that the subspaces spanned by $\mathbf{G}$ and by $\mathbf{H}$ are indeed orthogonally complementary.

✍ **Example 6.4** _____

Consider a (7,3) linear block code generated by

$$\mathbf{G} = \begin{pmatrix} 1 & 1 & 0 & 1 & 1 & 0 & 0 \\ 0 & 1 & 1 & 1 & 0 & 1 & 0 \\ 1 & 0 & 1 & 1 & 0 & 0 & 1 \end{pmatrix}. \tag{6.14}$$

All codewords generated by **G** are

Messages	Codewords	Messages	Codewords
000	0000000	100	1101100
001	1011001	101	0110010
010	0110010	110	1000110
011	1100011	111	0001111

○

～～✍

✍ **Example 6.5** _____

Consider a (7,4) linear block code generated by

$$\mathbf{G} = \begin{pmatrix} 1 & 1 & 0 & 1 & 0 & 0 & 0 \\ 0 & 1 & 1 & 0 & 1 & 0 & 0 \\ 1 & 1 & 1 & 0 & 0 & 1 & 0 \\ 1 & 0 & 1 & 0 & 0 & 0 & 1 \end{pmatrix}. \tag{6.15}$$

All codewords generated by **G** are tabulated below.

Messages	Codewords	Messages	Codewords
0000	0000000	1000	1101000
0001	1010001	1001	0111001
0010	1110010	1010	0011010
0011	0100011	1011	1001011
0100	0110100	1100	1011100
0101	1100101	1101	0001101
0110	1000110	1110	0101110
0111	0010111	1111	1111111

It is observed that among the 15 nonzero codewords, seven are of weight 3, seven are of weight 4, and one is of weight 7.

○

～～✍

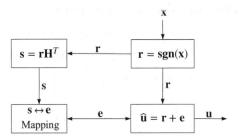

Figure 6.2 *Illustrating the principle behind decoding Hamming codes. Here,* **x** *denotes the received noisy vector*

6.3.1 Syndrome test

When a channel introduces an error vector **e** into a transmitted codeword **u**, the received vector $\mathbf{r} = \mathbf{u} + \mathbf{e}$ no longer strictly stays in the codeword subspace. Rather, it will have a nonzero projection **s** onto the orthogonal complementary subspace spanned by the row vectors of **H**, as shown by

$$\mathbf{s} = \mathbf{r}\mathbf{H}^T = \mathbf{e}\mathbf{H}^T. \tag{6.16}$$

The projection **s**, usually referred to as the *syndrome* of **r**, carries useful information for both error detection and correction. If $\mathbf{s} \neq \mathbf{0}$, the key issue is to determine the error vector **e**, which, once obtained, can be used for error correction to yield an estimator of **U**:

$$\hat{\mathbf{u}} = \mathbf{r} + \mathbf{e}. \tag{6.17}$$

Estimation of **e** from **s**, however, is not a simple issue since they *do not* have one-to-one correspondence. The reason is that vector **e** is n-dimensional while vector **s** is defined in its $(n - k)$-dimensional subspace. As a result of mapping **e** onto the $(n - k)$-dimensional subspace, a syndrome vector **s** can correspond to a number of possible error patterns. Nevertheless, we must establish an s-versus-e mapping table in order to implement error correction. The strategy is described below.

When transmitting a codeword through a noisy channel, theoretically the error vector **e** can be any binary vector in the n-space. However, the probability of a large number of bits in error is negligible. Therefore, we can focus on the 2^{n-k} most likely error patterns, which correspond to the zero bit or a small number of bits in error. With this constraint, the syndromes and the error patterns form a one-to-one mapping. The idea used to retrieve a linear block code from a noisy received data is illustrated in Figure 6.2.

In this flowchart, the received noisy vector **x** is one-bit quantized (hard decision) to a binary vector **r**, which is then used to determine the error pattern **e** for error correction. For illustration, let us consider an example.

✍ Example 6.6

Suppose a (6,3) code is generated by

$$\mathbf{G} = \begin{pmatrix} 1 & 1 & 0 & 1 & 0 & 0 \\ 1 & 0 & 1 & 0 & 1 & 0 \\ 0 & 1 & 1 & 0 & 0 & 1 \end{pmatrix}.$$

Establish a mapping between the error pattern and its syndrome.

Solution

Given $n = 6$ and $k = 3$, there are a total of $2^k = 8$ syndromes. Thus, we have to limit the number of error patterns to 8 for establishing a one-to-one mapping between **e** and **s**. We start from $\mathbf{e} = \mathbf{0}$, then continue to **e** of only one nonzero entry, and so on, until we get eight error patterns. The syndrome s is calculated by using

$$
\mathbf{s} = \mathbf{e}\mathbf{H}^T = \mathbf{e} \begin{pmatrix} 1 & 0 & 0 \\ 0 & 1 & 0 \\ 0 & 0 & 1 \\ 1 & 1 & 0 \\ 1 & 0 & 1 \\ 0 & 1 & 1 \end{pmatrix},
$$

with the mapping between e and s tabulated below.

$$
\begin{matrix}
0 & 0 & 0 & 0 & 0 & 0 & & 0 & 0 & 0 \\
1 & 0 & 0 & 0 & 0 & 0 & & 1 & 0 & 0 \\
0 & 1 & 0 & 0 & 0 & 0 & & 0 & 1 & 0 \\
0 & 0 & 1 & 0 & 0 & 0 & & 0 & 0 & 1 \\
0 & 0 & 0 & 1 & 0 & 0 & \rightarrow & 1 & 1 & 0 \\
0 & 0 & 0 & 0 & 1 & 0 & & 1 & 0 & 1 \\
0 & 0 & 0 & 0 & 0 & 1 & & 0 & 1 & 1 \\
1 & 0 & 0 & 0 & 0 & 1 & & 1 & 1 & 1 \\
\underbrace{}_{\mathbf{e}} & & & & & & & \underbrace{}_{\mathbf{s}}
\end{matrix}
\qquad (6.18)
$$

Note: When e iterates over all one-bit error patterns, s iterates over all the row patterns in $\mathbf{H}^T$. The only row vector that does not appear in $\mathbf{H}^T$ is $\mathbf{s} = (111)$. If this s is restricted to having been generated from two-bit error patterns, the possible error pattern can be $\mathbf{e} = (010010)$, (100001), or (001100). Thus, for $\mathbf{s} = (111)$, we may choose any one of them, say $\mathbf{e} = (100001)$, to complete the mapping table. By conditioning on one-bit errors, we are able to establish a one-to-one mapping between s and e. If the one-bit error condition is relaxed, each s can correspond to multiple e's. For example, $\mathbf{e} = (100100)$, (001110), and (010000) can produce the same syndrome ($\mathbf{s} = 010$). However, if $\mathbf{s} = (010)$ does occur, what we can assert is that the one-bit error pattern (010000) has the highest probability to happen, provided that the SNR is moderate to high.   ○

In the above example, the (6,3) linear block code is capable of correcting all six one-bit error patterns. But among a total of 15 possible two-bit error patterns, it can correct only one of them, leaving a large amount of ambiguity. We are, therefore, particularly interested in a subclass of linear codes, which are capable of correcting a complete set of error patterns with 0-to-t bits errors but nothing more. These types of codes are called *perfect codes*.

Definition 6.3.1 *Perfect code. A t-correcting q-ary code of length n and redundancy r is called a perfect code if the code parameters satisfy the following condition:*

$$
q^r = \sum_{i=0}^{t} \binom{n}{i} (q-1)^i. \qquad (6.19)
$$

✍ **Example 6.7** _____

Determine which of the following Hamming codes are perfect: (a) the (7,3) code, (b) the (7,4) code, and (c) the (15,11) code.

Solution

(a) For the (7,3) code, we have $r = 7 - 3 = 4$, implying that the number of possible syndromes is 2^4, which is less than

$$\sum_{i=0}^{2} \binom{7}{i} = 29,$$

but greater than

$$\sum_{i=0}^{1} \binom{7}{i} = 8. \tag{6.20}$$

Hence, the (7,3) code is not a perfect code. To conclude, there are 16 correctable error patterns in the (7,3) code. These 16 error patterns are sufficient to cover all 7 single-bit error patterns plus part of the two-bit error patterns.

(b) For the (7,4) code, we have $r = 3$, for which

$$\sum_{i=0}^{1} \binom{7}{i} = 2^r. \tag{6.21}$$

Thus, the (7,4) code is a perfect code.

(c) Turning to the (15,11) code, for which $r = 15 - 11 = 4$, we have

$$2^r = 16 = \sum_{i=0}^{1} \binom{15}{i}. \tag{6.22}$$

Hence, it is a perfect code. ○

An (n,k) code has 2^k valid codewords. Alternatively, a perfect code can be examined by arranging the total of 2^n n−tuples as an array, such that all the codewords are placed in the first row while all the correctable error patterns, called *coset leaders*, in the first column.

✍ Example 6.8

In Example 6.6, assume that the code vector is $\mathbf{u} = 011110$ whereas the received vector is $\mathbf{r} = 010110$. Determine the decoder's output.

Solution

Direct calculation yields $\mathbf{s} = \mathbf{r}\mathbf{H}^T = (001)$, which, from the mapping table above, leads to $\hat{\mathbf{e}} = (001000)$. It follows that

$$\hat{\mathbf{u}} = \mathbf{r} + \hat{\mathbf{e}} = 011110 \rightarrow \mathbf{m} = (110).$$

○

6.3.2   Error-correcting capability

Two concepts closely related to the code performance of Hamming codes are *Hamming weight* and *Hamming distance*. The Hamming weight $w(\mathbf{u})$ of a codeword $\mathbf{u}$ is defined as the number of nonzero elements in $\mathbf{u}$. Similarly, the Hamming distance between two Hamming codewords $\mathbf{u}$ and $\mathbf{v}$ is defined as the weight of their difference, given by $w(\mathbf{u} - \mathbf{v})$ or, equivalently, by $w(\mathbf{u} + \mathbf{v})$.

✍ **Example 6.9** ⎯⎯⎯

The Hamming weight of $\mathbf{u} = 100101101$ is $w(\mathbf{u}) = 5$. The Hamming distance between the vectors $\mathbf{u} = 100101101$ and $\mathbf{v} = 011110100$ is equal to

$$d(\mathbf{u},\ \mathbf{v}) = w(\mathbf{u} + \mathbf{v}) = 6.$$

〰〰〰✍

The error-correcting capability of a block code depends on its minimum distance. Let $\lfloor x \rfloor$ denote the integer part of $x \geq 0$. The *error-correcting capability* of linear codes is given by

$$t = \lfloor \frac{d_{\min} - 1}{2} \rfloor \text{ bits.} \tag{6.23}$$

That is, such a block code is capable of *correcting* all the error patterns of up to t-bit errors within a block, and probably part of error patterns with $(t + 1)$ errors. Its *error-detecting capability*, on the other hand, is given by

$$(d_{\min} - 1) \text{ bits} \tag{6.24}$$

meaning that the block code is capable of *detecting* all the error patterns of up to $(d_{\min} - 1)$-bit errors within a block.

The minimum distance of a code, designated by $d_{\min}$, is defined as the smallest value of the Hamming distance between all pairs in its codeword set C. Mathematically, we write

$$d_{\min} = \min\{d(\mathbf{u},\ \mathbf{v}) : \mathbf{u} \neq \mathbf{v}; \forall\ \mathbf{u},\ \mathbf{v} \in C\}. \tag{6.25}$$

Recall that the sum of any two codewords $\mathbf{u}$ and $\mathbf{v}$ is still a codeword, say $\mathbf{c}$, in the code C. Thus,

$$d(\mathbf{u},\ \mathbf{v}) = w(\mathbf{u} + \mathbf{v}) = w(\mathbf{c}), \tag{6.26}$$

suggesting that $d_{\min}$ can be alternatively defined as the minimum weight of all *nonzero* codewords in a code set. That is,

$$d_{\min} = \min\{w(\mathbf{c}) : \mathbf{c} \neq \mathbf{0},\ \mathbf{c} \in C\}. \tag{6.27}$$

6.3.2.1   *Two methods to determine $d_{\min}$*

In practice, two methods can be used to determine the minimum distance of a linear block code. The first method is to directly list all the codewords of a code C, and then determine their minimum weight.

The second method is based on the assertion that the minimum distance is equal to the smallest number of columns of $\mathbf{H}$ that sum to 0. In other words, $d_{\min}$ columns of $\mathbf{H}$ are linearly dependent. This can be proved as follows: Denote $\mathbf{H} = [\mathbf{h}_1,\ \mathbf{h}_2, \cdots,\ \mathbf{h}_n]$. Assume that $\mathbf{u}$ is the minimum-weight codeword with nonzero components located at position indexed by $(1), \cdots, (d_{\min})$. Since $\mathbf{u}\mathbf{H}^T = 0$, direct multiplication leads to

$$\mathbf{h}_{(1)} + \mathbf{h}_{(2)} + \cdots + \mathbf{h}_{(d_{\min})} = \mathbf{0},$$

which completes the proof.

6.3.2.2 *Error probability*

Consider the error performance of transmitting a Hamming coded sequence with code length n and error-correcting capability t over a binary symmetrical channel (BSC) with transit probability $\Pr\{0|1\} = \Pr\{1|0\} = p$. Then, by definition, the block (i.e., codeword) error probability can be shown to be upper-bounded by

$$P_w = \sum_{j=t+1}^{n} \binom{n}{j} p^j (1-p)^{n-j}$$

$$= 1 - \sum_{j=0}^{t} \binom{n}{j} p^j (1-p)^{n-j}. \tag{6.28}$$

The second line is obtained by using the binomial expansion of $(p+q)^n$ and noting that $p + q = 1$ in our case.

The average bit error can be also determined in a similar manner, yielding

$$P_b = \frac{1}{n} \left[\sum_{i=t+1}^{n} i \binom{n}{i} p^i (1-p)^{n-i} \right], \tag{6.29}$$

where the quantity inside the square brackets represents the average number of error bits in an n-bit codeword, which, via some skilful manipulation, can be simplified leading to

$$P_b = p - \frac{1}{n} \sum_{i=0}^{t} i \binom{n}{i} p^i (1-p)^{n-i}. \tag{6.30}$$

The proof is left to the reader as Problem 6.9. The second term represents the benefit from coding.

Note that the equality in (6.28) and (6.30) holds only for perfect codes since such codes can correct and only correct up to t-bit error patterns. Nonperfect codes can partly correct $(t + 1)$-bit error patterns. Thus, (6.28) and (6.30) only provide the upper-bound performance for nonperfect codes. In fact, the exact performance can be obtained for general linear block codes by including the part of error code patterns that can be corrected.

✍ Example 6.10

Calculate the improvement in probability of message (block) error relative to an uncoded transmission for a (24,12) double-error correcting linear block code. Assume that the coherent BPSK modulation is used such that the received SNR $\frac{E_b}{N_0} = 10\,\text{dB}$.

Solution

First, consider the case without coding. For AWGN channels, the channel transit probability is

$$p_u = Q\left(\sqrt{\frac{2E_b}{N_0}} \right) = Q(\sqrt{2 \times 10}) = 3.8721 \times 10^{-6},$$

whereby the error probability for an uncoded block message can be determined as

$$P_w^{(u)} = 1 - (1 - p_u)^{12} = 4.6464 \times 10^{-5}.$$

Next, consider the case with coding, for which the channel SNR becomes

$$\frac{E_c}{N_0} = \frac{k}{n} \frac{E_b}{N_0} = \frac{10}{2} = 5$$

and, thus, the channel transit error probability is equal to

$$p = Q\left(\sqrt{\frac{2E_c}{N_0}}\right) = Q(\sqrt{2 \times 5}) = 7.827 \times 10^{-4}.$$

From (6.28), it follows that the block error probability with coding is given by

$$P_w \leq \sum_{k=3}^{24} \binom{24}{k} p^k (1-p)^{24-k} = 9.5862 \times 10^{-7}.$$

From this example, it is clear that if all the symbol energy is invested on modulation, the block error probability will be 4.6464×10^{-5}. If, on other hand, half of the symbol energy is invested on modulation while another half on coding, the block error probability reduces to 9.5862×10^{-7}, gaining nearly two orders of improvement.

6.4   Cyclic codes

Cyclic codes constitute an important subclass of linear block codes. In this section, we focus on binary cyclic codes. For a more in-depth and comprehensive treatment of cyclic codes, the reader is referred to Chapter 4 of [7].

Definition 6.4.2 *An (n,k) linear block code is called a* cyclic code *if the cyclic shift (end-around shift) of a codeword is still a codeword. For example, if* $u = (u_1, u_2, \cdots, u_{n-1})$ *is a valid codeword, then* $\mathbf{u}^{(1)} = (u_{n-1}, u_0, u_1, \cdots, u_{n-2})$ *is also a valid codeword.*

A cyclic codeword is obtained from its generating polynomial via mapping:

$$u(x) = u_0 + u_1 x + \cdots + u_{n-1} x^{n-1} \rightarrow \mathbf{u} = (u_0, u_1, \cdots, u_{n-1}). \tag{6.31}$$

In what follows, the operation $v(x)|\text{modulo } p(x)$ is often simply denoted as $v(x)|p(x)$. In fact, the two notations are used interchangeably.

6.4.1   The order of elements: a concept in GF(q)

Consider a Galois field consisting of seven natural numbers, $\text{GF}(7) = \{0, 1, 2, 3, 4, 5, 6\}$, with modulo-7 addition and multiplication. It is easy to examine that the elements resulting from the modulo-7 addition or multiplication on any elements in $\text{GF}(7)$, indeed, remain in the same field. In general, different elements in a Galois field have different orders. The order of $\alpha \in \text{GF}(q)$ is generally defined as the smallest positive integer m, such that $\alpha^m = 1$. Returning to the $\text{GF}(7)$, we note that

$$\begin{array}{llll}
3^1 = 3 & 5^1 = 5 & & \\
3^2 = 2 & 5^2 = 4 & 2^1 = 2 \quad 4^1 = 4 & \\
3^3 = 6 & 5^3 = 6 & 2^2 = 4 \quad 4^2 = 2 & 6^1 = 6 \quad 1^1 = 1 \\
3^4 = 4 & 5^4 = 2 & 2^3 = 1 \quad 4^3 = 1 & 6^2 = 1 \\
3^5 = 5 & 5^5 = 3 & & \\
3^6 = 1 & 5^6 = 1 & &
\end{array} \tag{6.32}$$

Thus, by letting S_i denote the set of elements of order i in GF(7), we have $S_1 = \{1\}$, $S_2 = \{6\}$, $S_3 = \{2, 4\}$, $S_6 = \{3, 5\}$. In fact, the number of order-k elements in GF(q), denoted here by N_k, can be determined without checking all the elements one by one, as shown above. Instead, it can be evaluated by using

$$N_k \triangleq \text{size}(S_k) = \begin{cases} 0, & k \setminus (q-1) \\ \phi(k), & k|(q-1) \end{cases}, \tag{6.33}$$

where $k|(q-1)$ means "k divides $(q-1)$" while $k \setminus (q-1)$ means "k does not divide $(q-1)$," and $\phi(t)$ is the Euler ϕ function defined by

$$\phi(k) = k \prod_{p|k} \left(1 - \frac{1}{p}\right), \tag{6.34}$$

with the product taking over all *prime* numbers $1 < p \le k$ that divide k. The above expression indicates that the order must be a factor of $(q-1)$ including 1 and $(q-1)$ itself. For illustration, consider $k = 3$ in our case for which $q - 1 = 6$. It follows that

$$N_3 = \phi(3) = 3\left(1 - \frac{1}{3}\right) = 2, \tag{6.35}$$

leading to the same results as previously obtained. Very often, it is sufficient to assert the existence of the element of order $q - 1$ in a GF(q) rather than finding this element itself. Such an element is called a *primitive element* in the GF(q). Its importance lies in that, once such an element exists, we can use *its powers* to represents all the nonzero elements in the Galois field.

In the preceding example, the elements in the Galois field assume natural numbers. In algebraic coding, a Galois field is often constructed from a root of a *primitive polynomial*. A polynomial $p(x)$ is said to be *irreducible* if it has no factor other than 1 and itself. An irreducible polynomial $p(x) \in$ GF(q)[x] of degree m is said to be primitive if the *smallest* positive integer n for which $p(x)$ divides $x^n + 1$ is $n = q^m - 1$. In particular, a root of a primitive polynomial of degree $2^m - 1$ and its powers can be used to represent all the elements in the GF(2^m).

As an example, let us consider a GF(8) constructed from a *primitive* root of $p(x) = 1 + x + x^3$, denoted here by α. It is easy to see that $p(x)$ is a primitive polynomial. Then, α is of order 7. According to the closure property under multiplication, all the nonzero elements in the GF(8) are expressible in terms of its powers, which, when simplified using the relationships $\alpha^3 = 1 + \alpha$ and $\alpha^7 = 1$, leads to three types of representation for the nonzero elements.

$$\begin{pmatrix} \alpha^1 \\ \alpha^2 \\ \alpha^3 \\ \alpha^4 \\ \alpha^5 \\ \alpha^6 \\ \alpha^7 \end{pmatrix} = \begin{pmatrix} \alpha \\ \alpha^2 \\ 1 + \alpha \\ \alpha + \alpha^2 \\ 1 + \alpha + \alpha^2 \\ 1 + \alpha^2 \\ 1 \end{pmatrix} \rightarrow \begin{pmatrix} 010 \\ 001 \\ 110 \\ 011 \\ 111 \\ 101 \\ 100 \end{pmatrix}. \tag{6.36}$$

For this GF(8), the number of elements of order 7 is given by

$$N_7 = \phi(7) = 7\left(1 - \frac{1}{7}\right) = 6, \tag{6.37}$$

implying that α–α^6 are all of order 7 and any of them can be used a base to represent all the nonzero elements in the field.

It is clear that the roots of a primitive polynomial $p(x) \in \mathrm{GF}(q)$ of degree m are of order $q^m - 1$. They, along with 0 and 1, constitute a Galois field $\mathrm{GF}(q^m)$ whose nonzero elements, in turn, can be represented as the powers of a primitive root of $p(x)$.

Many polynomials used for the generation of algebraic codes have close connections to Galois fields in a certain way. The elements in the Galois field are divided into subclasses, whereby a generator and parity-check polynomials are constructed. The algebraic structures of the relevant Galois field or the ring of polynomials are implicitly imbedded into the code and, thus, dominate its performance.

6.4.2   Cyclic codes

For a cyclic code

$$\mathbf{u} = [u_0, \, u_1, \cdots, \, u_{n-1}], \tag{6.38}$$

its ith end-around shifted version

$$\mathbf{u}^{(i)} = [u(n - i), \, u(n - i + 1), \cdots, \, u(n - 1), \, u_0, \, u_1, \cdots u(n - i - 1)], \tag{6.39}$$

is still a codeword. For understanding cyclic codes and their design, it is often convenient to link a codeword to a polynomial through mapping

$$\mathbf{u} = [u_0, \, u_1, \cdots, \, u_{n-1}] \to u(x) = u_0 + u_1 x + \cdots + u_{n-1} x^{n-1}. \tag{6.40}$$

With this mapping, it becomes easier to add an algebraic structure to a codeword and exert manipulation on a code. For example, an end-around shift operation can be simply written as

$$\mathbf{u}^{(1)} \to x \, u(x)\big|_{\mathrm{modulo}(x^n + 1)}. \tag{6.41}$$

To show this, we proceed with simple multiplication, obtaining

$$x \, u(x) = u_0 x + u_1 x^2 + \cdots u_{n-1} x^n, \text{ or}$$
$$= (u_{n-1} + u_0 x + \cdots + u_{n-2} x^{n-1}) + (u_{n-1} x^n + u_{n-1}), \tag{6.42}$$

and thus

$$x \, u(x)\big|_{\mathrm{modulo}(x^n + 1)} = u_{n-1} + u_0 x + \cdots + u_{n-2} x^{n-1}, \tag{6.43}$$

which, after mapping, leads to

$$\mathbf{u}^{(1)} = (u_{n-1}, \, u_0, \cdots, \, u_{n-2}). \tag{6.44}$$

More generally, we have

$$\mathbf{u}^{(i)}(x) = x^i u(x) \text{ modulo } (x^n + 1), \tag{6.45}$$

which implies that

$$u^{(i)} = (u_{n-i}, \, u_{n-i+1}, \cdots, \, u_{n-1}, \, u_0, \cdots, \, u_{n-i-1}). \tag{6.46}$$

The multiplier x^i is a shift operator, shifting the codeword to the right by i positions. But the resulting polynomial has a degree higher than a standard code word; we, therefore, need a modulo-$(x^n + 1)$ operation.

✍ Example 6.11 _____

Given $\mathbf{u} = (1101)$, we may perform the 3rd end-around shift on it to obtain $\mathbf{u}^{(3)}$.

Solution

Noting that $n = 4$ and $u(x) = 1 + x + x^3$, we have

$$x^3 u(x) = x^3 + x^4 + x^6.$$

Dividing both sides by $(x^4 + 1)$, the remainder is

$$u^{(3)}(x) = x^3 + x^2 + 1 = 1 + x^2 + x^3.$$

Hence, $\mathbf{u}^{(3)} = (1011)$.

In modulo operation, two identities can be useful. Given two polynomials $f_1(x)$ and $f_2(x)$, let $r_1(x) = f_1(x)|_{\text{modulo}(g(x))}$ and $r_2 = f_2(x)|_{\text{modulo}(g(x))}$. Then, we have

$$[f_1(x) + f_2(x)]|_{\text{modulo}(g(x))} = r_1(x) + r_2(x),$$
$$[f_1(x) \cdot f_2(x)]|_{\text{modulo}(g(x))} = r_1(x) \cdot r_2(x). \tag{6.47}$$

The proof is very simple and is thus ignored here for brevity.

6.4.3 Generator, parity check, and syndrome polynomial

The generator and message polynomials for an $(n, \ k)$ cyclic code can be represented, respectively, as

$$g(x) = g_0 + g_1 x + \cdots + g_{n-k} x^{n-k},$$
$$m(x) = m_0 + m_1 x + \cdots + m_{k-1} x^{k-1}.$$

An (n, k) cyclic code has the following properties:

Property 6.4.1 *The generator polynomial $g(x)$ of an $(n, \ k)$ cyclic code is a factor of $x^n + 1$, that is,*

$$x^n + 1 = g(x)h(x).$$

Furthermore, $g(x)$ is a primitive polynomial of degree $(n - k)$.

Property 6.4.2 *All codewords can be expressed as the product of $g(x)$ and the message polynomial $m(x)$ of degree equal to k or less, as shown by*

$$u(x) = m(x)g(x). \tag{6.48}$$

Note that the codewords so obtained are not in systematic form.

Property 6.4.3 *The codeword generated by $u(x) = m(x)g(x)$ can be checked using $h(x)$ through*

$$u(x)h(x)|(x^n + 1) = 0 \rightarrow u(x)|g(x) = 0. \tag{6.49}$$

It indicates that $h(x)$ behaves in a similar way as the parity-check matrix H for a linear block code.

These results are the direct consequence of the fact that $(x^n + 1) = g(x)h(x)$ and $u(x) = m(x)g(x)$.

✍ **Example 6.12** _____

Consider the factorization

$$x^7 + 1 = (x^3 + x^2 + 1)(x^3 + x + 1)(x + 1). \tag{6.50}$$

A (7,4) cyclic code can be generated by using $g_1(x)$ or $g_2(x)$ defined by

$$g_1(x) = x^3 + x^2 + 1,$$
$$g_2(x) = x^3 + x + 1.$$

Suppose we choose $g_2(x)$ as the generator polynomial. Find the codeword for message $\mathbf{m} = [1011]$.

Solution

The check polynomial is $h_2(x) = (x+1)(x^3 + x^2 + 1) = 1 + x + x^2 + x^4$. Since $n = 7$ and $(n-k) = 3$, we have $k = 4$ and a total of 2^4 codewords. The codeword for $\mathbf{m} \to m(x) = 1 + x^2 + x^3$ is obtained as

$$u(x) = m(x)g_2(x) = 1 + x + x^2 + x^3 + x^4 + x^5 + x^6 \to \mathbf{u} = [1111111],$$

which, as expected, is not in a systematic form. Let us check

$$u(x)h_2(x) = u(x)(x+1)(x^3 + x^2 + 1) = (x^7 + 1)(x^3 + x^2 + 1)|(x^7 + 1) = 0,$$

confirming Property 6.4.3. ○

〰✍

6.4.4   Systematic form

The systematic form of an (n, k) cyclic code contains two portions, one for the message and the other for parity check. For portion 1 of an (n, k) cyclic code, it is easy to keep the message by simply shifting $m(x)$ to the rightmost, as shown by

$$x^{n-k}m(x). \tag{6.51}$$

The parity portion should be chosen to maintain the divisibility of a codeword by the generator. To this end, consider

$$x^{n-k}m(x)|g(x) = r(x). \tag{6.52}$$

Alternatively, we can write $x^{n-k}m(x) = q(x)g(x) + r(x)$, whereby the codeword

$$u(x) = r(x) + x^{n-k}m(x) = q(x)g(x), \tag{6.53}$$

is indeed in the systematic form and is divisible by $g(x)$.

The procedure described above for generating a systematic cyclic code can be summarized as follows:

- Step 1: Use the delay operator x^{n-k} to shift the message to the rightmost, yielding

$$x^{n-k}m(x) = m_0 x^{n-k} + m_1 x^{n-k+1} + \cdots + m_{k-1}x^{n-1}. \tag{6.54}$$

- Step 2: Determine the parity polynomial $r(x)$ by finding the remainder

$$r(x) = x^{n-k}m(x)|g(x). \tag{6.55}$$

- Step 3: Obtain the code polynomial as

$$u(x) = r(x) + x^{n-k}m(x)$$

$$= r_0 + r_1 x + \cdots + r_{n-k-1}x^{n-k-1} + m_0 x^{n-k} + \cdots + m_{k-1}x^{n-1}. \tag{6.56}$$

- Step 4: Collect the corresponding coefficients to give the codewords in systematic form:

$$\mathbf{u} = [r_0, \ r_1, \cdots, \ r_{n-k-1}, \ m_0, \cdots, \ m_{k-1}]. \tag{6.57}$$

✍ **Example 6.13**

With the same setting as in Example 6.12, determine the systematic cyclic code for the same message vector $\mathbf{m} = 1011$.

Solution

First, shift the message to the rightmost:

$$x^{n-k}m(x) = x^3(1 + x^2 + x^3) = x^3 + x^5 + x^8. \tag{6.58}$$

Next, determine the parity polynomial:

$$r(x) = x^{n-k}m(x)|g(x) = 1.$$

Combining the two, we obtain

$$u(x) = r(x) + x^3 m(x) = 1 + x^3 + x^5 + x^6, \tag{6.59}$$

which, mapped to a codeword, yields

$$\mathbf{u} = 100\underline{1011}, \tag{6.60}$$

with the second portion identical to the message.

○

✍ **Example 6.14**

Determine the generator and parity-check matrices for the code obtained in Example 6.13.

Solution

The row vectors of $\mathbf{G}$ are the codewords that correspond to the messages 1, x, x^2, and x^3. It follows that

$$\mathbf{G} = \begin{pmatrix} 1 & 1 & 0 & 1 & 0 & 0 & 0 \\ 0 & 1 & 1 & 0 & 1 & 0 & 0 \\ 1 & 1 & 1 & 0 & 0 & 1 & 0 \\ 1 & 0 & 1 & 0 & 0 & 0 & 1 \end{pmatrix}. \tag{6.61}$$

The codeword vectors corresponding to message polynomials 1, x, $x^2, \cdots$, x^k are called the *basis vectors*. The corresponding parity check is given by

$$\mathbf{H} = \begin{pmatrix} 1 & 0 & 0 & 1 & 0 & 1 & 1 \\ 0 & 1 & 0 & 1 & 1 & 1 & 0 \\ 0 & 0 & 1 & 0 & 1 & 1 & 1 \end{pmatrix}. \tag{6.62}$$

This is exactly the parity-check matrix of a Hamming code.

○

6.4.5   Syndrome and decoding

Consider an (n, k) cyclic code with generator $g(x)$ of degree $(n - k)$ and code polynomial $u(x)$. Suppose that the received polynomial is $v(x) = u(x) + e(x)$ with error pattern $e(x)$. Recall that a codeword $u(x)$ is "orthogonal" to $h(x)$. Thus, the remainder $s(x) = v(x)h(x)|(x^n + 1)$ can be used to examine the presence of a possible error pattern in the received signal, and hence the name syndrome polynomial. Using the fact that $x^n + 1 = g(x)h(x)$ and $u(x) = m(x)g(x)$, we have

$$s(x) = v(x)h(x)|(x^n + 1) = v(x)|g(x) \tag{6.63}$$

$$= e(x)|g(x), \tag{6.64}$$

which will be used for error correction. As a first step, use (6.64) to construct a syndrome versus error pattern mapping table. For one-bit errors, we may set $e(x) = x^i$, $i = 1, \cdots, n$ and calculate their syndromes by

$$s_i(x) = x^i|g(x). \tag{6.65}$$

A two-bit error pattern can be written similarly, in the form of $e(x) = x^i + x^j$, $i \neq j$, $i, j = 1, 2, \cdots, n$. By the same token, we can treat multiple bit errors. Just as for linear block codes, the maximum number of error patterns that can be corrected is upper-bounded by 2^{n-k}.

6.4.5.1   Decoding procedure

Decoding of a cyclic code consists of three steps.

- Step 1: syndrome computation;
- Step 2: associating a syndrome to an error pattern;
- Step 3: error correction by adding the error pattern to the received vector.

The direct implementation of the above scheme can lead to circuit complexity that grows exponentially with the code size and the number of errors to be corrected. A much simpler decoder is called the Meggitt decoder, which exploits the algebraic structure of cyclic codes [7].

✍ **Example 6.15**

Suppose that a (7,4) cyclic code is generated by $g(x) = 1 + x + x^3$. Associate the syndromes to the related error patterns.

Solution

There are seven possible single-error patterns and a total of 2^3 syndromes, implying that all the single-error patterns are correctable.

Error pattern $e(x)$	Syndrome $s(x)$	$(s_0\ s_1\ s_2)$
x^6	$1 + x^2$	(101)
x^5	$1 + x + x^2$	(111)
x^4	$x + x^2$	(011)
x^3	$1 + x$	(110)
x^2	x^2	(001)
x^1	x	(010)
x^0	1	(100)

6.5 Golay code

The Golay is a linear cyclic code directly derived from the factorization of $(x^{23} + 1)$, as shown by

$$x^{23} + 1 = (x + 1)g_1(x)g_2(x), \tag{6.66}$$

where

$$g_1(x) = 1 + x^2 + x^4 + x^5 + x^6 + x^{10} + x^{11},$$
$$g_2(x) = 1 + x + x^5 + x^6 + x^7 + x^9 + x^{11}, \tag{6.67}$$

both having the degree $m = 11$. We can use either of them to generate the $(23, 12)$ cyclic code, called the Golay code $(23, 12)$, named after its inventor.

Unlike the $(7, 4)$ cyclic code, the length of the Golay code $n = 23$ is not in the form of one less the power of 2. If we use a primitive polynomial of degree $m = 11$ to generate a code, the code size is $2^m - 1 = 2047$. Clearly, neither $g_1(x)$ nor $g_2(x)$ is a primitive polynomial. Then, what is their relationship to the GF(2048)? If α is a primitive root of $x^{2047} + 1$, all the nonzero elements in GF(2048) can be represented as $(\alpha, \alpha^2, \cdots, \alpha^{2047})$ for which $\alpha^{2047} = 1$. Note that $2047 = 89 \times 23$. Thus, the root $\beta = \alpha^{89}$ is of degree 23, that is, $\beta^{23} = 1$. The subset $(\beta, \beta^2, \cdots, \beta^{23})$ in the GF(2048) represents the 23rd roots of unity. We want to use them to form two factor polynomials. To this end, we calculate

$$\mathbf{a} = \beta^{2^\mathbf{k}} = \begin{pmatrix} \beta \\ \beta^2 \\ \beta^4 \\ \beta^8 \\ \beta^{16} \\ \beta^9 \\ \beta^{18} \\ \beta^{13} \\ \beta^3 \\ \beta^6 \\ \beta^{12} \\ \beta \end{pmatrix}, \quad \mathbf{b} = (\beta^5)^{2^\mathbf{k}} = \alpha^{(89*2^\mathbf{k})|_{\mathrm{mod}(2047)}} = \begin{pmatrix} \beta^5 \\ \beta^{10} \\ \beta^{20} \\ \beta^{17} \\ \beta^{11} \\ \beta^{22} \\ \beta^{21} \\ \beta^{19} \\ \beta^{15} \\ \beta^7 \\ \beta^{14} \\ \beta^5 \end{pmatrix}, \tag{6.68}$$

where $\mathbf{k}$ is a column vector defined by $\mathbf{k} = [0, 1, \cdots, 11]^T$. All the 23 roots are included in these two classes. We can use a different base for power calculation, such as β^3 and β^{13}, and so on, leading to the same result as the left column, while the bases β^7 and β^{11}, and so on, leading to the same as the right column. It is clear that the column entries repeat after the 12th element; we only use the first 11 entries for factorization. The resulting polynomial product is

$$(x + 1)g_1(x)g_2(x). \tag{6.69}$$

The GF(2048) has $\phi(23) = 22$ elements of degree 23.

6.6 BCH codes

So far, we have seen that the correction capability of general cyclic codes can only be determined by an enumerative list of the code weights after a code has been generated. A BCH code can have its correction capability designed by theory, with the term BCH to honor its inventors Hocquenghem [3], Bose and Ray-Chaudhuri [4]. To proceed, we need the following definitions:

Definition 6.6.1 *A minimal polynomial* $M(x)$ *of* α *is defined as the lowest order nonzero polynomial such that* $M(\alpha) = 0$.

If $M(x)$ is a minimal polynomial of $\alpha \in \mathrm{GF}(2^m)$, then all of its roots are given by α, α^q, $\alpha^{q^2}, \cdots, \alpha^{q^{m-1}}$. In other words, $M(x)$ is an mth-degree binary primitive polynomial that defines the $\mathrm{GF}(2^m)$, for which α is a primitive element. The generator of a BCH code is defined by a set of minimal polynomials. In particular, the generator polynomial $g(x)$ for a t-bit correcting BCH code of length $(2^m - 1)$ has

$$\alpha, \ \alpha^2, \cdots, \ \alpha^{2t} \tag{6.70}$$

as its roots. Let $M_i(x)$ be the *minimal* polynomial of α^i. Then,

$$\begin{aligned}
g(x) &= \mathrm{LCM}\{M_1(x), \ M_2(x), \cdots, \ M_{2t}(x)\} \\
&= \mathrm{LCM}\{M_1(x), \ M_3(x), \cdots, \ M_{2t-1}(x)\},
\end{aligned} \tag{6.71}$$

where LCM is the abbreviation for least common multiple.

6.6.1   Generating BCH codes

For $m \leq 3$ and $t < 2^{m-1}$, there always exists a binary BCH code of $n = 2^m - 1$ bits long and t-bit correcting capability. Let us illustrate the BCH code design through an example.

✍ **Example 6.16** ⸻

Design a binary BCH code of length 15 and correction capability $t = 3$ by using the primitive polynomial $p(x) = x^4 + x + 1$. Determine (a) the generator polynomial $g(x)$, (b) the number of codewords, and (c) codeword for message $m(x) = 1 + x^3 + x^4$.

Solution

From $p(x)$, we have $m = 4$ and the code size $n = 2^m - 1 = 15$. The $\mathrm{GF}(2^4)$ defined by $p(x)$ will be used for our design. More specifically, if α is a primitive root of $p(x)$, we have the following relationships:

$$\alpha^4 = \alpha + 1, \ \text{and} \ \alpha^{15} = 1.$$

Since $t = 3$, $g(x)$ includes $\alpha^i, i = 1, \ 2, \cdots, \ 2t$ as its roots. Each root, say α_i, has a corresponding conjugacy class, which constitutes a polynomial $M_i(x)$. In particular, we have

$$\alpha \rightarrow (\alpha, \ \alpha^2, \ \alpha^{2^2}, \ \alpha^{2^3}, \underbrace{\alpha^{2^4}, \ \alpha^{2^5}, \cdots}_{\text{Start to repeat}}) \rightarrow (\alpha, \ \alpha^2, \ \alpha^4, \ \alpha^8). \tag{6.72}$$

Given that $\alpha^{2^4} = \alpha$, the sequence repeats from the fifth element, and thus α is said to be of order 4. Since our concern is LCM, we just keep the first period, as shown on the right of the above expression. Observe that this conjugacy set contains two other roots, α^2 and α^4, implying that their polynomials are the same as $M_1(x)$.

In the same manner, we find that the two remaining roots α^3 and α^5 are of order 4 and 2, respectively, as given by

$$\begin{aligned}
\alpha^3 &\rightarrow (\alpha^3, \ \alpha^6, \ \alpha^{12}, \ \alpha^{24}) = (\alpha^3, \ \alpha^6, \ \alpha^{12}, \ \alpha^9), \\
\alpha^5 &\rightarrow (\alpha^5, \ \alpha^{10}).
\end{aligned}$$

It follows that

$$M_1(x) = (x+\alpha)(x+\alpha^2)(x+\alpha^4)(x+\alpha^8) = x^4 + x + 1,$$

$$M_3(x) = (x+\alpha^3)(x+\alpha^6)(x+\alpha^9)(x+\alpha^{12}) = x^4 + x^3 + x^2 + x + 1,$$

$$M_5(x) = (x+\alpha^5)(x+\alpha^{10}) = x^2 + x + 1.$$

In the above, we have used (6.72) for simplification. The computation can be further reduced if we note that for *binary* BCH and integers a_i and r, we have $(\sum_i a_i x^i)^{2r} = \sum_i (a_i x^i)^{2r}$ since the coefficients for the cross terms are all even numbers and, thus, vanish after modulo-2 operation. The generator polynomial is thus given by

$$g(x) = M_1(x)M_3(x)M_5(x) = x^{10} + x^8 + x^5 + x^4 + x^2 + x + 1,$$

from which we observe that $n - k = 10$, leading to $k = 5$. The number of codewords is then $2^k = 32$. To generate a systematic codeword for $m(x) = 1 + x^3 + x^4$, we calculate

$$x^{n-k}m(x) = x^{10} + x^{13} + x^{14},$$

$$x^{n-k}m(x)|g(x) = x + x^2 + x^5 + x^6 + x^7 + x^8,$$

which enables us to write the codeword

$$u(x) = x + x^2 + x^5 + x^6 + x^7 + x^8 + x^{10} + x^{13} + x^{14}.$$

In summary, the procedure for generating a binary BCH code is listed below:

- Step 1: Choose a primitive polynomial of degree m whereby the GF(2^m) is constructed.
- Step 2: Find the minimal polynomial $M_i(x)$ of α^i for $i = 1, 2, \cdots, 2t$.
- Step 3: Obtain $g(x) = \text{LCM}\{M_1(x), M_3(x), \cdots, M_{2t-1}(x)\}$.
- Step 4: Determine k from $(n-k)$, that is, the degree of $g(x)$.
- Step 5: Find the minimum distance $d_{\min} \geq 2t + 1$ by referring to the weight of $g(x)$.

6.6.2   Decoding BCH codes

Suppose a received polynomial $v(x) = u(x) + e(x)$ is a code polynomial $u(x)$ corrupted by error $e(x)$. Since $u(x)$ is capable of correcting $2t$ bit errors, it has roots $\alpha, \alpha^2, \cdots, \alpha^{2t}$. Thus, $S_i = v(\alpha^i) = e(\alpha^i)$, $1 \leq i \leq 2t$ can be used as syndromes for error correction. Explicitly, we write

$$S_i = \sum_{j=0}^{n-1} e_j x^j |_{x=\alpha^i}, \tag{6.73}$$

where v coefficients will be nonzero (equal to unity) if there are $v \leq 2t$ bits in error. Denote these error locations by $(j) = 1, \cdots, v$, and denote $x_\ell = \alpha^{(\ell)}$. We can further write

$$S_i = \sum_{\ell=1}^{v} x_\ell^i, \quad i = 1, 2, \cdots, v. \tag{6.74}$$

Directly solving these nonlinear simultaneous equations for error-bit locations x_ℓ is difficult. It is wise to consider, instead, the error locator polynomial

$$L(z) = \prod_{\ell=1}^{v}(1 + x_\ell z)$$

$$= \sigma_0 + \sigma_1 z + \cdots + \sigma_v z^v \tag{6.75}$$

whose roots $z = x_\ell^{-1} = \alpha^{n-(\ell)}$ carry information about the error locations (ℓ). The coefficients σ's are functions of x_ℓ's. In order for $L(z)$ to be useful, we represent σ's in terms of the known syndromes by appropriately combining (6.74) and (6.75). There are different ways to achieve this goal. Here, we only give the result by Peterson [8]:

$$\underbrace{\begin{pmatrix} S_1 & S_2 & \cdots & S_v \\ S_2 & S_3 & \cdots & S_{v+1} \\ \vdots & \vdots & \cdots & \vdots \\ S_v & S_{v+1} & \cdots & S_{2v-1} \end{pmatrix}}_{\mathbf{S}_v} \begin{pmatrix} \sigma_v \\ \sigma_{v-1} \\ \vdots \\ \sigma_1 \end{pmatrix} = \begin{pmatrix} S_{v+1} \\ S_{v+2} \\ \vdots \\ S_{2v} \end{pmatrix}, \quad v = 1, \cdots, t. \tag{6.76}$$

We outline the procedure for decoding BCH codes as follows:

- Step 1: Calculate the syndromes $S_i = v(x)|_{x=\alpha^i}$ for $i = 1, 2, \cdots, 2t$.
- Step 2: Determine the rank v of $\mathbf{S}_{2t}$, which indicates the number of bits in error. We begin by examining $\xi_t = \det(\mathbf{S}_t)$. If $\xi_t \neq 0$, then $v = t$ and stop. If $\xi_t = 0$, we delete the rightmost column and the bottom row of $\mathbf{S}_t$, and continue until we find the $v \times v$ matrix $\mathbf{S}_v$ such that $\det(\mathbf{S}_v) \neq 0$.
- Step 3: Solve the corresponding (6.75) for $\sigma_1, \cdots, \sigma_v$, whereby $L(x) = \prod_{\ell=1}^{v}(1 + \sigma_\ell x)$ is constructed.
- Step 4: Find the v roots of $L(x)$ by trying $x = \alpha^i$, $i = 1, \cdots, n$ in succession. Denote these roots by $x = \alpha^{(a)}$; then the error locations are $(\ell) = n - (a)$.

Note that the value $t = 2$ implies the minimum distance $d = 2t + 1 = 5$. Then, we have

$$g(x) = \text{LCM}\{M_1(x), M_2(x), M_3(x), M_4(x)\}. \tag{6.77}$$

It follows that the use of bases α, α^2, and α^4 leads to the same minimal polynomial. Thus, it suffices to use only $M_1(x) = x^5 + x^2 + 1$ and $M_3(x)$ to construct the LCM. The minimal polynomial resulting from using the base α^3 is

$$M_3(x) = (x - \alpha^3)(x - \alpha^9)(x - \alpha^{27})(x - \alpha^{81})(x - \alpha^{243}),$$

which, after using the GF(2^5) defined by $p(x)$ (i.e., $\alpha^5 = \alpha^2 + 1$ and $\alpha^{31} = 1$), leads to

$$M_3(x) = x^5 + x^4 + x^3 + x^2 + 1.$$

Hence, we obtain

$$g(x) = M_1(x)M_3(x) = x^{10} + x^9 + x^8 + x^6 + x^5 + x^3 + 1, \tag{6.78}$$

which produces a (31,21) code. The parity check matrix contains $n = 31$ columns and $2t$ rows, and is given by

$$\mathbf{H} = \begin{pmatrix} 1 & \alpha & \alpha^2 & \cdots & \alpha^{30} \\ 1 & \alpha^2 & \alpha^4 & \cdots & \alpha^{60} \\ 1 & \alpha^3 & \alpha^6 & \cdots & \alpha^{90} \\ 1 & \alpha^4 & \alpha^8 & \cdots & \alpha^{120} \end{pmatrix}. \tag{6.79}$$

with the ith column formed by the base α^{i-1}.

A primitive polynomial over GF(q) of degree m can always be factorized in GF(q^m). For example, for GF(2^3),

$$x^3 + x^2 + 1 = (x + a)(x + a^2)(x + a^4)$$
$$x^3 + x + 1 = (x + a^3)(x + a^5)(x + a^6). \tag{6.80}$$

If β is a root of a primitive polynomial of degree m, then β^q, β^{q^2}, β^{q^3}, $\cdots$ are also its roots.

✍ Example 6.17

In Example 6.16, suppose that the received polynomial is given by $v(x) = 1 + x + x^2 + x^4 + x^8$ with two bit errors. Decode the received signal and see whether the original message can be recovered.

Solution

First calculate the syndromes

$$S_1 = v(a) = 1 + a + a^2 + a^4 + a^8 = 1$$
$$S_2 = v(a^2) = 1 + a^2 + a^4 + a^8 + a^{16} = 1$$
$$S_3 = v(a^3) = 1 + a^3 + a^6 + a^{12} + a^{24} = 0$$
$$S_4 = v(a^4) = 1 + a^4 + a^8 + a^{16} + a^{32} = 1$$
$$S_5 = v(a^5) = 1 + a^5 + a^{10} + a^{20} + a^{40} = 1$$
$$S_6 = v(a^6) = 1 + a^6 + a^{12} + a^{24} + a^{48} = 0, \tag{6.81}$$

where the final simplification is achieved with the help of $a^4 = a + 1$. Next check that

$$\det(\mathbf{S}_3) = \begin{pmatrix} 1 & 1 & 0 \\ 1 & 0 & 1 \\ 0 & 1 & 1 \end{pmatrix} = 0.$$

Then, proceed to examine

$$\det(\mathbf{S}_2) = \begin{pmatrix} 1 & 1 \\ 1 & 0 \end{pmatrix} = 1 \neq 0,$$

thus enabling the assertion that the number of errors equals 2. Next, locate the error bits by solving the following equations:

$$\begin{pmatrix} S_1 & S_2 \\ S_2 & S_3 \end{pmatrix} \begin{pmatrix} \sigma_2 \\ \sigma_1 \end{pmatrix} = \begin{pmatrix} S_3 \\ S_4 \end{pmatrix} \rightarrow \begin{pmatrix} 1 & 1 \\ 1 & 0 \end{pmatrix} \begin{pmatrix} \sigma_1 \\ \sigma_2 \end{pmatrix} = \begin{pmatrix} 0 \\ 1 \end{pmatrix}.$$

The solution is $\sigma_1 = \sigma_2 = 1$, whereby

$$\sigma(x) = x^2 + x + 1,$$

which has roots a^5 and a^{10} in GF(2^4) defined by $p(x) = x^4 + x + 1$. Hence, the error polynomial is $e(x) = x^5 + x^{10}$, which, when added to the received polynomial, results in the codeword

$$u(x) = v(x) + e(x) = 1 + x + x^2 + x^4 + x^5 + x^8 + x^{10}.$$

6.7 Convolutional codes

In previous study, algebraic structures are embedded into a code through a parity-check submatrix, field structures, or ring polynomials. In this section, a code structure is embedded through filters. A convolutional encoder can be considered to be an FIR filter having a single input and multiple output ports. The output is the convolution of the input with the filter's impulse responses; thus the name *convolutional codes*. An example is shown in Figure 6.3, where two storage units are used to form two FIR filters for the encoder. As a convention, the oldest data is on the rightmost of the shift register, while the newest data is on the leftmost. The encoder takes *one* bit at a time from the input message sequence to produce the two output bits.

In general, a convolutional encoder can consist of multiple (say K) k-tuple registers to form n FIR filters in parallel, such that it shifts k bits at a time to the right and is fed with a new k-bit byte from its input sequence. In so doing, a k-bit message byte produces n coded output bits and, thus, the ratio k/n is called the *code rate*. The number of registers K represents how many output bytes are influenced by a current input byte, and hence the term *constraint length*. Correspondingly, the quantity $M = (K - 1)$ is called the *memory depth*.

Just like FIR filters, a convolutional encoder is an FIR filter with multiple outputs and, thus, can be represented in different ways, such as generator polynomials, graphes, impulse responses, or delay operators, in the z-transform domain. A discrete symbol input endows a convolutional encoder with Markovian properties, which can be characterized by a state diagram, trellis diagram, and transfer function. Various filter representations for an encoder are used to study the generation of convolutional codes. The Markovian properties are exploited by using dynamic programming for efficient decoding. This is the basic idea behind the Viterbi algorithm. The transfer function is used to study the distance performance.

6.7.1 Examples

Let us consider the $\frac{1}{2}$-encoder shown in Figure 6.3. The constraint length is $K = 3$. The graph shown therein is one way to represent a convolutional encoder. The corresponding impulse response representation is given by

$$g_1 = [1\ 0\ 1]: \ \text{low order} \ \rightarrow \ \text{high order},$$

$$g_2 = [1\ 1\ 1]: \ \text{low order} \ \rightarrow \ \text{high order}.$$

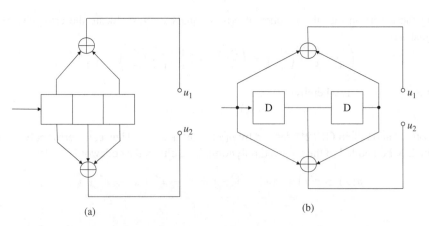

(a)   (b)

Figure 6.3 *Two usual ways to show a convolutional encoder (rate $= \frac{1}{2}$, $K = 3$)*

Alternatively, it can be represented as two generator polynomials

$$g_1(x) = 1 + x^2,$$
$$g_2(x) = 1 + x + x^2,$$

which are used to generate the first and second code symbol, respectively. The FIR filter is essentially a shift delay line. Thus, the two polynomials, when represented in terms of a delay operator D, are given by

$$g_1(D) = 1 + D^2,$$
$$g_2(D) = 1 + D + D^2.$$

They have the same form as the generator polynomials but different physical significance. Thus, the two representations will be used interchangeably.

6.7.2 Code generation

If the input is 1011, determine the coded output sequence. Note that the newest data in the input is placed on the leftmost:

$$1011$$
$$\text{new} \leftarrow \text{old}$$

6.7.2.1 *Method of filters*

We can directly work on the filters to generate the results. Since the filter's output is the convolution of its impulse response with its input, we need to reverse the input message as 1101 for calculation. Suppose that the initial stages of the encoder are set to be zeros, and the input sequence is forced to be truncated by being appended with three zeros. The equivalent input is 101100. This method is based on the impulse response and we directly work on the encoder diagram.

The message sequence right-shifts to the memory of the encoder by one bit at a time and the filters perform modulo-2 operation on the contents in the memory. This is equivalent to moving the window of width K, in reverse order, over the message sequence and performing the same filtering operations within the window. For example, the first window is 100, with which $u_1 = 1 + 0 = 1$ and $u_2 = 1 + 0 + 0 = 1$, leading to result 11 inserted on the first row of the following table. The second window is 010, resulting in output 01 on row 2, and so on.

Window	Output $[u_1, u_2]$
00110<u>100</u>	11
0011<u>010</u>0	01
001<u>101</u>00	00
00<u>110</u>100	10
0<u>011</u>0100	10
<u>001</u>10100	11

Finally, we put the output data together to obtain the coded sequence 110100101011.

6.7.2.2 *The method of polynomials*

When using polynomial multiplication for code generation, we need to keep the message in the original order 1011, with the newest bit on the left. Write the message polynomial $m(x) = 1 + x^2 + x^3$ and multiply it with

$g_1(x)$ and $g_2(x)$, respectively, to produce outputs from the two filters:

$$v_1(x) = m(x)g_1(x) = (1 + x^2 + x^3)(1 + x^2) = 1 + x^3 + x^4 + x^5,$$
$$v_2(x) = m(x)g_2(x) = (1 + x^2 + x^3)(1 + x + x^2) = 1 + x + x^5. \tag{6.82}$$

Add *all* the missing terms with zero coefficients, organize both $v_1(x)$ and $v_2(x)$ in ascending order, and interlace their coefficients yielding the codeword 110100101011; the result is identical to that obtained by treating the encoder as filters.

The interlacing can be done, alternatively, through polynomial operations, as shown by

$$v(x) = v_1(x^2) + xv_2(x^2) = 1 + x + x^3 + x^6 + x^8 + x^{10} + x^{11}, \tag{6.83}$$

whose coefficients give the same output sequence 110100101011. Here, The use of x^2 to replace x in $v_i(x)$ aims to create room between its symbols to accommodate another sequence, whereas multiplying $v_2(x^2)$ with x shifts $v_2(x^2)$ one bit to the right to allow for interlacing with $v_1(x^2)$. The same sequence can be produced by inputting 1101 to the filters!

6.7.3   Markovian property

Convolutional encoders have properties not shared by usual FIR filers. Unlike usual filters with real or complex signals as their input, the input to an encoder is *discrete* symbols and the filtering operation is done in a finite field. As discrete symbols are shifted in a constrained-K convolutional encoder, its $K - 1$ registers constitute a finite-state machine with its contents transiting from one state to another, under the control of each incoming symbol. The *state* of an (n, k, K)-encoder is defined as the contents of the *M=(K-1)* rightmost states of the shift register. Consider the example in Figure 6.3 again. When a bit 1 is shifted in the encoder initialized with zeros, graphically shown below as

$$\xrightarrow{\;1\;} \boxed{1\,|\,0\,|\,0}\,,$$

the encoder state switches from the current state 00 to the next state 10 driven by the newly input bit 1. More intuitively, we write

$$\boxed{00} \longrightarrow \boxed{10}.$$

The state transition only indicates the contents in the encoder memory. For a complete description, it still needs to include the encoder's outputs that capture the topological structure (or simply, connections) defined by the filters of the encoder. This can be done by

$$\boxed{00} \xrightarrow{1/11} \boxed{10}.$$

This symbolic representation means that with a newly entered bit 1, the encoder state will transit from 00 to 10 while producing an encoder output 11, which is also known as *branch output* in the literature.

Property 6.7.1 *A K-constrained binary convolutional code is of 2^{K-1} states. Each state can transit to two possible states under the control of newly input bit.*

Property 6.7.2 *Two branches emanating from the same node re-merge again at a certain node after K time slots.*

The filter structure, together with the modulo 2 addition, endows a convolutional encoder with Markovian properties. These properties are captured by the state transition and the trellis diagrams.

The state diagram of the encoder in Figure 6.3 is shown in Figure 6.4, where, as a convention, dashed lines represent input equal to bit 1 while solid lines represent input equal to bit 0. The state diagram of an encoder shows the transition of the encoder from one state to another. But it does not show the state evolution over time. This goal is achieved by using a trellis diagram, as illustrated in Figure 6.5. It is observed that paths emanating from a node re-merge at the same node after K symbol intervals.

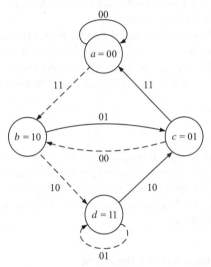

Figure 6.4 *State diagram of the encoder in Figure 6.3. Dashed lines represent input = 1, whereas solid lines represent input = 0*

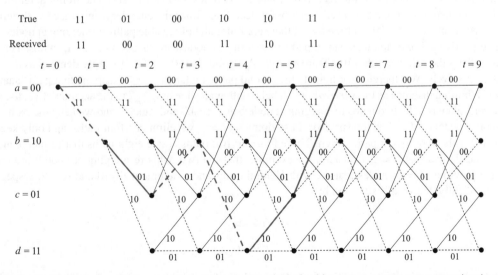

Figure 6.5 *Trellis diagram of the encoder in Figure 6.3, with dashed line representing input equal to bit 1 and solid line representing bit 0*

The encoded sequence is modulated and then sent over a noisy channel. The receiver performs the reverse operation: demodulation and decoding. Because of noise corruption and possible bandlimited characteristics of the channel, demodulation may introduce decision errors. The question is: what is the best way to recover the original message sequence?

A powerful tool for decoding a convolutional code is the Viterbi algorithm. It exploits the Markov property of convolutional codes and always eliminates paths that are unlikely to be a code sequence whenever possible, thereby significantly reducing the computational complexity. The idea comes from the following observation: any two paths in the trellis merge to the same state, and one of them can always be eliminated in the search for an optimum path. The reason is that the present state summarizes the encoder history, and only through it the previous states can influence the future output.

The decoding performance is closely related to the minimum free distance, or simply *free distance*, of the code, which, in the context of convolutional codes, is defined as the minimum distance in the set of arbitrarily long paths that diverge and re-merge. The free distance of a convolutional code, denoted by d_{free}, can be strictly derived from its transfer function, as will be addressed shortly. The number of bits that can be corrected is closely related to d_{free} by

$$t = \lfloor \frac{d_{\text{free}} - 1}{2} \rfloor, \tag{6.84}$$

where $\lfloor q \rfloor$ denotes the operation of taking the integral part of $q > 0$.

The contribution of Viterbi elegantly represents the constraints as a limited number of feasible path patterns, thereby significantly reducing the computation. Forney argued that the Viterbi algorithm is the maximum likelihood (ML) decoder based on a hard decision estimator $\{\hat{s}_k\}$. In particular, his argument is based on the fact that $p(\mathbf{c}|\hat{\mathbf{s}})$ is a binomial distribution.

6.7.4   Decoding with hard-decision Viterbi algorithm

We use the encoder in Figure 6.3 as an example to illustrate the procedure for trellis-based convolutional decoding. The encoder converts the message 1011 encoded into output sequence 110100101011. Suppose that, due to errors in detection, the received sequence becomes 110000111011. The trellis diagram of the encoder, sketched in Figure 6.5, is used as a tool for decoding. Since the encoder is initialized with zeros, the trellis starts from state $a = 00$. At time $t = 3$, there are four pairs of feasible paths re-merging at nodes a, b, c, and d, respectively. For example, a pair of paths re-merging at node a are "aaaa" and "abca," whose Hamming distances from the received bits 110000 are $d = [2, \ 0, \ 0]$ and $d = [0, \ 1, \ 2]$, or simply denoted as $d = 2$ and $d = 3$, respectively. We, therefore, eliminate the second path of a larger $d = 3$, keep the first path "aaaa" of a shorter Hamming distance as the *survivor*, and label it with underline as $\underline{d = 2}$ to the accumulative distance of a survivor. Likewise, the two paths re-merging at node b are "aaab" and "abcb", and we keep the path "abcb" of a shorter distance $d = 1$ as the survivor. In so doing, at $t = 3$ we eliminate four paths, and only keep four survivors for the subsequent path search. These four survival paths create eight paths that re-merge at nodes a, b, c, and d, again, at $t = 4$. Consider, for example, that the two paths re-merging at node b are incident from the previous ($t = 3$) nodes a and c with Hamming distance increment $d = 0$ and $d = 2$, respectively. We symbolically denote this situation by

$$a \xrightarrow{0} b$$
$$c \xrightarrow{2} b.$$

We check the two survivors terminating at nodes a and c at $t = 3$ are "aaaa: $d = 2$" and "abdc: $d = 2$," both having the same accumulative distance $d = 2$. By adding the distance increment to the previous survival paths, we obtain a complete description of paths at $t = 4$, as shown by

$$\underbrace{\text{aaaa}}_{2} \xrightarrow{0} \text{b} : \underline{d = 2}$$

$$\underbrace{\text{abdc}}_{2} \xrightarrow{2} \text{b} : d = 4.$$

We use the same technique to determine the four survivors for the next search, and continue. Tabulating all the information we obtain yields

Pair@	$t = 3$	$t = 4$	$t = 5$	$t = 6$
a	aaaa : $\underline{d = 2}$	$\underbrace{\text{aaaa}}_{2} \xrightarrow{2} \text{a} : d = 4$	$\underbrace{\text{abdca}}_{2} \xrightarrow{1} \text{a} : d = 3$	$\underbrace{\text{abcbca}}_{3} \xrightarrow{2} \text{a} : d = 5$
	abca : $d = 3$	$\underbrace{\text{abdc}}_{2} \xrightarrow{0} \text{a} : \underline{d = 2}$	$\underbrace{\text{abcbc}}_{2} \xrightarrow{1} \text{a} : \underline{d = 3}$	$\underbrace{\text{abcbdc}}_{2} \xrightarrow{0} \text{a:} \boxed{d = 2}$
b	aaab : $d = 4$	$\underbrace{\text{aaaa}}_{2} \xrightarrow{0} \text{b} : \underline{d = 2}$	$\underbrace{\text{abdca}}_{2} \xrightarrow{1} \text{b} : d = 3$	$\underbrace{\text{abcbca}}_{3} \xrightarrow{0} \text{b} : \underline{d = 3}$
	abcb : $\underline{d = 1}$	$\underbrace{\text{abdc}}_{2} \xrightarrow{2} \text{b} : d = 4$	$\underbrace{\text{abcbc}}_{2} \xrightarrow{1} \text{b} : \underline{d = 3}$	$\underbrace{\text{abcbdc}}_{2} \xrightarrow{2} \text{b} : d = 4$
c	aabc : $d = 5$	$\underbrace{\text{abcb}}_{1} \xrightarrow{1} \text{c} : \underline{d = 2}$	$\underbrace{\text{aaaab}}_{2} \xrightarrow{2} \text{c} : d = 4$	$\underbrace{\text{abcbcb}}_{3} \xrightarrow{1} \text{c} : d = 4$
	abdc : $\underline{d = 2}$	$\underbrace{\text{abdd}}_{2} \xrightarrow{1} \text{c} : d = 3$	$\underbrace{\text{abcbd}}_{2} \xrightarrow{0} \text{c} : \underline{d = 2}$	$\underbrace{\text{aaaabd}}_{2} \xrightarrow{1} \text{c} : \underline{d = 3}$
d	aabd : $d = 5$	$\underbrace{\text{abcb}}_{1} \xrightarrow{1} \text{d} : \underline{d = 2}$	$\underbrace{\text{aaaab}}_{2} \xrightarrow{0} \text{d} : \underline{d = 2}$	$\underbrace{\text{abcbcb}}_{3} \xrightarrow{1} \text{d} : d = 4$
	abdd : $\underline{d = 2}$	$\underbrace{\text{abdd}}_{2} \xrightarrow{1} \text{d} : d = 3$	$\underbrace{\text{abcbd}}_{2} \xrightarrow{2} \text{d} : d = 4$	$\underbrace{\text{aaaabd}}_{2} \xrightarrow{1} \text{d} : \underline{d = 3}$

At $t = 6$, the path with the minimum Hamming distance $d = 2$ from the received sequence is "abcbdca," which is marked with a box for ease of identification. This path is chosen as the estimate of true coded sequence with message 101100, and is exactly the same as the original message appended with two zeros. It is clear that regardless of two bit errors in the received sequence, the convolutional code has the capability of complete correction.

Let us summarize the procedure for Viterbi decoding:

- Step 1: Draw a trellis diagram for the given convolutional code, and start from the all-zero initial state (initially clearing the registers). Label each branch with input bits and output branch words.
- Step 2: For each time step, say t_i, calculate the Hamming distance of each branch (its branch word) from the corresponding received symbol, and label this distance at the end of the branch.

- Step 3: After $i \geq K$, there are always two surviving paths that merge at each node. Calculate their *cumulative* Hamming distance accumulated back to time $t = 0$. Then, between the two surviving paths, eliminate the one with the larger Hamming distance. Thus, after $i \geq K$, only 2^{K-1} paths survive.
- Step 4: Proceed in the same way until t reaches an appropriately preset time instant, say $t = L$, and stop. Among the 2^{K-1} survivors, choose the one with minimum cumulative distance as the optimal path.
- Step 5: Determine the decoding output sequence from the input labeled on the optimal path.

It remains to determine an appropriate length L for convolutional decoding. Unlike a block code, a convolutional code does not have a particular block size. For ease of decoding, convolutional codes are often forced into a block structure by *periodic truncation*. Namely, append *M zero bits* to the end of the input data sequence for the purpose of clearing or flushing the encoder shift register. A side effect is the reduction of the effective code rate. There are a number of factors that affect the choice of the value of L. On one hand, finite decoding depth causes additional errors, called *truncation errors*. It was shown by Forney (1973) that the truncation error is negligible if $L > 5.8K$. On the other hand, L represents the time delay in decision making; a larger L implies a larger time delay. Furthermore, a larger L implies the need for a larger register memory. More importantly, increased L implies an exponentially increasing decoding complexity and storage requirement. The storage requirement in the Viterbi algorithm is roughly equal to $(4 \sim 5)K2^{K-1}$. Decoding complexity, decision delay, and error performance constitute the tradeoff in the design of a practical Viterbi algorithm (VA) decoder. As a tradeoff, L can be chosen to be sightly greater than $5.8K$.

✍ Example 6.18

For example, a rate-$\frac{1}{2}$, $K = 9$ convolutional code is used in IS-95 sync and page channels:

$$g_1(x) = 1 + x + x^2 + x^3 + x^5 + x^7 + x^8,$$
$$g_2(x) = 1 + x^2 + x^3 + x^4 + x^8,$$

which has the same free distance $d_{\text{free}} = 12$ as the one with generator $g = (561, 753)$. The speech data of $9.6\,\text{kb/s}$, after coding, is converted to a bit rate of $19.2\,\text{kb/s}$. Each coded bit is then spread by a 64-bit Walsh–Hadamard vector, ending up with a chip rate of $1.2288\,\text{Mb/s}$. The spread coded sequence is transmitted using BPSK.

○

〜〜〜✍

6.7.5   Transfer function

The *minimum distance* between any two adversary paths of a convolutional code is a critical parameter to determine the code's correction capability. It is also known as the *free distance*. Just as the case of linear block codes, the minimum distance, when measured equivalently with respect to the all-zero path, is the minimum Hamming weight of all possible paths. The minimum distance of a code is dictated by the code's constraint length and topological connections inherent in its transfer function. Such a transfer function is derivable from the state transition diagram of the code.

Let us show how to derive the transfer function through a particular $\frac{1}{3}$-convolutional code defined by $(D + D^2,\, 1 + D,\, 1 + D + D^2)$ with state diagram given in Figure 6.6. The procedure is detailed as follows.

1. To *convert* the state diagram to the form of a linear system, we split the all-zero node into two nodes, with one representing the input port (denoted by a) and the other representing the output port (denoted by e). Further, we remove the self-loop at the all-zero state a since it has no contributions to the distance properties of the code. The resulting diagram is shown in Figure 6.7.

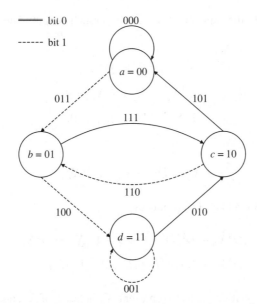

Figure 6.6 *State diagram for the rate-1/3 encoder*

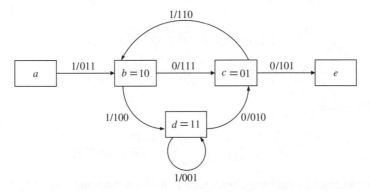

Figure 6.7 *Transfer diagram for the example*

2. Label each branch with D^κ, where the exponent κ is set equal to the number of bit one in the relevant branch codeword.

3. Label state a by X_a, state b by X_b, and so on.

Following this procedure, we obtain an equivalent linear system model of the rate-$\frac{1}{3}$ convolutional code depicted in Figure 6.7.

We need to determine the transfer function of the linear system defined in Figure 6.7. In a way similar to the application of Kirhoff's current law to a linear circuit, we obtain "current" equations for all the nodes as follows:

$$X_b = D^2 X_a + D^2 X_c,$$
$$X_c = D^3 X_b + D X_d,$$
$$X_d = D X_b + D X_d,$$
$$X_e = D^2 X_c, \qquad (6.85)$$

where the *degree* of D equals the number of 1's in the branch output. The first three equations can be written asy

$$X_b - D^2 X_c = D^2 X_a, \tag{6.86}$$

$$D^3 X_b - X_c + DX_d = 0, \tag{6.87}$$

$$DX_b + (D - 1)X_d = 0, \tag{6.88}$$

$D^2 \times (6.88) - (6.87)$ results in

$$X_c = D(1 + D - D^2)X_d, \tag{6.89}$$

which, when inserted into (6.86) and (6.87), produces

$$D^3 X_b - D^2[D - D^2(D - 1)]X_d = D^2 X_a,$$
$$D^3 X_b + (D - [D - D^2(D - 1)])X_d = 0. \tag{6.90}$$

Eliminate X_b from the above equations so as to represent X_d in terms of X_a, and then insert the resulting X_d into (6.89) to obtain X_c. It turns out that

$$X_d = \frac{D^5 X_a}{1 - D + D^2 - D^3 - D^5 - D^7 + D^8},$$
$$X_c = \frac{D^5 - D^7(D - 1)}{1 - D + D^2 - D^3 - D^5 - D^7 + D^8} X_a. \tag{6.91}$$

Insert X_c into (6.85), yielding

$$T(D) = \frac{D^7 + D^9 - D^{10}}{1 - D + D^2 - D^3 - D^5 - D^7 + D^8}. \tag{6.92}$$

The free distance is indicated by the lowest order term in the numerator and, thus, $d_{\text{free}} = 7$.

The transfer function can be made to provide more information than just the free distance by introducing a counter for input bit "1" on each branch, denoted by M, and a counter to denote the number of branches, denoted by A, that need to connect to the all-zero state in the state equations in (6.85). Besides the free distance, this enhanced transfer function, denoted here by $T(D, A, M)$, provides detailed information for each possible path, with reference to the all-zero path. In general, $T(D, A, M)$ can be expanded in the form

$$T(D, A, M) = \sum_{i=d_{\text{free}}}^{\infty} \varphi_i D^i A^{\alpha_i} M^{\beta_i}, \tag{6.93}$$

where the summands indicate that there are φ_i paths, each of distance i, from the all-zero path; and each of these paths spans over α_i branches and is driven by a source sequence of β_i bit ones relative to the all-zero paths. The transfer function $T(D, A, M)$ is also very useful in the calculation of the pairwise error probability (PEP) between any two paths that start from the same node and re-merge later. The use of $T(D, A, M)$ is very flexible, and its argument can be set to unity if the corresponding information is not of interest. For the determination and other details of $T(D, A, M)$, the reader is referred to [9, 10].

Table 6.1 *Rate-$\frac{1}{2}$ convolutional codes with optimal distance spectrum profiles where n_i denotes the number of paths with Hamming weight $d_{free} + i$. (Adapted from Johannesson and Stahl [11], Copyright (1999) IEEE, with permission)*

M	$g_1(D)$	$g_2(D)$	d_{free}	# of paths			Equivalent generators
				n_0	n_1	n_2	
2	5	7	5	1	2	4	
3	17	13	6	1	3	5	$(17, 15)$
4	31	27	7	2	3	4	$(23, 35)$
5	77	45	8	2	3	8	$(77, 51)$
6	147	135	10	12	0	53	$(163, 135)$
7	313	275	10	1	6	13	$(323, 275)$
8	751	557	12	10	9	30	$(457, 755)$
9	1337	1475	12	1	8	8	$(1755, 1363)$
10	2457	2355	14	19	0	80	$(3645, 2671)$

6.7.6 Choice of convolutional codes

In choosing a convolutional code for practical use, some considerations must be taken into account.

The first is catastrophic error propagation. In coding, *catastrophic error propagation* refers to the phenomenon in which a finite number of decoding errors can cause an infinite number of decoded bit errors. A criterion for diagnosing a convolutional catastrophic encoder is to check whether all of its generators have a common polynomial factor of degree at least 1. As an example, let us consider the encoder defined by

$$g_1(D) = 1 + D,$$
$$g_2(D) = 1 + D^3.$$

Since

$$g_2(D) = (1 + D)(1 + D + D^2), \tag{6.94}$$

it includes $g_1(D)$ as a factor. Hence, the encoder can cause catastrophic error propagation.

The second consideration is to ensure a necessary free distance to meet the requirement of error correction capability. The free distance of a code usually increases with its constraint length. Nevertheless, codes with a same constraint length often have different free distances, thus raising the issue of searching for good codes. Research has been carried out along this line, resulting in abundant results, part of which is tabulated below. The equivalent generator polynomials are also included in the table. When represented in binary form, they provide exactly the same polynomial coefficients as the corresponding optimal pair, except that they are in reverse order.

In some applications such as the construction of turbo codes, systematic codes are preferred. A *systematic* rate-$\frac{k}{n}$ convolutional code is one in which the input k-tuple appears as part of the n-tuple output codeword produced by the k-tuple input. Most of the convolutional codes shown above are nonsystematic. For linear block codes, any nonsystematic code can be linearly transformed into a systematic one of the *same* block distance properties. The same assertion is *not* true of convolutional codes. Making a convolutional code systematic, in general, *reduces* the maximum possible free distance for a given constraint length and rate. One advantage of a systematic code is that it can never be catastrophic.

✍ Example 6.19

In GSM, the input rate for speech transmission mode on a full-rate channel is equal to 13 kps, in blocks of 260 bits. Each block consists of 182 bits of class 1 and 78 unprotected bits of class 2. The first 50 bits in class 1 are protected by 3 parity bits, and 4 zero tailing bits are added before convolutional coding. After coding, the block size becomes $[(50 + 3) + 132 + 4] \times 2 + 78 = 456$ bits. The rate-$\frac{1}{2}$ code generator polynomials are given by

$$g_1(D) = D^4 + D^3 + 1,$$
$$g_2(D) = D^4 + D^3 + D + 1.$$

6.7.7   Philosophy behind decoding strategies

At the transmitter, the encoded sequence is modulated before transmission over a noisy channel. It is therefore natural, in traditional thinking, to perform reverse operation at the receiver by bit-by-bit detection before decoding. Let $\mathbf{b}$ denote the message vector and $\mathbf{G}$ denote the corresponding coding matrix. Assume, for simplicity, that the coded vector $\mathbf{Gb}$ is converted into a bipolar vector

$$\mathbf{u} = 2\mathbf{Gb} - 1 \triangleq \varphi(\mathbf{b}), \tag{6.95}$$

before being BPSK-modulated for transmission over an AWGN channel, and assume a coherent receiver is used. Then, in baseband form, the received vector is expressible as

$$\mathbf{x} = \sqrt{E_c}\varphi(\mathbf{b}) + \mathbf{n}, \tag{6.96}$$

where $\mathbf{n} \sim \mathcal{N}(\mathbf{0}, \sigma_n^2 \mathbf{I})$, and E_c is the coded bit energy at the receiver which has taken into account the channel attenuation, thus allowing us to write the SNR at the receiver as

$$\rho = E_c/\sigma_n^2. \tag{6.97}$$

To simplify notation, we may divide both sides of (6.96) by $\sqrt{E_c}$ to obtain

$$\mathbf{y} = \underbrace{\varphi(\mathbf{b})}_{\mathbf{u}} + \mathbf{e}, \tag{6.98}$$

where $\mathbf{e} \sim \mathcal{N}(\mathbf{0}, \rho^{-1}\mathbf{I})$. Thus, the PDF of $\mathbf{y}$ is given by

$$f(\mathbf{y}|\mathbf{u}) = (2\pi/\rho)^{-n/2} \exp(-\rho\|\mathbf{y} - \mathbf{u}\|^2/2). \tag{6.99}$$

Given the received vector $\mathbf{y}$, decoding the message vector $\mathbf{b}$ in the ML framework is essentially an integer programming problem, defined by

$$\min_{\mathbf{b}:\mathbf{u}=\varphi(\mathbf{b})} \|\mathbf{y} - \mathbf{u}\|^2. \tag{6.100}$$

This is, in nature, a nonlinear regression problem, with nonlinearity defined by the code structures $(\mathbf{Gb})|_{\text{mod } 2}$, which, along with the integer constraint on $\mathbf{b}$, determines that decoding cannot be solved by simply using the least-squares method. There are two different strategies to handle the nonlinearity involved in the constraint.

$$\min_{\mathbf{b}:\mathbf{u}=\varphi(\mathbf{b})} \|\mathbf{y} - \mathbf{u}\|^2 \rightarrow \begin{cases} \text{directly exploiting code structures for : ML decoding} \\ \min_{\mathbf{b}:\mathbf{u}=\varphi(\mathbf{b})} \|\text{sgn}(\mathbf{y}) - \mathbf{u}\|^2 : \text{ hard-decision decoding} \end{cases} .$$

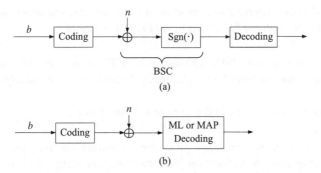

Figure 6.8 *Distinction between hard and soft decoding*

The second strategy is a two-step procedure with the essential idea of using $\text{sgn}(\mathbf{y})$ to replace $\mathbf{y}$ to simplify the decoding:

$$\text{step 1: } \mathbf{y} = \mathbf{u} + \mathbf{e} :\rightarrow \hat{\mathbf{u}} \longrightarrow \text{step 2: } \mathbf{Gb} = \hat{\mathbf{c}} :\rightarrow \hat{\mathbf{b}}$$

$$\boxed{\text{bit-by-bit detection}} \longrightarrow \boxed{\text{error correction}} \tag{6.101}$$

where the operation $\hat{\mathbf{c}} = (\hat{\mathbf{u}} + 1)/2$ converts the bipolar binary vector $\hat{\mathbf{u}}$ back to a unipolar one. The distinction between the hard and soft decoding is illustrated in Figure 6.8.

6.7.7.1 Hard decoding

Along the first line of thought, we treat the received samples as if they were independent and thus conduct bit-by-bit detection first, leaving the code structure to be exploited in the second step of correction. Binary decision on received samples $\mathbf{y}$ corresponds to the operation $\text{sgn}[\mathbf{y}]$. Hard-decision-based optimization is equivalent to finding the path m for which

$$\Lambda_m(\text{sgn}[\mathbf{y}]) = \sum_{i=1}^{\ell} \frac{\rho_i}{2} \|\text{sgn}[\mathbf{y}_i] - \mathbf{u}_i^{(m)}\|^2 \tag{6.102}$$

is minimized. In BSC, the transition probabilities $\Pr\{1 \rightarrow -1\}$ and $\Pr\{-1 \rightarrow 1\}$ are equal, say equal to $p < 1/2$. Suppose that $\mathbf{u}^{(m)}$ and $\text{sgn}(\mathbf{y})$ differ in d_m bits. Since the channel noise is white, binary decisions for successive bits in the demodulator are independent. Thus, for a *given* code sequence along path m, we may use the binomial rule to express the likelihood metric over BSC as

$$\Pr(\mathbf{y}|\mathbf{u}^{(m)}, d_m) = p^{d_m}(1 - p)^{n\ell - d_m}, \tag{6.103}$$

which, after taking the logarithm, produces

$$\log \Pr(\mathbf{y}|\mathbf{u}^{(m)}, d_m) = -d_m \underbrace{\log[(1 - p)/p]}_{>0} + n\ell \log(1 - p). \tag{6.104}$$

This expression indicates that ML decoding on BSC is identical to searching for the path with minimum Hamming distance on a trellis diagram. As such, we can write

$$\max_m \{\log \Pr(\mathbf{y}|\mathbf{u}^{(m)}, d_m)\} \rightarrow \min_m \{d_m\}. \tag{6.105}$$

Property 6.7.3   *Hard-decision Viterbi decoding along a trellis with the minimum cumulative Hamming distance criterion represents the optimal ML decoding on the BSC channel.*

The problem is that, in converting an AWGN channel to its BSC counterpart with hard decision, some information has been lost and cannot be remedied any more by determining the relatively inferior performance of hard-decision decoding.

Summarizing the above analysis, we can draw our assertions in order.

Property 6.7.4   *Each step in the two-step procedure, standing on its own right, is optimal. The sample-by-sample detection leads to a single-bit ML estimator if focusing solely on one sample. The second step focuses on the minimum Hamming distance, also representing an ML decoder but valid only for the BSC. However, treating a single decoding problem in two separate steps is incapable of achieving the optimal estimation.*

It is easy to imagine that hard decision, though simple, introduces more errors than it should. The drawback of hard decision is threefold. First, it treats a single optimization problem in two separate steps. The resulting solution is not optimal. Second, it fails to provide the estimation reliability, such as its confidence interval. For example, the sample values $y_i = 0.001$ and $y_i = 3.24$ end up with the same estimate $\hat{u}_i = 1$ regardless of their significant difference in reliability.

The use of a hard decision to replace the exact data $\mathbf{y}$ in decoding will inevitably cause information loss. The situation can be improved if we use multilevel quantization, instead; the resulting improved schemes are usually known as *soft-decision decoding*. If we use the hard-decision decoding as a reference, then the improvement (in SNR) with soft-decision decoding is as tabulated below.

Number of levels	SNR improvement (dB)
2 (hard decision)	0
8	2
∞ (i.e., Yusing $\mathbf{y}$)	2.2

The SNR gain is achieved at the cost of increased storage requirement in the decoder. Soft-decision decoding can be easily used along with convolutional decoding, but it is generally not used in decoding block codes because of its difficulty in implementation. If the scheme of κ-bit soft decision is employed, we partition the entire signal range $(-\sqrt{E_c}, \sqrt{E_c})$ into 2^κ subintervals, in much the same way as is done in uniform quantization.

6.7.7.2   Soft decision

Refer to (6.98) and focus on a single sample, denoted by $y_i = u_i + e_i$. Using the chain rule, we write

$$P(u_i|y_i) = f(u_i|y_i)dy_i = \frac{f(y_i|u_i)P(u_i)}{f(y_i)}dy_i, \tag{6.106}$$

where u_i takes values of 1 or -1 with equal likelihood, with probability $P(u_i = 1) = P(u_i = -1) = 1/2$. Then, the likelihood ratio can be calculated as

$$L = \frac{P(u_i = 1|y_i)}{P(u_i = -1|y_i)} = \frac{f(y_i|u_i = 1)}{f(y_i|u_i = -1)} = e^{2\rho y_i}, \tag{6.107}$$

which, along with the total probability $P(u_i = 1|y_i) + P(u_i|y_i) = 1$, leads to

$$P(u_i = 1|y_i) = e^{2\rho y_i}(1 + e^{2\rho y_i})^{-1},$$

$$P(u_i = -1|y_i) = (1 + e^{2\rho y_i})^{-1}. \tag{6.108}$$

With these expressions and the fact that $u_i = \pm 1$, we can follow the definition to determine the conditioned expectation of u_i as

$$\mathbb{E}[u_i|y_i] = P(u_i = 1|y_i) - P(u_i = -1|y_i) = \tanh(\rho y_i), \tag{6.109}$$

which can be used as a soft estimate of u_i. The decision is done here by $\tanh(\cdot)$, which behaves similar to the signum function in neural networks wherein a neuron outputs soft decisions to the next layer. This soft output not only provides a decision but also indicates its reliability. A decision output close to ± 1 implies a very high reliability, whereas a decision output close to zero points to a very unreliable estimate. In contrast, a hard decision provides information of one-bit quantization, namely only the sign information.

6.7.7.3 *ML decoding of convolutional codes*

In the first step of the two-step decoding procedure, each of the received samples is used separately to estimate its corresponding coded bit as if these samples were independent. ML decoding is a one-step procedure, directly estimating the source sequence from the received signal. Before proceeding, let us define the notations.

$\mathbf{u}^{(m)}$	the code sequence along path m
$\mathbf{y}$	the received sequence
$u_{ij}^{(m)}$	the jth bit in the i true branch word along path m
y_{ij}	the jth bit in the ith received branch word
L	constraint length
ℓ	the depth into the trellis, that is, the total number of branches in $\mathbf{y}$
m	the path index in a trellis
$r = \frac{k}{n}$	code rate

The mth path sequence can be explicitly written as $\mathbf{u}^{(m)} = [\mathbf{u}_1^{(m)}, \cdots, \mathbf{u}_\ell^{(m)}]$, where $\mathbf{u}_i^{(m)} = [u_{i1}^{(m)}, \cdots, u_{in}^{(m)}]$ denotes the ith n-bit word associated with the ith branch of path m. Correspondingly, we can partition $\mathbf{y}$ into ℓ segments each of n entries, such that $\mathbf{y} = [\mathbf{y}_1, \cdots, \mathbf{y}_\ell]$.

There are multiple feasible paths that are likely to have been responsible for the generation of the received sequence. The decoding problem is, therefore, to decide which path is the most likely one. This is done by virtue of the maximum log posterior probability, as shown by

$$\hat{m} = \max_{\text{all possible path } m} \log \Pr\{\mathbf{u}^{(m)}|\mathbf{y}\}, \tag{6.110}$$

where the log MAP can be further written as

$$\log \Pr\{\mathbf{u}^{(m)}|\mathbf{y}\} = \underbrace{\log \Pr\{\mathbf{y}|\mathbf{u}^{(m)}\}}_{\text{log likelihood } \Lambda_m(\mathbf{y})} + \underbrace{\log \Pr\{\mathbf{u}^{(m)}\}}_{prior}. \tag{6.111}$$

The expression of the log likelihood term depends on the particular channel environment.

For exposition, let us assume an AWGN channel for which the ith segment of the received signal $\mathbf{y}$ is expressible as

$$\mathbf{y}_i = \mathbf{u}_i + \mathbf{e}_i \text{ with } \mathbf{e}_i \sim \mathcal{N}(\mathbf{0}, \rho^{-1}\mathbf{I}).$$

Thus, the log likelihood of the feasible path m is equal to

$$\Lambda_m(\mathbf{y}) = \log \prod_{i=1}^{\ell} P(\mathbf{y}_i|u_i^{(m)}) = \sum_{i=1}^{\ell} \frac{\rho_i}{2}\|\mathbf{y}_i - \mathbf{u}_i^{(m)}\|^2 + (\text{terms independent of } m)$$

$$= \sum_{i=1}^{\ell} \rho_i \mathbf{y}_i^T \mathbf{u}_i^{(m)} + (\text{terms independent of } m). \tag{6.112}$$

It is clear that for AWGN channels, the log likelihood principle leads to the minimum distance criterion, as shown by line 1 of the above expression. This minimum distance criterion is equivalent to the criterion of maximum correlation that measures the *similarity* between the received vector and the codeword sequence of path m, as shown in the last line. In a practical system, the analog received samples y_i in $\mathbf{y}$ can be replaced by their multilevel quantized versions to save memory. Inserting the above into (6.110) and dropping the terms of relevance to m yields

$$\hat{m} = \arg \max_{m} \left\{ \left(\sum_{i=1}^{\ell} \rho_i \mathbf{y}_i^T \mathbf{u}_i^{(m)} \right) + \log \Pr\{\mathbf{u}^{(m)}\} \right\}, \tag{6.113}$$

implying that ML decoding of a convolutional code is to choose, among all the feasible candidates, the path that most resembles the received vector $\mathbf{y}$.

For a transmitted sequence of $n\ell$ bits long, the number of possible paths in the trellis is $2^{n\ell}$. For large $n\ell$, the comparison of the log likelihood metrics for 2^{ℓ} possible paths with brute force is extremely time consuming or even impossible. The Viterbi algorithm, based on a two-step procedure, enables exploiting the Markovian properties of convolutional codes to avoid the enumeration of all the possible paths, thus considerably reducing the computational complexity. Furthermore, its performance can be improved by adopting soft decisions.

6.7.8   Error performance of convolutional decoding

We are interested in the error performance of convolutional decoding. Recall that the decoding process is based on a trellis structure, and convolutional codes have a large number of feasible paths. We can imagine that it is extremely difficult to determine the exact error performance associated with decoding a convolutional code. For simplicity, binary PSK signaling is assumed. Without loss of generality, we further assume that the transmit sequence is along the all-zero path. When using the Viterbi algorithm, the all-zero path and its adversary constitute a competing pair; both paths start from the all-zero state at $t = 0$ and re-merge with the all-zero state some time after $t \geq (2^{K-1} - 1)$. An error event occurs when the metric of the adversary path exceeds its all-zero counterpart. The basic method is to determine the PEP between two competing paths with which to upper bound the error probability.

6.7.8.1   *ML decoding*

For PEP calculation, making a decision between two competing paths is similar to that between two signal points in a constellation. As such, we can argue that the PEP between the two competing paths depends only on their distance in bits, which is denoted here by d. The PEP turns out to be

$$P_2(d) = Q\left(d\sqrt{\frac{2g^2 E_c}{N_0}}\right), \tag{6.114}$$

where g stands for the channel gain, and E_c denotes the channel bit energy. With this PEP, we can bound the error probability of convolutional decoding to yield

$$P_e \leq \sum_{d=d_{\text{free}}}^{\infty} \varphi_d P_2(d), \tag{6.115}$$

where φ_d, denoting the number of paths that differ d bits from the all-zero path, can be determined from the expansion of the transfer function

$$T(D, M) = \sum_{d=d_{\text{free}}}^{\infty} \varphi_d D^d M^{f(d)}. \tag{6.116}$$

For details, the reader is referred to Proakis [9].

6.7.8.2   *Hard-decision-based decoding*

A hard decision converts an AWGN channel to a BSC for which the rule of binomial distribution is applied. As such, the PEP between two competing paths with d-bit difference can be shown to be [9]

$$P_2(d) = \begin{cases} \sum_{k=(d+1)/2}^{d} \binom{d}{k} p^k (1-p)^{d-k}, & d = odd \\ \\ \sum_{k=d/2}^{d} [1 - \frac{1}{2}\delta(k - \frac{d}{2})] \binom{d}{k} p^k (1-p)^{d-k}, & d = even \end{cases} \tag{6.117}$$

where $\delta(k)$ is the Dirac delta function which equals 1 when $k = 0$ and zero otherwise. With the above PEP, the error probability can be upper-bounded by

$$P_e < \sum_{d=d_{\text{free}}}^{\infty} \varphi_d P_2(d). \tag{6.118}$$

✍ **Example 6.20** _____

Given a convolutional code with $d_{\text{free}} = 5$ and the transition error probability over a BSC $p = 0.01$, determine $P_2(d)$ if hard-decision VA decoding is used.

Solution

From the given condition, $d = 5$ and hence

$$P_2(d)|_{d=5} = \sum_{k=3}^{5} \binom{5}{k} (0.01)^k (0.99)^{5-k}$$

$$= 9.8506 \times 10^{-6}.$$

6.7.8.3   *Coded BPSK in flat fading*

In the previous two subsubsections, we investigated the error performance of convolutional decoding on AWGN channels. Let us turn to its PEP performance in flat fading. Denote the channel vector by $\mathbf{h} = [h_1, \cdots, h_\ell]^T$. Define $\mathbf{H} = \text{diag}\{\mathbf{h}\}$, by which the $\ell \times 1$ received vector can be written as

$$\mathbf{x} = \sqrt{E_c}\, \mathbf{Hu} + \mathbf{n}, \tag{6.119}$$

where E_c denotes the coded bit energy and $\mathbf{n} \sim \mathcal{N}(\mathbf{0}, \sigma_n^2 \mathbf{I})$. Given that the code vector $\mathbf{u}_1$ is transmitted, we would like to know the probability that a decision is made in favor of $\mathbf{u}_2$. This happens when

$$\|\mathbf{x} - \sqrt{E_c}\mathbf{Hu}_1\|^2 > \|\mathbf{x} - \sqrt{E_c}\mathbf{Hu}_2\|^2. \tag{6.120}$$

For BPSK signaling, $\mathbf{u}_1^T \mathbf{H}^T \mathbf{Hu}_1 = \mathbf{u}_2^T \mathbf{H}^T \mathbf{Hu}_2 = \sum_{i=1}^{\ell} |h_i|^2$. The above expression can be simplified to

$$\xi \triangleq \mathbf{x}^T \mathbf{H}(\mathbf{u}_1 - \mathbf{u}_2) < 0. \tag{6.121}$$

The decision variable ξ can be further written as

$$\xi = \sqrt{E_c}\mathbf{u}_1^T \mathbf{H}^T \mathbf{H}(\mathbf{u}_1 - \mathbf{u}_2) + \mathbf{n}^T \mathbf{H}(\mathbf{u}_1 - \mathbf{u}_2) < 0. \tag{6.122}$$

This is a binary decision problem whose error probability is determined by the SNR. Let d denote the Hamming distance between $\mathbf{u}_1$ and $\mathbf{u}_2$. The entries of $(\mathbf{u}_1 - \mathbf{u}_2)$ are zeros except at the d places that take values of 2 or -2; we denote the corresponding channel coefficients by $h_{(i)}$, $i = 1, \cdots, d$. Thus, we can find the SNR yielding

$$\gamma = \frac{E_c \left(\sum_{i=1}^{d} |h_{(i)}|^2 \right)^2}{\sigma_n^2 \sum_{i=1}^{d} |h_{(i)}|^2} = \frac{rE_b \sum_{i=1}^{d} |h_{(i)}|^2}{\sigma_n^2}, \tag{6.123}$$

which enables us to obtain the PEP

$$P_e = Q\left(\sqrt{\frac{rE_b \sum_{i=1}^{d} |h_{(i)}|^2}{\sigma_n^2}} \right), \tag{6.124}$$

with r denoting the code rate. By averaging over the channel gains $\{h_{(i)}\}$, we can determine the average PEP for various fading channels.

Property 6.7.5 *For BPSK signaling, the PEP is independent of the codewords.*

From the discussion above, coding brings in an increase in minimum distance and, thus, a better error performance. The power of a code is usually measured in terms of the reduction in E_b/N_0 required by a coded system to achieve a specific error probability, say 10^{-5}, as compared to its uncoded counterpart with the same modulation and channel conditions.

6.8   Trellis-coded modulation

Up to this point, we have treated modulation and coding separately as if there were no connection between them. This is also the traditional strategy to handle the two topics, where coding is simply viewed as a means to correct errors made in demodulation. In 1974, J.L. Massey suggested a new notion of considering modulation and coding as two aspects of a same entity [12]. In 1982, Ungerboeck implemented the thought of Messey by

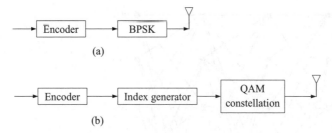

Figure 6.9 *Comparing the philosophy of (a) a conventional encoder and (b) TCM*

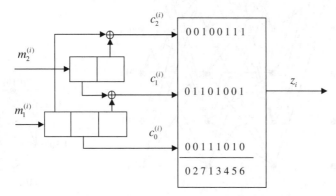

Figure 6.10 *TCM encoder with 8-PSK*

inventing trellis-coded modulation (TCM) [13]. In classical coding, a coded binary output is *directly* mapped into complex symbols for transmission. In TCM, a coded binary output is used as *indexes* for selecting the signal points from an *augmented* constellation. The philosophical difference between the two is illustrated in Figure 6.9. By augmented constellation, we mean a constellation with more signal points than necessary for modulation.

Consider Ungerboeck's 8-state 8PSK code of rate $\frac{1}{2}$, sketched in Figure 6.10. The encoder takes a two-bit input during the ith interval, which is denoted as

$$\mathbf{m}^{(i)} = [m_2^{(i)}, m_1^{(i)}],$$

while its coded output is organized in the order

$$\mathbf{c}^{(i)} = [c_2^{(i)}, c_1^{(1)}, c_0^{(0)}].$$

The redundancy of the TCM is reflected in the fact the coded symbols are selected from an augmented constellation. In this illustrating example, QPSK is sufficient for representing two-bit messages. Yet, an 8PSK constellation is used for TCM.

The trellis diagram of the TCM code is depicted in Figure 6.11, where the eight possible states of the encoder are listed on its rightmost column. Each of these states is determined by the contents of the two registers of the encoder in the following manner. Suppose the contents are given by

$$\boxed{m_2}\ \boxed{b_1}$$
$$\boxed{m_1}\ \boxed{a_2}\ \boxed{a_1}.$$

(6.125)

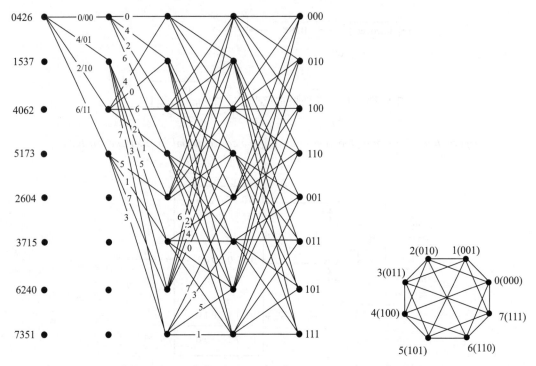

Figure 6.11   *Trellis diagram for the TCM encoder with 8PSK*

where $[m_1 m_2]$ is the current input to the encoder. The current state of the encoder is defined as

$$s_1 = [b_1 a_2 a_1].$$

Driven by input $[m_2 m_1]$, the encoder transits from its current state s_1 to the next state s_2, as shown by

$$\underbrace{[b_1 a_2 a_1]}_{s_1} \overset{m_2 m_1}{\rightarrow} \underbrace{[m_2 m_1 a_2]}_{s_2}.$$

In the leftmost column of Figure 6.11, each node is labeled with its four possible outcomes. For example, state 110 is labeled with 5173, meaning that with different inputs, the encoder in state 011 can produce four possible coded outputs, that is, 101, 001, 111, 011, which, in turn, guides the encoder to select the signal points 5, 1, 7, and 3 from the 8PSK constellation, respectively. There are four branches emanating from each node, which, in clockwise order, correspond to two-bit inputs 00, 01, 10, *and*11, respectively. For a more complete characterization, each branch is labeled with its corresponding encoder output $\mathbf{c}^{(i)}$ and input $\mathbf{m}^{(i)}$, in the form of $\mathbf{c}^{(i)}/\mathbf{m}^{(i)}$, where the superscript (i) denotes the ith interval. For example, the branch input/output of the third branch emanating from state 100 (4062) is easily found to be

$$[100] \overset{2/11}{\rightarrow} [110],$$

which simply means that, driven by input 11, the encoder transits from state 100 to state 110, producing coded output 010 that selects signal point 2.

✍ Example 6.21

Suppose that an input sequence is applied to the TCM encoder shown in Figures 6.10 and 6.11, in the order as sketched below.

$$01101011 \longrightarrow \boxed{\text{TCM Encoder}}$$

$$\text{new} \longrightarrow \text{old}$$

Assuming that the encoder is initialized with zeros, find the evolution of the state and the corresponding coded output phase sequence.

Solution

$$\boxed{000} \xrightarrow{6/11} \boxed{110} \xrightarrow{7/10} \boxed{101} \xrightarrow{4/10} \boxed{100} \xrightarrow{0/01} \boxed{010}$$

where boxes of three-bit content are used to signify the states. Let θ_i denote the phase corresponding to the ith signal point. Then, the phase evolution is

$$\theta_6 \rightarrow \theta_7 \rightarrow \theta_4 \rightarrow \theta_0.$$

It is observed from Figure 6.11 that there are a number of paths that start from an all-zero node and re-merge at another all-zero node after sketching over L_{rem} branches. We label these paths with their corresponding output coded phase sequences $\theta_1, \theta_2, \cdots, \theta_{L_{\text{rem}}}$ and the Hamming weight of their input sequences. The competing paths for each L_{rem} are tabulated in order.

1. $L_{\text{rem}} = 2$: Only one possible erroneous sequence, that is, 24.
2. $L_{\text{rem}} = 3$: five possible erroneous sequences. They are
   412(1), 436(2), 204(2), 652(2), 676(3).
3. $L_{\text{rem}} = 4$: 17 possible erroneous sequences. They are
   4104(2), 4532(2), 4516(3), 4344(3), 4772(3), 4756(4);
   2012(2), 2036(3), 2664(3), 2252(3), 2276(4);
   6504(3), 6132(3), 6116(4), 6744(4), 6372(4), 6356(5).

For example, notation 4104(2) signifies a four-branch path with coded phase sequence 4104 and source weight equal to 2. Among all the paths with different L_{rem}, path 676 has the minimum distance from its all-zero counterpart. To calculate their Euclidean distance, note that the separation between signal points θ_0 and θ_7 is equal to $d_{07} = 2R \sin \frac{\pi}{8}$, with R denoting the radius. Likewise, the distance between θ_0 and θ_6 can be determined as $d_{06} = \sqrt{2}\, R$. Hence, the squared free distance is

$$d_{\text{free}}^2 = (2 + 2 + 4\sin^2(\pi/8))R^2 = 4.585R^2. \tag{6.126}$$

In contrast, the uncoded free distance is defined by the separation between two adjacent points of QPSK, and is thus given by $d_0^2 = \sqrt{2}\, R$. Given that both the QPSK and 8PSK constellations have the same average energy, the coding gain can be directly calculated as

$$\mathcal{G} = \frac{d_{\text{free}}^2}{d_0^2} = 4.585/2 = 2.29 \ (3.60 \text{ dB}). \tag{6.127}$$

During trellis encoding, as the coded index sequence moves along a feasible path selected by the message input, the modulated signal jumps from one point to another on the augmented signal constellation following a trajectory guided by the coded indices. The Hamming metric in the coded indices is mapped to the Euclidean distance in the signal constellation. Although Hamming distance is of relevance, it is the Euclidean distance between signal points that eventually determines the system error performance. TCM offers a much better performance than directly transmitting a coded bit sequence via BPSK signaling. More importantly, such improvement is achieved without the need for bandwidth expansion and increased transmit power. The drawback of TCM lies in its decoding complexity.

6.9   Summary

In this chapter, we have studied various linear block codes and convolutional codes. Linear block codes have their structures defined in fields and rings, with a solid foundation in modern abstract algebra. Convolutional codes, on the other hand, are formed by finite impulse filters with or without feedback loops; yet their operations are also defined in finite fields. We further investigated the issue of trellis-coded modulation, with its code structures built on an augmented signal constellation of the relevant modulation scheme. Given that various structures have been used for channel encoding, an interesting question arises: is there any optimal structure for coding?

Shannon has demonstrated that channel capacity-approaching codes should be sufficiently long and as random as possible. From Shannon's demonstration, it seems that a good code has no structure, which, however, poses a difficulty in decoding. Thus, there is a balance between the structures of a code and its randomness. With a code structure in mind, good, long codes should have a distance distribution that minimizes the occurrence of low-weight codes rather than pursuing a large minimum distance. Channel coding is more an art than a science. A masterpiece relies on inspiration and novel philosophy.

Problems

Problem 6.1   In $GF(13)$, pursue the following issues.

(a) Evaluate 5^{841}.
(b) Determine the order of 3.
(c) Identify a primitive element.

Problem 6.2   Determine the product $c = a \cdot b$ of two elements $a = 1 + x^2$ and $b = 1 + x + x^2$ in the $GF(2^3)$ constructed by using the primitive polynomial $p(x) = 1 + x^2 + x^3$ in $GF(2)$.

Problem 6.3   Channel coding is implemented by inserting certain algebraic structures into a information sequence and, nearly without exception, structures of this type have their definition and operations in the Galois fields.

(a) Provide your understanding to justify the rationale of doing so.
(b) Does this happen coincidently because of a powerful tool available in the theory of abstract algebra that can be borrowed for the use in channel encoding, or is this structure itself indeed an optimal choice?
(c) Question for discussion: Is there other better structures or frameworks for coding? Describe your imagination and justification.

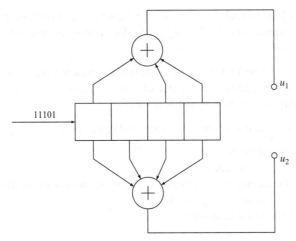

Figure 6.12 *Encoder for Problem 6.12. The oldest data goes first to the encoder*

Problem 6.4 Determine and compare the correcting capability of the three convolutional codes defined by

$$\text{code 1}: \begin{cases} g_1(D) = 1 + D + D^2 \\ g_2(D) = 1 + D^2 \end{cases}$$

$$\text{code 2}: \begin{cases} g_1(D) = 1 + D + D^2 \\ g_2(D) = D + D^2 \end{cases}$$

$$\text{code 3}: \begin{cases} g_1(D) = 1 + D + D^2 \\ g_2(D) = 1 + D \end{cases}$$

Problem 6.5 Suppose a primitive polynomial $x^4 + x + 1$ is used to generate a GF(16) with modulo-2 addition and multiplication. Determine the degrees of all non-zero elements in this field.

Problem 6.6 Suppose a (9, 5) systematic cyclic code is generated by using the primitive polynomial $p(x) = x^4 + x^3 + 1$. (1) List all of its single-bit error patterns. (2) Determine the codeword for message $\mathbf{m} = [11001]$.

Problem 6.7 A cyclic code is usually generated by a primitive polynomial, say $p(x)$. (1) What is the role played by the Galois field, which is defined by $p(x)$, in the code generation? (2) What would happen if the polynomial $g(x) = x^3 + 1$ is used to generate a (7, 4) code?

Problem 6.8 Refer to the encoder shown in Figure 6.12. Determine the encoder output in response to the input message sequence

$$\mathbf{m} = (1\,0\,1\,1\,1),$$

with the oldest bit shown on the leftmost.

Problem 6.9 Suppose an n-bit linear block code of t-bit error-correcting capability is transmitted over a binary symmetric channel with transit probability p. Show that the average bit error probability is given by

$$P_b \leq p - \frac{1}{n} \sum_{i=0}^{t} i \binom{n}{i} p^i (1-p)^{n-i}.$$

Problem 6.10 A $(24, 12)$ linear block code, capable of correcting double errors, is transmitted over an AWGN channel with noncoherent binary FSK signaling. Determine the average bit error probability if the channel SNR is $E_b/N_0 = 5\,\text{dB}$.

Problem 6.11 A GSM system employs a rate-$\frac{1}{2}$ 2^4-state convolutional code used in GSM with generator polynomials $g(D) = [31, 33]$. (1) Show that the free distance of this code is 7. (2) Determine the number of paths that have the Hamming weight of 7, 8, and 9, respectively.

Problem 6.12 A rate-$\frac{1}{2}$ convolutional encoder is shown in Figure 6.13.

(a) Draw the state diagram of the encoder.
(b) Draw the trellis diagram of the encoder.
(c) Suppose the received sequence is $\mathbf{z} = [10\ 11\ 00\ 01\ 11]$. Show how to use a hard-decision Viterbi algorithm to estimate the original message sequence.
(d) Find the transfer function of the encoder.

Problem 6.13 A rate-1/3 convolutional encoder is shown in Figure 6.14. (1) Draw its trellis diagram. (2) Determine the original message if the received sequence is 11000001110100.

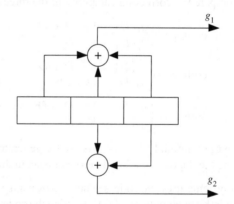

Figure 6.13   *Convolutional encoder for Problem 6.12*

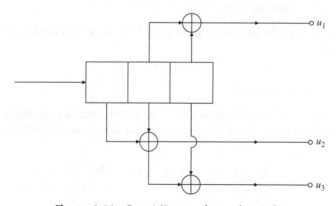

Figure 6.14   *Rate-1/3 convolutional encoder*

Table 6.2   *For Problem 6.14; elements in GF(2^4), which is generated by $P(x) = 1 + x + x^4$*

Power Rep.	Polynomial Rep.	Vector Rep.
0	0	(0000)
1	1	(1000)
α	α	(0100)
α^2	α^2	(0010)
α^3	α^3	(0001)
α^4	$1 + \alpha$	(1100)
α^5	$\alpha + \alpha^2$	(0110)
α^6	$\alpha^2 + \alpha^3$	(0011)
α^7	$1 + \alpha + \alpha^3$	(1101)
α^8	$1 + \alpha^2$	(1010)
α^9	$\alpha + \alpha^3$	(0101)
α^{10}	$1 + \alpha + \alpha^2$	(1110)
α^{11}	$\alpha + \alpha^2 + \alpha^3$	(0111)
α^{12}	$1 + \alpha + \alpha^2 + \alpha^3$	(1111)
α^{13}	$1 + \alpha^2 + \alpha^3$	(1011)
α^{14}	$1 + \alpha^3$	(1001)

Table 6.3   *For Problem 6.14: minimum polynomials in GF(2^4) generated by $P(x) = 1 + x + x^4$*

Conjugate roots	Minimum polynomial
0	x
1	$x + 1$
$\alpha, \alpha^2, \alpha^4, \alpha^8$	$x^4 + x + 1$
$\alpha^3, \alpha^6, \alpha^9, \alpha^{12}$	$x^4 + x^3 + x^2 + x + 1$
α^5, α^{10}	$x^2 + x + 1$
$\alpha^7, \alpha^{11}, \alpha^{13}, \alpha^{14}$	$x^4 + x^3 + 1$

Problem 6.14 (BCH code construction). Let α be a primitive element of GF(2^4) generated by the primitive polynomial $P(x) = 1 + x + x^4$. The elements in this field and their minimum polynomials are shown in Tables 6.2 and 6.3. Construct a BCH code of length 15 that can correct at most three errors. Determine the generator polynomial and the parity-check matrix.

Problem 6.15 (Code BCH construction). Find the generator polynomials of all the primitive BCH codes of length 15. Assume that the GF(2^4) is generated by the primitive polynomial $P(x) = 1 + x + x^4$.

Problem 6.16 (BCH decoding). Consider the primitive triple-error-correction BCH code constructed in Problem 6.14. Suppose that the received polynomial is $r(x) = x^3 + x^5 + x^{12}$. Decode $r(x)$.

Problem 6.17 Refer to the trellis coded modulator in Figure 6.11.

(a) Find the shortest path that starts from an all-zero node and re-merges at an all-zero node later and determine its distance from the all-zero path.

(b) Suppose a binary input sequence is applied to the encoder, as shown by

1000111001

where the leftmost entry represents the oldest data. Determine coded phase output sequence if the encoder is initialized with zeros.

References

1. C.E. Shannon, "A mathematical theory of communication," *Bell Syst. Tech. J.*, vol. 27, pp. 379–423 (Part I), pp. 623–656 (Part II), 1948.
2. R.W. Hamming, "Error detecting and error correcting codes," *Bell Syst. Tech. J.*, vol. 29, pp. 147–160, 1950.
3. A. Hocquenghem, "Codes correcteurs d'erreurs," *Chiffres*, vol. 2, pp. 147–156, 1959.
4. R.C. Bose and D.K. Ray-Chandhuri, "On a class of error correcting binary group codes," *Inf. Control*, vol. 3, pp. 68–79, 1960.
5. I.S. Reed and G. Solomon, "Polynomial codes over certain finite fields," *J. Soc. Ind. Appl. Math.*, vol. 8, pp. 300–304, 1960.
6. P. Elias, "Coding for noisy channels," *IRE Conv. Rec.*, Part 4, pp. 37–47, 1955.
7. S. Lin and D.J. Costello, Error Control Coding, Upper-Saddle River, NJ: Prentice-Hall, 2004.
8. W.W. Peterson, "Encoding and error-correction procedures for the Bose-Chaudhuri codes," *IRE Trans. Inf. Theory*, vol. 6, pp. 459–470, 1960.
9. J.G. Proakis, *Digital Communications*, 3rd ed., New York: McGraw-Hill, 1996.
10. B. Sklar, *Digital Communications: Fundamentals and Applications*, 2nd ed., Upper Saddle River, NJ: Prentice Hall, 2001.
11. R. Johannesson and P. Stahl, "New rate 1/2, 1/3, and 1/4 binary covolutional encoders with optimal distance profile," *IEEE Trans. Inf. Theory*, vol. 45, no. 5, pp. 1653–1658, 1999.
12. J.L. Massey, "Coding and modulation in digital communication," *Proceedings of 1974 International Zurich Seminar on Digital Communication*, Zurich, Switzerland, pp. E2(1)–E2(4), March 1974.
13. G. Ungerboeck, "Trellis-coded modulation with redundant signal sets, part I: introduction," *IEEE Commun. Mag.*, vol. 25, no. 2, pp. 5–21, 1987.
14. F.J. MacWilliamsQ1 and N.J.A. Sloane, *The Theory of Error-Correcting Codes*, Amsterdam, The Netherlands: North-Holland, 1977.
15. G.D. Forney Jr., "The Viterbi algorithm," *Proc. IEEE*, vol. 61, pp. 268–278, 1973.
16. G.D. Forney Jr., *Concatenated Codes*, Cambridge, MA: MIT Press, 1966.
17. R.G. Gallager, "Low density parity check codes," *IRE Trans. Inf. Theory*, vol. 8, pp. 21–28, 1962.
18. M. Zivkovic, "A table of primitive binary polynomials," *website by Google*, 1985.
19. M. Zivkovic, "Table of primitive binary polynomials: II," *Math. Comput.*, vol. 63, no. 207, pp. 301–306, 1994.
20. G. Ungerboeck, "Channel coding with multilevel phase signals," *IEEE Trans. Inf. Theory*, vol. 16, no. 1, pp. 55–67, 1982.
21. M. Cedervall and R. Johannesson, "A fast algorithm for computing distance spectrum of convolutional codes," *IEEE Trans. Inf. Theory*, vol. 35, no. 6, pp. 1146–1159, 1989.

7

Diversity Techniques

Abstract

The error probability of coherent communication systems in AWGN drops exponentially with increased signal-to-noise ratio, γ, in accordance with the Q-function $Q(\sqrt{\gamma}\,)$. When operating in a random field, multipath fading randomizes γ, thereby considerably degrading the average error performance. Indeed, a random field has a mechanism to damage the transmission reliability. Yet, there must be another mechanism therein for its solution. A possible remedy is to acquire multiple copies of the transmitted signal to smoothen its randomness. This can be done, for example, by placing a number of antennas over the random field. Such multiple copies can be combined in different ways, resulting in various diversity combining techniques, such as maximum ratio combining, equal-gain combining, selection combining, and post-detection combining. In this chapter, the reader is exposed to various diversity techniques and related mathematical skills used thereto.

♣

7.1 Introduction

For the motivation of diversity techniques, let us examine the influence of multipath fading on transmission reliability. When a signal is transmitted over a randomly scattering field, the received signal is a superposition of phased components arriving from different random paths. The consequence of the phaser summation is the random fluctuation in the received signal amplitude, usually called *fading*, as described in Chapter 3. Depending on the signal bandwidth relative to the channel coherent bandwidth, fading channels can be classified into two categories: frequency-flat fading and frequency-selective fading. In this chapter, we focus on flat fading channels, which are characterized by a single random channel gain.

To understand the influence of random fading gain, let us compare the error performance of an MPSK system in AWGN and in flat Rayleigh fading. In its baseband form, the received signal over the kth symbol interval is given by

$$x_k = \begin{cases} \sqrt{E_s}\, e^{j\phi_k} + n_k, & \text{AWGN} \\ g\sqrt{E_s}\, e^{j\phi_k} + n_k, & \text{Flat fading} \end{cases}, \tag{7.1}$$

Wireless Communications: Principles, Theory and Methodology, First Edition. Keith Q.T. Zhang.
© 2016 John Wiley & Sons, Ltd. Published 2016 by John Wiley & Sons, Ltd.
Companion Website: www.wiley.com/go/zhang7749

where ϕ_k denotes the MPSK symbol, E_s denotes the symbol energy, and $n_k \sim \mathcal{CN}(0, N_0/2)$ is the AWGN component with N_0 denoting its one-sided power spectral density. For Rayleigh fading, the channel gain g varies according to $g \sim \mathcal{CN}(0, \sigma_g^2)$. Throughout this chapter, we denote the average transmit SNR by ρ_s, and denote the average received SNR by $\overline{\gamma}$ or ρ_r. For both channels, ρ_s is the same given by $\rho_s = E_s/N_0$. However, ρ_r is different for the two channels. In particular, $\rho_r = \rho_s$ for the AWGN channel, whereas $\rho_r = \sigma_g^2 \rho_s$ for the flat fading channel. For a fair comparison of the error performance over the two channels, we assume $\sigma_g^2 = 1$.

From studies in Chapter 4, the error performance of various modulation schemes in AWGN takes the form $P_e = a\, Q(\sqrt{b\rho_s})$, which is exact for BPSK and very accurate for MPSK and square M-QAM over moderate or large SNR regimes. The constants a and b depend on the modulation scheme in use. For a flat fading channel, the influence of the channel gain must be taken into account. The conditional error performance depends on the channel gain g and is given by $P_e = a \cdot Q(\sqrt{b|g|^2 \rho_s})$, which, when averaged over the Gaussian-distributed fading gain g_k, enables us to determine the *average* probability of error

$$\overline{P}_e = \frac{a}{2}\left(1 - \sqrt{\frac{b\rho_r}{2 + b\rho_r}}\right) \tag{7.2}$$

as proved subsequently in Section 7.7.1. For coherent BPSK, the bit error rate (BER) can be further written as

$$\mathrm{BER} = \begin{cases} Q(\sqrt{2\rho_r}), & \text{AWGN} \\ \frac{1}{2}\left(1 - \sqrt{\frac{\rho_r}{1+\rho_r}}\right), & \text{Rayleigh} \end{cases} \tag{7.3}$$

Here, we deliberately denote the error performance in the two channels by the same notation BER for comparison. We are in particular concerned with their behavior over the large SNR regime, which is of practical interest. After a simple calculation, their slope for large SNRs can be shown to be

$$\frac{d\,\mathrm{BER}}{d\,\rho_r} = \begin{cases} -\frac{1}{\sqrt{4\pi\rho_r}}e^{-\rho_r}, & \text{AWGN} \\ -\frac{1}{4\rho_r}, & \text{Rayleigh} \end{cases} \tag{7.4}$$

The conclusion is that, in AWGN channels, the probability of bit error drops *exponentially* with increase in average SNR ρ_r. Its counterpart for Rayleigh fading channels drops only *linearly* with the average SNR. To be more intuitive, when a coherent BPSK system operates in AWGN, it only needs an SNR of 6.7 dB to achieve the error performance of BER $= 10^{-3}$. When operating in flat Rayleigh fading, however, it requires an average SNR of 24 dB to maintain the same performance.

Multipath scattering constitutes a random field. With multipath fading, the channel conditions can deteriorate into unreliable reception of digital signals. In random fields, there exists a mechanism to damage reliable transmission and, yet, there must be another mechanism for remedy. Random motion of mobile units tends to spatially decorrelate the received signal, making it possible to acquire multiple independent copies of the same transmitted signal by placing a number of antennas over spatially well-separated spots in the random field. Combining these independent signal copies through an appropriate *diversity* technique, we can significantly enhance the detection performance.

We use the following symbols throughout this chapter.

E_s	Symbol energy
N_0	One-sided power spectral density in watt/Hz
$N_0/2$	Two-sided noise power spectral density, used interchangeably with σ_n^2
$\rho_s = E_s/N_0$	Average transmit signal to noise ratio (SNR)
g_i	Channel gain at branch i

$\alpha_i = |g_i|$ Fading-gain magnitude at branch i
$\Omega_i = \alpha_i^2 E_s$ Received symbol energy
$\gamma_i = \alpha_i^2 E_s / N_0$ Instantaneous SNR at branch i
γ SNR for a single branch system
ρ_i or $\overline{\gamma}_i$ Denotes $\mathbb{E}[\gamma_i]$, that is, the average received SNR at branch i
$\overline{\gamma}_o$ Equal average branch SNR for i.i.d. systems; that is, $\mathbb{E}[\gamma_i] = \overline{\gamma}_o, \forall i$
$\overline{\gamma}_{MRC}, \overline{\gamma}_{EGC}, \overline{\gamma}_{SC}$ Average output SNR for MRC, EGC, and SC, respectively
σ_n^2 The noise variance at each receive antenna

7.2   Idea behind diversity

Multipath fading has a profound impact on the transmission reliability. As studied in the preceding chapters, the error performance of a coherent communication system in an AWGN channel takes the form of the Q-function, as shown by $P_e = a\, Q(\sqrt{b\gamma})$. In AWGN channels, the SNR γ is a fixed number. With multipath fading, however, γ is a random variable and, theoretically, fluctuates between zero and infinity. Thus, γ can be very small, though with a small probability. It is the events of small γ that damage the system reliability. The idea to overcome the fading effect in γ is to acquire several, say m, independent or weakly correlated samples of γ and combine them in some way to form a diversity system. The reason is simple; for a sufficiently large m, the likelihood of all these samples *simultaneously dropping* to small values is negligible.

For a more intuitive understanding of diversity combining, let us show how multiple samples can ensure a more stable SNR. Consider m i.i.d. samples $\gamma_i, i = 1, \cdots, m$ acquired from independent Rayleigh fading channels. Each instantaneous SNR, up to a scaling factor, is distributed as $\gamma_i \sim \chi^2(2)$. An effective scheme for a more stable SNR is to combine these samples to form

$$\gamma_c = \gamma_1 + \cdots + \gamma_m \sim \chi^2(2m),$$

which, according to Chapter 2, follows the $\chi^2(2m)$ distribution with mean $2m$ and variance $4m$. As such,

$$\frac{\gamma_c}{2m} \to 1 + O(m^{-1}),$$

implying that the normalized chi-square variable is "frozen" to a fixed number as m increases, thereby removing the effect of multipath fluctuation. This is exactly the concept of diversity combining. Though introduced here through a Rayleigh fading channel and a special combining scheme, the principle of diversity combining is generally applicable.

Diversity combining can be carried out in the space, time, frequency, and polarization domains, leading to space diversity, time diversity, frequency diversity, and so on. Channel coding can be also considered a form of temporal diversity, and is, thus, sometime called *code diversity*. In this chapter, without loss of generality, we expound the principle using space diversity and, in particular, concentrate on diversity reception with multiple antennas. There are three typical ways to combine the signals from different branches, as listed below:

- selection combining (SC),
- maximum ratio combining (MRC),
- equal-gain combining (EGC).

The schemes MRC and EGC require the knowledge of channel phase information and, thus, are used along with coherent detection. On the contrary, SC is usually used along with noncoherent detection to take advantage of implementation simplicity. Two metrics are commonly used to evaluate the performance improvement of a combining scheme over the conventional nondiversity receiver; they are the improvement factor in average SNR, and error performance enhancement.

7.3   Structures of various diversity combiners

For ease of illustration, consider the use of L antennas to receive an MPSK signal through flat fading channels. The received signal at the kth antenna, where $k = 1, \cdots, L$, can be represented as

$$r_k(t) = \alpha_k A \cos[\omega_c(t - \tau_k) + \phi_m + \theta_k] + n_k(t), \ 0 \leq t < T_s, \tag{7.5}$$

where A is the transmit signal amplitude, T_s is the symbol duration, and ϕ_m is the information symbol. The fading channel is characterized by the fading magnitude α_k, uniformly distributed random phase θ_k, and time delay τ_k. The transmit symbol energy is therefore equal to $E_s = A^2 T_s/2$. The received signal is further corrupted by the AWGN $n_k(t)$ with mean zero and variance $N_0/2$. Assume that both α_k and θ_k remain unchanged over one symbol period. The first step in coherent diversity reception is the time alignment among all the received signals at the L receive antennas according to $\tau_{\max} = \max\{\tau_1, \cdots, \tau_L\}$. The aligned signal $r_k(t)$ is then applied to the quadratic receiver, producing outputs from its I and Q branches which, when combined, produces a single complex variable

$$x_k = g_k \sqrt{E_s} \, e^{j\phi_m} + e_k, \tag{7.6}$$

where $g_k = \alpha_k \, e^{j\theta_k}$ is the complex channel gain of branch k. Throughout this chapter, the magnitude of channel gain g_k is denoted as $|g_k|$ or α_k, interchangely. The noise component e_k now consists of two independent I/Q channels, thus doubling its variance to give $e_k \sim \mathcal{CN}(0, N_0)$. For a more general baseband modulated signal, we can replace $e^{j\phi_m}$ with a complex symbol $s_m = a + jb$. It is easy to check that the *instantaneous* SNR remains the same no matter whether evaluated using the real model $r_k(t)$ or its complex counterpart x_k, and is given by

$$\gamma_k = \frac{\alpha_k^2 E_s}{N_0}. \tag{7.7}$$

Different branches may have a different noise variance; but they can be calibrated to have the same variance. Given L noisy copies of the transmitted signal ϕ_m, we may linearly combine them with weighting factors $w_k, k = 1, \cdots, L$ to produce a combiner's output

$$x = \sum_{k=1}^{L} w_k^* x_k = \sqrt{E_s} \mathbf{w}^\dagger \mathbf{g} \, e^{j\phi_m} + \mathbf{w}^\dagger \mathbf{e}, \tag{7.8}$$

where the superscripts $*$, T, and $\dagger$ signify complex conjugate, transpose, and Hermitian transpose, respectively, and the vectors $\mathbf{g}$, $\mathbf{w}$, and $\mathbf{e}$ are defined by

$$\mathbf{g} = [g_1, \cdots, g_L]^T, \ \mathbf{w} = [w_1, \cdots, w_L]^T, \ \mathbf{e} = [e_1, \cdots, e_L]^T. \tag{7.9}$$

The choice of w's depends on the combining scheme to be used. See Figure 7.1 for illustration.

Besides linear diversity combining, there are nonlinear combining techniques such as post-detection combining, which eliminate the need of the channel state information at the receiver and are, thus, widely used along with differential MPSK and noncoherent MFSK. In what follows, we will first investigate linear combining, and revisit noncoherent combining schemes later.

7.3.1   MRC

Among all possible linear combiners given in (7.8), the most natural way is to choose the weighting vector $\mathbf{w}$ such that the instantaneous output SNR

$$\gamma = \frac{E_s |\mathbf{w}^\dagger \mathbf{g}|^2}{\mathbb{E}|\mathbf{w}^\dagger \mathbf{e}|^2} \tag{7.10}$$

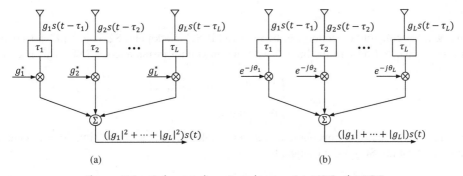

Figure 7.1 *Coherent diversity schemes. (a) MRC. (b) EGC*

is maximized. We can directly calculate the expectation in the denominator to obtain $\mathbb{E}|\mathbf{w}^\dagger\mathbf{e}|^2 = (\mathbf{w}^\dagger\mathbf{w})N_0$, and then employ the Schwartz inequality to upper-bound the numerator, yielding

$$\gamma \le \frac{E_s(\mathbf{w}^\dagger\mathbf{w})(\mathbf{g}^\dagger\mathbf{g})}{N_0(\mathbf{w}^\dagger\mathbf{w})} = \underbrace{(E_s/N_0)}_{\rho_s}\mathbf{g}^\dagger\mathbf{g}. \tag{7.11}$$

The maximum value is attained when $\mathbf{w}$ is proportional to $\mathbf{g}$. The resulting combiner is known as the maximum ratio combiner (MRC). Without loss of generality, we choose $\mathbf{w}_{\text{opt}} = \mathbf{g}$ for simplicity. The operation $\mathbf{g}^\dagger\mathbf{g}$ implies that the receiver perfectly knows the gains of the channels, and performs two operations: that is, co-phasing among different branches, and matching the magnitudes of their channel gains for the maximal instantaneous SNR at the MRC output. In summary, there are four steps to implement the MRC: temporal alignment, co-phasing, channel gain matching, and combining. Combining (7.7) and (7.11) enables the assertion

$$\gamma_{\text{MRC}} = \gamma_1 + \cdots + \gamma_L. \tag{7.12}$$

Namely, the instantaneous output SNR of the MRC is equal to the sum of instantaneous SNRs at all of its input branches.

7.3.2 EGC

To acquire the channel gains $\{g_k = \alpha_k \exp(\theta_k)\}$, the diversity combining system needs to estimate their magnitudes and phases. Phase estimation is relatively easy to implement, by simply using a phase-locked loop. Therefore, the receiver structure can be considerably simplified if the combiner's tap weights only employ the channel phase information but setting their magnitudes to an arbitrary real constant $a > 0$, so that we can write

$$w_k = a\,e^{j\theta_k},\ k = 1, \cdots, L. \tag{7.13}$$

The term EGC is derived from the fact that all the weights have the same magnitude equal to a. It is straightforward to determine the output instantaneous SNR of the EGC; the result is

$$\gamma_{\text{EGC}} = \frac{E_s(\alpha_1 + \cdots + \alpha_L)^2}{\mathbb{E}[|e_1|^2 + \cdots + |e_L|^2]} = \frac{E_s}{LN_0}(\alpha_1 + \cdots + \alpha_L)^2. \tag{7.14}$$

Without the need for channel-magnitude information, EGC takes three steps for its implementation: temporal alignment, co-phasing, and combining.

7.3.3   SC

The requirement for coherent detection can be removed if SC is adopted. SC is just like a switch, with its output terminal connected only to the branch with the maximum instantaneous SNR. This is equivalent to setting the weights to

$$w_k = \begin{cases} 1, & \text{if } k = \arg\max_i\{\gamma_i\} \\ 0, & \text{otherwise} \end{cases}.$$

(7.15)

An SC is usually followed by a noncoherent receiver to take advantage of the simplicity in detection. The instantaneous SNR of the SC output is equal to

$$\gamma_{\text{SC}} = \max\{\gamma_1, \cdots, \gamma_L\}.$$

(7.16)

7.4   PDFs of output SNR

The PDF of output SNR, or equivalently its CHF, is necessary for the performance evaluation of a diversity combiner for the detection of a digitally modulated signal in a fading environment. Fading channels can follow various possible distributions. Because of space limitation, we only focus on some commonly encountered fading environments.

7.4.1   MRC

The output SNR of MRC has been derived in (7.11). Its PDF is environment-dependent. We identify three cases, as addressed below.

7.4.1.1   *Rayleigh fading*

For Rayleigh fading, the channel gains are jointly Gaussian distributed, as shown by $\mathbf{g} \sim \mathcal{CN}(\mathbf{0}, \mathbf{R}_g)$. Let us begin with the simplest case of i.i.d. Rayleigh fading so that $\mathbf{R}_g = \sigma_g^2 \mathbf{I}$. We rewrite the instantaneous SNR at branch in (7.7) as

$$\gamma_i = \rho_s |g_i|^2 = \frac{\rho_s \sigma_g^2}{2} \underbrace{(2\sigma_g^{-2}|g_i|^2)}_{\chi^2(2)} = \frac{\rho_r}{2}\chi^2(2).$$

(7.17)

Since g_i is a complex Gaussian variable with two dimensions, each of variance $\sigma_g^2/2$, the product inside the parentheses is the power summation of two independent normalized Gaussian variables and, hence, follows the $\chi^2(2)$ distribution. From statistical theory, it is well known that the mean and variance of $\chi^2(\nu)$ are ν and 2ν, respectively. With these properties in mind, it is easy to check $\rho_r = \mathbb{E}[\gamma_i]$, which represents the average SNR at each branch. It is also known that the sum of independent chi-square variates with *equal* weights is still a chi-square whose degrees of freedom (DFs) equal the sum of their individual DFs. This is the semiclosure property of chi-square distributions. As such, we have

$$\gamma_{\text{MRC}} = \frac{\rho_r}{2}[\chi_1^2(2) + \cdots + \chi_L^2(2)] \sim \frac{\rho_r}{2}\chi^2(2L).$$

(7.18)

Its PDF, when explicitly expressed, is given by

$$f_{\gamma_{\text{MRC}}}(x) = \frac{1}{(L-1)!\rho_r}\left(\frac{x}{\rho_r}\right)^{L-1} e^{-x/\rho_r}.$$

(7.19)

This is the PDF of the order-L MRC with i.i.d. Rayleigh fading branches.

Next, we turn to the general Rayleigh fading with an arbitrary Hermitian channel covariance matrix $\mathbf{R}_y$, as encountered in many practical diversity systems with limited antenna spacing, electromagnetic coupling, or nonhomogenous scattering. In this general case, it is more convenient to characterize γ_{MRC} in the CHF domain. By definition, we can write

$$\phi_{\gamma_{\mathrm{MRC}}}(z) = \mathbb{E}[e^{z\gamma_{\mathrm{MRC}}}] = \det(\mathbf{I} - z\rho_s\mathbf{R}_g)^{-1}. \tag{7.20}$$

With this CHF, we can explicitly represent the PDF of γ_{MRC}. For ease of illustration, we assume that the rank of $\mathbf{R}_g$ is $L_0 \leq L$ with distinct eigenvalues $\lambda_i, i = 1, \cdots, L_0$. We can, thus, further represent it in fractional form

$$\phi_{\gamma_{\mathrm{MRC}}}(z) = \prod_{i=1}^{L_0} (1 - z\rho_s\lambda_i)^{-1} = \sum_{i=1}^{L_0} \frac{a_i}{1 - z\rho_s\lambda_i}, \tag{7.21}$$

with coefficients a' determined by

$$a_k = \prod_{i=1, i\neq k}^{L_0} (1 - z\rho_s\lambda_i)^{-1}\big|_{z=(\rho_s\lambda_k)^{-1}}. \tag{7.22}$$

Note that each term in the above CHF corresponds to a $\chi^2(2)$ distribution, which, after inverse Fourier transform, leads to an explicit PDF

$$f_{\gamma_{\mathrm{MRC}}}(x) = \sum_{i=1}^{L_0} \frac{1}{\rho_s\lambda_i} e^{-x/(\rho_s\lambda_i)}. \tag{7.23}$$

In deriving the PDF (7.23), we assumed $\mathbf{R}_g$ of distinct eigenvalues. For a general case with eigenvalues of higher order, the treatment is similar but needs the use of the residue theorem.

7.4.1.2 Rician fading

For correlated Rician fading, the channel gain vector $\mathbf{g} \sim \mathcal{CN}(\mathbf{u}_g, \mathbf{R}_g)$. Correspondingly, γ_{MRC} has the CHF

$$\phi_{\gamma_{\mathrm{MRC}}}(z) = \mathbb{E}[e^{z\gamma_{\mathrm{MRC}}}] = \det(\mathbf{I} - z\mathbf{R}_g)^{-1} \exp(\mathbf{u}_g^\dagger[z^{-1}\mathbf{I} - \mathbf{R}_g]^{-1}\mathbf{u}_g). \tag{7.24}$$

Unlike the case of Rayleigh fading, there is no explicit expression for the PDF of a generic γ_{MRC} in Rician fading. However, the PDF of γ_{MRC} can be easily determined if all the branches suffer i.i.d. Rician fading with each following the distribution $g_i \sim \mathcal{CN}(\mu, \sigma_g^2)$. In this case, we directly obtain

$$\gamma_{\mathrm{MRC}} = \rho_s(|g_1|^2 + \cdots + |g_L|^2) \sim \frac{\rho_s\sigma_g^2}{2}\chi^2(2L, \lambda), \tag{7.25}$$

where the non-centrality parameter is given by $\lambda = 2L|\mu|^2/\sigma_g^2$.

7.4.1.3 Nakagami fading

For Nakagami fading with fading parameter m, the CHF is given by

$$\phi_\gamma(s) = \det(\mathbf{I} - s\mathbf{A})^{-m}, \tag{7.26}$$

where $A(i, j) = \sqrt{R_g(i, j)}$.

7.4.2 EGC

Generally speaking, EGC is analytically difficult to handle due to the lack of appropriate tools in mathematics. As a consequence, there is no generic closed-form solution for the PDF of γ_{EGC}. We therefore just focus on the case of independent branches. For the performance analysis of EGC, it is easier to consider the square root of γ_{EGC} than γ_{EGC} itself, although they are equivalent. It follows from (7.14) that the square-root SNR of EGC is given by

$$\zeta = \sqrt{\gamma_{\text{EGC}}} = \sqrt{\rho_s/L}\,(\alpha_1 + \cdots + \alpha_L), \tag{7.27}$$

where $\rho_s = E_s/N_0$. By definition, the CHF of ζ is given by $\phi_\zeta(\omega) = \mathrm{E}[e^{j\omega\zeta}]$. Thus, it can be obtained from the joint CHF of individual branches. To proceed, we assume independent Nakagami fading branches with $\alpha_i \sim \mathcal{NK}(m_i, \sigma_i^2)$, where m_i and σ_i^2 denote their fading parameters and average power, respectively. For details of Nakagami distribution, the reader is referred to Chapter 2. By evaluating the cosine and sine transforms of the PDF of α_i, respectively, and using formulas (14) and (19) of [1] for their simplification, we obtain the CHF of α_i:

$$\phi_{\alpha_i}(z) = {}_1F_1\left(m_i, \frac{1}{2}; \frac{\sigma_i^2 z^2}{4m_i}\right) + z\frac{\Gamma(m_i + \frac{1}{2})}{\Gamma(m_i)}\sqrt{\frac{\sigma_i^2}{m_i}}\,{}_1F_1\left(m_i + \frac{1}{2}; \frac{3}{2}; \frac{\sigma_i^2 z^2}{4m_i}\right). \tag{7.28}$$

Thus, replacing z with $j\omega\sqrt{\rho_s/L}$, denoting

$$\eta_i = \frac{\sigma_i^2 \rho_s}{Lm_i} = \frac{\overline{\gamma}_i}{Lm_i}, \tag{7.29}$$

and using the independent branches assumption, we can evaluate the CHF of ζ, which is defined by $\phi_\zeta(\omega) = \mathrm{E}[e^{j\omega\zeta}]$, yielding

$$\phi_\zeta(\omega) = \prod_{i=1}^{L}\left[{}_1F_1\left(m_i; \frac{1}{2}; -\frac{\eta_i\omega^2}{4}\right) + j\omega\sqrt{\eta_i}\,\frac{\Gamma(m_i + \frac{1}{2})}{\Gamma(m_i)}\right.$$
$$\left. \times\,{}_1F_1\left(m_i + \frac{1}{2}; \frac{3}{2}; -\frac{\eta_i\omega^2}{4}\right)\right]. \tag{7.30}$$

For the special case $m = 1$, the above expression reduces to the joint CHF for EGC in independent Raleigh fading, as shown by

$$\phi_\zeta(\omega) = \prod_{i=1}^{L}\left[{}_1F_1\left(1; \frac{1}{2}; -\frac{\overline{\gamma}_i\omega^2}{4}\right) + j\omega\sqrt{\frac{\pi\overline{\gamma}_i}{4}}\,\exp\left(-\frac{\overline{\gamma}_i\omega^2}{4}\right)\right]. \tag{7.31}$$

7.4.3 SC

We next show how to obtain the PDF of γ_{SC}. For simplicity, we assume, again, independent Rayleigh fading channels with average received SNR $\{\rho_i\}$. The cumulative distribution of an order-L SC output SNR is then equal to

$$F_{\gamma_{\text{SC}}}(x) = \Pr\{\cap_{\ell=1}^{L}(\gamma_\ell < x)\}$$
$$= \prod_{\ell=1}^{L}\Pr(\gamma_\ell < x)$$
$$= \prod_{\ell=1}^{L}(1 - e^{-x/\rho_i}). \tag{7.32}$$

For i.i.d. channels, with equal average received SNR ρ_r, the above expression is reducible to

$$F_{\gamma_{\mathrm{SC}}}(x) = (1 - e^{-x/\rho_r})^L, \tag{7.33}$$

whereby the PDF can be determined as

$$\begin{aligned} f_{\gamma_{\mathrm{SC}}}(x) &= \frac{dF_{\gamma_{\mathrm{SC}}}(x)}{d\,x} \\ &= \frac{L}{\rho_r}(1 - e^{-x/\rho_r})^{L-1} e^{-x/\rho_r}. \end{aligned} \tag{7.34}$$

For generalized SC, we need a joint PDF of m largest ordered variables. The relevant PDFs can be found in Refs [2, 3]. Generalized SC always combines the branches with the largest instantaneous SNRs and, thus, has a much better performance. The cost is the need for coherent detection.

7.5 Average SNR comparison for various schemes

Given the randomness of multipath fading channels, the output SNRs obtained above are random variables. It is therefore more intuitive to compare their ensemble average. For illustration, we assume that all the branches are subject to Nakagami fading, as shown by $|g_i| \sim \mathcal{NK}(m_i, \sigma_i^2)$, and further assume that the AWGN component at each branch has the same variance, equal to $\sigma_n^2 = N_0$. Then, the instantaneous output SNR at the three combiners can be written as

$$\begin{aligned} \gamma_{\mathrm{MRC}} &= \frac{E_s}{N_0}(|g_1|^2 + \cdots + |g_L|^2), \\ \gamma_{\mathrm{EGC}} &= \frac{E_s(|g_1| + \cdots + |g_L|)^2}{LN_0}, \\ \gamma_{\mathrm{SC}} &= \frac{E_s}{N_0}\max\{|g_1|^2, \cdots, |g_L|^2\}. \end{aligned} \tag{7.35}$$

Let us show how to determine the mean value of these instantaneous output SNRs.

7.5.1 MRC

From Chapter 2, the assumption $|g_i| \sim \mathcal{NK}(m_i, \sigma_i^2)$ implies that

$$|g_i|^2 \sim \frac{\sigma_i^2}{2m_i}\chi^2(2m_i),$$

which, when used along with the property of $\mathbb{E}[\chi^2(m)] = m$ and the fact that $E_s\sigma_i^2/N_0$ is the average SNR $\overline{\gamma}_i$ at branch i, enables the assertion

$$\mathbb{E}[\gamma_{\mathrm{MRC}}] = \frac{E_s}{N_0}\sum_{i=1}^{L}\mathbb{E}[|g_i|^2] = (\overline{\gamma}_1 + \cdots + \overline{\gamma}_L). \tag{7.36}$$

In other words, the average output SNR of MRC is equal to the sum of individual SNRs at all of its branches. Note that the conclusion is derived without the need of independent branches assumption. For the special case with i.i.d. branches, all the branches have the same average SNR, given by $\bar{\gamma}_1 = \cdots = \bar{\gamma}_L = \bar{\gamma}_o$. We have $\mathbb{E}[\gamma_{\mathrm{MRC}}] = L\bar{\gamma}_o$ and the SNR gain of MRC is then equal to

$$\mathbb{E}[\gamma_{\mathrm{MRC}}]/\bar{\gamma}_o = L \tag{7.37}$$

which is proportional to the diversity order L. For MRC, correlated and uncorrelated branches lead to the same results of average SNR.

7.5.2   EGC

Next we consider EGC with *independent* Nakagami fading branches $|g_i| \sim \mathcal{NK}(m_i, \sigma_i^2)$, since the general case with correlated Nakagami fading branches is difficult to handle. From Chapter 2, we know that $|g_i| \sim \frac{\sigma_i}{\sqrt{2m_i}}\chi(2m_i)$, with the first two moments given by

$$\mathbb{E}[|g_i|] = \frac{\sigma_i}{\sqrt{m_i}} \frac{\Gamma(m_i + \frac{1}{2})}{\Gamma(m_i)}, \text{ and } \mathbb{E}[|g_i|^2] = \sigma_i^2. \tag{7.38}$$

With these properties, we are in position to determine the average SNR gain of the EGC. To this end, expanding the total EGC output power

$$\xi \overset{\Delta}{=} (|g_1| + \cdots + |g_L|)^2 = \sum_{i=1}^{L} |g_i|^2 + \sum_{i=1}^{L} \sum_{j=1, j\neq i}^{L} |g_i||g_j|, \tag{7.39}$$

and using the independent branch assumption, we can take the expectation term by term to obtain

$$\mathbb{E}[\xi] = \sum_{i=1}^{L} \sigma_i^2 + \sum_{i=1}^{L} \sum_{j=1, j\neq i}^{L} \frac{\sigma_i \sigma_j}{\sqrt{m_i m_j}} \frac{\Gamma(m_i + \frac{1}{2})\Gamma(m_j + \frac{1}{2})}{\Gamma(m_i)\Gamma(m_j)}. \tag{7.40}$$

Inserting this result into the second line of (7.35) and noting that $E_s \sigma_i^2/N_0 = \bar{\gamma}_i$, we obtain

$$\mathbb{E}[\gamma_{\mathrm{EGC}}] = \frac{1}{L} \sum_{i=1}^{L} \bar{\gamma}_i + \frac{1}{L} \sum_{i=1}^{L} \sum_{j=1, j\neq i}^{L} \sqrt{\frac{\bar{\gamma}_i \bar{\gamma}_j}{m_i m_j}} \frac{\Gamma(m_i + \frac{1}{2})\Gamma(m_j + \frac{1}{2})}{\Gamma(m_i)\Gamma(m_j)}. \tag{7.41}$$

This is the general result for EGC in Nakagami fading with independent branches.

Some special cases are of practical interest. In particular, for the case of i.i.d. Nakagami fading branches with $m_1 = \cdots = m_L = m$ and $\bar{\gamma}_1^2 = \cdots = \bar{\gamma}_L^2 = \bar{\gamma}_o$, the above expression reduces to

$$\mathbb{E}[\gamma_{\mathrm{EGC}}] = \bar{\gamma}_o \left[1 + \frac{(L-1)}{m} \frac{\Gamma^2\left(m + \frac{1}{2}\right)}{\Gamma^2(m)} \right], \tag{7.42}$$

whereby the combining gain is obtained as

$$\frac{\mathbb{E}[\gamma_{\mathrm{EGC}}]}{\bar{\gamma}_o} = \left[1 + \frac{(L-1)}{m} \frac{\Gamma^2\left(m + \frac{1}{2}\right)}{\Gamma^2(m)} \right]. \tag{7.43}$$

For Rayleigh fading, the expression (7.43) can be further simplified to give

$$\frac{\gamma_{\mathrm{EGC}}}{\bar{\gamma}_o} = 1 + \frac{\pi}{4}(L-1). \tag{7.44}$$

7.5.3   SC

Let us turn to SC with i.i.d. Rayleigh branches, each of average SNR $\overline{\gamma}_o$, for which, as shown in (7.34), the output instantaneous SNR at the SC is

$$f_{\gamma_{SC}}(x) = \frac{L}{\overline{\gamma}_0}(1 - e^{x/\overline{\gamma}_o})^{L-1}e^{-x/\overline{\gamma}_o}. \tag{7.45}$$

The average SNR is then equal to

$$\overline{\gamma}_{SC} = \mathbb{E}[\gamma_{SC}] = \int_0^\infty x f_{\gamma_{SC}}(x)\, dx. \tag{7.46}$$

Two methods can be used to determine $\overline{\gamma}_{SC}$. The first method is quite straightforward; it directly applies the binomial expansion to the power factor in $f_{\gamma_{SC}}(x)$, and then integrates the results term by term, leading to

$$\overline{\gamma}_{SC} = L\int_0^\infty \sum_{k=0}^{L-1}(-1)^k \binom{L-1}{k}\frac{x}{\overline{\gamma}_o}e^{-(k+1)x/\gamma_0}\, dx$$

$$= L\overline{\gamma}_o \sum_{k=0}^{L-1}(-1)^k \binom{L-1}{k}\frac{1}{(k+1)^2}. \tag{7.47}$$

This is the exact result but not the simplest.

   The second approach that leads to the most elegant result is due to Brennan [4]. Using this approach, we first change the integration variable by setting $y = 1 - \exp(-x/\overline{\gamma}_o)$. After simplification and expanding $-\log(1 - y) = \sum_{k=1}^\infty y^k/k$, we obtain

$$\overline{\gamma}_{SC} = L\overline{\gamma}_o \int_0^1 y^{L-1}\sum_{k=1}^\infty \left(\frac{y^k}{k}\right)\, dy$$

$$= L\overline{\gamma}_o \sum_{k=1}^\infty \frac{1}{k(k+L)}$$

$$= \overline{\gamma}_o \left(\sum_{k=1}^\infty \frac{1}{k} - \sum_{k=1}^\infty \frac{1}{k+L}\right). \tag{7.48}$$

The key step here is to recognize that the first sum is expressible as $\sum_{k=1}^L k^{-1} + \sum_{i=1}^\infty (i+L)^{-1}$, showing that the second summation is only part of the first. As such, we obtain

$$\overline{\gamma}_{SC} = \sum_{k=1}^L \frac{1}{k}, \tag{7.49}$$

whereby the average SNR gain is determined as

$$\frac{\overline{\gamma}_{SC}}{\overline{\gamma}_o} = \sum_{k=1}^L \frac{1}{k}. \tag{7.50}$$

We summarize the average SNR gains for various combining schemes as follows:

$$\frac{\gamma_{\text{MRC}}}{\overline{\gamma}_o} = L,$$

$$\frac{\gamma_{\text{EGC}}}{\overline{\gamma}_o} = 1 + (L-1)\pi/4,$$

$$\frac{\gamma_{\text{SC}}}{\overline{\gamma}_o} = \sum_{k=1}^{L} \frac{1}{k}. \tag{7.51}$$

The second and third lines are applicable to only to i.i.d. Rayleigh fading. The first line is applicable to i.i.d. Nakagami fading as well. Specifically, the average SNR gain of MRC is equal to the number of branches. EGC is slightly inferior. SC is the worst among the three, in the sense that adding the Lth branch increases the average SNR by only $1/L$. Thus, for large L, further increasing the number of branches in selection diversity yields a diminishing return. It is therefore suggested to use the order not exceeding $L = 5$.

7.6   Methods for error performance analysis

The use of the average SNR as a performance index provides only a rough idea to compare various combining schemes. A more accurate performance metric is the probability of bit error or symbol error. The methodology used to analyze diversity combining systems is very flexible, depending upon the combining and modulation schemes to be used and the operational fading environment. There is no single best approach universally applicable to all the scenarios. The common methods fall into three categories.

1. Classical chain rule (CCR) in probability
2. Method of characteristic function (CHF)
3. The combination of the two methods.

The idea behind the CCR method is to directly exploit the existing formulas for the error performance of various modulation schemes in AWGN, as summarized in Table 4.3. The output SNR of many combiners takes the form of a sum of SNRs at individual branches. This motivates the use of its CHF to convert a convolution operation in finding the PDF of the combiner output SNR into a simple product in the Fourier transform domain. This is the idea of the CHF method.

7.6.1   The chain rule

Since error-probability formulas of various modulation schemes in AWGN are available in the literature, it is advantageous to directly use them as the *conditional* error probability for deriving the combiner's average error performance. These formulas, as shown in Table 4.3, are usually a function of the SNR γ, taking the form of $P_e(\gamma)$. For fading channels, γ is a random variable, dependent on the fading channel gains. Thus, the average error probability is obtained by averaging $P_e(\gamma)$ over the PDF of γ, denoted here by $f_\gamma(\gamma)$, leading to

$$\overline{P}_e = \int_0^\infty P_e(x) f_\gamma(x) d\,x. \tag{7.52}$$

The idea here is to use the chain rule $P(AB) = P(A|B)P(B)$ for calculating the marginal probability $P(A) = \mathbb{E}_B[P(AB)]$. The question is: under what circumstances is this chain rule applicable to a diversity system? Recall that $P_e(\gamma)$ has been derived for a single-antenna nondiversity system. Thus, to justify the use of the chain rule, the combiner output must have the *same functional form* as that of the single-antenna system. It is easy to see that linear diversity combiners such as MRC and EGC satisfy this condition because their output

$$x = (\mathbf{w}^\dagger \mathbf{g})s + \mathbf{w}^\dagger \mathbf{n} \tag{7.53}$$

is of exactly the same form as its counterpart in an AWGN channel except for variations in the amplitude and noise components. Once the chain rule is justified for use, what we need is to determine the PDF of the combiner's output SNR γ for a given fading environment, and then use it to ensemble-average the conditional probability of the error, $P_e(\gamma)$, for the corresponding signaling scheme in AWGN to obtain the average P_e.

However, one must be cautious in the use of the chain rule to post-detection diversity reception of differential MPSK signals or to noncoherent diversity MFSK systems; discussion of the relevant issues is postponed to a later section.

7.6.2 The CHF method

In diversity reception, a combiner's output SNR γ can be a complicated function of the SNRs γ_i at individual branches, making it extremely difficult to directly determine its PDF. In this case, the CHF method is often simpler and more effective because it is capable of converting a convolution in the probabilistic domain into a product in the CHF domain. In fact, the idea of CHF is also useful in the performance evaluation of noncoherent diversity schemes, as shown in the subsequent sections.

7.7 Error probability of MRC

MRC usually requires coherent detection. Recall that the conditional symbol error rate of most coherent signaling schemes takes the form

$$P_e(\gamma) = \alpha \, Q(\sqrt{\beta \gamma}),$$

where α and β are two modulation-scheme-dependent parameters. This expression is exact to BPSK and coherent BFSK, while it provides an accurate approximation to higher level modulation schemes in the regime of moderate or large SNR, as shown in Table 4.3. Thus, what we need is to calculate the expectation $\overline{P}_e = \alpha \, \mathbb{E}_\gamma[Q(\sqrt{\beta \gamma})]$.

7.7.1 Error performance in nondiversity Rayleigh fading

Let us begin with the error performance of MRC on nondiversity fading channels, which often serves as a benchmark for comparison.

Without loss of generality, let us consider a coherent BPSK system operating on a flat Rayleigh fading channel characterized by a random gain of $g \sim \mathcal{CN}(0, \sigma_g^2)$. We need to average the conditional bit-error probability

$$P_e(\gamma) = Q(\sqrt{2\gamma}) \tag{7.54}$$

over the instantaneous received SNR defined by $\gamma = |g|^2 E_b/N_0$, where E_b is the bit energy, and $N_0/2$ is the variance of the AWGN component. The average received SNR is thus equal to $\overline{\gamma}_o = \sigma_g^2 E_b/N_0$. To determine the *average* error probability $\overline{P}_e = \mathbb{E}_\gamma[P_e(\gamma)]$, we need the PDF of γ. From the property of chi-square distribution, we can assert that γ is distributed as $\chi^2(2)$, as shown by

$$\gamma \sim (\overline{\gamma}_o/2)\chi^2(2), \tag{7.55}$$

whose PDF of is simply given by

$$f_\gamma(z) = \frac{1}{\overline{\gamma}_o} e^{-z/\overline{\gamma}_o}. \tag{7.56}$$

By invoking the results of (Ref. [1], p. 177(8)), or Formulas 6.283(1) and 8.250(1) of Gradshteyn and Ryzhik [5] to work out the integral

$$\int_0^\infty Q(\sqrt{2qt}) e^{-pt} dt = \frac{1}{2p}\left[1 - \sqrt{\frac{q}{p+q}}\right], \tag{7.57}$$

we obtain

$$\overline{P}_e = \int_0^\infty Q(\sqrt{2z})\left(\frac{1}{\overline{\gamma}_o} e^{-z/\overline{\gamma}_o}\right) dz$$

$$= \frac{1}{2}\left(1 - \sqrt{\frac{\overline{\gamma}_o}{1+\overline{\gamma}_o}}\right). \tag{7.58}$$

For general coherent signaling schemes other than BPSK with an approximate conditional symbol error probability $P_e(\gamma) = \alpha\, Q(\sqrt{\beta\gamma})$, we obtain

$$\overline{P}_e = \alpha\mathbb{E}_\gamma[Q(\sqrt{\beta\gamma})]$$

$$= \frac{\alpha}{2}\left(1 - \sqrt{\frac{\beta\overline{\gamma}_o}{2+\beta\overline{\gamma}_o}}\right). \tag{7.59}$$

The above results can be proved *alternatively* by using the CHF method. Let us first rewrite the Q-function by change of variable, as follows:

$$Q(\sqrt{\beta\gamma}) = \int_{\sqrt{\beta\gamma}}^\infty \frac{1}{\sqrt{2\pi}} e^{-x^2/2} dx$$

$$= \frac{1}{2}\int_{\beta\gamma}^\infty \underbrace{\frac{1}{\sqrt{\pi}}\left(\frac{y}{2}\right)^{-1/2} e^{-y/2}}_{\chi^2(\frac{1}{2})} dy$$

$$= \frac{1}{2}\left[1 - \int_0^{\beta\gamma} f(y)dy,\right]. \tag{7.60}$$

where $f(y)$ denotes the PDF of $\chi^2(\frac{1}{2})$. With this expression, it follows that

$$\overline{P}_e = \frac{1}{2} - \frac{1}{2\beta\overline{\gamma}_o}\left[\int_0^\infty \left(\int_0^u f(y)dy\right) e^{-\frac{u}{\beta\overline{\gamma}_o}} du\right]. \tag{7.61}$$

We recognize that the double integral inside the square brackets is essentially the Laplace transform of the integral of a chi-square PDF with transform variable $p = \frac{1}{\beta\overline{\gamma}_o}$. The result should be the CHF of $\chi^2(\frac{1}{2})$ multiplied by p, as shown by $p(1+p)^{-1/2}$. Hence,

$$\overline{P}_e = \frac{\alpha}{2}[1 - (1+p)^{-1/2}], \tag{7.62}$$

which is identical to (7.59).

Comparison of (7.54) and (7.58) indicates that fading channels significantly degrade the error performance of digital communication system from an exponentially dropping slope, defined by the Q-function, to a much lower rate defined by a square-root function.

7.7.2   MRC in i.i.d. Rayleigh fading

The method used above for a nondiversity coherent system can be extended to MRC with i.i.d. Rayleigh-faded branches. Assume each branch of distribution $g_i \sim \mathcal{CN}(0, \sigma_g^2)$. Then, from (7.11), the output SNR is given by

$$\gamma_{\mathrm{MRC}} = \rho_s(|g_1|^2 + \cdots + |g_L|^2). \tag{7.63}$$

Since $|g_i|^2 \sim \frac{1}{2}\sigma_g^2\chi^2(2)$, the use of the semiclosure property enables us to assert that $\gamma_{\mathrm{MRC}} \sim \frac{1}{2}\overline{\gamma}_o\chi^2(2L)$ with $\overline{\gamma}_o$ denoting the average received SNR at each branch, that is

$$\overline{\gamma}_o = \rho_s\sigma_g^2. \tag{7.64}$$

Hereafter, we drop the subscript from γ_{MRC} and denote $z = \frac{1}{2}\chi^2(2L)$ for notational simplicity. We need to average the conditional symbol error performance, which is assumed, again, to take the form

$$P_e(\gamma) = \alpha\, Q(\sqrt{z\beta\overline{\gamma}_o}\,).$$

To this end, we represent the PDF of z explicitly as

$$f_z(z) = \frac{p^L}{(L-1)!}z^{L-1}e^{-pz}, \; z \geq 0, \tag{7.65}$$

where $p = 1$ is a dump variable for subsequent use. It follows that

$$\overline{P}_e = \frac{p^L}{(L-1)!}\int_0^\infty \underbrace{\alpha\, Q(\sqrt{\beta\overline{\gamma}_o z}\,)}_{P_e(z)}z^{L-1}e^{-pz}\, dz. \tag{7.66}$$

The integral here does not seem easy to handle. We temporarily ignore the factor z^{L-1} in the integrand and just focus on the Fourier transform of the Q-function. The use of (7.57) enables us to write

$$\psi(p) = \int_0^\infty Q(\sqrt{\beta\overline{\gamma}_o z}\,)e^{-pz}\, dz$$

$$= \frac{1}{2p}(1 - \sqrt{\beta\overline{\gamma}_o/(p + \beta\overline{\gamma}_o)}\,). \tag{7.67}$$

The two integrands in (7.66) and (7.67) differ only by the factor z^{L-1}, thus suggesting us to relate $\overline{P}_e$ to (7.67) by $(L-1)$ times differentiating the latter to yield

$$\overline{P}_e = -\frac{\alpha}{2}\frac{(-p)^L}{(L-1)!}\frac{d^{L-1}}{d\,p^{L-1}}[p^{-1}\psi(p)]|_{p=1}$$

$$= -\frac{\alpha}{2}\frac{(-p)^L}{(L-1)!}\frac{d^{L-1}}{d\,p^{L-1}}(p^{-1}[1 - \sqrt{\beta\rho_r/(p + \beta\rho_r)}\,])|_{p=1}. \tag{7.68}$$

We further invoke Leibnitz's rule for differentiation to simplify, yielding

$$\overline{P}_e = \frac{\alpha}{2}\left[1 - \mu - \mu\sum_{k=1}^{L-1}\frac{1}{k!}\frac{1\cdot 2\cdots(2k-1)}{2^k}(1-\mu^2)^k\right]$$

$$= \frac{\alpha}{2}\left[1 - \mu - \mu\sum_{k=1}^{L-1}\binom{2k}{k}\left(\frac{1-\mu^2}{4}\right)^k\right], \tag{7.69}$$

where

$$\mu = \sqrt{\beta\overline{\gamma}_o/(2+\beta\overline{\gamma}_o)}. \tag{7.70}$$

By appropriately setting α and β, we can obtain the average symbol error performance for various coherent signaling schemes; see Chapter 4 for details. For MPSK, for example, we set $\alpha = 2$ and $\beta = 1 - \cos(2\pi/M)$, while for BPSK we set $\alpha = 1$, and $\beta = 2$.

The formula (7.69) can be shown to be identical to the well-known expression

$$\overline{P}_e = \left[\frac{1}{2}(1-\mu)\right]^L\sum_{k=0}^{L-1}\binom{L-1+k}{k}\left[\frac{1}{2}(1+\mu)\right]^k. \tag{7.71}$$

The equivalence of the two formulas can be verified by numerical examination. For example, by setting $\alpha = 1$ and $L = 1$, both formulas lead to the traditional result (7.59). It is also easy to check that for $L = 2$ both formulas lead to the same result, namely

$$\overline{P}_e = \frac{1}{4}(2 - 3\mu + \mu^3). \tag{7.72}$$

The asymptotic behavior of (7.71) is of practical importance. Consider BPSK for which $\beta = 2$. As $\overline{\gamma}_o$ tends to infinity, $(1+\mu) = 2$ and $(1-\mu) = \frac{1}{2\overline{\gamma}_o}$, which, when used along with the equality

$$\sum_{k=0}^{L-1}\binom{L-1+k}{k} = \binom{2L-1}{L},$$

lead to an asymptotic expression for BPSK

$$\overline{P}_e = \left(\frac{1}{4\overline{\gamma}_o}\right)^L\binom{2L-1}{L}. \tag{7.73}$$

It reveals that for $\overline{\gamma}_o \gg 1$, the error performance in decibels improves linearly with L.

✍ Example 7.1

Compare the error performance of a nondiversity receiver and an order-3 MRC. Assume coherent BPSK and i.i.d. Rayleigh fading channels, each of average SNR $\overline{\gamma}_o = 5\,\text{dB}$.

Solution

In this problem, we are given $\overline{\gamma}_o = 5\,\text{dB} = 3.16$, which, when inserted into (7.70), produces $\mu = 0.87$. Hence, we have

$$\frac{1}{2}(1-\mu) = 0.065,$$

$$\frac{1}{2}(1+\mu) = 0.935.$$

It follows from (7.58) and (7.68) that

$$\overline{P}_e = \begin{cases} 2.48545 \times 10^{-3}, & L = 3 \\ 0.064, & L = 1 \end{cases}.$$

Clearly, the performance improvement brought by the MRC exceeds an order of magnitude. ◯

7.7.3 MRC in correlated Rayleigh fading

When operating in correlated Rayleigh fading, the channel gain vector of MRC can be written as $\mathbf{g} \sim \mathcal{CN}(\mathbf{0}, \mathbf{R}_g)$, for which the combiner's output SNR is given by $\gamma_{\mathrm{MRC}} = \rho_s \mathbf{g}^\dagger \mathbf{g}$. In what follows, we drop the subscript MRC for simplicity and assume, again, that the conditional symbol error probability takes the form $P_e(\gamma) = \alpha Q(\sqrt{\beta \gamma})$. The procedure for finding the average error performance is the same as the i.i.d. case. The only difficulty lies in the PDF of γ. To proceed, assume $\mathbf{R}_g$ of distinct eigenvalues λ_i. According to (7.23), the PDF of γ is equal to

$$f_\gamma(x) = \sum_{i=1}^{L} \frac{1}{\overline{\gamma}_i} e^{-x/\overline{\gamma}_i}, \tag{7.74}$$

where $\overline{\gamma}_i = \rho_s \lambda_i$ denotes the average SNR at the virtual independent branch i defined by the ith eigen channel of $\mathbf{R}_g$, which corresponds to eigenvalue λ_i. Inserting this PDF into (7.59), it follows that

$$\overline{P}_e = \int_0^\infty \alpha Q(\sqrt{\beta x}) f_\gamma(x)\, dx$$

$$= \frac{\alpha}{2} \sum_{i=1}^{L} \left(1 - \sqrt{\frac{\beta \overline{\gamma}_i}{2 + \beta \overline{\gamma}_i}}\right). \tag{7.75}$$

In the above, all the poles of the channel matrix were assumed to be simple. Yet, the methodology is applicable to the general case with poles of multiplicity. Let us consider an example for illustration.

✍ **Example 7.2** _____

Consider the use of an order-3 MRC for the detection of M-ary PSK signals through independent fading branches, such that $g_i \sim \mathcal{CN}(0, \sigma_g^2), i = 1, 2$ and $g_3 \sim \mathcal{CN}(0, 2\sigma_g^2)$. Determine its average probability of symbol error.

Solution

For large SNR, the average probability of symbol error can be calculated by averaging $P_e(\gamma) = Q[\sqrt{2\gamma}\,\sin(\pi/M)]$ over the instantaneous SNR $\gamma = \rho_s(|g_1|^2 + |g_2|^2 + |g_3|^2)$. For the given condition, we can assert that

$$\gamma = \rho_s \sigma_g^2 \underbrace{\left(\frac{1}{2}\chi^2(2) + \frac{1}{2}\chi^2(2) + \chi^2(2)\right)}_{z}.$$

As such, by denoting $\overline{\gamma}_o = \rho_s \sigma_g^2$, we may write the average error performance as

$$\overline{P}_e = \mathbb{E}_z[Q(\sqrt{2z\overline{\gamma}_o}\,\sin(\pi/M)].$$

Averaging over γ, or equivalently over z, leads to the same result; but the latter is simpler. To find the PDF of z, we write its CHF in fractional form, yielding

$$\phi_z(s) = (1-s)^{-2}(1-2s)^{-1}$$
$$= -(1-s)^{-2} - 2(1-s)^{-1} + 4(1-2s)^{-1},$$

which, when inverse-Fourier-transformed, leads to the PDF of z, given by

$$f_z(x) = 2e^{-x/2} - 2e^{-x} - xe^{-x}.$$

Using it to average the conditional error probability, we obtain

$$\overline{P}_e = (1-\mu_1) - (1-\mu_2) - \frac{1}{4}(1-\mu_2)^2(2+\mu_2)$$
$$= 2\mu_2 - \mu_1 - \mu_2^3,$$

where

$$\mu_1 = \sqrt{\frac{2\overline{\gamma}_o\sin^2(\pi/M)}{1+2\overline{\gamma}_o\sin^2(\pi/M)}}, \quad \mu_2 = \sqrt{\frac{\overline{\gamma}_o\sin^2(\pi/M)}{1+\overline{\gamma}_o\sin^2(\pi/M)}}.$$

7.7.4 P_e for generic channels

The previous analysis of the MRC error performance heavily relied on a property of the Q-function that $Q(\sqrt{\gamma})$, when averaged over a $\chi^2(2m)$-distributed γ of integer-valued m, always leads to a closed-form expression. Such a requirement on γ is generally not satisfied except for Rayleigh fading channels. Nevertheless, the average error performance of MRC operating on a more general fading channel is always expressible as a onefold integral, as long as the CHF of γ exists.

As shown in Table 4.3 of Chapter 4, the conditional error probability for most coherent signaling schemes takes the form of $Q(\sqrt{\gamma})$, or is expressible in terms of exponential integrals. As such, we can always represent the conditional error performance of MPSK and square MQAM as

$$P_e(\gamma) = \frac{1}{\pi}\left[a\int_0^{\psi_1}\exp\left(-\frac{\alpha\gamma}{\sin^2\theta}\right)d\theta - b\int_0^{\psi_2}\exp\left(-\frac{\beta\gamma}{\sin^2\theta}\right)d\theta\right], \tag{7.76}$$

where a, b, ψ_1, and ψ_2 are modulation-scheme-dependent parameters. Taking MPSK as an example, $a = 1$, $\alpha = \sin^2(\pi/M)$, $\psi_1 = \pi - (\pi/M)$, and $b = 0$. To find the values of these parameters for other modulation schemes, the reader is referred to Chapter 4. Thus, the ensemble average of $P_e(\gamma)$ over γ corresponds to taking the Fourier transform of γ. It turns out that

$$\overline{P}_e = \frac{a}{\pi}\int_0^{\psi_1}\phi_\gamma(s)|_{s=-\alpha/\sin^2\theta}\,d\theta - \frac{b}{\pi}\int_0^{\psi_2}\phi_\gamma(s)|_{s=-\beta/\sin^2\theta}\,d\theta, \tag{7.77}$$

where $\phi_\gamma(s) = \mathbb{E}_\gamma[e^{s\gamma}]$ denotes the CHF of γ.

7.8 Error probability of EGC

The error performance of EGC is the most difficult to determine among the three types of diversity combiners. We only consider two scenarios: BPSK-EGC systems with three independent Rayleigh faded branches [6], and general EGC for detection of arbitrary coherent signals in various fading environments with an explicit joint channel CHF [7]. The former case leads to a closed-form solution, whereas the latter takes an integral form solution.

7.8.1 Closed-form solution to order-3 EGC

For the case of order-3 EGC for detection of BPSK signals in flat Rayleigh fading, we can derive a closed-form solution by resorting to neither the chain rule nor the CHF method. Rather, we directly work on a decision variable. Let us begin with the received signal at the kth antenna given by

$$y_k(t) = \alpha_k \sqrt{E_b} \sqrt{2/T_b} \cos[\omega_c t + \theta_m(t) + \varphi_k] + n_k(t), \qquad (7.78)$$

where ω_c signifies the carrier frequency in radians/second, θ_m of value 0 or π denotes the antipodal information phase with symbol duration T_b and bit energy E_b, and $n_k(t)$ is the AWGN component of mean zero and variance $N_k/2$. The fading channel is characterized by a Rayleigh faded magnitude α_k of power $\overline{\alpha_k^2}$ and a uniformly distributed random phase $\varphi_k \in [0, 2\pi)$. After demodulation, co-phasing, and equal-gain combining, the EGC output is determined to yield

$$\xi = \pm\underbrace{(x_1 + \cdots + x_L)}_{x} + \underbrace{(w_1 + \cdots + w_L)}_{w}, \qquad (7.79)$$

where $w_k \sim \mathcal{N}(0, N_k/2)$ and $x_k = \alpha_k \sqrt{E_b}$ are Rayleigh-distributed variables with power $\Omega_k = \mathbb{E}[x_k^2] = \overline{\alpha_k^2} E_b$. The random variables w_k and x_k are assumed independent from branch to branch. The positive and negative signs in front of the first term correspond to the message bit $\theta_m = 0$ and $\theta_m = \pi$, respectively. As such, the decision rule is given by

$$\xi \begin{cases} > 0, & \hat{\theta}_m = 0 \\ < 0, & \hat{\theta}_m = \pi \end{cases}. \qquad (7.80)$$

To determine the average BER $\overline{P}_e$ associated with this binary decision, we need to seek the PDF of the summation of L independent Rayleigh variables in (7.79). This implies an intractable $(L-1)$-fold integration if brute force is used. We, therefore, consider working on the CHF domain to convert the multiple-fold convolution in the probabilistic domain to a product in the CHF domain. The resulting CHF is directly connected to P_e, as revealed by the Gil–Pelaez lemma. From (7.79), it is easy to write the CHF of ξ as

$$\phi_\xi(v) = \mathbb{E}[e^{jv\xi}] = \phi_x(v)\phi_w(v). \qquad (7.81)$$

The CHF expression for a Rayleigh sum has been given in (7.31) and, with a slight modification, we can obtain $\phi_x(v)$ for x. Here, we allow for fading branches of unequal noise variances. The CHF of noise is given by

$$\phi_{w_k}(v) = \exp\left(-\frac{N_k v^2}{4}\right). \qquad (7.82)$$

Denoting $N_0 = (N_1 + \cdots + N_L)/L$ and $\rho_k = \frac{\Omega_k}{N_0} = \overline{\alpha_k^2}(E_b/N_0), k = 1, \cdots, L$, and using the property of the confluent (i.e., Kummer's) hypergeometric function $_1F_1(a; b; z)$ [1], we obtain

$$\phi_\xi(v) = e^{-\frac{(\Omega_1 + \cdots + \Omega_L + LN_0)v^2}{4}} \prod_{k=1}^{L} \left\{ {}_1F_1\left(-\frac{1}{2}; \frac{1}{2}; \frac{\Omega_k v^2}{4}\right) + j\sqrt{\frac{\pi \Omega_k}{4}} \, v \right\}. \tag{7.83}$$

Relating the average error probability $\overline{P}_e$ to the CHF via the Pelaez lemma presented in Chapter 2, we get

$$\overline{P}_e = \Pr\{\xi < 0\}$$

$$= \frac{1}{2} - \frac{1}{2\pi} \int_{-\infty}^{\infty} \frac{\Im\{\phi_\xi(v)\}}{v} \, dv. \tag{7.84}$$

For $L = 3$, the integral has a closed-form expression. Inserting (7.83) into (7.84) and invoking Formula 7.622(1) of [5] to simplify, we obtain

$$\overline{P}_e = \frac{1}{2} - \mathcal{F}(\rho_1, \rho_2, \rho_3, 3\kappa) - \mathcal{F}(\rho_2, \rho_3, \rho_1, 3\kappa) - \mathcal{F}(\rho_3, \rho_1, \rho_2, 3\kappa)$$

$$+ \frac{\pi}{4}\sqrt{\frac{\rho_1 \rho_2 \rho_3}{\rho_1 + \rho_2 + \rho_3 + 3\kappa)^3}}, \tag{7.85}$$

where $\kappa = 1$ for BPSK and $\kappa = 2$ for coherent BFSK, and $\mathcal{F}$ is defined as

$$\mathcal{F}(x, y, z, c) = \frac{1}{2}\sqrt{\frac{x(x + y + c)(x + z + c)}{(x + y + z + c)^3}}$$

$$\times \, {}_2F_1\left(-\frac{1}{2}, -\frac{1}{2}; \frac{1}{2}; \frac{yz}{(x + y + c)(x + z + c)}\right). \tag{7.86}$$

This is a closed-form solution for order-3 EGC expressible in terms of the Gauss hypergeometric function $_2F_1$. Recently, Nadarajah *et al.* [8] have pointed out that $_2F_1$ in the above expression can be represented as an elementary function by invoking [9]

$$_2F_1\left(-\frac{1}{2}, -\frac{1}{2}; \frac{1}{2}; c\right) = \sqrt{1 - c} + \sqrt{c}\arcsin(\sqrt{x}). \tag{7.87}$$

As such, $\overline{P}_e$ can be further simplified to

$$\overline{P}_e = \frac{1}{2} - \frac{1}{2}(\rho_1 + \rho_2 + \rho_3 + 3\kappa)^{-1}\sum_{i=1}^{3}\sqrt{\rho_i(\rho_i + 3\kappa)} + \frac{\sqrt{\rho_1 \rho_2 \rho_3}}{2(\rho_1 + \rho_2 + \rho_3 + 3\kappa)^{3/2}}$$

$$\times \left[\frac{\pi}{2} - \varphi(\rho_1, \rho_2, \rho_3) - \varphi(\rho_2, \rho_3, \rho_1) - \varphi(\rho_3, \rho_1, \rho_2)\right], \tag{7.88}$$

where

$$\varphi(\rho_i, \rho_j, \rho_k) = \arcsin\sqrt{\frac{\rho_i \rho_j}{(\rho_i + \rho_k + 3\kappa)(\rho_j + \rho_k + 3\kappa)}}. \tag{7.89}$$

In the above expression, the formula for $L = 2$ and $L = 1$ can be obtained by letting $\rho_3 = 0$ and $\rho_2 = \rho_3 = 0$, respectively, leading to

$$\overline{P}_e = \begin{cases} \frac{1}{2}\left(1 - \frac{\sqrt{\rho_1(\rho_1 + 2\kappa)} + \sqrt{\rho_2(\rho_2 + 2\kappa)}}{\rho_1 + \rho_2 + 2\kappa}\right), & L = 2 \\[4mm] \frac{1}{2}\left(1 - \sqrt{\frac{\rho_1}{\rho_1 + \kappa}}\right), & L = 1 \end{cases}. \tag{7.90}$$

On commenting the challenges facing the analysis of EGC, Brennan notes that "the characteristic function of a Rayleigh variable is not expressible in an immediately useful form." He, therefore, remarks in his seminal paper [4] that, unfortunately, the integral for the bit error rate "is quite as frightful as it appears; numerous workers – going back to Lord Rayleigh himself – have tried to express in terms of tabulated functions, but with no success if $L \geq 3$." Today, closed-form solutions for EGC are available only up to $L = 3$. The general case for $L \geq 4$ is still open, awaiting research efforts in the future.

7.8.2   General EGC error performance

The Gil-Pelaez lemma used in the preceding subsection is associated with a binary decision and is, thus, inapplicable to higher level modulation. For the analysis of the average symbol error rate of a generic EGC, we must go back to the chain rule

$$\overline{P}_e = \int_0^\infty P_e(\gamma) p_\gamma(\gamma) d\gamma, \tag{7.91}$$

where the SNR equals $\gamma = (\alpha_i + \cdots + \alpha_L)^2 E_s / (L N_0)$. To avoid the square operation in γ, we work on its square root

$$\zeta = (\alpha_1 + \cdots + \alpha_L)\sqrt{E_s/N_0} \tag{7.92}$$

and rewrite

$$\overline{P}_e = \int_0^\infty P_e(x) f_\zeta(x) \, dx. \tag{7.93}$$

where the PDF and CHF of ζ are denoted by $f_\zeta(x)$ and $\phi_\zeta(\omega)$, respectively. This is the inner product of two functions. We, thus, invoke Parseval's theorem to represent it in the CHF domain, yielding

$$\overline{P}_e = \frac{1}{2\pi} \int_{-\infty}^\infty \underbrace{\mathrm{FT}[P_e(\zeta)]}_{G(\omega)} \phi_\zeta^*(\omega) d\omega, \tag{7.94}$$

where FT is the short form of Fourier transform, and the CHF of ζ is given in (7.30). The conditional error probability for generic EGC with arbitrary square M-QAM and some coherent MPSK, when expressible in terms of ζ, takes the form

$$P_e(\zeta) = a \, Q(\sqrt{b}\,\zeta) + c \, Q^2(\sqrt{b}\,\zeta). \tag{7.95}$$

The coefficients a, b, and c for different modulations are given in Table 7.1.

Table 7.1   *Parameter setting for various schemes*

Schemes	Coefficients		
	a	b	c
BPSK	1	2	0
Coherent BFSK	1	1	0
QPSK	2	1	1
MQAM	$4\left(1 - \frac{1}{\sqrt{M}}\right)$	$\frac{3}{M-1}$	$4\left(1 - \frac{1}{\sqrt{M}}\right)$

The Fourier transform of $P_e(\zeta)$ given in (7.95) can be represented as a single expression [10]:

$$G(\omega) = \frac{1}{\sqrt{2\pi\,b}}\left[(a-c)\,{}_1F_1\left(1;\frac{3}{2};-q\right) + \frac{c}{\sqrt{2}}\,{}_1F_1\left(1;\frac{3}{2};-\frac{q}{2}\right)e^{-q/2}\right]$$

$$+ j\left\{\frac{a}{2\omega}\left(1-e^{-q}\right) - \frac{c}{4\omega}\left[1-e^{-z} - \frac{2q}{\pi}\,{}_1F_1^2\left(1;\frac{3}{2};-\frac{q}{2}\right)\right]\right\}, \tag{7.96}$$

where $q = \omega^2/(2b)$, and the confluent hypergeometric function ${}_1F_1$ can be calculated by using

$$\,{}_1F_1(a;b;z) = \sum_{n=0}^{\infty} \frac{(a)_n}{(b)_n}\frac{z^n}{n!} \tag{7.97}$$

with $(a)_n = a(a+1)\cdots(a+n-1)$ and $(a)_0 = 1$. Inserting (7.96) and (7.30) into (7.94) allows us to numerically determine the probability of symbol error for EGC with independent branches in various fading environments.

7.8.3 Diversity order of EGC

It is interesting to find the diversity order of EGC for large SNR. Physically, the diversity order of a communication system is the slope at which its error probability, in a log scale, to fall off with increased SNR. We have more to say about diversity order in the chapter of MIMO channels. For simplicity, we assume BPSK signaling. Let us consider an EGC with i.i.d. Nakagami faded branches, each of fading parameter m and average power Ω, so that we can write its instantaneous SNR as $\xi_i \sim \sqrt{\Omega/(2m)}\,\chi(2m)$. Nakagami suggests approximating the sum $\xi = \sum_{i=1}^{L}\xi_i$ by a single chi-variable $\sqrt{\hat{\Omega}/(2\hat{m})}\,\chi(2\hat{m})$, such that [11]

$$\hat{m} = mL,$$

$$\hat{\Omega} = L^2\Omega\left(1 - \frac{1}{5m}\right). \tag{7.98}$$

Hence, we can write the EGC instantaneous SNR as

$$\gamma = \frac{E_s\xi^2}{2LN_0} \sim \frac{1}{2\hat{m}}\underbrace{\frac{E_s\hat{\Omega}}{LN_0}}_{\overline{\gamma}}\chi^2(2\hat{m}). \tag{7.99}$$

It is easy to determine its CHF as

$$\phi_\gamma(z) = \left(1 + z\frac{\hat{\Omega}}{\hat{m}}\frac{E_s}{LN_0}\right)^{-\hat{m}}$$

$$= \left[1 + z\overline{\gamma}\frac{(1-\frac{1}{5m})}{m}\right]^{-mL}, \tag{7.100}$$

where $\overline{\gamma} = \Omega\,E_s/N_0$ is the average SNR at each branch.

We insert this approximate CHF into the Gaussian integral form of the Q-function and invoke the mean value theorem of integration, yielding the average bit error probability

$$\overline{P}_b = \mathbb{E}_\gamma[Q(\sqrt{b\gamma}\,)] = \frac{1}{\pi} \int_0^{\pi/2} \mathbb{E}\left[e^{-\frac{b\gamma}{2\sin^2\theta}}\right] d\theta$$

$$= \left(1 - \frac{1}{M}\right) \left[1 + \frac{1 - \frac{1}{5m}}{m} \frac{b\overline{\gamma}}{\sin^2\theta_0}\right]^{-mL}, \tag{7.101}$$

where $\theta_0 \in (0, \frac{\pi}{2})$ is m-dependent. Clearly, an EGC with L branches approaches the diversity order of L.

7.9   Average error performance of SC in Rayleigh fading

Selection combining can be further classified into two categories: pure SC and threshold SC. Pure SC always selects the branch with the highest instantaneous SNR as the output branch. In threshold SC, the output terminal retains its connection to the previously selected branch until its instantaneous SNR drops below a prescribed level (threshold). In this section, we focus on the pure SC. The study of SC heavily relies on the PDFs of ordered random variables, which, for the i.i.d. scenarios, have been intensively addressed in Refs [3, 12].

7.9.1   Pure SC

Consider the use of an order-L SC for noncoherent detection of a differential BPSK or BFSK signal with i.i.d. Rayleigh fading branches. For differential BPSK and noncoherent BFSK, the conditional bit error rate is given by $P_b = \frac{1}{2}\exp(-bx)$, with $b = 1$ for the former and $b = \frac{1}{2}$ for the latter, while the PDF of the SC output SNR has been obtained in (7.34). Thus, by using the chain rule, the average error probability of SC can be determined by

$$\overline{P}_b = \int_0^\infty P_b(x) f_{\gamma_{\mathrm{SC}}}(x)\, dx$$

$$= \int_0^\infty \frac{1}{2} e^{-bx} (L/\overline{\gamma}_o)(1 - e^{-x/\overline{\gamma}_o})^{L-1}\, e^{-x/\overline{\gamma}_o}\, dx$$

$$= L \sum_{k=0}^{L-1} (-1)^k \binom{L-1}{k} \frac{1}{2b\overline{\gamma}_o + 2(k+1)}. \tag{7.102}$$

It is simple, yet fails to intuitively indicate the diversity order of $\overline{P}_b$.

To find the diversity order offered by an SC of L branches, we employ the method of mathematical induction to rewrite (7.102) as

$$\overline{P}_b = \frac{\psi(L)}{\prod_{k=0}^{L-1}(2b\overline{\gamma}_o + 2k + 2)}, \tag{7.103}$$

where the coefficient $\psi(L)$ depends on the number of antennas, L, by

$$\psi(L) = L \cdot 2^{L-1} \sum_{k=0}^{L-1} (-1)^k \binom{L-1}{k} \frac{L!}{k+1}. \tag{7.104}$$

For $L = 1, 2, 3, and\ 4$, the coefficient ψ takes the value of 1, 4, 24, and 192, respectively.

Property 7.9.1 *Clearly, just as the MRC, as the average SNR approaches infinity, the average error probability of SC with i.i.d. Rayleigh fading branches is also on the order of $\overline{\gamma}_o^{-L}$, implying that the SC achieves the diversity order L.*

If coherent MPSK or square M-QAM, signaling is used alongside an order-L SC instead, and the average symbol-error rate can be calculated by using (7.57) to yield

$$\overline{P}_e = \int_0^\infty a\, Q(\sqrt{b\,x}\,)\frac{L}{\overline{\gamma}_o}(1 - e^{-x/\overline{\gamma}_o})^{L-1}\, e^{-x/\overline{\gamma}_o}\, dx$$

$$= \frac{aL}{\overline{\gamma}_o}\sum_{k=0}^{L-1}(-1)^k \binom{L-1}{k}\int_0^\infty Q(\sqrt{b\,x}\,)\exp\left(-\frac{(k+1)x}{\overline{\gamma}_o}\right)\, dx$$

$$= \frac{aL}{2}\sum_{k=0}^{L-1}\frac{(-1)^k}{k+1}\binom{L-1}{k}\left[1 - \sqrt{\frac{b\overline{\gamma}_o}{b\overline{\gamma}_o + 2(k+1)}}\right]. \tag{7.105}$$

This expression is exact for BPSK, while it provides an accurate approximation for QPSK and M-QAM in moderate and large SNR regimes. The choice of coefficients a and b depends on the modulation scheme to be used.

✍ Example 7.3 _____

It is of practical interest to compare the error performance of MRC, EGC, and SC. Let us consider the three combiners operating in Rayleigh fading with $L = 3$ i.i.d. branches for the detection of BPSK signals. For fairness, SC also adopts coherent detection. The error performance of the three combiners is calculated by

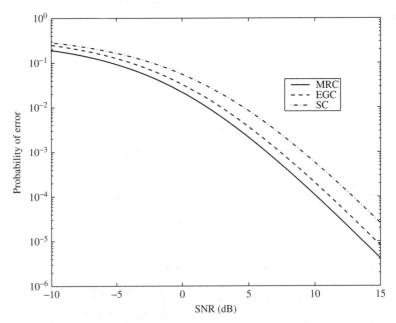

Figure 7.2 *Performance comparison between MRC, EGC, and SC for the detection of coherent BPSK signals in Rayleigh fading with L = 3 i.i.d. branches*

using (7.71), (7.88), and (7.105), respectively. When $\overline{\gamma}_o = 0\,\text{dB}$, their BER is $0.0218, 0.0227$, and 0.0554, respectively. For $\overline{\gamma}_o = 5\,\text{dB}$, the corresponding values become $0.0021, 0.0037$, and 0.0084. For more intuitive comparison, their performance is depicted in Figure 7.2.

7.9.2 Generalized SC

In a strict sense, pure SC only picks up the branch with the maximum instantaneous SNR as its output. This idea can be extended to combining the $N < L$ branches with the highest output SNRs for a better performance. In other words, a generalized SC scheme is the integration of the conventional SC with MRC or EGC.

Let $\gamma_{(1)}, \cdots, \gamma_{(L)}$ be the instantaneous branch SNRs in descending order, satisfying $\gamma_{(1)} > \gamma_{(2)} > \cdots > \gamma_{(L)}$, and the N branches with largest SNR are maximal-ratio combined, so that the instantaneous combiner output SNR is given by

$$\gamma = \sum_{i=1}^{N} \gamma_{(i)}. \tag{7.106}$$

Assuming i.i.d. Rayleigh branch fading, the PDF of γ for $N = 2$ and 3 can be explicitly expressed as [12]

$$f_\gamma(x) = \begin{cases} L(L-1)\frac{1}{\gamma_0}e^{-\gamma/\gamma_0}\left[\frac{\gamma}{2\gamma_0} + \sum_{k=1}^{L-2}\binom{L-2}{k}\frac{(-1)^k}{k}\left(1-e^{-\frac{k\gamma}{2\gamma_0}}\right)\right], \\ \qquad\qquad N = 2 \\ \frac{L(L-1)(L-2)}{2}\frac{1}{\gamma_0}e^{-\gamma/\gamma_0}\left[\frac{\gamma^2}{6\gamma_0^2} + \sum_{k=1}^{L-3}\binom{L-3}{k}\frac{(-1)^k}{k^2}\left(\frac{k\gamma}{\gamma_0} - 3(1-e^{-\frac{k\gamma}{3\gamma_0}})\right)\right], \\ \qquad\qquad N = 3. \end{cases} \tag{7.107}$$

Thus, for the case $N = 2$, we obtain the average symbol error rate

$$\overline{P}_e = \int_0^\infty a\,Q(\sqrt{b\gamma}\,)f_\gamma(\gamma)d\gamma$$

$$= \frac{a}{2}\binom{L}{2}\left\{\left(1 - \frac{1}{\sqrt{1+\alpha}} - \frac{\alpha}{2(1+\alpha)\sqrt{1+\alpha}}\right) + 2\sum_{k=1}^{L-2}\binom{L-2}{k}(-1)^k V(k)\right\}, \tag{7.108}$$

where

$$\alpha = 1/(b\gamma_0),$$

$$V(k) = \frac{1}{2+k} - \frac{1}{k\sqrt{1+\alpha}} + \frac{2}{k(k+2)\sqrt{1+0.5\alpha(k+2)}}. \tag{7.109}$$

For $N = 3$, it yields

$$\overline{P}_e = \frac{a}{2}\binom{L}{3}\left[1 - \mu\sum_{k=0}^{2}\binom{2k}{k}\left(\frac{1-\mu^2}{4}\right)^k + 3\sum_{k=1}^{L-3}\binom{L-3}{k}(-1)^k G(k)\right], \tag{7.110}$$

where

$$G(k) = \frac{k-3}{k^2}(1-\mu) - \frac{\mu}{2k}(1-\mu^2) + \frac{9}{k^2(k+3)}\left(1 - \left[1+\alpha\left(1+\frac{k}{3}\right)\right]^{-1/2}\right) \tag{7.111}$$

and

$$\mu = \sqrt{b\gamma_0/(1 + b\gamma_0)}.$$ (7.112)

Before concluding this subsection, we remark that the generalized SC, though of theoretical interest, is not a practical scheme. Just as the MRC, the generalized SC requires channel state information for all the branches. Yet, it achieves neither the error performance of MRC nor the simplicity of the traditional SC.

7.10   Performance of diversity MDPSK systems

In the remainder of this chapter, *M*-ary DPSK is further abbreviated as MDPSK for ease of description. The performance evaluation for noncoherent diversity reception of MDPSK signals is quite complicated since an error event is defined usually as the intersection of two product decision variables. In this section, we describe two methods to analyze the average *symbol* error performance for the post-detection combining of MPSK signals, both assuming i.i.d. Rayleigh faded branches. One method is based on linear prediction [13], while the other relies on directly finding the distribution of the instantaneous phase angle [14]. To further determine the average bit error rate of the same system with Gray-coded mapping, the reader is referred to [13, 15].

7.10.1   Nondiversity MDPSK in Rayleigh fading

Let us begin with the average error performance of nondiversity MDPSK systems in flat Rayleigh fading, aiming to derive a general solution in closed form [16]. Let g_k and g_{k-1} denote the two successive samples of the Rayleigh flat fading process. For simplicity, we assume that the information phase is e^{j0} and the transmitted energy E_s has been absorbed into g_k so that $\sigma_g^2 = \mathbb{E}[|g_k|^2]$ represents the average received signal power. As such, the two successive received samples can be written as

$$x_{k-1} = g_{k-1} + n_{k-1},$$
$$x_k = g_k + n_k,$$ (7.113)

where $n_k \sim \mathcal{CN}(0, \sigma_n^2)$ is the AWGN component. Denote the autocorrelation of the fading process and the average received SNR by

$$c_g = \mathbb{E}[g_k g_{k-1}^*]/\sigma_g^2,$$
$$\overline{\gamma} = \sigma_g^2/\sigma_n^2,$$ (7.114)

respectively. Then, the average received signal power and cross-correlation are given by

$$\sigma_x^2 = \mathbb{E}(|x_k|^2) = (1 + \overline{\gamma})\sigma_n^2,$$
$$c_x = \frac{\mathbb{E}[x_k x_{k-1}^*]}{\sigma_x^2} = \frac{\overline{\gamma} c_g}{1 + \overline{\gamma}}.$$ (7.115)

For differential decoding, we use the product $x_k x_{k-1}^*$ as a decision variable for which the correct decision region is given by

$$-\frac{\pi}{M} < \arg(x_{k-1}^* x_k) < \frac{\pi}{M}.$$ (7.116)

This sectorial region can be equivalently represented as the intersection $(u_1 > 0) \cap (u_2 < 0)$ where

$$u_1 = \Im\left\{ x_{k-1}^* x_k \exp\left(\frac{j\pi}{M} \right) \right\},$$

$$u_2 = \Im\left\{ x_{k-1}^* x_k \exp\left(-\frac{j\pi}{M} \right) \right\}. \tag{7.117}$$

The difficulty lies in the fact that both decisions involve two random variables in quadratic form of x_k and x_{k-1}. We take two steps to circumvent the difficulty. First normalize u's with $|x_{k-1}|$ to obtain two new random variables

$$v_1 = \Im\left\{ \frac{x_{k-1}^* x_k}{|x_{k-1}|} \exp\left(\frac{j\pi}{M} \right) \right\},$$

$$v_2 = \Im\left\{ \frac{x_{k-1}^* x_k}{|x_{k-1}|} \exp\left(-\frac{j\pi}{M} \right) \right\}. \tag{7.118}$$

The normalization has no influence on the decision region, so that the correct decision can be written as

$$P_c = \Pr\{(v_1 > 0) \cap (v_2 < 0)\}. \tag{7.119}$$

Next, we use one-step linear prediction to represent x_k by projecting x_k onto the past sample x_{k-1}, such that

$$x_k = c_x x_{k-1} + e_k. \tag{7.120}$$

The prediction error e_k is orthogonal to x_{k-1} and has the variance

$$\sigma_e^2 = (1 - |c_x|^2)\sigma_x^2. \tag{7.121}$$

Inserting (7.120) into (7.118) and rearranging yields

$$v_1 = |c_x x_{k-1}| \sin\left(\frac{\pi}{M} \right) + \frac{\Im\left\{ x_{k-1}^* e_k \exp\left(\frac{\pi}{M} \right) \right\}}{|x_{k-1}|},$$

$$v_2 = -|c_x x_{k-1}| \sin\left(\frac{\pi}{M} \right) + \frac{\Im\left\{ x_{k-1}^* e_k \exp\left(-\frac{\pi}{M} \right) \right\}}{|x_{k-1}|}. \tag{7.122}$$

Here, we have placed c_x inside the absolute signs for the following reason. The power spectra of most fading processes of practical interest, such as that described by the Jakes' model, are of even symmetry about the frequency, which, when transformed to the time domain, implies a correlation of real value. Thus, we may assume that c_g is real. It follows from (7.115) that c_x is also real, thus justifying its absorption inside the absolute signs. Clearly, conditioned on x_{k-1}, variables v_1 and v_2 are jointly Gaussian distributed with the mean vector and covariance matrix given by

$$\mathbf{m}_v = \begin{pmatrix} |c_x x_{k-1}| \sin(\pi/M) \\ -|c_x x_{k-1}| \sin(\pi/M) \end{pmatrix}, \quad \mathbf{R}_v = \frac{\sigma_e^2}{2} \begin{pmatrix} 1 & \cos(2\pi/M) \\ \cos(2\pi/M) & 1 \end{pmatrix}, \tag{7.123}$$

respectively. Theoretically, one can use this joint PDF to determine the conditional correct decision probability. But the correlation between v_1 and v_2 makes the calculation quite complicated. Fortunately, $\mathbf{R}_v$ has two very special eigenvectors, which enable the decorrelation of v's to obtain a new pair of variables to work on:

$$z_1 = (v_1 + v_2)/\sqrt{2},$$
$$z_2 = (v_1 - v_2)/\sqrt{2}. \tag{7.124}$$

Using simple inequality operations, it is easy to verify that the set $(v_1 > 0) \cap (v_2 < 0)$ is the same as $(-z_2 < z_1 < z_2) \cap (0 < z_2 < \infty)$. It also follows that

$$z_1 \sim \mathcal{N}(0, \sigma_1^2), \ z_2 \sim \mathcal{N}(\sqrt{2}\,|c_x x_{k-1}| \sin(\pi/M), \sigma_2^2), \tag{7.125}$$

with variances given by

$$\sigma_1^2 = \frac{\sigma_e^2}{2}(1 + \cos(2\pi/M)),$$

$$\sigma_2^2 = \frac{\sigma_e^2}{2}(1 - \cos(2\pi/M)), \tag{7.126}$$

respectively. Denoting $r = |x_{k-1}|$, the correct decision probability conditioned on r is given by

$$P_c(r) = \int_0^\infty f_{z_2}(z_2) \int_{-z_2}^{z_2} f_{z_1}(z_1)\,dz_1 dz_2, \tag{7.127}$$

which can be simplified as a onefold integral involving the Q-function. For Rayleigh fading, r is distributed as

$$f_R(r) = \frac{2r}{\sigma_x^2} \exp\left(-\frac{r^2}{\sigma_x^2}\right), \tag{7.128}$$

which is used to average $P_c(r)$ and simplify, leading to an expression for the average symbol error probability in closed form:

$$\overline{P}_e = 1 - \overline{P}_c = \frac{M-1}{M} + \frac{|c_x|\tan(\pi/M)}{\xi(c_x)}\left[\frac{1}{\pi}\arctan\left(\frac{\xi(c_x)}{|c_x|}\right) - 1\right], \tag{7.129}$$

where $\xi(c_x) = \sqrt{1 - |c_x|^2 + \tan^2(\pi/M)}$. This is an exact and general formula for nondiversity MDPSK in flat Rayleigh fading.

We consider a few special cases of practical interest. For binary DPSK, $M = 2$, and the above formula reduces to

$$\overline{P}_e = \frac{1}{2}\left(1 - \frac{\overline{\gamma}}{(1+\overline{\gamma})}|c_g|\right), \tag{7.130}$$

which is identical to the result by Simon and Alouini [17]. Two extreme cases are of interest. The first is the large SNR behavior. As $\overline{\gamma} \to \infty$, we have $|c_x| \to |c_g|$, whereby the above general formula is simplified to

$$\overline{P}_e = \frac{M-1}{M} + \frac{|c_g|\tan(\pi/M)}{\xi(c_g)}\left[\frac{1}{\pi}\arctan\left(\frac{\xi(c_g)}{|c_g|}\right) - 1\right] > 0, \tag{7.131}$$

implying the presence of an error floor for MDPSK in fast fading. For slow fading for which $|c_g| = 1$, the above result reduces to $P_e = 0$, indicating a vanishing error floor as $\overline{\gamma} \to \infty$. The next is the behavior in slow fading, for which $|c_g| = 1$ and, thus, $c_x = \overline{\gamma}/(1+\overline{\gamma})$. The general formula becomes

$$\overline{P}_e = \frac{M-1}{M} - \frac{1}{\pi}\sqrt{\frac{\overline{\gamma}^2 \sin^2(\pi/M)}{1 + 2\overline{\gamma} + \overline{\gamma}^2 \sin^2(\pi/M)}}\left[\pi - \arccos\left(\frac{\overline{\gamma}}{1+\overline{\gamma}}\cos(\pi/M)\right)\right]. \tag{7.132}$$

At this point, it remains to answer whether the method of chain rule is applicable to post-detection combining of MDPSK signals.

7.10.2   Remarks on use of the chain rule

We have used the chain rule for MRC and EGC. A rule of thumb for the suitability of the chain rule is that the decision variable obtained at a combiner output must be in the same functional form as the one that has been used for deriving the existing conditional probability of error. Thus, one must be cautious in using the chain rule for cases with nonlinearity, such as those encountered in post-detection combining MDPSK or noncoherent MFSK signals.

Consider an L-diversity product receiver for the detection of differential MPSK signal $\phi_k = \theta_k - \theta_{k-1}$. Denote the kth received signal, gain, and noise samples from branch i by $y_i(k)$, $g_i(k)$, and $n_i(k)$, respectively. The decision variable is given by

$$\xi_L = \sum_{\ell=1}^{L} \underbrace{[g_i(k)e^{j\theta_k} + n_i(k)]}_{y_i(k)}\underbrace{[g_i(k-1)e^{j\theta_{k-1}} + n_i(k-1)]^*}_{y_i^*(k-1)}, \tag{7.133}$$

which, when viewed from the information symbol $e^{j\theta_k}$, is clearly not in the same form as that with nondiversity reception:

$$\xi_1 = [g(k)e^{j\theta_k} + n(k)][g(k-1)e^{j\theta_{k-1}} + n(k-1)]^*.$$

Thus, the performance formulas derived for ξ_1 cannot serve as a basis for the performance evaluation of ξ_L.

The unique feature of the noncoherent differential MPSK diversity system is the use of the previous sample as a reference for symbol detection. To analyze its performance, we need to handle two difficult issues: random variable products and their summation. Therefore, linear prediction and CHF are expected to be two useful tools. Let us illustrate the two methods by focusing on the case for the detection of binary DPSK signals on i.i.d. Rayleigh fading channels. The results, though derived for binary DPSK, are equally applicable to MDPSK with Gary-coded mapping [13, 15].

7.10.3   Linear prediction to fit the chain rule

Clearly, a noncoherently received MPSK signal, in its original form, is not justified for the use of the chain rule. The situation changes, however, if we rewrite the received signal in a linear-prediction format.

Assume $n_i(k) \sim \mathcal{CN}(0, \sigma_n^2)$ and i.i.d. Rayleigh faded branches, such that two adjacent samples are distributed as

$$\begin{bmatrix} g_i(k) \\ g_i(k-1) \end{bmatrix} \sim \mathcal{CN}\left(\mathbf{0}, \sigma_g^2 \begin{bmatrix} 1 & c_g \\ c_g^* & 1 \end{bmatrix}\right), \tag{7.134}$$

where σ_g^2 and c_g are the fading variance and the lag-1 correlation coefficient, respectively. Further, denote $\overline{\gamma}_o = \sigma_g^2/\sigma_n^2$. Given $y_i(k-1)$, it follows from linear prediction theory that $y_i(k) = ay_i(k-1) + e_i(k)$, where the one-step prediction coefficient $a = R_y(0)R_y^{-1}(1) = c_g\overline{\gamma}_o/(1+\overline{\gamma}_o)$ and $e_i(k) \sim \mathcal{CN}(0, \sigma_e^2)$ with prediction-error power $\sigma_e^2 = (\sigma_g^2 + \sigma_n^2)(1 - |a|^2)$. Then, conditioned on $y_i(k-1)$, we can write

$$y_i(k)y_i(k-1)^* = a|y_i(k-1)|^2 + w_i(k),$$

where $w_i(k) \sim \mathcal{CN}(0, \sigma_e^2|y_i(k-1)|^2)$. Thus, given $\{y_i(k-1), i = 1, \cdots, L\}$, the post-detection combiner's output decision variable is given by

$$\xi = a\underbrace{\sum_{i=1}^{L} |y_i(k-1)|^2}_{\xi} + \underbrace{\sum_{i=1}^{L} w_i(k)}_{w}, \tag{7.135}$$

which is now in the same form as what we have for coherent combining. With this modification, the chain rule is applicable. It implies that we can use the conditional symbol error probability of the corresponding coherent modulation for calculating the average performance of the post-detection combining MDPSK system. Thus, given $\{y_i(k-1), i = 1, \cdots, L\}$, the instantaneous SNR at the post-detection combiner (PDC) output can be determined as

$$\gamma_{\text{PDC}} = \frac{[|a| \sum_{i=1}^{L} |y_i(k-1)|^2]^2}{\mathbb{E}[|w|^2]} = \frac{|a|^2 \sum_{i=1}^{L} |y_i(k-1)|^2]}{\sigma_e^2}. \tag{7.136}$$

As such, for a conditional symbol error probability taking the form of $P_e(\gamma) = \alpha Q(\sqrt{\beta\gamma})$, the average symbol rate with PDC is equal to

$$\overline{P}_e = \alpha \mathbb{E}_{\gamma_{\text{PDC}}}[Q(\sqrt{\beta\gamma_{\text{PDC}}})]. \tag{7.137}$$

From (7.136), it is easy to see that $\gamma_{\text{PDC}} \sim \frac{1}{2}|a|^2(1 - |a|^2)^{-1}\chi^2(2L)$. Thus, we can imagine that the formula for the average symbol rate of the MDPSK PDC system with i.i.d. Rayleigh faded branches takes the same form as (7.71) for MRC, except that μ is defined differently as $\mu = |c_g|\overline{\gamma}_o/(1 + \overline{\gamma}_o)$. It seems that this linear prediction approach was first used by Kam [13].

✍ **Example 7.4** —————————————————————————————

As an example, consider the average error performance of the post-detection detection of a binary DPSK signal with $L = 2$ independent Rayleigh faded branches.

Solution

For a binary signaling scheme, we only consider the in-phase component for which the noise power reduces by half. For the case of $L = 2$, the average BER becomes

$$\overline{P}_b = \frac{1 + \overline{\gamma}_o(1 - |c_g|)}{2(1 + \overline{\gamma}_o)}.$$

○

〰✍

At this point, the reader may ask whether the linear prediction method can be extended to the case of Rician fading. The answer is no. The reason is that for Rician fading, the linear prediction error $e(k)$, though Gaussian, is of nonzero mean. Thus, the conditional probability of error is no longer simply in the form of a Q-function.

7.10.4   Alternative approach for diversity MDPSK

In the preceding subsection, we described a linear prediction method for determining the average symbol error performance for post-detection combining MDPSK signals. In this subsection, we describe another approach.

Let $\mathbf{y}_i = [y_i(k), y_i(k-1)]^T$, $\mathbf{g}_i = [g_i(k), g_i(k-1)]^T$, and $\mathbf{n}_i = [n_i(k), n_i(k-1)]^T$ denote the received vector, gain vector, and noise vector, respectively. Assuming that the information phase of 0 is sent, we can write

$$\mathbf{y}_i = \mathbf{g}_i + \mathbf{n}_i \tag{7.138}$$

and construct a product decision variable $D = \sum_{i=1}^{L} \mathbf{g}_i^\dagger \mathbf{g}_i$ for which the correct decision region is

$$-\frac{\pi}{M} < \arg(D) < \frac{\pi}{M}.$$

To proceed, assume i.i.d. Rayleigh faded branches, such that $\mathbf{g}_i \sim \mathcal{CN}(\mathbf{0}, \mathbf{R}_g)$ and $\mathbf{n}_i \sim \mathcal{CN}(\mathbf{0}, \sigma_n^2\mathbf{I})$. As before, let σ_g^2 and c_g denote the variance of channel gain $g_i(k)$ and its temporal correlation coefficient, respectively, and let $\overline{\gamma}_o = \sigma_g^2/\sigma_n^2$ denote the average received SNR. The covariance matrix of $\mathbf{y}_i$ is then given by $\mathbf{R}_y = \mathbf{R}_g + \sigma_n^2 \mathbf{I}$, and its inverse $\mathbf{S} = \mathbf{R}_y^{-1}$ equals

$$\mathbf{S} = \begin{pmatrix} S_{11} & S_{12} \\ S_{21} & S_{22} \end{pmatrix} = \begin{pmatrix} (1+\overline{\gamma}_o)\sigma_n^2 & c_g\overline{\gamma}_o\sigma_n^2 \\ c_g^*\overline{\gamma}_o\sigma_n^2 & (1+\overline{\gamma}_o)\sigma_n^2 \end{pmatrix}^{-1}. \tag{7.139}$$

Let φ denote the phase of S_{12}. The distribution of the phase θ of D is determined to be [18]

$$f(\theta) = \frac{L(L+1)\cdots(2L-1)(\det \mathbf{S})^L}{2\sqrt{\pi}\,\Gamma(2L+\frac{3}{2})(|S_{12}|\cos(\theta-\varphi)+S_{11})^{2L}}$$

$$\times\ {}_2F_1\left(2L, L-\frac{1}{2}; L+\frac{3}{2}; \frac{|S_{12}|\cos(\theta-\varphi)-S_{11}}{|S_{12}|\cos(\theta-\varphi)+S_{11}}\right), \tag{7.140}$$

with which the SER of the MDPSK diversity system can be evaluated in terms of a onefold integral

$$\overline{P}_e = 1 - \int_{-\pi/M}^{\pi/M} f(\theta)d\theta. \tag{7.141}$$

The reader is referred to [14] for more details.

7.11   Noncoherent MFSK with diversity reception

Next we turn to a noncoherent MFSK diversity system with post-detection combining. Assume that L-diversity is employed for reception, for which the M output variables are given by

$$\xi_1 = \sum_{k=1}^{L} |g_k(1) + n_k(1)|^2,$$

$$\xi_m = \sum_{k=1}^{L} |n_k(m)|^2,\ m = 2,\cdots,M, \tag{7.142}$$

where $n_k(m) \sim \mathcal{CN}(0, \sigma_n^2)$ denote the AWGN components at the mth detector of the kth diversity receiver. For simplicity, we assume that all ξ's have been normalized so that the noise variance $\sigma_n^2 = 1$. Then, the instantaneous SNR is expressible as

$$\gamma = \frac{\sum_{k=1}^{L} |g_k(1)|^2}{\sigma_n^2} = \mathbf{g}^\dagger\mathbf{g}. \tag{7.143}$$

The method used for handling single-branch noncoherent MFSK described in Chapter 4 is also applied here to determine the correct-detection probability conditioned on $z = \xi_1$, obtaining

$$P_c(z) = \Pr\{\cap_{i=2}^{M}(\xi_i < z|z)\}$$

$$= [\Pr\{\xi_i < z|z\}]^{M-1},\ i \neq 1. \tag{7.144}$$

The difference is that, with L antennas, we have $\xi_i \sim \frac{1}{2}\chi^2(2L), i = 2,\cdots,M$, which, along with 2.321(2) of [5], leads to

$$\Pr\{\xi_i < z|z\} = \int_0^\infty \frac{u^{L-1}}{\Gamma(L)}e^{-u}\,du$$

$$= 1 - e^{-z}\sum_{k=0}^{L-1}\frac{z^k}{k!}. \tag{7.145}$$

It follows that

$$P_c(z) = \left[1 - e^{-z} \sum_{k=0}^{L-1} \frac{z^k}{k!}\right]^{M-1}. \tag{7.146}$$

The next step is to average $P_c(z)$ over z. The barrier is the power in the expression. We, therefore, use the binomial expansion to represent the right-hand side to obtain

$$P_c(z) = 1 + \sum_{m=1}^{M-1} \binom{M-1}{m} (-1)^m e^{-mz} \left[\sum_{k=0}^{L-1} \frac{z^k}{k!}\right]^m. \tag{7.147}$$

We next apply the multinomial expansion to the power with square brackets, yielding

$$P_c(z) = 1 + \sum_{m=1}^{M-1} (-1)^m (M-1)_{m-1} \sum_{m_0 + \cdots + m_{L-1}} \alpha\, z^p e^{-mz}, \tag{7.148}$$

where $(M-1)_k = (M-1)(M-2)\cdots(M-k)$ and $(M-1)_0 = 1$. The second summation is taken over all possible L-tuples $\mathbf{m} = (m_0, \cdots, m_{L-1})$ such that $m_0 + \cdots + m_{L-1} = m$ for $m_k \in \{0, 1, \cdots m\}$. Both α and p are functions of the L-tuples $\mathbf{m}$, defined by

$$\alpha = \left[\prod_{k=0}^{L-1} m_k!(k!)^{m_k}\right]^{-1},$$

$$p = \sum_{k=1}^{L-1} k m_k. \tag{7.149}$$

Taking the ensemble average of $P_c(z)$ requires the evaluation of $\mathbb{E}_z[z^p E^{-mz}]$. We recognize that $\mathbb{E}_z[e^{sz}]$ is the CHF of z, and is denoted here by $\phi_z(s)$. A skill used here is the use of the relationship

$$\frac{d^p \phi_z(s)}{d\, s^p} = \mathbb{E}\left[\frac{d^p e^{sz}}{d\, s^p}\right] = \mathbb{E}[z^p e^{sz}], \tag{7.150}$$

which, when inserted into $P_c(z)$ above, results in the average symbol error probability

$$\overline{P}_e = \sum_{m=1}^{M-1} (-1)^{m+1} (M-1)_{m-1} \sum_{m_0 + \cdots + M_{L-1} = m} \alpha \frac{d^p \phi_z(s)}{ds^p}\bigg|_{s=-m}. \tag{7.151}$$

Recall that $z = \xi_1 = \sum_{k=1}^{L} |g_k(1) + n_k(1)|^2$. To account for arbitrary fading channels g_k, we take two steps to determine its CHF. Conditioned on g_k, z follows the noncentral chi-square distribution, as shown by $z|_\mathbf{g} \sim \frac{1}{2}\chi^2(2L, 2\gamma)$ with CHF

$$\phi_z(s|\gamma) = (1-s)^{-L} \exp\left(\frac{s\gamma}{1-s}\right). \tag{7.152}$$

Thus, we may represent $\phi_z(s)$ as

$$\phi_z(s) = (1-s)^{-L} \phi_\gamma(u)\big|_{u=s/(1-s)}, \tag{7.153}$$

where $\phi_\gamma(u) = \mathbb{E}[e^{u\gamma}]$ is the CHF of the diversity channel γ. We may combine the above to write

$$\overline{P}_e = \sum_{m=1}^{M-1} (-1)^{m+1} (M-1)_{m-1} \sum_{m_0 + \cdots + m_{L-1} = m} \alpha \left[\frac{d^p (1-s)^{-L} \phi_\gamma(u)\big|_{u=s/(1-s)}}{ds^p}\right]_{s=-m}. \tag{7.154}$$

This is a generic formula applicable to arbitrary fading channels as long as they have CHFs.

As an example, consider a Rician fading channel. Its normalized channel gain vector is distributed as $\mathbf{g} \sim \mathcal{CN}_L(\mu_g, \mathbf{R}_g)$, for which we have

$$\phi_\gamma(u) = \mathbb{E}[e^{u\mathbf{g}^\dagger\mathbf{g}}] = \det(\mathbf{I} - u\mathbf{R}_g)^{-1} \exp(\mu_g^\dagger(u^{-1}\mathbf{I} - \mathbf{R}_g)^{-1}\mu_g). \qquad (7.155)$$

With it is inserted into (7.154), we can obtain the average performance of Rician fading channels.

✍ Example 7.5

Suppose that a diversity system with $L = 2$ antennas is used for noncoherent reception of BPSK signal ($M = 2$) in Rician fading with mean vector and covariance matrix given by $\mu = \mu[1, 1]^T$ and $\mathbf{R}_g = \sigma_g^2\mathbf{I}$. Assume AWGN of variance $\sigma_n^2 = 1$, so that σ_g^2 and $|\mu|^2$ represent the SNRs due to the diffused and line-of-sight components, respectively. Through calculation, we find that

$$\overline{P}_e = (2 + \sigma_g^2)^{-1}\left[\frac{4 + 3\sigma_g^2}{2 + \sigma_g^2} + \frac{2\mu^2}{2 + \sigma_g^2}\right]\exp\left(\frac{2\mu^2}{2 + \sigma_g^2}\right).$$

In this example, if we further assume $\mu = 0$, the result reduces to the case for order-2 diversity noncoherent reception of BFSK signals in i.i.d Rayleigh fading:

$$\overline{P}_e = \frac{4 + 3\rho}{(2 + \rho)^3}, \qquad (7.156)$$

where $\rho = \sigma_g^2/\sigma_n^2$. $\overline{P}_e$ tends to zero as $\rho \to \infty$, and tends to 1/2 as $\rho \to 0$, consistent with our intuition. ○

〰〰✍

7.12   Summary

Each combining technique, standing alone, has its own merits. As general remarks, MRC is the best among the three schemes, in terms of error performance. This is achieved, however, at the cost of requiring both the phase and gain information of the random channels. The SC is at the other extreme, being the worst in error performance and the simplest in implementation. SC is usually used together with noncoherent detection. EGC runs in between, requiring only the channel phase information. Yet, its performance is only lightly inferior to the MRC by about 2 dB.

Methods used to analyze diversity combining seem very flexible and diverse. However, there is an underlying thought behind them. Coherent combining schemes, such as MRC and EGC, possess two fundamental characteristics: coherence and combining. The powerful means to exploit coherence is the chain rule, while the powerful means to handle the instantaneous SNR summation due to combining is the CHF method. The bridge linking the MRC to CHF is the Weinstein's integral representation of Q-function, whereas its counterpart for EGC is the Parseval's theorem, which transforms the chain rule into an inner product in the Fourier domain. Wisely combining the chain rule and CHF methods is the basic idea to tackle the performance issue of various coherent combiners.

For the particular case of MRC in Rayleigh fading, the PDF of its output instantaneous SNR takes the form of $\chi^2(2m)$ or their linear combination. Closed-form solutions always exist thanks to the property that $Q(\sqrt{x})$ has a simple Fourier transform and the error performance of MRC is simply its derivatives. The same assertion is true for MRC on Nakagami channels with integral fading parameters.

The treatment of noncoherent combining MDPSK is more complicated because, in general, the chain rule is no longer applicable. Nevertheless, for post-detection combining of binary DPSK, there is always a closed-form solution. In this case, there is only one product decision variable, and its CHF possesses a unique pole of multiplicity, ensuring the use of the residue theorem for a closed-form solution.

We finish this chapter by including Refs. [21–45] for additional reading.

Problems

Problem 7.1 Consider a three-branch diversity system. Suppose the instantaneous SNR at the three branches is -17, -15, and $-13\,\text{dB}$, respectively. Determine the instantaneous output SNR if the following combining scheme is to be used:

(a) SC
(b) MRC

Problem 7.2 Consider an order-2 diversity receiver with *independent* branch instantaneous SNRs $\gamma_i, i = 1, 2$. Assume that γ_i follows the probability density function $f_{\gamma_i}(x) = \frac{1}{\overline{\gamma}_i} e^{-x/\overline{\gamma}_i}$, where the average SNRs at the two branches are given by $\overline{\gamma}_1 = 1/4$ and $\overline{\gamma}_2 = 1/2$, respectively. Determine the average output SNR if a pure selection combining is used.

Problem 7.3 Consider an order-3 MRC for the reception of BPSK signals. The three branches have gains $\mathbf{g} = [g_1, g_2, g_3]^T$, such that $\mathbf{g} \sim \mathcal{CN}(\mathbf{0}, \mathbf{C}_g)$ with correlation matrix $\mathbf{C} = [1\ 0.8\ 0.6; 0.8\ 1\ 0.8; 0.6\ 0.8\ 1]$. Determine the system error performance.

Problem 7.4 Consider a receiver that employs order-L pure selection combining, for which the instantaneous SNRs, $\gamma_i, i = 1, \cdots, L$, at different branches are independent, and each of them has the same probability density function: $f_{\gamma_i}(x) = \frac{1}{\rho_r} e^{-x/\rho_r}$, with $\rho_r = \mathbb{E}[\gamma_i]$ denoting the average SNR at each branch.

(a) Show that the probability density function of the SC output SNR is given by

$$f_{\gamma_{\text{SC}}}(x) = \frac{L}{\rho_r}(1 - e^{-x/\rho_r})^{L-1} e^{-x/\rho_r}.$$

(b) If the output SNR, γ_{SC}, of the SC is required to be not less than ρ_r with probability 0.89, determine the minimum diversity order L.
(c) If the average output of the SC is not required to exceed $2\rho_r$, determine the minimum diversity order L.

Problem 7.5 An MRC diversity receiver employs three antennas with independent branch gains g_1, g_2, and g_3, so that the received signal at branch i is given by

$$x_i(t) = g_i s(t - \tau_i) + n_i(t),$$

where $s(t)$ is a BPSK signal of symbol energy $E_s = 0.2$ Watts; $g_1 \sim \mathcal{CN}(0, 1)$, $g_2 \sim \mathcal{CN}(0, 1)$, and $g_3 \sim \mathcal{CN}(0, 2)$ denote the three channel gains; and $n_i \sim \mathcal{CN}(0, 0.1)$ is the AWGN component. (a) Derive a closed-form expression for the bit error performance P_b. (b) Calculate P_b for the given conditions.

Problem 7.6 Consider a diversity system of order L, and assume that all branches are independent and have the same average SNR $\overline{\gamma}_o = -15\,\text{dB}$. Determine the value of L required for maintaining the average output SNR not less than $-10\,\text{dB}$ for the three combining schemes (a) MRC, (b) EGC, and (c) SC.

Problem 7.7 A diversity system has two branches, each having an average SNR $\overline{\gamma}_o = 5\,\text{dB}$. Suppose we expect the probability of instantaneous output SNR $\gamma < -5\,\text{dB}$ is less than 0.5%. Determine an appropriate combining scheme.

Problem 7.8 Suppose that an EGC of L branches is to be used in i.i.d. Rayleigh fading and each branch has the average SNR of $\overline{\gamma}_o = 8\,\text{dB}$. We expect less than 0.5% of the time over which the instantaneous output SNR γ is less than $0\,\text{dB}$. Determine the minimum value of L that meets the requirement.

Problem 7.9   The SNR improvement factor of the threshold selection scheme depends on the choice of the threshold Γ_T, as shown by

$$\frac{\overline{\gamma}}{\overline{\gamma}_o} = 1 + \frac{\Gamma_T}{\overline{\gamma}_o} \exp\left(-\frac{\Gamma_T}{\overline{\gamma}_o}\right).$$

Determine the value of Γ_T for which the improvement factor is maximized.

Problem 7.10   Consider the detection of a coherent MPSK signal using an order-L MRC with independent Rayleigh faded gains $g_i \sim \mathcal{CN}(0, \sigma_i^2)$ in an AWGN channel of symbol energy E_s and one-sided noise power spectral density N_0 such that $2E_s/N_0 = 1$. Derive the probability of symbol error for this MRC.

Problem 7.11   Suppose an order-L MRC receiver is used for the detection of a coherent BPSK and a non-coherent BFSK signals on independent Nakagami channels. The instantaneous SNR γ_i at branch i possesses fading parameter m_i and mean $\mathbb{E}[\gamma_i] = b_i$. Assume that the ratio m_k/b_k is equal for all the branches. Show that the average bit error probability for the two systems is given, respectively, by

$$P_{eC} = \frac{1}{2\sqrt{\pi}} \frac{\Gamma(m_t + 1/2)}{\Gamma(m_t + 1)} \left(\frac{2m_t}{2m_t + b_t}\right) {}_2F_1\left(\frac{1}{2}, m_t, m_t + 1; \frac{2m_t}{2m_t + b_t}\right)$$

$$P_{eN} = \frac{1}{2}\left(1 + \frac{b_t}{2m_t}\right)^{-m_t}$$

where $m_t = m_1 + \cdots + m_L$ and $b_t = b_1 + \cdots + b_L$.
Al-Hussaini and Al-Bassiounip [19].

Problem 7.12   An L-diversity system operates in an i.i.d. Rayleigh fading environment such that each branch has identical average SNR equal to $\overline{\gamma}$. Suppose we employ $m < L$ branch to form a GSC. Show that the average SNR gain of the GSC is given by

$$g = m + \frac{m}{m+1} + \frac{m}{m+2} + \cdots + \frac{m}{L}.$$

Kong and Milstein [2].

Problem 7.13   Consider a PDC for detecting binary DPSK signals with two i.i.d. Rayleigh faded branches. The received signal in branch i is $x_i(k) = g_i(k)e^{j\theta_k} + n_i(k)$, where $n_i(k) \sim \mathcal{CN}(0, \sigma_n^2)$, and the branch gain is of power σ_g^2 and correlation coefficient c_g. Denote matrix $\mathbf{\Omega} = [0\ 1; 1\ 0]$ and vector $\mathbf{x}_i = [x_i(k), x_i(k-1)]^T$. Then, $\mathbf{x}_i \sim \mathcal{CN}(\mathbf{0}, \mathbf{R}_x)$ and $\mathbf{R}_x = (\sigma_n^2 \mathbf{I} + \mathbf{R}_g)$, with $\mathbf{R}_g$ denoting the covariance matrix of two consecutive samples of each branch. The decision variable is $\xi = \Re\{\sum_{i=1}^{L} \mathbf{x}_i(k)^\dagger \mathbf{\Omega} \mathbf{x}_i(k-1)\}$, and an erroneous event occurs when $\xi < 0$. According to the Gil–Pelaez lemma, the BER $\Pr\{\xi < 0\}$ is directly related to the CHF of ξ by

$$P_e = -\mathrm{Res}\left\{\frac{\phi_\xi(s)}{s}, \text{poles in } \Re(s) < 0\right\},$$

where the CHF of ξ can be easily determined as

$$\phi_\xi(s) = \mathbb{E}[e^{s\xi}] = \det(\mathbf{I} - s\mathbf{\Omega R}_x)^{-L} = (1 - s\lambda_1)^{-L}(1 - s\lambda_2)^{-L},$$

with λ_1 and λ_2 denoting the eigenvalues of $\mathbf{\Omega R}_x$.

(a) Show that $\mathbf{R}_x = \mathbf{\Omega}(\sigma_n^2 \mathbf{I} + \mathbf{R}_g)$ has two eigenvalues: one negative and one positive.
(b) Use the residue theorem to show that the BER of the system is given by

$$P_e = (1+\alpha)^{-2} + 2\alpha(1+\alpha)^{-3},$$

where

$$\alpha = \frac{1 + \overline{\gamma}_o(1 + |c_g|)}{1 + \overline{\gamma}_o(1 - |c_g|)}.$$

Problem 7.14 The average bit error of using an order-L MRC for the detection of BPSK signals on i.i.d. Rayleigh fading channels is given in (7.71). Prove this formula by invoking the integral identity [20]

$$\frac{1}{\pi} \int_0^{\pi/2} \left(\frac{\sin^2\theta}{\sin^2\theta + \alpha} \right)^m d\theta = \left(\frac{1-\mu}{2} \right)^{m-1} \sum_{k=0}^{m-1} \binom{m-1+k}{k} \left(\frac{1+\mu}{2} \right)^k,$$

where α is a constant and $\mu = \sqrt{\alpha/(1+\alpha)}$.

Problem 7.15 Use results obtained in Problem 7.13 to show that, when $|c_g| < 1$, there always exists an error floor in the error performance of a noncoherent differential PSK diversity system, regardless of its average SNR. However, increase in diversity order can reduce the error floor. In particular, show that

(a) When $|c_g| < 0$, $\displaystyle\lim_{\overline{\gamma}_o \to \infty} P_e = \frac{1}{4}(1 - |c_g|)^2(2 + |c_g|)$.

(b) When $|c_g| = 1$, $P_e = \frac{2 + 3\overline{\gamma}_o}{4(1 + \overline{\gamma}_o)^3}$.

Problem 7.16 Consider an L-diversity reception of DPSK signals, with the kth received sample at branch i denoted as $y_i(k) = g_i(k)s_k + n_i(k)$. Denote $\overline{\gamma}_o = \sigma_g^2/\sigma_n^2$ and $c_g = \mathbb{E}[g_i(k)g_i^*(k-1)]/\sigma_g^2$. Define

$$\xi = \Re \left\{ \sum_{i=1}^{L} y_i(k)y_i^*(k-1)e^{j\psi} \right\},$$

where $\psi \neq \pm\pi/2$ is a fixed angle. Show that

$$\Pr\{\xi < 0\} = \left(\frac{1-\mu}{2} \right)^L \sum_{i=0}^{L-1} \binom{L-1+k}{k} \left(\frac{1+\mu}{2} \right)^k,$$

where

$$\mu = \left\{ 1 + \left[\left(\frac{\overline{\gamma}_o + 1}{|c_g|\overline{\gamma}_o} \right)^2 - 1 \right] \cos^{-2}\psi \right\}^{-1/2}.$$

Kam [13].

References

1. A. Erdelyi, Ed., *Tables of Integrals Transforms*, New York: McGraw-Hill, 1954.
2. N. Kong and L.B. Milstein, "Average SNR of a generalized diversity selection combining scheme," *IEEE Commun. Lett.*, vol. 3, no. 3, pp. 57–59, 1999.
3. E.B. Manoukian, *Mathematical Nonparametric Statistics*, New York: Gordon and Beach Science Publishers, 1986.
4. D.G. Brennan, "Linear diversity combining techniques," *Proc. IRE*, vol. 47, pp. 1075–1102, 1959.
5. I.S. Gradshteyn and I.M. Ryzhik, *Table of Integrals, Series, and Products*, New York: Academic Press, 1980.
6. Q.T. Zhang, "Probability of error for equal-gain combiners over Rayleigh channels: some closed-form solutions," *IEEE Trans. Commun.*, vol. 45, no. 3, pp. 270–273, 1997.
7. A. Annamalai, C. Tellumbura, and V.K. Bhargava, "Equal-gian diversity receiver performance in wireless channels," *IEEE Trans. Commun.*, vol. 48, no. 10, pp. 1732–1745, 2000.

8. S. Nadarajah and S. Kotz "On the closed form solutions for the probability of error for EG combiners," *IEEE Commun. Lett.*, vol. 11, no. 2, pp. 132–133, 2007.

9. A P Prudnikov, Y.A. Brychkov, and O.I. Marichev, *Integrals and Series*, vol. 3, Amsterdam: Gordon and Breach Science Publishers, 1986.

10. A. Annamalai, C. Tellambura, and V.K. Bhargava, "Equal-gain diversity receiver performance in Wireless channels," *IEEE Trans. Commun.* vol. 48, no. 10, pp. 1732–1746, 2000.

11. M. Nakagami, "The m-distribution: a general formula of intensity distribution of rapid fading," in *Statistical Methods in Radio Wave Propagation*, W.C. Hoffman, Ed. New York: Pergamon, 1960.

12. T. Eng, N. Kong, and L.B. Milstein, "Comparison of diversity combining techniques for Rayleigh-fading channels," *IEEE Trans. Commun.*, vol. 44, no. 9, pp. 1117–1129, 1996.

13. P.Y. Kam, "Bit error probabilities of MDPSK over the nonselective Rayleigh fading channel with diversity reception," *IEEE Trans. Commun.*, vol. 39, no. 2, pp. 220–224, 1991.

14. Q.T. Zhang and X.W. Cui, "New performance results for MDPSK with noncoherent diversity in fast Rayleigh fading," *IEEE Commun. Lett.*, vol. 10, no. 8, pp. 605–607, 2006.

15. P.J. Lee, "Computation of the bit error rate of coherent M-ary PSK with Gray code bit mapping," *IEEE Trans. Commun.*, vol. 34, no. 5, pp. 488–491, 1986.

16. Q.T. Zhang and X.W. Cui, "A closed-form expression for symbol-error rate of M-ary DPSK in fast fading" *IEEE Trans. Commun.*, vol. 53, no. 7, pp. 1085–1087, 2005.

17. M.K. Simon and M. Alouini, "Average bit-error probability performance for optimum diversity combining of noncoherent FSK over Rayleigh fading channels," *IEEE Trans. Commun.*, vol. 51, no. 4, pp. 566–569, 2003.

18. K.S. Miller, *Multidimensional Gaussian Distribution*, New York: John Wiley & Sons, Inc., 1964.

19. E.K. Al-Hussaini and A.A.M. Al-Bassiounip, "Performance of MRC diversity systems for the detection of signals with Nakagami fading," *IEEE Trans. Commun.*, vol. 33, no. 12, pp. 1315–1319, 1985.

20. M.K. Simon and M.S. Alouini, *Digital Communication Over Fading Channels: A Unified Approach to Performance Analysis*, New York: John Wiley & Sons, Inc., 2001.

21. J.P. Imhof, "Computing the distribution of quadratic forms in normal variables," *Biometrika*, vol. 48, vols. 3-4, pp. 419–426, 1961.

22. J.G. Proakis, *Digital Communications*, 3rd ed., New York: McGraw-Hill, 1995.

23. F.S. Weinstein, "Simplified relationships for the probability distribution of the phase of a sine wave in narrow-band normal noise," *IEEE Trans. Inf. Theory*, vol. 20, pp. 658–661, 1974.

24. J.W. Craig, "A new, simple, and exact resolution for calculating the performance for error for two-dimensional signal constellations," *MILCOM'91*, pp. 571–575, 1991.

25. R.F. Pawula, "Relations between the Rice Ie-function and Marcum Q-function with applications to error rate calculations," *Electron. Lett.*, vol. 31, no. 20, pp. 1717–1719, 1995.

26. Q.T. Zhang, "Exact analysis of post-detection combining for DPSK and NFSK systems over arbitrary correlated Nakagami channels," *IEEE Trans. Commun.*, vol. 46, no. 11, pp. 1459–1467, 1998.

27. Q.T. Zhang, "Maximal-ratio combining over Nakagami fading channels with an arbitrary branch covariance matrix," *IEEE Trans. Veh. Technol.*, vol. 48, no. 4, pp. 1141–1150, 1999.

28. R.K. Mallik, P. Gupta, and Q.T. Zhang, "Minimum selection GSC in independent Rayleigh fading," *IEEE Trans. Veh. Technol.*, vol. 54, no. 3, pp. 1013–1021, 2005.

29. Q.T. Zhang and H.G. Lu, "A general analytical approach to multi-branch selection combing over various spatially correlated fading channels," *IEEE Trans. Commun.*, vol. 50, no. 7, pp. 1066–1073, 2002.

30. K. Dietze Jr., C.B. Dietrich, and W.L. Stutzman, "Analysis of a two-branch maximal ratio and selection diversity system with unequal SNRs and correlated inputs for a Rayleigh fading channel," *IEEE Trans. Wireless Commun.*, vol. 1, no. 2, pp. 274–281, 2002.

31. Y. Chen and C. Tellambura, "Performance analysis of L-branch equal gain combiners in equally correlated Rayleigh fading channels," *IEEE Commun. Lett.*, vol. 8, no. 3, pp. 150–152, 2004.

32. S. Loyka and G. Levin, "Diversity-multiplexing tradeoff via asymptotic analysis of large MIMO systems," *IEEE ISIT'07*, June 2007, Nice, France.

33. G.K. Karagiannidis, "Moments-based approach to the performance analysis of equal gain diversity in Nakagami-m fading," *IEEE Trans. Commun.*, vol. 52, no. 5, pp. 685–690, 2004.

34. A. Annamalai, "Error rates for Nakagami-m fading multichannel reception of binary and M-ary signals," *IEEE Trans. Commun.*, vol. 49, no. 1, pp. 58–68, 2001.

35. S. Liu, J. Cheng, and N.C. Beaulieu, "Asymptotic error analysis of diversity schemes on arbitrarily correlated Rayleigh channels," *IEEE Trans. Commun.*, vol. 58, no. 5, pp. 1351–1355, 2010.

36. Y. Ma and Q.T. Zhang, "Accurate evaluation for MDPSK with noncoherent diversity," *IEEE Trans. Commun.*, vol. 50, no. 7, pp. 1189–1200, 2002.

37. S. Chennakeshu and J.B. Anderson, "Error rates for Rayleigh fading multichannel reception of MPSK signals," *IEEE Trans. Commun.*, vol. 43, no. 2/3/4, pp. 338–346, 1995.

38. Q.T. Zhang, "Error performance of noncoherent MFSK with L-diversity on correlated fading channels," *IEEE Trans. Wireless Commun.*, vol. 1, no. 3, pp. 531–539, 2002.

39. F. Patenaude, J.H. Lodge, and J.Y. Chouinard, "Noncoherent diversity reception over Nakagami-fading channels," *IEEE Trans. Commun.*, vol. 46, no. 8, pp. 985–991, 1998.

40. N.C. Beaulieu and A.A. Abu-Dayya, "Analysis of equal gain diversity on Nakagami fading channels," *IEEE Trans. Commun.*, vol. 39, no. 2, pp. 225–234, 1991.

41. R.F. Pawula, "A new formula for MDPSK symbol error probability," *IEEE Commun. Lett.*, vol. 2, pp. 271–272, 1998.

42. M.A. Smadi and V.K. Prabhu, "Postdetection EGC diversity receivers for binary and quaternary DPSK systems over fading channels," *IEEE Trans. Veh. Technol.*, vol. 54, no. 3, pp. 1030–1036, 2005.

43. G.M. Vitetta, U. Mengeli, and D.P. Taylor, "Error probability with incoherent diversity reception of FSK signals transmitted over fast Rician fading channels," *IEEE Trans. Commun.*, vol. 46, no. 11, pp. 1443–1447, 1998.

44. M.J. Barrett, "Error probability for optimal and suboptimal quadratic receivers in rapid Rayleigh fading channels," *IEEE J. Sel. Areas Commun.*, vol. 5, no. 2, pp. 302–304, 1987.

45. M. Brahler and M.K. Varanasi, "Asymptotic error probability analysis of quadratic receivers in Rayleigh-fading channels with applications to a unified analysis of coherent and non-coherent space-time receivers," *IEEE Trans. Inf. Theory*, vol. 47, no. 6, pp. 2383–2399, 2001.

8

Processing Strategies for Wireless Systems

Abstract

In this chapter, we examine the evolution of processing strategies used for wireless communications, aiming to seek their underlying thoughts that may ignite inspiration for the future.

In this chapter, we re-examine the philosophical issue of processing strategies for wireless communications initiated in Chapter 1. Various physical channels encountered in wireless communications are imperfect in the sense that they introduce additive noise, distortion, and interference, therefore requiring various signal processing strategies for remedy.

8.1 Communication problem

For exposition, let us start from a particular received signal model. Suppose that a unipolar information-bearing bit vector $\mathbf{b}$ is coded, modulated, and precoded for transmission over a communication channel defined by matrix $\mathbf{H}$. Let $\mathbf{C}$, $\mathbf{\Pi}$, $\mathbf{M}$, and $\mathbf{Q}$ denote the matrices for coding, interleaving, modulation, and precoding operations, respectively, and let j denote $j = \sqrt{-1}$. The received signal in baseband form is then expressible as

$$\mathbf{y} = \alpha \mathbf{HQM\Pi}(2\mathbf{Cb} - 1)_{\mathrm{mod}\,2} + \mathbf{n}, \tag{8.1}$$

where $\mathbf{n} \sim \mathcal{CN}(\mathbf{0},\, \sigma_n^2 \mathbf{I})$ denotes the noise vector, and the constant factor α is used to scale the transmitted power. In the above signal model, we assume that the dimensions of all matrices and vectors are compatible, so that their operations are meaningful. For example, the modulation matrix $\mathbf{M}$ for 2^m-QAM takes the form

$$\mathbf{M} = \begin{bmatrix} \mathbf{v} & j\mathbf{v} & \mathbf{0} & \mathbf{0} & \cdots & \mathbf{0} & \mathbf{0} \\ \mathbf{0} & \mathbf{0} & \mathbf{v} & j\mathbf{v} & \mathbf{0} & \cdots & \mathbf{0} \\ \vdots & & \cdots & & & & \vdots \\ \mathbf{0} & \mathbf{0} & \mathbf{0} & \cdots & \mathbf{0} & \mathbf{v} & j\mathbf{v} \end{bmatrix}, \tag{8.2}$$

where the row vector $\mathbf{v} = [2^{m-1}, 2^{m-2}, \cdots, 1]$. For QPSK, vector $\mathbf{v}$ reduces to $v = 1$.

Wireless Communications: Principles, Theory and Methodology, First Edition. Keith Q.T. Zhang.
© 2016 John Wiley & Sons, Ltd. Published 2016 by John Wiley & Sons, Ltd.
Companion Website: www.wiley.com/go/zhang7749

The purpose of communication is twofold. First, given the resources of system bandwidth and total power, and a particular channel, we need to design subsystems $\mathbf{C}$, $\mathbf{\Pi}$, $\mathbf{Q}$, and $\mathbf{M}$ to fulfill the system data rate and other performance requirements. This is a design issue. The second issue is, given an operational environment defined by matrix $\mathbf{H}$ and a received vector $\mathbf{y}$, to retrieve the original message vector $\mathbf{b}$ as reliable as possible under a certain optimum criterion such as the maximum *a posteriori* (MAP) probability. Assuming an AWGN channel with one-sided power spectral density N_0 and up to a constant factor, the objective function is expressible in terms of the PDF of $\mathbf{y}$, as

$$f(\mathbf{y}) = \exp\left(-\|\mathbf{y} - \mathbf{HQM\Pi C}(\mathbf{b})\|^2/N_0\right), \tag{8.3}$$

where $\|\cdot\|$ represents the length of a vector. Upon the reception of $\mathbf{y}$, our task is to find $\hat{\mathbf{b}}$ such that

$$\hat{\mathbf{b}} = \arg\max_{\{\mathbf{b}\}} f(\mathbf{b}|\mathbf{y}). \tag{8.4}$$

Because $\mathbf{b}$ are discrete symbols, this optimization is essentially an integer programming problem, which involves highly nonlinear operators, making it extremely difficult to handle. Closed-form solutions are almost impossible, and even the use of numerical search for the optimal solution is computationally prohibitive.

The signal model is even more complicated if K users share a same physical channel. For example, the received vector for a code-division multiple-access (CDMA) system takes the form

$$\mathbf{y}_{cdma} = \sum_{k=1}^{K} \mathbf{H}_k\mathbf{Q}_k\mathbf{M}_k\mathbf{S}_k\mathbf{\Pi}_k\mathbf{C}_k(\mathbf{b}_k) + \mathbf{n}, \tag{8.5}$$

where $\mathbf{S}_k = (\mathbf{I} \otimes \mathbf{s}_k)$, with column vector $\mathbf{s}_k$ representing the spreading code for user k. Other matrices with subscript k are defined in a similar manner to those in (8.1).

The algebraic structure of channel coding penetrates all the functional blocks of a communication system, forming a superstructure that belongs to the *entire* system. The constraints imposed by this superstructure greatly narrow down the region for the search of the message bit vector $\mathbf{b}$.

8.2   Traditional strategy

If we carefully examine the historical trajectory of communications in tackling the aforementioned challenge, we find that the dominant strategy used in the past was to split the global optimization problem into a number of localized suboptimization problems, so that the design of each functional block could be isolated from others. For example, one could introduce an equalizer to combat ISI, introduce spreading codes to prevent multiuser interference (MUI), and transmit along eigen beams of MIMO to eliminate inter-antenna interference (IAI), and finally perform decoding separately. These subsystem processing schemes are listed in Table 8.1.

The traditional strategy is characterized by *isolated* design and *isolated* processing. A typical flowchart is shown below.

$$\boxed{\text{Removal of channel-induced interference}} \longrightarrow \boxed{\text{Symbol detection}} \longrightarrow \boxed{\text{Decoding}}$$

Table 8.1   *Examples to illustrate traditional strategies*

Phenomenon	Source	Local Solutions	Remarks
ISI	Dispersive channels	Equalizer or OFDM	
MUI	Multiuser channels	User-space partition	Orthogonal principle
IAI	Correlated MIMO channels	Precoder, space-time codes	

The resulting solution is a local suboptimum, with the advantage of simplicity in architecture and computation.

8.3 Paradigm of orthogonality

The pillar of the traditional strategy of isolation is the paradigm of orthogonality, which is the simplest and most effective technique to prevent the occurrence of interference. The paradigm of orthogonality enables multiple signals to share a single time-dispersive channel without incurring interference. Partitioning the channel in frequency as orthogonal subchannels is intuitively simple and appealing, but not always feasible.

Interference exists in many communication systems, often being a dominant factor that degrades their error performance. The physical intuition of interference generation is simple, lying in that *multiple* symbols squeeze into a *lower* dimensional space. These symbols can come from a data stream, multiple users, or multiple antennas. The solution is also intuitive, and we must increase the space dimension to create sufficient "room" to allow these symbols to transmit on their exclusive orthogonal subchannels. In the history of communications, the paradigm of orthogonality is repeatedly employed in various scenarios to combat interference, conscientiously or otherwise. Very often, the construction of orthogonal subchannels for transmission corresponds to the singularity decomposition of a physical channel, and the orthogonal subchannels are defined by their singular vectors. This principle are equally applicable to both temporally dispersive channels and spatially correlated MIMO channels, but requiring the channel station information at the transmitter (CSIT). Nevertheless, the paradigm of orthogonality finds its applications in many practical scenarios without CSIT. For a sequentially transmitted symbol sequence over a time-invariant ISI channel, the channel matrix is of a Toeplitz structure and its singular vectors asymptotically approach the Fourier vectors. The resulting system operates on orthogonal subcarriers without the need for CSIT. For parallel data transmission over a MIMO channel, Alamouti wisely organized message symbols with redundancy in a spatially orthogonal pattern, which is now known as the *Alamouti's codes*. The orthogonality of the Alamouti space-time codes is convertible to a set of *virtual* orthogonal subchannels, when viewed at the receivers. In a multiple access environment, symbols from a number of users are transmitted in parallel, usually assumed to compete for accessing an AWGN channel. The AWGN channel is characterized by its identity channel matrix, leaving many degrees of freedom in choosing a set of basis functions. Different sets of basis functions lead to different multiple-access schemes such as TDMA, CDMA, and OFDMA.

Today, the idea of orthogonal channels has been extended to the concept of interference alignment. This design philosophy considerably simplifies the transceiver structures, *isolating* the processing of one symbol from the others to implement a symbol-by-symbol or a user-by-user decision. Figure 8.1 outlines various applications of the paradigm of orthogonality addressed or to be addressed in this book.

8.4 Turbo processing principle

Orthogonal design is not always feasible, especially when CSIT is not available or a channel matrix is not Toeplitz. The advent of the turbo principle represents a big move from the traditional philosophy of isolative design, allowing for a joint treatment of two subsystems including the codec and one other component in an iterative manner.

Turbo codes and the turbo processing principle (or simply the turbo principle) are two facets that constitute Berrou's unified framework. It is the turbo principle that makes concatenated codes a practical coding scheme for communications. Apart from its key role as a powerful means for decoding, the turbo principle opens a new horizon for us to ponder over the design framework for wireless communication.

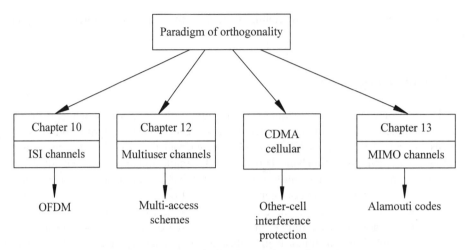

Figure 8.1 *Paradigm of orthogonality with applications*

The message vector **b** takes discrete values. The estimation problem in (8.4) falls into the framework of integer optimization, which usually has no closed-form solution and requires tedious calculations even when a numerical solution is sought. The difficulty is further aggravated if we observe the nonlinear operation $(\mathbf{Cb})_{\mathrm{mod2}}$ involved in the signal model.

Clearly, an algebraic coding structure, once imbedded into the transmitted data, belongs not only to the coding/decoding subsystems but to the entire communication system. It is also clear that for the best estimation, global integer optimization should be used, although it is mathematically intractable, especially till a couple of decades ago due to limited computing power. It is, therefore, common practice for a traditional transceiver to split the estimation into two separate steps of detection and decoding, at the cost a poorer decoding error performance. With the advent of low-cost, powerful computing devices, global optimization has become the trend for a better solution. Turbo decoding was born out of this transition.

Turbo decoding is not a global optimization yet. It represents a joint local optimal strategy by simultaneously considering constraints imposed by a few functional blocks, typically two. The elegance of the turbo principle lies in that it implements integer optimization via iteration, thereby wisely avoiding any analytical treatment of discrete-symbol constraints. The two blocks to be processed should be relatively independent. The commonly used combinations are as follows:

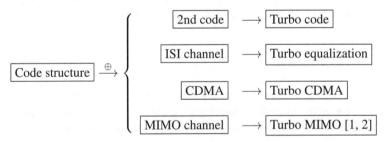

where the symbol $\xrightarrow{\oplus}$ means "incorporated into." In turbo code, a second code is artificially designed so that the turbo principle is applicable. The resulting contaminated code, no matter whether in series or parallel, possesses a much better performance. In all other cases, the second structure inherently exists in a communication environment, which, when exploited and used alongside the code structure, considerably improves the overall system performance.

The turbo principle represents a novel thought in solving estimation problems in wireless communications that involve nonlinear discrete constraints. With increased computing capability, its future evolution is likely to cover more functional blocks to approach global optimization for the best performance. The evolution of signal processing strategies for wireless communications can be outlined as follows:

$$\boxed{\text{Block-by-block process}} \longrightarrow \boxed{\text{Combined turbo/block process}} \longrightarrow \boxed{\text{Global optimization}}$$

Problems

Problem 8.1   Describe the basic idea behind the turbo principle. Is it possible to extend the turbo principle to a mechanism of three devices?

Problem 8.2   In a system using the interleave-division multiple access (IDMA) scheme, the received vector is given by

$$\mathbf{x}_{\text{IDMA}} = \sum_{k=1}^{K} \mathbf{H}_k \mathbf{Q}_k \mathbf{M}_k \mathbf{\Pi}_k \mathbf{R}_k \mathbf{C}_k \mathbf{b}_k + \mathbf{n},$$

where $\mathbf{R}_k$ is a repetition code matrix. Compare it with the CDMA signal model given in (8.5) and find their difference in philosophy. Offer your comments.

References

1. M. Sellathural and S. Haykin, "Turbo-BLAST for wireless communications: theory and experiments," *IEEE Trans. Signal Process.*, vol. 50, no. 10, pp. 2538–2546, 2002.
2. S. Haykin and M. Sellathural, "Turbo-MIMO for wireless communications," *IEEE Commun. Mag.*, vol. 42, no. 10, pp. 48–53, 2004.

9

Channel Equalization

Abstract

Channel equalization represents a *defensive* strategy against temporally dispersive channels. It created a splendid history of data transmission by enabling the data rate on a twisted line telephone to increase from 1.2 to 19.2 kb/s shortly in two decades from 1970s to 1980s. When a pulse sequence is transmitted over a dispersive channel, the pulse waveforms are spread over time, producing an overlap between the overall channel impulse response and its translates. All the dispersive symbols compete for use of the same channel, leading to a phenomenon referred to as inter-symbol interference (ISI). The influence of ISI can be minimized based on different criteria, resulting in different types of channel equalizers, which are studied in this chapter. A beautiful picture exists that links channel equalization as an engineering practice to the power spectral decomposition theory in stochastic processes.

♣

9.1 Introduction

For digital communication in a high-data-rate environment, the propagation is usually modeled as a frequency-selective channel. The channel behaves like a filter, which can be described by a channel matrix $\mathbf{H}$ in the discrete-time domain or, equivalently, by a nonideal bandlimited frequency transfer function $H(f)$ in the frequency domain. This filter causes the temporal spillover of signal waveforms. The consequence of competing for the use of the finite channel frequency resource by consecutive symbols is inter-symbol interference (ISI). A basic means to combat ISI is channel equalization.

The classical work of Nyquist [1] laid an important foundation for channel equalization. By assuming a noise-free environment, Nyquist derived a sufficient condition for completely removing the ISI of all the adjacent symbols with the current one at the sampling instant. Inspired by the pioneering work of Nyquist [1], research work has been done intensively on channel equalization for four decades, falling into two categories, that is, linear and nonlinear equalization. The simplest linear scheme is to directly decorrelate the channel-induced signal correlation to enable decisions made on a symbol-by-symbol basis, resulting in the so-called zero-forcing equalizer (ZFE) [2]. The price to pay for this simple structure of ZFE is noise amplification. The resulting signal-to-noise ratio (SNR) is proportional to the harmonic means of the powers of singular values of the channel matrix and is, thus, dominated by the smallest singular value. At very high data rates, most of multipath channels have very spread-out eigenvalues, forcing ZFE to work in poor conditions. To overcome the drawback of ZFE, the MMSE equalizer minimizes the overall effect of the noise and ISI. The idea is to make the overall frequency response as flat as possible, so that the corresponding impulse response looks like a delta function. The dominance of the channel's smallest eigenvalue is partly compensated by a compensating term in the denominator of the output SNR of an MMSE equalizer. However, the effect of such compensation gradually vanishes as the average transmit SNR increases.

Both ZFE and MMSE equalizers are of harmonic mean type, thus leaving a considerable amount of ISI component at their output. Part of this ISI residual can be further removed by using the previously detected symbols; the idea of using previous decisions to cope with the ISI problem was first introduced by Austin [3], The optimal DFE usually implementable in the form of a decision feedback equalizer (DFE) either in the time [4] or in the hybrid time-frequency domain [5]. The optimal DFE receiver was obtained by Monsen [6] under the MMSE criterion, while Price [7] derived the same result through spectral factorization. The joint transmitter and receiver optimization with MMSE was solved by Salz [8], also by resorting to spectral factorization; while the problem of optimal MMSE DFE receiver with QAM signaling was tackled in Ref. [9]. Belfiore and Park [10] investigated MMSE DFE receivers from a different prospective of linear prediction, showing that the DFE was realizable as an optimum linear equalizer followed by an estimator of a random distortion sequence. By assuming no errors in previous symbol decisions, it can be shown that the jointly optimized feed-forward filter (FFF) and feedback filter (FBF) are capable of completely removing all the ISI. This property is logically not surprising since the assumption is more demanding than the conclusion.

Besides the MMSE criterion, other criteria are also used in the derivation of channel equalization for better error performance, resulting in various maximum *a posteriori* probability (MAP) receivers [11–17] and maximum likelihood symbol sequence detectors [18]. Nonetheless, the MMSE equalizer seems more attractive in practice because of its trade-off in the system performance and implementation complexity.

The operation of DFE relies on the assumption of error-free decisions on the previous symbols and an infinitely long feedback loop. Any errors in detected symbols can cause severe error propagation. If channel state information is available at the transmitter (CSIT), it is shown that error propagation can be eliminated [19] by implementing the feedback loop as a Timlinson–Harashima precoder (THP) [20, 21] at the transmitter.

9.2   Pulse shaping for ISI-free transmission

For a digital communication system that transmits a sequence of pulse waveforms over a dispersive channel, we need to find out what kind of waveforms can be free from ISI. To proceed, let $p(t)$, $c(t)$, and $q(t)$ denote the pulse-shaping filter, channel impulse response, and receiver filter, respectively. The overall impulse response is thus given by

$$u(t) = p(t) \star c(t) \star q(t) \tag{9.1}$$

with $\star$ denoting the operator of convolution. Thus, when a symbol sequence $\{s_i, i = 0, \pm 1, \pm 2, \dots \}$ of symbol duration T_s is transmitted over the channel, the receiver filter output is expressible as

$$x(t) = \sum_{i=-\infty}^{\infty} s_i u(t - iT_s) + n(t), \tag{9.2}$$

where $n(t)$ denotes the AWGN with mean zero and noise variance Σ_n^2. Denote the symbol power and the average transmit SNR, respectively, by

$$\sigma_s^2 = \mathbb{E}[|s_i|^2]$$

$$\gamma_0 = \frac{\sigma_s^2}{\sigma_n^2}. \tag{9.3}$$

Suppose that we are interested in the symbol s_k and sample $x(t)$ at $t = kT_s$ to produce information about s_k:

$$x(kT_s) = s_k u(0) + \underbrace{\sum_{i=-\infty, i \neq k}^{\infty} s_i u((k-i)T_s)}_{\text{ISI}} + n(kT_s). \tag{9.4}$$

Clearly, apart from the desired signal component given in the first term, the channel impairment introduces an ISI component shown in the second term. Ideally, the overall impulse response $u(k)$ free of ISI should be designed such that

$$u(k) = \begin{cases} 1, & k = 0 \\ 0, & k \neq 0 \end{cases}, \tag{9.5}$$

which is a discrete delta function and thus has a constant Fourier transform. On the other hand, $u(k)$ is a sampled version of $u(t)$. From the sampling theorem, the Fourier spectrum of $u(k)$ equals the superposition of the spectrum of $u(t)$ with its shifted versions, as shown by

$$\mathcal{F}\{u(k)\} = \sum_{i=-\infty}^{\infty} U\left(f - \frac{i}{T_s}\right). \tag{9.6}$$

It follows that

$$\sum_{i=-\infty}^{\infty} U\left(f - \frac{i}{T_s}\right) = \text{constant}. \tag{9.7}$$

The overall frequency transfer function $U(f)$ satisfying this condition takes the form of a raised cosine, and the corresponding waveform $u(t)$ is usually referred to as the *Nyquist waveform*, named after its inventor Nyquist. Since we are only concerned with the sampling instants $t = kT_s$, Nyquist shaping guarantees that the adjacent symbols have no interference with the current symbol at the *sampling instant*.

Nyquist shaping is important; however, it only guarantees transmission free of ISI but does not guarantee achieving a maximum signal-to-interference-plus-noise ratio at the receiver output. ISI is indeed a major issue in a bandlimited communication system, but it is not the only consideration. Noise amplification must be considered as well, in order to obtain a good system error performance. Various equalization techniques, which we will study, represent the efforts along this direction.

9.3 ISI and equalization strategies

For most practical situations, a channel is either time-varying or unknown, making it impossible to employ a fixed raised-cosine type of receiver filter to completely eliminate ISI. Denote the overall filter by

$$h(t) = p(t) \star c(t). \tag{9.8}$$

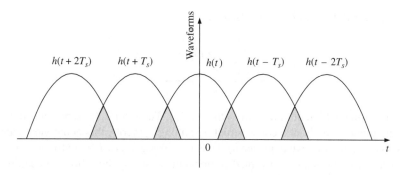

Figure 9.1   *Overlap among adjacent waveforms, which is the source of ISI*

Without loss of generality, normalize $h(t)$ so that $h(0) = 1$. To get insight into the way of constructing a receive filter $g_o(t)$, we focus on the detection of symbol s_0 since the detection of other symbols is similar. Represent the channel output as

$$x(t) = s_0 h(t) + \sum_{i=-\infty, i\neq 0}^{\infty} s_i h(t - iT) + n(t). \tag{9.9}$$

If the duration of $h(t)$ is less than T_s, there is no interference among symbols and the receiver can simply use $g_o(t) = h(t)$ for correlation. For most practical applications, $h(t)$ has a long tail, causing a spillover of a number, say m, of its adjacent shifted versions $h(t \pm iT_s), i = 1, 2, \cdots, m$. The situation of ISI caused by waveform spill-over is shown in Figure 9.1. Without certain algebraic (e.g., orthogonality or coding) structures added to the channel waveform, the overlapped portion of $h(t)$ is difficult to recover.

The overall channel response is continuous in time, possibly of infinite length, while symbol detection is conducted once every T_s seconds. Then, how would the corresponding receiver look like? There are two ways to derive the receiver structure. The first approach is based on some advanced mathematical tools, mainly calculus of variation and spectral factorization [6, 8]. The resulting equalizer is preceded with a filter, $h^*(-t)$, to match the overall channel impulse, so that the filter output SNR at the sampling instants $t = iT_s$ is maximized. This arrangement agrees with our physical intuition. The samples from the matched filter are then applied to a tap-delayed line equalizer, which is designed under certain criteria, such as ZFE, MMSE, and DFE. The second, and more natural, way to introduce equalization is a geometric method based on orthogonal projection, whereby an equivalent channel model and its corresponding equalizer are determined all at once. We follow the second approach, and expound its principle in the context of ZFE.

9.4   Zero-forcing equalizer

Without loss of generality, let us focus on symbol s_0, which is carried by the waveform $h(t)$. Part of the waveform $h(t)$ is corrupted by its shifted versions that carry other symbols. Zero-forcing equalization is closely related to the notion of orthogonal decomposition of $h(t)$. We thus begin with orthogonal projection before proceeding.

9.4.1   Orthogonal projection

Assuming an order-$(2m + 1)$ linear smoothing filter (i.e., interpolator), we can decompose $h(t)$ into two components, that is, the portion $h_c(t)$ that is expressible as a linear combination of its shifted versions, and its *orthogonal* complement denoted by $e_o(t)$. That is,

$$h(t) = \underbrace{\sum_{i=-m, i \neq 0}^{m} c_i h(t - iT_s)}_{h_c(t)} + e_o(t), \tag{9.10}$$

where c_i is the projection coefficient of $h(t)$ onto $h(t - iT_s)$. The term $h_c(t)$ represents the component *corrupted* by adjacent symbols. Without a coded structure among symbols, this component cannot be used for the retrieval of s_0. Geometrically, we consider $h_c(t)$ as the projection of $h(t)$ onto the space spanned by its shifted versions, leaving $e_o(t)$ as an uncorrupted component satisfying the following condition:

$$e_o(t) \perp \text{span}\{h(t - jT_s), j = \pm 1, \ldots, \pm m\}. \tag{9.11}$$

The situation is shown in Figure 9.2.

The orthogonality here implies that $e_o(t) \perp h(t - jT_s)$ for all $j \neq 0$, namely

$$\underbrace{\int_{-\infty}^{\infty} [h(t) - \sum_{i \neq 0} c_i h(t - iT_s)]}_{e_o(t)} h^*(t - jT_s) dt = 0, \tag{9.12}$$

which can be simplified to

$$\sum_{i=-\infty, i \neq 0}^{\infty} r_{j-i} c_i = r_j, j = \pm 1, \ldots, \pm m, \tag{9.13}$$

where r's denote the cross correlation between $h(t)$ and its shifted versions, given by

$$r_{j-i} = \int_{-\infty}^{\infty} h(t - iT_s) h^*(t - jT_s) dt. \tag{9.14}$$

To find the projection coefficients c's, denote vectors

$$\mathbf{c} = [c_{-m}, \ldots, c_{-1}, c_1, \ldots, c_m]^T,$$

$$\mathbf{r}_{\backslash 0} = [r_{-m}, \ldots, r_{-1}, r_1, \ldots, r_m]^T \tag{9.15}$$

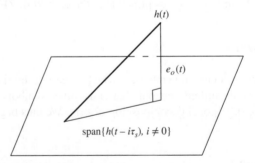

Figure 9.2 *Geometric significance of $e_o(t)$*

and matrix

$$
\mathbf{R}_{\backslash 0} = \begin{pmatrix}
r_0 & r_{-1} & \cdots & r_{m+1} & r_{-m-1} & \cdots & r_{-2m} \\
r_1 & r_0 & \cdots & r_{-m+2} & r_{-m} & \cdots & r_{-2m+1} \\
\vdots & & \vdots & & & \vdots & \\
r_{m-1} & r_{m-2} & \cdots & & & & r_{-m-1} \\
r_{m+1} & r_m & \cdots & & & & r_{-m+1} \\
\vdots & & & & & & \vdots \\
r_{2m} & r_{2m-1} & \cdots & & & \cdots & r_0
\end{pmatrix}.
\tag{9.16}
$$

It follows from (9.13) that the projection coefficients c's are equal to

$$
\mathbf{c} = \mathbf{R}_{\backslash 0}^{-1}\mathbf{r}_{\backslash 0}, \text{ with } m \to \infty.
\tag{9.17}
$$

Note that the above covariance matrix is not a Toeplitz matrix; but it can be easily generated from a Toeplitz one by deleting the $(m+1)$th row and the $(m+1)$th column.

The projection error $e_o(t)$ can be considered as the output of the interpolation error filter; namely

$$
e_o(t) = h(t) - \sum_{i \neq 0} c_i h(t - iT_s) = \sum_{i=-\infty}^{\infty} a_i h(t - iT_s),
\tag{9.18}
$$

where the interpolation filter coefficients a's are defined by

$$
a_i = \begin{cases} -c_i, & i \neq 0 \\ 1, & i = 0 \end{cases} \text{ or, in vector form,}
$$

$$
\mathbf{a} = [\ldots, -c_{-1}, 1, -c_1, \ldots]^T.
\tag{9.19}
$$

As such, when defined in the z-transform domain, the interpolation error filter is given by

$$
A(z) = \sum_{i=-\infty}^{\infty} a_i z^{-i}, \text{ and}
$$

$$
e_o(t) = A(z)[h(t)],
\tag{9.20}
$$

where $A(z)$ is treated as a delay operator performing on $h(t)$, such that $z^{-i}h(t) = h(t - iT_s)$.

9.4.2   ZFE

The strategy of ZFE employs only the uncorrupted component for symbol retrieval. To this end, it uses $e_o^*(-t)$ as the local reference to correlate the incoming signal $x(t)$. ZFE is, in essence, an interpolation-error filter. Using the relationships (9.12)–(9.13) and the fact that

$$
\langle h(t), e_o(t) \rangle = \langle h_c(t) + e_o(t), e_o(t) \rangle = \|e_o\|^2,
\tag{9.21}
$$

we can see that

$$
\langle h(t - iT_s), e_o(t) \rangle = \|e_o\|^2 \delta(i),
\tag{9.22}
$$

which, when expressed explicitly in terms of (9.14), gives an expression in discrete form as

$$
\sum_{i=-\infty}^{\infty} r_{j-i} a_i = \|e_o\|^2 \delta(j).
\tag{9.23}
$$

This equation provides another way to determine the ZFE coefficients a's apart from (9.17). A ZFE filter is usually of infinite length. However, it can be approximately implemented with finite taps, say $\mathbf{a}_m = [a_{-m}, \ldots, a_{-1}, 1, a_1, \ldots, a_m]^T$. It should be emphasized that the equivalent channel filter $R_h(z)$ in Figure 9.5 cannot be truncated since it represents the mechanism inside the black box. Let $\mathbf{R}_m$ denote the Toeplitz channel correlation matrix with $[r_0, r_{-1}, \ldots, r_{-2m}]^T$ as its first row and with $[r_0, r_1, \ldots, r_{2m}]^T$ as its first column. We may rewrite the ZFE filter defined in (9.23) in vector form to yield

$$\mathbf{a}_m = [a_{-m}, \ldots, 1, \ldots, a_m]^T = \mathbf{R}_m^{-1}\mathbf{d}_m, \tag{9.24}$$

with $\mathbf{d}_m$ denoting the all-zero column $(2m+1)$ vector except for the $(m+1)$th entry of unity. Note that we usually need to normalize the result calculated from the right-hand side to ensure $a_0 = 1$. The $(2m+1)$-tap ZFE produces error power $\mathbf{a}_m^\dagger \mathbf{R}_m \mathbf{a}_m = \mathbf{d}_m^\dagger \mathbf{R}_m^{-1} \mathbf{d}_m$, which is greater than the interpolation error power $\|e_o\|^2$ but approaches the latter as $m \to \infty$.

It is often more convenient to describe an equalizer in the z-transform domain. Denote the z-transform of r_i by $R_h(z)$. Taking the z-transform of both sides of (9.24), we obtain the interpolation-error filter

$$A(z) = \frac{\|e_o\|^2}{R_h(z)}. \tag{9.25}$$

Since the numerator is a scaling factor, we simply employ $A(z) = 1/R_h(z)$ as the equalizer in subsequent discussion.

A basic block diagram of ZFE is sketched in Figure 9.3. The output of the correlator is sampled at every T_s seconds, producing symbol information for $s_0, s_1, \ldots$, respectively. In particular, the output sample at time instant $t = 0$ is given by

$$y(t)\|_{t=0} = \langle x(t), e_o^*(t) \rangle$$

$$= s_0 \underbrace{\int_{-\infty}^{\infty} |e_o(t)|^2 dt}_{\|e_o\|^2} + \underbrace{\int_{-\infty}^{\infty} n(t)e_o^*(t)dt}_{w}$$

$$= s_0\|e_o\|^2 + w, \tag{9.26}$$

where $w \sim \mathcal{CN}(0, \sigma_n^2\|e_o(t)\|^2)$. Clearly, all the ISI from symbols other than s_0 has been completely removed. Its signal output is the interpolation error of $s_0 h(t)$. It follows that the output SNR is given by

$$\gamma_{ZFE} = \|e_0\|^2(\gamma_0). \tag{9.27}$$

The channel interpolation-error power $\|e_o\|^2$ depends only on the channel power spectrum through its harmonic mean, as will be shown in the next subsection.

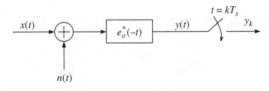

Figure 9.3   *Another way to see the zero-forcing equalizer*

✍ **Example 9.1** _____

Consider transmission, at symbol rate $1/T_s$, over a channel with impulse response $h(t) = \sqrt{2\alpha} \exp(-\alpha t)$, $t > 0, \alpha > 0$. Determine the equivalent channel filter $R_h(z)$ and the ZFE filter $A(z)$. If the channel parameter α is set such that $\alpha T_s = 0.5$, find the filter coefficients of the resulting ZFE.

Solution

For $\tau > 0$, the channel correlation function is found to be

$$r_h(\tau) = \int_0^\infty h(t+\tau)h(t)dt = 2\alpha \int_0^\infty e^{-\alpha t}e^{-\alpha(t+\tau)}dt = e^{-\alpha \tau}. \tag{9.28}$$

It is easy to examine that $r_h(-\tau) = r_h(\tau)$. Thus, we can simply write

$$r_h(\tau) = e^{-\alpha|\tau|}.$$

Physically, α represents the 3-dB bandwidth of the overall channel. Sampling $r_h(\tau)$ at the symbol rate and denoting

$$\omega_0 = \alpha T_s, r = e^{-\omega_0}, \tag{9.29}$$

we obtain a discrete correlation sequence $\{r_h(k) = r^k, \forall k\}$. The equivalent channel filter can, thus, be determined as

$$R_h(z) = \sum_{n=-\infty}^\infty r_h(n)z^{-n} = \sum_{n=0}^\infty r^n z^{-n} + \sum_{m=1}^\infty r^m z^m$$

$$= \frac{1}{1 - rz^{-1}} + \frac{r}{1 - rz}$$

$$= \frac{1 - r^2}{(1 - rz^{-1})(1 - rz)}, \quad \text{for } |r| < |z| < |r|^{-1}. \tag{9.30}$$

Up to a constant factor, its reciprocal defines the ZFE filter

$$A(z) = rz + (1 + r^2) + rz^{-1},$$

which has only three nonzero tap coefficients. In particular, for $\omega_0 = 0.5$ and after normalization, these three taps are $a_1 = a_{-1} = 0.4434$ and $a_0 = 1$.   ○

〰✍

9.4.3   Equivalent discrete ZFE receiver

From the previous discussion, we see that the ZFE conducts continuous-time convolution before the sampler and operates in the discrete-time domain after that. We are interested in deriving a discrete ZFE model.

Referring to Figure 9.3 and $x(t)$ given in (9.9), we can write the equalizer's output as

$$y(t) = \int_{-\infty}^\infty x(\tau)e_o^*(\tau - t)d\tau = y_s(t) + w(t),$$

where the signal component $y_s(t)$ is defined as

$$y_s(t) = \int_{-\infty}^\infty \left(\sum_i s_i h(t - iT_s)\right)\left(\sum_j a_j h^*(\tau - jT_s - t)\right) d\tau, \tag{9.31}$$

which, when sampled at $t = kT_s$, produces output samples

$$y_s(t)|_{t=kT_s} = s_k\|e_o\|^2$$

$$= \sum_j a_j \sum_i s_i \int_{-\infty}^{\infty} h(\tau - iT_s)h^*(\tau - kT_s - jT_s)\, d\tau$$

$$= \sum_j a_j \underbrace{\sum_i s_i r_{k+j-i}}_{u_{k+j}}. \tag{9.32}$$

The above expression indicates that the signal output from the ZFE is the result of the operator $A(z)R_h(z)$ operating on the input symbols:

$$y_s(k) = A(z)R_h(z)s_k, \tag{9.33}$$

where the z-transform $A(z)$ has been defined in (9.20) and $R_h(z)$ is the z-transform of r_i defined just above (9.25). The expression (9.32) reveals a structure of ZFE, which consists of a front-end filter matched to the overall channel response $h(t)$, a sampler at the symbol rate, and an equalizer $A(z)$. The equivalent channel model is shown in Figure 9.4a. Figure 9.4b shows its discrete counterpart, which is capable of producing statistically sufficient samples. It is easy to verify by direct calculation that the two equivalent channel diagrams lead to the same results. Since discrete samples are much easier to handle via various signal processing techniques, we will adopt them in the subsequent analysis. The noise filter $f^*(z^{-1})$ is derivable from the factorization of $R_h(z)$ through the Fejer–Riesz theorem [23], as shown by

$$R_h(z) = f(z)f^*(z^{-1}). \tag{9.34}$$

The decomposition is unique if we impose a causal constraint on $f(z)$ that all of its poles are inside the unit circle. Although derived in the context of ZFE, the models shown in Figure 9.4 are equally applicable to the MMSE and DFE equalizers.

Summarizing the above discussion, we can write a discrete model for the ZFE, yielding

$$y_k = A(z)u_k = A(z)[R_h(z)s_k + w_k], \tag{9.35}$$

which offers a basis to construct a discrete equivalent model for the ZFE. One such model based on (9.35) is shown in Figure 9.5.

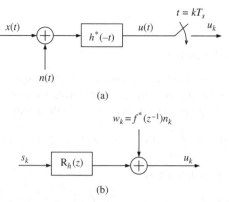

(a)

(b)

Figure 9.4 *Matched filter to generate statistically sufficient samples and its equivalent diagram: a) original subsystem with a matched filter followed by a sampler; b) an equivalent discrete-time model.*

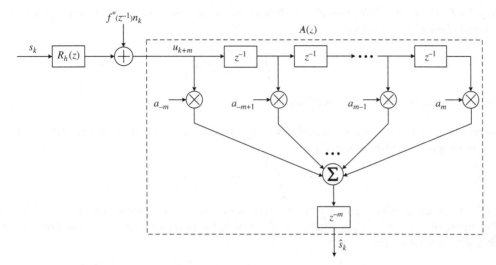

Figure 9.5 *Equivalent block diagram of the ZFE receiver*

Property 9.4.1 *The ZFE $A(z)$ indeed decorrelates the signal components in u_k, resulting in*

$$y_k = s_k + f^{-1}(z)n_k, \qquad (9.36)$$

thus satisfying the ISI-free condition.

The proof is trivial since $A(z) = 1/R_h(z) = 1/[f(z)f^*(z^{-1})]$. From the above expression, we see that ZFE is infeasible when channel frequency response has singularities. In this case, the ZFE produces no signal output; that is, the output SNR is zero. We also see that the ZFE decorrelates the signal component but colors the noise. It remains to determine the noise power in y_k.

Property 9.4.2 *Let $\tilde{S}_h$ denote the aliased version of the power spectrum of $h(t)$ due to sampling the matched filter output at the symbol rate. The noise power in the ZFE output is given by*

$$\sigma_{n,\text{ZFE}}^2 = \sigma_n^2 \int_{-1/2}^{1/2} \tilde{S}_h^{-1}(f)df. \qquad (9.37)$$

An explicit expression of $\tilde{S}_h$ will be derived shortly in (9.40). The noise output at the ZFE can be regarded as passing noise n_k through the discrete filter $f^{-1}(z)$. Since $|f^{-1}(z)|^2 = R_h(z)$, we have

$$\sigma_{n,\text{ZFE}}^2 = \sigma_n^2 \int_{-1/2}^{1/2} R_h^{-1}(z)|_{z=e^{j\omega}}df,$$

where $\omega = 2\pi f$ denotes the angular frequency in radian/sec. Thus, here and hereafter, we need to distinguish the two usages of symbol f. As a variable in ω or an integration variable, it stands for the frequency in Hz whereas denoting a function such as $f(z)$, it represents a filter. The proof is complete if we note the integrand $R_h(z)$, when setting $z = e^{j2\pi f}$, is just the power spectrum $\tilde{S}_h(f)$ in accordance with the Wiener–Khinchin theorem. Thus, the symbols $R_h(z)|_{z=e^{j\omega}}$ and $\tilde{S}_h(f)$ will be used interchangeably in this chapter.

Property 9.4.3 *The output SNR at the ZFE is dictated by the harmonic mean of the aliased power spectrum of the overall channel, as shown by*

$$\gamma_{\text{ZFE}} = \gamma_0 \left[\int_{-1/2}^{1/2} \tilde{S}_h^{-1}(f) df \right]^{-1}, \tag{9.38}$$

where γ_0 is the average transmit SNR defined in (9.3).

There are two ways to determine the folded power spectrum of the overall channel. Recall that $r_k = r(kT_s)$ results from sampling the channel correlation

$$r(\tau) = \int_{-\infty}^{\infty} h(t)h(t - \tau) dt \tag{9.39}$$

at $\tau = kT_s$. The Fourier transform of $r(\tau)$ is the power spectrum of $h(t)$, denoted here by $S_h(f)$. Recall that sampling theory of $r(\tau)$ causes the periodic repetition of $S_h(f)$ in the frequency domain. Thus, the Fourier transform of r_k takes the form

$$\tilde{S}_h(f) = \frac{1}{T_s} \sum_{m=-\infty}^{\infty} S_h\left(f + \frac{m}{T_s} \right) = \frac{1}{T_s} \sum_{m=-\infty}^{\infty} \left| H\left(f + \frac{m}{T_s} \right) \right|^2, \tag{9.40}$$

with $H(f)$ denoting the Fourier transform of $h(t)$. The second way is straightforward, directly calculating $\tilde{S}_h(f)$ from the z-transform of r_k by using

$$\tilde{S}_h(f) = R_h(z)|_{z=e^{j\omega}} = \sum_{i=-\infty}^{\infty} r_i z^{-i}|_{z=e^{j\omega}}. \tag{9.41}$$

✍ **Example 9.2**

We continue from Example 9.1. Find the equivalent block diagram for the channel with a ZFE, and determine the output noise power.

Solution

As obtained in Example 9.1, the equivalent channel filter is

$$R_h(z) = \frac{1 - r^2}{(1 - rz^{-1})(1 - rz)}$$

$$= f(z)f^*(z^{-1}), \quad |r| < |z| < 1/|r|,$$

where $r = \exp(-\alpha T_s)$. The equivalent noise filer is given by

$$f^*(z^{-1}) = \frac{(1 - r^2)^{1/2}}{1 - rz}.$$

The ZFE output noise power is equal to

$$P_n = \sigma_n^2 \int_{-1/2}^{1/2} R_h^{-1}(z)|_{z=e^{j\omega}} df$$

$$= \frac{\sigma_n^2}{1 - r^2} \int_{-1/2}^{1/2} (1 + r^2 - 2r \cos \omega) df$$

$$= \sigma_n^2 (1 + r^2)/(1 - r^2).$$

Thus, as $r \to 0$, that is, as the channel bandwidth is much wider than the symbol rate, $P_n \to \sigma_n^2$, implying no noise amplification. On the other extreme, as $r \gg 1$, that is, as the channel bandwidth is much smaller than the symbol rate, the output noise power is amplified to infinity.

In practice, a truncated ZFE $A_m(z)$ may be used instead. In this case, the residual ISI should be found from $A_m(z)R_h(z)s_k$, by all the terms other than s_k. The output noise power is evaluated by

$$P_n = \sigma_n^2 \int_{-1/2}^{1/2} |A_m(z)R_h(z)|^2_{z=e^{j\omega}} \, df.$$

9.5 MMSE linear equalizer

The equivalent diagram for the ZFE consists of a receiver filter $h^*(-t)$ matched to the overall channel response, a sampler, and a discrete ZFE filter $A(z)$ which is designed for the complete removal of ISI induced by the channel, certainly at the cost of noise amplification. The MMSE equalizer has the same structure, but is designed to minimize the overall effect of the ISI and AWGN components using the MMSE criterion.

Let $u(t)$ denote the matched filter output in response to $x(t)$ given in (9.9). Then, we can write

$$u(t) = h(-t) \star x(t) = h(-t) \star \sum_{i=-\infty}^{\infty} s_i h(t - iT_s) + \underbrace{h(-t) \star n(t)}_{e(t)}. \tag{9.42}$$

The output noise component $e(t)$ from the matched filter is colored noise with zero mean and covariance:

$$r_e(\tau) = \mathbb{E}[e(t)e^*(t - \tau)] = \sigma_n^2 r_h(-\tau), \tag{9.43}$$

where $r_h(\tau)$ is the channel correlation defined in (9.39). Sample $u(t)$ at $t = kT_s$, and let $\mathbf{u}$ denote the $2m \times 1$ sample vector centered at time index zero. Then, in vector form, we can rewrite (9.42) as

$$\mathbf{u} = \mathbf{R}_h \mathbf{s} + \mathbf{e} \quad \text{where} \quad \mathbf{e} \sim \mathcal{CN}(\mathbf{0}, \sigma_n^2 \mathbf{R}_h), \tag{9.44}$$

and $\mathbf{R}_h = [r_h(i, j)]$ denotes the $(2m + 1)$-by-$(2m + 1)$ channel covariance matrix.

With the MMSE criterion for channel equalization, the linear estimator of symbol s_0 is $\hat{s}_0 = \mathbf{a}^\dagger \mathbf{u}$, and we need to find the tap weight vector $\mathbf{a}$ of the MMSE equalizer that minimizes the mean error power

$$J(\mathbf{a}) = \mathbb{E}[|s_0 - \mathbf{a}^\dagger \mathbf{u}|^2]. \tag{9.45}$$

Intuitively, the equalizer $\mathbf{a}$ should be chosen such that the error process is orthogonal to $\mathbf{u}$, that is

$$\mathbb{E}[\mathbf{u}(s_0 - \mathbf{a}^\dagger \mathbf{u})^*] = 0. \tag{9.46}$$

By directly working out $\mathbb{E}[\mathbf{u}\mathbf{u}^\dagger]$ and $\mathbb{E}[s_0^* \mathbf{u}]$, we obtain $(\gamma_0^{-1}\mathbf{I} + \mathbf{R}_h)\mathbf{a} = \mathbf{d}$, which leads to the MMSE equalizer

$$\mathbf{a} = (\gamma_0^{-1}\mathbf{I} + \mathbf{R}_h)^{-1}\mathbf{d}, \tag{9.47}$$

where $\mathbf{d} = [0, \ldots, 0, 1, 0, \ldots, 0]^T$ is the signal of delta type. Inserting (9.47) into (9.45), we obtain the minimum mean squared error as

$$J_{\min} = \sigma_s^2 - \mathbf{a}^\dagger \, \mathbb{E}[\mathbf{u}\mathbf{u}^\dagger]\mathbf{a} = \sigma_s^2[1 - \mathbf{d}^\dagger(\gamma_0^{-1}\mathbf{I} + \mathbf{R}_h)^{-1}\mathbf{R}_h\mathbf{d}]. \tag{9.48}$$

Property 9.5.1 *The infinite-length MMSE equalizer is determined by the channel correlation through*

$$A(z) = \frac{1}{R_h(z) + \gamma_0^{-1}}. \tag{9.49}$$

The proof is straightforward if we take the z-transform of Equation (9.47) and treat $\mathbf{d}$ as the delta function.

Property 9.5.2 *For a $(2m + 1)$-long MMSE equalizer, if we denote the eigenvalues of $\mathbf{R}_h$ by λ_i, then the minimum mean squared error is approximately given by*

$$J_{\min} = \sigma_n^2 \sum_{i=0}^{2m} (\lambda_i^2 + \gamma_0^{-1})^{-1}. \tag{9.50}$$

In the extreme case with $m \to \infty$, we have

$$J_{\min}^{\infty} = \lim_{m \to \infty} J_{\min} = \sigma_n^2 \int_{-1/2}^{1/2} \frac{1}{\gamma_0^{-1} + \tilde{S}_h(f)} df, \tag{9.51}$$

where $\tilde{S}_h(f)$ denotes the folded power spectrum of the channel impulse response $h(t)$ and is numerically related to the channel transfer function $H(f)$ by

$$\tilde{S}_h(f) = \frac{1}{T_s} \sum_{k=-\infty}^{\infty} \left| H\left(f + \frac{k}{T_s}\right)\right|^2 \tag{9.52}$$

or directly evaluated from $R_h(z)$ by setting $z = e^{j\omega}$.

Note that $J_{\min}$ represents the error power between the symbol and its linear estimate, containing both the noise and residual ISI components. Since from (9.48) the signal power in the estimated symbol equals $(\sigma_s^2 - J_{\min})$, we have the following assertion:

Property 9.5.3 *The average SNR at the MMSE equalizer output is given by*

$$\gamma = \frac{\mathbf{d}^{\dagger}(\gamma_0^{-1}\mathbf{I} + \mathbf{R}_h)^{-1}\mathbf{R}_h\mathbf{d}}{1 - \mathbf{d}^{\dagger}(\gamma_0^{-1}\mathbf{I} + \mathbf{R}_h)^{-1}\mathbf{R}_h\mathbf{d}}. \tag{9.53}$$

When the length of equalizer tends to infinity asymptotically, we have

$$\gamma_{\infty} = \frac{1 - J_{\min}^{\infty}}{J_{\min}^{\infty}} = \gamma_0 \left[\int_{-1/2}^{1/2} \frac{1}{\tilde{S}_h(f) + \gamma_0^{-1}} df\right]^{-1} - 1.$$

The MMSE equalizer tries to make the overall frequency response as flat as possible, so that the corresponding impulse response is like a delta function. Hence, it performs better than its ZFE counterpart in terms of SNR.

✍ **Example 9.3** ———

To illustrate the MMSE linear prediction-error power, let us consider a rational power $\tilde{S}_h(\omega) = \frac{\sigma_e^2}{|1 - a\, z^{-1}|^2}\big|_{z=\exp(j\omega)}$, which physically represents the power spectrum of an AR(1) process driven an i.i.d. process $e_k \sim \mathcal{CN}(0, \sigma_e^2)$. Calculate the total power and harmonic mean of this spectrum, and compare the result with an order-1 linear prediction filter output.

Solution

Denote $\alpha = |a|$, and simplify the power spectrum to

$$\tilde{S}_h(\omega) = \frac{\sigma_e^2}{1 + \alpha^2 - 2\alpha \cos \omega}.$$

First, determine the total power, which is equal to

$$\sigma_x^2 = \frac{1}{2\pi} \int_{-\pi}^{\pi} \tilde{S}_h(\omega) d\omega = \frac{1}{2\pi} \int_{-\pi}^{\pi} \frac{\sigma_e^2}{(1 + \alpha^2) - 2\alpha \cos \omega} d\omega.$$

By virtue of the Integral Formula (341) of the *CRC Handbook* [24], we have

$$\sigma_x^2 = \frac{\sigma_e^2}{\pi(1 - \alpha^2)} \tan^{-1} \left[\frac{1 + \alpha}{1 - \alpha} \tan^{-1} \frac{x}{2} \right]_{x=-\pi}^{\pi} = \frac{\sigma_e^2}{1 - \alpha^2}.$$

We next turn to the linear prediction-error power σ_e^2. For an AR(1) process, its partial correlation is numerically equal to the model coefficient. Hence, according to linear prediction theory, we obtain

$$\sigma_e^2 = \sigma_x^2(1 - \alpha^2).$$

Finally, we determine the harmonic mean by evaluating

$$H_m = \left[\frac{1}{2\pi} \int_{-\pi}^{\pi} \tilde{S}_h^{-1}(\omega) d\omega \right]^{-1} = \frac{\sigma_e^2}{1 + \alpha^2} < \sigma_e^2.$$

The harmonic mean represents the smoothed-estimate error power, and is clearly smaller than its linear prediction counterpart, as expected.

9.6   Decision-feedback equalizer (DFE)

From the previous study, we can fully or partly remove the ISI component by using the ZF or MMSE criterion. By using the MMSE criterion, we can achieve the optimal tradeoff between the ISI and noise components, so that the SNR is maximized. We also observe that, in doing so, the noise-plus-ISI residue becomes correlated. A natural question is, therefore: can we exploit such correlation to further reduce residual noise component? The answer is in the affirmative by exploiting the nonlinearity in the decision device, resulting in a so-called DFE with its block diagram sketched in Figure 9.6. The equivalent channel model for DFE is the same as that for ZFE. A DFE is composed of a forward transversal filter, which is designed to provide a smoothed estimate of the current symbol, and a decision-feedback filter, which serves as a linear predictor of the current

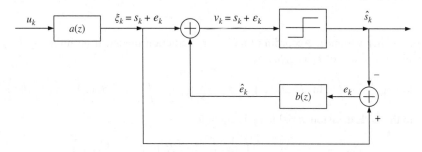

Figure 9.6   *Block diagram for DFE*

noise component by past noise samples [25]. It is the nonlinear symbol decision that makes it possible to separate the past noise samples from the pre-decision variables. Again, the DFE can be designed under either the MMSE or the ZF criterion, and here we focus on the MMSE criterion since the treatment of the latter is similar. As shown in the preceding section, the MMSE forward filter should be chosen to be

$$[s_k - a(z)u_k] \perp u_k,$$ (9.54)

implying that

$$\mathbb{E}[\{s_k - a(z)u_k\}u_k^*] = 0,$$ (9.55)

which, after some manipulation, leads to

$$a(z) = \frac{1}{R_h(z) + \gamma_0^{-1}}.$$ (9.56)

Here, we intentionally use $a(z)$ to denote the z-transform of the forward-filter coefficients of the DFE to distinguish from its counterpart, $A(z)$, for the ZFE. Just as was done in DFE, $R_h(z)$ is factored into $R_h(z) = f(z)f^*(z^{-1})$, so that the noise in the equivalent channel model is $f^*(z^{-1})n(t)$.

The output from an MMSE transversal filter $a(z)$ is given by

$$\xi_k = a(z)[R_h(z)s_k + f^*(z)n_k]$$
$$= s_k + \underbrace{[a(z)R_h(z) - 1]s_k + a(z)f^*(z)n_k}_{e_k}.$$ (9.57)

The process $\{e_k\}$ is the combined noise and ISI residue in the forward filter's output, which, when inserted with (9.56), can be explicitly represented as

$$e_k = -\underbrace{\frac{\gamma_0^{-1}}{R_h(z) + \gamma_0^{-1}}s_k}_{\text{ISI}} + \underbrace{\frac{f^*(z)}{R_h(z) + \gamma_0^{-1}}n_k}_{\text{noise}}.$$ (9.58)

The second and the third terms on the last line represent the residual ISI and noise components, respectively; they are mutually uncorrelated. Their power spectral density can be considered as the result of passing the signal and noise sequences through the corresponding filters. Thus, the use of the Wiener–Khinchin theorem enables us to yield their combined power spectral density

$$S_e(\omega) = \frac{\gamma_0^{-2}\sigma_s^2 + R_h(z)\sigma_n^2}{[R_h(z) + \gamma_0^{-1}]^2} = \frac{\sigma_n^2}{R_h(z) + \gamma_0^{-1}}|_{z=j\omega}.$$ (9.59)

Clearly, the combined error process e_k is a correlated process. Thus, its past samples, if available, can be used to predict its current sample value. We can remove this predicted error component from the forward filter's output ξ_k to produce a new decision variable v_k

$$v_k = s_k + \epsilon_k$$ (9.60)

with error ϵ_k of smaller variance. According to Wiener prediction theory, if an infinitely long predictor is used, the residual error power of ϵ_k is given by

$$\sigma_\epsilon^2 = \exp\left(\frac{1}{2\pi}\int_{-\pi}^{\pi}\ln S_e(\omega)d\omega\right) = \sigma_n^2 \exp\left(-\frac{1}{2\pi}\int_{-\pi}^{\pi}\ln\left\{\tilde{S}_h(\omega) + \gamma_0^{-1}\right\}d\omega\right).$$ (9.61)

Thus, the SNR in the DFE decision variable v_k is equal to

$$\gamma_{\text{DFE}} = \frac{\sigma_s^2}{\sigma_\epsilon^2} = \gamma_0 \exp\left(\frac{1}{2\pi}\int_{-\pi}^{\pi}\ln\left\{\tilde{S}_h(\omega) + \gamma_0^{-1}\right\}d\omega\right).$$ (9.62)

This is the maximum signal-to-ISI residual plus noise power ratio achievable at the decision-device input of the DFE, assuming ideal decision making without errors.

Equation (9.61) shows the minimum power of noise plus ISI residual achievable by the DFE. The actual ISI and noise residual power P_n is usually larger because of the error propagation in the feedback loop. To find the actual P_n, approximately represent the past samples as the difference between the decision output $\hat{s}_k$ and ξ_k to yield

$$e_i = \xi_i - \hat{s}_i, \ i = k-1, k-2, \cdots. \tag{9.63}$$

The feedback filter $b(z)$ is a linear predictor of e_k, based on its past samples $\{e_i\}$. It takes the form of $b(z) = \sum_{i=1} b_i z^{-i}$, with coefficients chosen to minimize the linear prediction-error power of

$$\epsilon_k = e_k - b(z)e_k = (1 - b(z))e_k. \tag{9.64}$$

In the ideal situation, $\hat{s}_i = s_i$, $i = k-1, k-2, \cdots$, and the variance of ϵ_k achieves its lower bound (9.61). This is roughly true for high SNR γ_0. For low to moderate SNRs, we have $\hat{s}_i = s_i + e_s$. Assuming that BPSK signaling is used with error and correct detection probabilities denoted by P_e and P_c, respectively, it is easy to examine that

$$\mathbb{E}[\hat{s}_i] = 2P_e + 0P_c = 2P_e,$$
$$\mathbb{E}[|\hat{s}_i|^2] = 4P_e + 0P_c = 4P_e, \tag{9.65}$$

which implies that $\hat{s}_i$ is a bias estimator with variance $V[\hat{s}_i] = 4P_e(1 - P_e)$. If its dc component is blocked, the prediction error output from $1 - V(z)$ contains two parts. One part is produced by $\xi_i - s_i$, and the other by e_s; namely $e_i = (\xi_i - s_i) + e_s$. Thus, we can write the output noise power as

$$P_n = \sigma_n^2 \exp\left(\int_{-1/2}^{1/2} \log(\tilde{S}_h(f) + \gamma_0^{-1})df\right) + 4P_e(1 - P_e)\int_{-1/2}^{1/2} \frac{1}{\tilde{S}_h(f) + \gamma_0^{-1}}df. \tag{9.66}$$

The second term on the right represents the excessive power introduced by symbol-decision error propagation.

9.7 SNR comparison and error performance

We are now in position to summarize the output SNR of various equalizers obtained in Equations (9.38), (9.54), and (9.62), as shown in Table 9.1. Both the ZFE and MMSE equalizers belong to the category of harmonic mean, but the latter with a correction term γ_0^{-1} in the denominator to avoid singularity and thus always ensuring the feasibility of MMSE equalizers. Since the geometric mean is greater than its harmonic counterpart, DFE performs better than ZFE.

Table 9.1 *Output SNR of various equalizers*

Type of equalizer	Output SNR	Relation to channel spectrum
ZFE	$\gamma_0\left[\int_{-1/2}^{1/2} \tilde{S}_h^{-1}(f)df\right]^{-1}$	Harmonic mean
MMSE	$\gamma_0\left[\int_{-1/2}^{1/2} \frac{1}{\tilde{S}_h(f)+\gamma_0^{-1}}df\right]^{-1} - 1$	Harmonic mean
DFE	$\gamma_0 \exp\left(\frac{1}{2\pi}\int_{-\pi}^{\pi} \ln\{\tilde{S}_h(\omega) + \gamma_0^{-1}\}d\omega\right)$	Geometric mean

With the SNR given above, we can calculate the error performance of an equalizer with arbitrary modulation schemes, using the formulas obtained in Chapters 4 and 8.

It is also interesting to consider the channel capacity of the bandlimited channel $h(t)$ without CSIT. Let $H(f)$ denote the Fourier transform of $h(t)$ and assume that $h(t)$ is of unit energy for convenience. As shown in Problem 9.10, the channel capacity without CSIT is given by

$$C = \int_{-\infty}^{\infty} \ln(1 + \gamma_0 |H(f)|^2) df \qquad (9.67)$$

nats per channel use. Recall that $\ln(1 + x)$ is a concave function of x. Thus, we have

$$C \le \ln\left(1 + \gamma_0 \int_{-\infty}^{\infty} |H(f)|^2\right) = \ln(1 + \gamma_0),$$

which corresponds to the channel capacity of the AWGN channel.

Property 9.7.1 *The capacity without CSIT of a bandlimited channel is less than its counterpart of the AWGN channel.*

Also note that $\ln(1 + e^x)$ is convex in x, which enables us to assert

$$C \ge \ln\left[1 + \rho_s \exp\left(\int_{-\infty}^{\infty} |H(f)|^2 df\right)\right].$$

The exponential function inside the logarithm represents the geometric mean of the channel power spectrum, corresponding to the case of a receiver with MMSE equalization.

Property 9.7.2 *The mutual information achieved by the receiver with MMSE equalization is less than the capacity of the corresponding bandlimited channel. The same assertion is applied to receivers with ZFE.*

9.8   An example

We would like to show that the z-transform of the correlation of the channel response can be factorized. As an example, consider a channel with impulse response $h(t) = \sqrt{2\alpha} \exp(-\alpha t), t > 0, \alpha > 0$, and find the MMSE of the DFE and linear-prediction-based equalizers with infinite length.

The channel correlation function can be easily determined as

$$r_h(\tau) = \int_0^{\infty} h(t+\tau)h(t)dt = 2\alpha \int_0^{\infty} e^{-\alpha t} e^{-\alpha(t+\tau)} dt = e^{-\alpha \tau}, \tau > 0, \qquad (9.68)$$

and, furthermore, $r_h(-\tau) = r_h(\tau)$. Thus, we have

$$r_h(\tau) = e^{-\alpha|\tau|} \overset{FT}{\to} S_h(f) = \frac{2\alpha}{\alpha^2 + (2\pi f)^2}. \qquad (9.69)$$

Clearly, α represents the 3 dB bandwidth in radians/second. Sampling $r_h(\tau)$ at the rate T_s, we obtain a correlation sequence $r_h(k) = \{e^{-\alpha|kT_s|}, k = \ldots, -1, 0, 1, \ldots\}$, the z-transform of which is given by

$$R_h(z) = \sum_{k=\infty}^{\infty} r_h(k)z^{-k} = \sum_{k=0}^{\infty} e^{-\alpha T_s k} z^{-k} + \sum_{k=-1}^{-\infty} e^{-\alpha T_s(-k)} z^{-k}$$

$$= \frac{z^{-1}(e^{-\omega_0} - e^{\omega_0})}{(1 - z^{-1}e^{-\omega_0})(1 - z^{-1}e^{\omega_0})}$$

$$= \frac{1 - e^{-2\omega_0}}{(1 - e^{-\omega_0}z^{-1})(1 - e^{-\omega_0}z)}$$

$$= f(z)f(1/z), \ e^{-\omega_0} < |z| < e^{\omega_0}, \tag{9.70}$$

where $\omega_0 = \alpha T_s$, and the last line is in a factorization form with factor given by

$$f(z) = \frac{(1 - e^{-2\omega_0})^{1/2}}{1 - e^{-\omega_0}z^{-1}}, \tag{9.71}$$

Physically, ω_0 represents the 3-dB channel bandwidth, in radians/second, normalized by the sampling rate. Alternatively, we may directly work on the Fourier transform of $h(t)$, which is given by

$$H(f) = \sqrt{2\alpha}/(\alpha + j2\pi f), \tag{9.72}$$

whereby the sampled power spectrum, with aliasing, can be written as

$$\tilde{S}_h(f) = f_s \sum_{k=-\infty}^{\infty} |H(f - kf_s)|^2 = \sum_{k=-\infty}^{\infty} \frac{2\alpha f_s}{\alpha^2 + 4\pi^2(f - kf_s)^2}$$

$$= \sum_{k=-\infty}^{\infty} \frac{2\omega_0}{\omega_0^2 + 4\pi^2(v - k)^2}, \tag{9.73}$$

where $f_s = 1/T_s$ is the sampling rate, and $v = fT_s$ is the normalized frequency. To avoid severe aliasing, T_s should be chosen such that $\omega_0 < \pi$.

Denote $\mathbf{x} = [0, 1 \ldots, m-1]T_s$, and $\mathbf{R}_m = toeplitz(\exp(-\alpha\mathbf{x}))$. Then the order-$m$ linear prediction error power equals

$$\sigma_m^2 = \frac{\det(\mathbf{R}_{m+1})}{\det(\mathbf{R}_m)}. \tag{9.74}$$

The harmonic mean and geometric mean of the power spectrum are given, respectively, by

$$\sigma_{hm}^2 = \left[\int_{-1/2}^{1/2} \tilde{S}_h^{-1}(v)dv\right]^{-1},$$

$$\sigma_{ge}^2 = \exp\left(\int_{-1/2}^{1/2} \ln \tilde{S}_h(v) \, dv\right). \tag{9.75}$$

From Figures 9.7 and 9.8, it is observed that for all the three equalizers, the power of residual ISI plus noise component rapidly reduces to σ_n^2 as the normalized channel bandwidth exceeds 0.5. The condition of normalized bandwidth $\alpha/(2\pi R_s) > 0.5$ is equivalent to sampling the channel at lower than the Nyquist rate. The aliasing of channel spectra guarantees the existence of an effective channel equalizer to completely remove the ISI component, consistent with the assertion by Nyquist. As the normalized bandwidth becomes smaller than 0.5, the channel spectrum and its shifted versions are well separated, and there are singularities and near-singularities in the sampled spectrum $R_h(z)|_{z=\exp(j\omega)}$. The consequence is noise amplification, as shown in the earlier part of the curves. It is over this region that the MMSE and DFE equalizers outperform their ZFE counterpart.

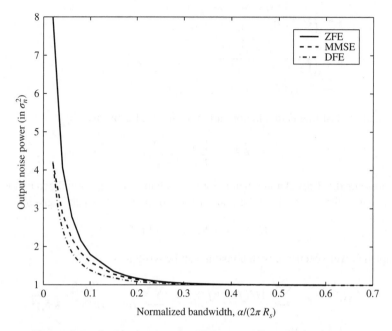

Figure 9.7   *Residual noise-plus-ISI power at the equalizer output*

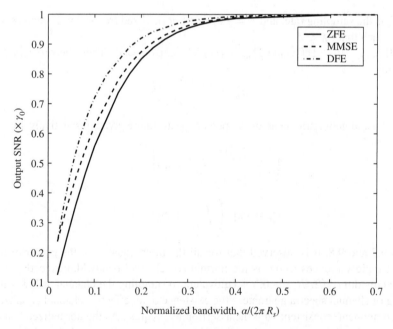

Figure 9.8   *Comparison of the output SNR of various equalizers*

9.9 Spectral factorization

In the above discussion, we often required to factorize a power spectrum for the construction of an equalizer, and showed that spectrum factorization was feasible through some particular examples. The question is how to factorize a general power spectrum.

Lemma 9.9.1 *For a zero-mean, WSS, discrete-time random process with a finite positive power spectrum* $\Phi(z)|_{z=e^{j\omega}} > 0$, *there exists factorization such that*

$$\Phi(z) = \sigma_e^2 g(z)g^*(z^{-1}), \tag{9.76}$$

where $g(z)$ *is a causal* ($g_i = 0$ *for* $k < 0$), *monic* ($g_0 = 1$), *and minimum phase filter (all of its poles and zeros are on or outside the unit circle), and* σ_e^2 *is directly related to the geometric mean of the power spectrum, as shown by*

$$\sigma_e^2 = \exp\left(\frac{1}{2\pi}\int_{-\pi}^{\pi} \ln \Phi(e^{j\omega})d\omega\right) \tag{9.77}$$

and

$$g(z) = \exp\left(\sum_{k=1}^{\infty} c_k z^{-k}\right),$$

$$g^*(z^{-1}) = \exp\left(\sum_{k=-\infty}^{-1} c_k z^{-k}\right), \tag{9.78}$$

where

$$c_k = \frac{1}{2\pi}\int_{-\pi}^{\pi} \ln\left[\Phi(e^{j\omega})\right] \exp\left(jk\omega\right)d\omega. \tag{9.79}$$

A filter $g(z)$ defined above is said to be canonical.

For the case of a rational power spectrum $\Phi(z)$, the factorization is very simple. Let us illustrate this by an example.

✎ Example 9.4 ────────────────────────────────

Given a rational power spectrum $\Phi(e^{j\omega}) = \frac{1.04+0.4\cos\omega}{1.25+\cos\omega}$, find its factorization.

Solution

It is easy to show that

$$\Phi(e^{j\omega}) = \frac{(z + 0.2)(z + 0.2)^*}{(z + 0.5)(z + 0.5)^*}\Big|_{z=\exp(j\omega)},$$

whereby we can determine

$$g(z) = \frac{1 + 0.2z^{-1}}{1 + 0.5z^{-1}}.$$

It is clear that a rational spectrum can be generated by driving a canonical filter with a white sequence of mean zero and variance σ_e^2. ○

9.10   Summary

Different equalization techniques are rooted in different methodologies. The transmitted waveforms, when passing through a temporally dispersive channel, are partly overlapped, causing the phenomenon of ISI. A simple methodology is to remove the overlapped portion, leading to the technique of ZFE. The drawback of ZFE is that its output SNR can be very small if the overall channel response has singularities or near-singularities in its Fourier spectrum. A remedy is the use of an MMSE channel equalizer. Part of the residual ISI component in ZFE or MMSE equalizers can be further eliminated by exploiting the estimated symbols from the decision device output at the receiver. Surprisingly, the output SNR of an equalizer is uniquely determined by the channel spectrum characteristics. In particular, the output SNR of ZFE and MMSE equalizers is uniquely determined by the harmonic mean of the channel power spectrum, while its DFE counterpart is determined by its geometric mean, revealing an elegant picture of channel equalization.

Channel equalization is a fully developed area. Readers are referred to [26]–[34] for additional reading.

Problems

Problem 9.1   Show that the overall impulse response of the ZFE meets the requirement of the Nyquist waveform.

Problem 9.2   Show that the overall impulse response of the MMSE equalizer does not meet the requirement of the Nyquist waveform.

Problem 9.3   Prove that the output power of interpolation-error filter is determined by the harmonic mean of the aliased channel power spectrum $\tilde{S}_h(f)$:

$$\|e_o\|^2 = \int_{-1/2}^{1/2} \tilde{S}_h^{-1}(f)df.$$

Hint: Consider $\|e_o\|^2$ as the output power of the interpolation-error filter driven by a random sequence with correlation $\{r_k\}$.

Problem 9.4   Suppose a pulse sequence $x(t) = \sum_i s_i p(t - iT_s)$ is transmitted over a two-path channel $c(t) = \delta(t) + \frac{1}{2}\delta(t - T_s)$, and is corrupted by AWGN $n(t)$ of mean zero and variance σ_n^2 at the receiver front-end. The shaping pulse $p(t)$ is defined by

$$p(t) = \begin{cases} \sin(\pi f_s t), & 0 \le t < T_s \\ 0 & \text{elsewhere} \end{cases}.$$

The receiver matches the overall channel response, followed by a sampler at symbol rate $f_s = 1/T_s$ and a ZFE.

(a) Determine an equivalent discrete model for the channel.
(b) Determine the tap-weight coefficients of the ZFE.
(c) Find the maximum SNR achievable by the ZFE.

Problem 9.5   In Problem 9.4, suppose that a five-tap ZFE is used instead.

(a) Determine its coefficients.
(b) Find the residual ISI component.
(c) Determine the signal-to-ISI ratio.
(d) Determine the output noise power.

Problem 9.6   In Problem 9.5, if we increase the truncated ZFE to 21 taps, check the residual ISI components again and show that $A(z)R_h(z) \approx 1$.

Problem 9.7   A pulse sequence $x(t) = \sum_{i=-\infty}^{\infty} s_i p(t - iT_s)$ is transmitted over a channel $c(t) = \delta(t) + 0.8\delta(t - T_s) + 0.6\delta(t - 2T_s)$. Assume that the shaping waveform $p(t)$ is a unit rectangular pulse defined over $[0, T_s)$, the random symbols s_i are of mean zero and variance E_s, and the channel output is corrupted by AWGN $n(t) \sim \mathcal{CN}(0, \sigma_n^2)$. Determine

(a) an equivalent discrete model for the overall channel;
(b) the ZFE filter response; and
(c) the output SNR of the ZFE.

Problem 9.8   The response of an equivalent channel filter is given by

$$R_h(z) = 0.5z^2 + 0.8z + 1 + 0.8z^{-1} + 0.5z^{-2}.$$

Show that the corresponding ZFE is infeasible.

Problem 9.9   Consider a QPSK sequence of symbol rate $1/T_s$ transmitted over a channel with impulse response $h(t) = \sqrt{2\alpha} \exp(-\alpha t), t > 0, \alpha > 0$. Assuming that a ZFE is used at the receiver and $\gamma_0 = 5$ dB, determine the system symbol error rate for $\alpha T_s = 0.5$ and 0.8.

Problem 9.10   Consider the channel capacity of a bandlimited channel $h(t)$ without assuming CSIT. Show that the channel capacity is given by

$$C = \int_{-\infty}^{\infty} \ln\left(1 + \gamma_0 |H(f)|^2\right) df,$$

nats per channel use, where γ_0 is the transmit SNR and $H(f)$ is the Fourier transform of $h(t)$.

Problem 9.11   A two-path channel has the impulse response $h(t) = 1 + a\delta(t - \tau)$, where $a = 0.8$ and $\tau = 10^{-5}$ s. Determine the channel capacity without CSIT.

Problem 9.12   Suppose a Gaussian signal sequence $\{s_1, s_2, \dots\}$ from a single user is transmitted over a random channel with impulse response $p(t) = \sum_{i=1}^{L} h_i \delta(t - \tau_i)$, so that the channel output is

$$y(t) = s_1 p(t - \tau_1) + s_2 p(t - \tau_2) + n(t). \tag{9.80}$$

Let $\gamma_{21} = \int p_1(t)p_2(t)dt / \|\mathbf{h}\|^2$ denote the cross correlation between $p_1(t) = p(t - \tau_1)$ and $p_2(t) = p(t - \tau_2)$, where $\|\mathbf{h}\|^2 = \sum_{i=1}^{L} |h_i|^2$. Show that given h_1 and h_2, the system mutual information is equal to

$$C = \log \det \left[\mathbf{I} + \frac{\sigma_s^2}{\sigma_n^2} \|\mathbf{h}\|^2 \begin{pmatrix} 1 + |\gamma_{21}|^2 & \gamma_{21}\sqrt{1 - |\gamma_{21}|^2} \\ \gamma_{21}\sqrt{1 - |\gamma_{21}|^2} & 1 - |\gamma_{21}|^2 \end{pmatrix} \right]. \tag{9.81}$$

Further show that the above expression is maximized when $\gamma_{21} = 0$.

Problem 9.13   The analysis of diversity combining is traditionally based on the assumption of flat fading channels. In this problem, study the average error performance of a coherent diversity system with DFE to operate in a frequency-selective Rayleigh fading environment. Suppose a dual-branch MRC diversity combiner is used for reception of a QPSK pulse sequence $x(t) = \sum_{i=-\infty}^{\infty} s_i p(t - iT_s)$ transmitted over independent Rayleigh selective fading channels with branch channel impulse response $c_k(t) = a_{k1}\delta(t - T_s) + a_{k2}\delta(t) + a_{k3}\delta(t + T_s)$ for channels $k = 1, 2$, respectively. Assume that the shaping waveform $p(t)$ is a unit rectangular pulse defined over $[0, T_s)$, the QPSK symbols are of symbol energy E_s, the channel output is corrupted by AWGN $n(t) \sim \mathcal{CN}(0, \sigma_n^2)$, and channel coefficients are mutually independent with $a_{ki} \sim \mathcal{CN}(0, \sigma_{ki}^2)$. Determine the average error probability of the system.

References

1. H. Nyquist, "Certain topics in telegraph transmission theory," *Trans. AIEE (Commun. Electron.)*, vol. 47, pp. 617–644, 1928.
2. D.W. Tufts, "Nyquist's problem-the joint optimization of transmitter and receiver in pulse amplitude modulation," *Proc. IEEE*, vol. 53, pp. 248–260, 1965.
3. M. Austin, "Decision-feedback equalization for digital communication over dispersive channels," *MIT Research Lab Electronics*, 371.
4. J.M. Cioffi, G.P. Dudevoir, M.V. Eyuboglu, and G.D. Forney, "MMSE decision-feedback equalizers and coding, Part I: equalization results, Part II: coding results," *IEEE Trans. Commun.*, vol. 43, pp. 2582–2604, 1995.
5. D. Falconer and S. Ariyavisitakul, "Broadband wireless using single carrier and frequency domain equalization," *Proc. Int. Symp. Wireless Pers. Multimedia Commun.*, vol. 1, pp. 27–36, 2002.
6. P. Monsen, "Feedback equalization for fading dispersive channels," *IEEE Trans. Inf. Theory*, vol. IT-17, pp. 56–64, 1971.
7. R. Price, "Nonlinearly feedback-equalized PAM vs. capacity from noisy filter channels," *International Conference on Communications*, 1972, pp. 22–12–22–16.
8. J. Salz, "Optimum mean-square decision feedback equalization," *Bell Syst. Tech. J.*, vol. 52, no. 8, 1973.
9. D. Falconer and G. Foschini, "Theory of minimum mean-square-error QAM systems employing decision feedback equalization," *Bell Syst. Tech. J.*, vol. 52, no. 10, 1973.
10. C.A. Belfiore and J.H. Park Jr., "Decision feedback equalization," *Proc. IEEE*, vol. 67, pp. 1143–1156, 1979.
11. M.R. Aaron and D.W. Tufts, "Intersymbol interference and error probability," *IEEE Trans. Inf. Theory*, vol. 12, pp. 26–34, 1966.
12. R.W. Chang and J.C. Hancock, "On receiver structures for channels having memory," *IEEE Trans. Inf. Theory*, vol. IT-12, pp. 463–468, 1966.
13. K. Abend and B.D. Fritchman, "Statistical detection for communication channels with intersymbol interference," *Proc. IEEE*, vol. 58, pp. 778–785, 1970.
14. R.A. Gonsalves, "Maximum-likelihood receiver for digital data transmission," *IEEE Trans. Commun. Technol.*, vol. COM-16, pp. 392–398, 1968.
15. G. Ungerboeck, "Nonlinear equalization of binary signals in Gaussian noise," *IEEE Trans. Commun. Technol.*, vol. COM-22, pp. 624–636, 1974.
16. K. Yao, "On the minimum average probability of error expression for binary pulse communication systems with intersymbol interference," *IEEE Trans. Inf. Theory*, vol. 18, pp. 363–378, 1972.
17. E. Shamash and K. Yao, "On the structure and performance of a linear decision feedback equalizer based on the minimum error probability criterion," *Record International Conference on Communication, ICC'74, Minneapolis, MN*, June 17–19, 1974.
18. G.D. Forney, "Maximum likelihood sequence estimation of digital sequences in the presence of intersymbol interference," *IEEE Trans. Inf. Theory*, vol. 18, pp. 363–378, 1972.
19. C. Windpassinger, R.F.H. Fischer, T. Vencel, and J.B. Huber, "Precoding in multiantenna and multiuser communication," *IEEE Trans. Commun.*, vol. 3, no. 4, pp. 1305–1316, 2004.
20. M. Timlinson, "New automatic equaliser employing modulo arithmetic," *Electron. Lett.*, vol. 7, no. 5, pp. 138–139, 1971.
21. H. Harashima and H. Miyakawa, "Matched-transmission technique for channels with intersymbol interference," *IEEE Trans. Commun.*, vol. 20, no. 4, pp. 774–780, 1972.
22. L. Brandenburgh and A. Wyner, "Capacity of the Gaussian channel with memory: a multivariate case," *Bell Syst. Tech. J.*, vol. 53, pp. 745–779, 1974.
23. A. Papoulis, *Signal Analysis*, New York: McGraw-Hill, 1977.
24. W.H. Beyer, *CRC Standard Mathematical Tables*, 28th ed., Boca Raton, FL: CRC Press, 1981.
25. S.U.H. Qureshi, "Adaptive equalization," *Proc. IEEE*, vol. 73, no. 9, pp. 1349–1387, 1985.
26. J.W. Smith, "The joint optimization of transmitted signal and receiving filter for data transmission systems," *Bell Syst. Tech. J.*, vol. 44, pp. 2363–2392, 1965.
27. D. Messerschmitt, "A geometric theory of intersymbol interference. Part I: zero-forcing and decision-feedback equalization," *Bell Syst. Tech. J.*, vol. 52, no. 9, pp. 1483–1519, 1973.

28. A. Salazar, "Design of transmitter and receiver filters for decision feedback equalization," *Bell Syst. Tech. J.*, vol. 53, no. 3, 1974.

29. N. Al-Dhahir and J. M. Cioffi, "MMSE decision-feedback equalizers: finite-length results," *IEEE Trans. Inf. Theory*, vol. IT-41, pp. 961–975, 1995.

30. D. Falconer, S.L. Ariyavisitakul, A. Benyamin-Seeyar, and B. Eidson, "Frequency domain equalization for single-carrier broadband wireless systems," *IEEE Commun. Mag.*, Vol. 40, no. 4, pp. 58–66, 2002.

31. H. Sari, G. Karam, and I. Jeanclaude, "Frequency-domain equalization of mobile radio and terrestrial broadcast channels," *Proceedings of Globecom'94, San Francisco, CA, Nov.-Dec.* 1994, pp. 1–5.

32. V. Aue, G.P. Fettweis, and R. Valenzuela, "Comparison of the performance of linearly equalized single carrier and coded OFDM over frequency selective fading channels using the random coding technique," *Proceedings of ICC'98*, pp. 753–757.

33. Y. Zhu and K.B. Letaief, "Single carrier frequency domain equalization with noise prediction for broadband wireless systems," *Proceedings of IEEE Global Telecommunications Conf (Globecom'04)*, vol. 5, pp. 3098–3102, Nov. 2004.

34. R.M. Gray, "On the asymptotic eigenvalue distribution of toeplitz matrices," *IEEE Trans. Inf. Theory*, vol. IT-18, no. 6, pp. 725–730, 1972.

10

Channel Decomposition Techniques

Abstract

In this chapter, we re-examine the nature of channel equalization in a more fundamental framework of channel decomposition. When a pulse sequence is transmitted over a temporally dispersive channel, the spread pulse waveforms overlap with each other, causing inter-symbol interference (ISI). Mathematically, it is the nondiagonal channel matrix that causes ISI among adjacent symbols. Therefore, the key to avoiding ISI is to diagonalize the channel matrix. Various channel equalizers use different techniques for channel decomposition. For example, the Tomlinson–Harashima equalizer has its roots in QR decomposition, whereas orthogonal frequency-division multiplexing (OFDM) is rooted in singular decomposition. Channel decomposition represents an early strategy in tackling impaired channels whereby to implement the philosophy of symbol-by-symbol detection.

♣

10.1 Introduction

The radical reason for the generation of inter-symbol interference (ISI) lies in the nondiagonal structure of a temporally dispersive channel matrix. It is through this nondiagonal channel matrix that adjacent symbols interfere with each other. Various channel equalization techniques described in the preceding chapter represents *passive* and *defensive* strategies for combating ISI, in the sense that the transmitter arbitrarily transmits a message pulse sequence without taking the channel characteristics into consideration and the receiver tries to find remedy once ISI is introduced. The thought behind this strategy lies in the philosophy of treating a nondiagonal channel matrix as an impairment.

A channel structure, ideal or nonideal, is a resource. A proactive strategy is to exploit such a structure to enhance signal transmission. This strategy can lead to a linear receiver or a nonlinear receiver. The latter includes the early maximum likelihood receiver by Forney [1], the Viterbi detector, and the more recent turbo equalizer, which we have more to say in Chapter 11. Linear receivers are preferred in past few decades because of their simplicity and reasonable error performance, and they are the focus of this chapter.

Wireless Communications: Principles, Theory and Methodology, First Edition. Keith Q.T. Zhang.
© 2016 John Wiley & Sons, Ltd. Published 2016 by John Wiley & Sons, Ltd.
Companion Website: www.wiley.com/go/zhang7749

From a mathematical point of view, the nature of the proactive strategy with a linear receiver is equivalent to the orthogonal decomposition of a nondiagonal channel matrix using classical techniques such as QR, eigen, or geometric mean decomposition (GMD). The resulting design is a precoder at the transmitter and a post-processor at the receiver to avoid the generation of ISI. Without ISI, detection can be done simply on a symbol-by-symbol basis.

10.2 Channel matrix of ISI channels

A communication channel behaves like a filter, which, in the discrete-time domain, can be described by a channel matrix $\mathbf{H}$, or in the frequency domain by a frequency transfer function $H(f)$. In its baseband form, the received signal is given by

$$\mathbf{y} = \mathbf{Hs} + \mathbf{n}, \tag{10.1}$$

where the signal vector $\mathbf{s}$ is transmitted through the discrete channel matrix $\mathbf{H}$ before arriving at the receiver and corrupted by AWGN $\mathbf{n} \sim \mathcal{CN}(\mathbf{0}, \sigma_n^2\mathbf{I})$. Denote the symbol power by σ_s^2, and denote the average transmit signal-to-noise ratio by $\gamma_0 = \sigma_s^2/\sigma_n^2$. In a practical system, the channel matrix $\mathbf{H}$ represents the overall system equivalent impulse response resulting from the convolution of the transmit shaping filter, the channel filter, and the receiver matched filter. For a temporal disperse channel, $\mathbf{H}$ is not diagonal, thereby unavoidably introducing ISI. The channel is usually slowly time-varying, as compared to the data rate. We can thus treat it as a time-invariant filter. In particular, if we model the channel as a tap-delay line filter of length L, we can write the channel matrix as

$$\mathbf{H} = \begin{pmatrix} h_0 & 0 & \cdots & & 0 & 0 \\ h_1 & h_0 & 0 & \cdots & & 0 \\ h_2 & h_1 & h_0 & & \cdots & 0 \\ \vdots & \vdots & & & \vdots & \\ h_{L-1} & h_{L-2} & \cdots & & & h_0 \end{pmatrix}. \tag{10.2}$$

The nondiagonal structure of the channel matrix is the source of ISI. This matrix has the special structure of equal diagonal elements, and is uniquely determined by the first column, that is, the tap-coefficient vector of the channel filter $\mathbf{h} = [h_0, h_1, \cdots, h_{L-1}]^T$. Such a matrix pattern, usually known as the Toeplitz structure, possesses many elegant properties and is, thus, often exploited in adaptive signal processing such as the Durbin–Levision algorithm.

10.3 Idea of channel decomposition

Since the radical reason for the occurrence of ISI is the nondiagonal structure of the channel matrix, the central idea to prevent ISI is the decorrelation of the channel matrix.

The traditional ZFE can be considered as the simplest example to carry out this idea. It simply takes the general pseudo-inverse of the channel as its equalizer, at the cost of noise amplification. A more appropriate strategy is to decompose the channel matrix $\mathbf{H}$ into matrix factors, such that

$$\mathbf{H} = \mathbf{A}\mathbf{\Phi}\mathbf{B}. \tag{10.3}$$

Then, the inverse of matrix $\mathbf{B}$ is used to precode the information vector to generate the transmitted vector $\mathbf{x} = \mathbf{B}^{-1}\mathbf{s}$, whereas $\mathbf{A}^{-1}$ is used for equalization at the receiver, as illustrated in Figure 10.1. To prevent

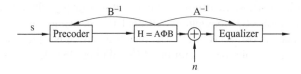

Figure 10.1   *Illustrating the idea of channel decomposition principle*

Table 10.1   *Precoder-and-equalizer design for various decompositions*

Equalizer	Precoder B^{-1}	Equalizer A^{-1}	Foundation	Required Info.
ZFE	N/A	$H^{\#}$	Pseudo-inverse	No CSIT
THP	R^{-1}	$Q^{\dagger}$	QR	CSIT
GMD	P^{-1}	$Q^{\dagger}$	Geometric mean	CSIT
SVD	$V^{\dagger}$	$U^{\dagger}$	Singular	No CSIT

noise amplification, matrix $\mathbf{A}$ requires to be unitary. The reason is that an AWGN vector, after processing by a unitary matrix, retains the same property unchanged. The matrix $\boldsymbol{\Phi}$ is either of a diagonal pattern or a triangular structure to facilitate symbol detection at the receiver.

Possible candidates for factoring $\mathbf{H}$ are Cholesky, singular value, and geometric mean decompositions:

$$\mathbf{H} = \begin{cases} \mathbf{QR} & \text{QR decomposition (QR)} \\ \mathbf{U\Lambda V^{\dagger}} & \text{Singular value decomposition (SVD)} \\ \mathbf{QRP} & \text{Geometric mean decomposition (GMD)} \end{cases}, \tag{10.4}$$

where matrices $\mathbf{P}$, $\mathbf{Q}$ and $\mathbf{U}$ are unitary matrices, $\boldsymbol{\Lambda}$ is a diagonal matrix, and $\mathbf{R}$ is an upper triangular matrix. With these notations, (10.1) becomes

$$\mathbf{y} = \begin{cases} \mathbf{Hs} + \mathbf{n}, & \text{Original} \\ \mathbf{QRs} + \mathbf{n}, & \text{QR} \\ \mathbf{U\Lambda V^{\dagger}s} + \mathbf{n}, & \text{SVD} \\ \mathbf{QRPs} + \mathbf{n}, & \text{GMD} \end{cases}. \tag{10.5}$$

These expressions, alongside Figure 10.1, suggest possible precoders and equalizers for combating ISI. The results are tabulated in Table 10.1, where we see that some channel decomposition techniques require channel state information (CSI) at the transmitter (CSIT).

10.4   QR-decomposition-based Tomlinson–Harashima equalizer

Let us start by briefly reviewing the principle of QR decomposition (QR). QR decomposition of the channel matrix corresponds to orthogonalizing the column vectors of $\mathbf{H} = [\mathbf{h}_1, \mathbf{h}_2, \cdots, \mathbf{h}_N]$ using the Gram–Schmidt procedure. The first base $\mathbf{q}_1$ is obtained by normalizing $\mathbf{h}_1$ with its length $r_{11} = \|\mathbf{h}_1\|$. As such, we have $\mathbf{h}_1 = r_{11}\mathbf{q}_1$. The second base $\mathbf{q}_2$ is obtained by normalizing the component that is orthogonal to $\mathbf{q}_1$, with its length r_{22}. As such

$$\mathbf{h}_2 = r_{12}\mathbf{q}_1 + r_{22}\mathbf{q}_2. \tag{10.6}$$

Proceeding in the same way, we can represent the channel matrix as

$$\mathbf{H} = [r_{11}\mathbf{q}_1, r_{12}\mathbf{q}_1 + r_{22}\mathbf{q}_2, \cdots, r_{1N}\mathbf{q}_1 + \cdots + r_{NN}\mathbf{q}_N]$$

$$= \underbrace{[\mathbf{q}_1, \cdots, \mathbf{q}_N]}_{Q} \underbrace{\begin{pmatrix} r_{11} & r_{12} & r_{13} & \cdots & r_{1N} \\ & r_{22} & r_{23} & \cdots & r_{2N} \\ & & \cdots & & \vdots \\ & & & & r_{NN} \end{pmatrix}}_{\mathbf{R}}. \tag{10.7}$$

It is clear that $\mathbf{Q}$ is the unitary matrix with its columns defined by the base vectors and $\mathbf{R}$ is the upper triangular matrix with its column entries representing the projection coefficients on the corresponding bases. Note that $r_{ii}, i = 1, \cdots, N$ are in general unequal.

To implement an equalizer based on (10.7), we normalize $\mathbf{R}$ with the diagonal matrix $\mathrm{diag}(\mathbf{R})$ formed by the diagonal entries of the former, so that the resulting upper diagonal matrix is of unit diagonal entries. For reason that will become clear shortly, we denote this normalized upper triangular matrix by $(\mathbf{I} + \tilde{\mathbf{R}})$. Hence, we can write

$$[\mathrm{diag}(\mathbf{R})]^{-1}\mathbf{R} = \mathbf{I} + \tilde{\mathbf{R}}, \tag{10.8}$$

where the (i, j)th entry of $\tilde{\mathbf{R}}$ is defined by

$$\tilde{R}(i, j) = \begin{cases} R(i, j)/R(i, i), & j > i; \\ 0, & j \le i. \end{cases} \tag{10.9}$$

The use of (10.8) enables us to rewrite the QR decomposition of the channel matrix as

$$\mathbf{H} = \mathbf{Q}[\mathrm{diag}(\mathbf{R})](\mathbf{I} + \tilde{\mathbf{R}}). \tag{10.10}$$

The above expression provides a basis for the implementation of the Tomlinson–Harashima precoder (THP) and equalizer [2, 3]. Following the idea of Figure 10.1, $(\mathbf{I} + \tilde{\mathbf{R}})^{-1}$ is used as the precoder for preprocessing the signal vector $\mathbf{s}$, so that the transmitted vector is $\mathbf{x} = (\mathbf{I} + \tilde{\mathbf{R}})^{-1}\mathbf{s}$. The received signal, thus, becomes

$$\mathbf{y} = \mathbf{Hx} + \mathbf{n} = \mathbf{Q}\mathrm{diag}(\mathbf{R})\mathbf{s} + \mathbf{n}, \tag{10.11}$$

which, after equalized by the unitary matrix $\mathbf{Q}^{-1}$ at the receiver, leads to the decision vector

$$\mathbf{z} = \mathrm{diag}(\mathbf{R})\mathbf{s} + \mathbf{n}. \tag{10.12}$$

With precoder $(\mathbf{I} + \tilde{\mathbf{R}})^{-1}$ at the transmitter and equalizer $\mathbf{Q}^{-1}$ at the receiver, the channel is diagonalized, completely preventing the occurrence of ISI. The use of the unitary equalizer $\mathbf{Q}^{-1}$ at the receiver prevents noise amplification on one hand and preserves the noise properties on the other. Thus, we can use the same notation $\mathbf{n}$ for the noise component.

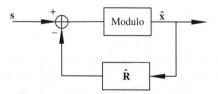

Figure 10.2   *Implementation of THP equalizer with a feedback loop*

In reality, direct implementation of matrix inversion is quite involved. It is common practice to implement $\mathbf{x} = (\mathbf{I} + \tilde{\mathbf{R}})^{-1}\mathbf{s}$ with a feedback loop involving a modulo operator to prevent the possible divergence in symbol magnitudes. The resulting approximation is

$$\hat{\mathbf{x}} = (\mathbf{I} + \tilde{\mathbf{R}})^{-1}(\mathbf{s} + \alpha\mathbf{e}), \tag{10.13}$$

where $\mathbf{e}$ is the error vector due to the modulo operation, and α is a constant depending on the modulation scheme in use [2, 3]. An illustrative diagram is sketched in Figure 10.2. The same type of modulo-adder is used in the receiver to recover the original data. The resulting equalizer is called the *Tomlinson–Harashima equalizer*.

At this point, it is interesting to compare the DFE and the THP equalizer. The former assumes no CSIT and only uses the previously detected symbols to remove the ISI residuals. Its good performance relies on error-free assumption on previously detected symbols and an infinitely long feedback loop. Should any errors occur in previously detected symbols, errors can propagate. If channel state information is available at the transmitter, it is shown that error propagation can be eliminated [4] by implementing the feedback loop as a THP [2, 5] at the transmitter.

10.5   The GMD equalizer

Geometric-mean decomposition [6] represents another technique to prevent ISI. The channel GMD is given by

$$\mathbf{H} = \mathbf{QRP}, \tag{10.14}$$

where $\mathbf{Q}$ and $\mathbf{P}$ are unitary matrices and $\mathbf{R}$ is an upper triangular matrix with identical diagonal entries. For the determination of these matrices, the reader is referred to [7]. With $\mathbf{P}^\dagger$ as the precoder at the transmitter and $\mathbf{Q}^\dagger$ as the equalizer at the receiver, the decision vector from the equalizer output turns out to be

$$\mathbf{y} = \mathbf{Q}^\dagger(\mathbf{HP}^\dagger\mathbf{s} + \mathbf{n}) = \mathbf{Rs} + \mathbf{n}. \tag{10.15}$$

The triangular structure of matrix $\mathbf{R}$ completely removes the ambiguity caused by ISI, enabling symbol-by-symbol decisions in an upward manner.

One basic feature of GMD is that the upper triangular matrix $\mathbf{R}$ has identical diagonal entries with value equal to the geometric mean of all nonzero singular values of $\mathbf{H}$. Recall that both the ZFE and the linear MMSE equalizer belong to the type of harmonic mean, in the sense that their output signal-to-interference-plus-noise ratio (SINR) is mainly dominated by the smallest singular value of the channel. In particular, for a discrete channel matrix $\mathbf{H}$ of K nonzero singular values $\{\alpha_k, k = 0, \cdots, N - 1\}$, it is easy to show that the output SINRs of ZFE and linear MMSE equalizer are given by

$$\gamma_{\text{ZFE}} = \gamma_0 \left[\frac{K}{\sum_{k=1}^{K} |\alpha_k|^{-2}} \right],$$

$$\gamma_{\text{MSE}} = \gamma_0 \left[\frac{K}{\sum_{k=1}^{K} \left(\gamma_0^{-1} + |\alpha_k|^2\right)^{-1}} \right],$$

respectively. The gap between the MMSE and the ZFE vanishes as γ_0 approaches infinity. The deterioration in the smallest singular channel is evened out by the GMD. The GMD evenly distributes the total channel gain by assigning to each diagonal entry of $\mathbf{R}$ the geometric mean of all the nonsingular values, thereby effectively preventing the drawback of noise amplification during symbol detection. Given this advantage, GMD equalizers have found applications in many other scenarios, such as MIMO [8, 9] and multiple access [10].

10.6   OFDM for time-invariant channel

The QR- and GMD-based techniques described above require CSIT. We also note that in both techniques, the Toeplitz structure of time-invariant channels is not exploited. A powerful means to exploit such a structure is the singular value decomposition.

Suppose that the channel impulse response is time-invariant and has the structure of a tap-delayed line with $L-1$ nonzero tap coefficients $\mathbf{h} = [h_0, h_1, \cdots, h_{L-1}]^T$. When an N-vector $\mathbf{s}$, where $N > L$, is transmitted through the channel, the channel matrix $\mathbf{H}$ is of a Toeplitz structure, with the augmented vector $\mathbf{h}_a = [\mathbf{h}, \mathbf{0}_{1 \times (N-L)}]^T$ as its first column. Symbolically, write $\mathbf{H} = Toeplitz(\mathbf{h}_a)$.

All the methods described above require the knowledge of SCIT to design a precoder for use in the transmitter, and such prior information is acquired at considerable cost. The need for a precoder in the ISI-free system can be removed if the Toeplitz structure of a time-invariant channel is exploited. That is, asymptotically as $N \to \infty$, the matrix of the eigenvectors of $\mathbf{H}$ approaches the Fourier matrix [11].

10.6.1   Channel SVD

The Toeplitz matrix $\mathbf{H}$ has a good asymptotic property; that is, as $N \to \infty$, its SVD approaches

$$\mathbf{H} \to \mathbf{F}^\dagger \mathbf{\Lambda} \mathbf{F}, \tag{10.16}$$

where $\mathbf{F}$ is the Fourier transform matrix, and $\mathbf{\Lambda} = \mathrm{diag}(\alpha_0, \cdots, \alpha_{N-1})$ with the diagonal entries equal the channel frequency response at the corresponding discrete frequencies [12]. The advantage of such a decomposition enables the use of a CSI-free matrix $\mathbf{F}^\dagger$ at the transmitter to eliminate the source of ISI generation. For a Toeplitz channel matrix, this ideal property is approximately attainable for a large block size N.

However, another matrix $\mathbf{H}_c$ is also generated from $\mathbf{h}_a$ but holding the above ideal SVD for any finite block size N. It takes the form

$$\mathbf{H}_c = \begin{bmatrix} h_0 & 0 & 0 & 0 & \cdots & 0 & h_L & \cdots & h_2 & h_1 \\ h_1 & h_0 & 0 & \cdots & & 0 & 0 & h_L & \cdots & h_2 \\ h_2 & h_1 & h_0 & \cdots & & 0 & 0 & 0 & \cdots & h_3 \\ \vdots & \vdots & \vdots & \vdots & \vdots & \vdots & \vdots & \vdots & & \\ h_{L-2} & & \cdots & h_0 & \cdots & 0 & 0 & \cdots & \cdots & h_{L-1} \\ h_{L-1} & h_{L-2} & \cdots & h_1 & \cdots & 0 & 0 & \cdots & \cdots & 0 \\ 0 & h_{L-1} & \cdots & h_2 & \cdots & 0 & 0 & \cdots & \cdots & 0 \\ \vdots & \vdots & \vdots & \vdots & \vdots & \vdots & \vdots & \vdots & & \\ 0 & 0 & \cdots & 0 & \cdots & h_L & & \cdots & \cdots & h_0 \end{bmatrix}.$$

This matrix is generated from its first column $\mathbf{h}_a$ by cyclic shifting. In particular, the vector $\mathbf{h}_a$ in the first column is downward 1 symbol cyclic shifted to produce column 2, and downward 2 symbol cyclic shifted to produce column 3, and so on. Symbolically, we write $H_c = \mathrm{cir}(\mathbf{h}_a)$.

Property 10.6.1 *Let $\mathbf{F}$ denote the normalized $N \times N$ Fourier matrix such that $\mathbf{F}^\dagger \mathbf{F} = \mathbf{F}\mathbf{F}^\dagger = \mathbf{I}$. The circulant matrix $\mathbf{H}_c$ possesses a very special eigen structure defined by the Fourier transform of $\mathbf{h}$. In particular, $\mathbf{H}_c$ can be eigen-decomposed as*

$$\mathbf{H}_c = \mathbf{F}^\dagger \mathbf{\Lambda} \mathbf{F}, \tag{10.17}$$

where $\Lambda = diag(\alpha_0, \cdots, \alpha_{N-1})$, *with* α_m *defined by the discrete Fourier transform of the channel vector*

$$\alpha_m = \sum_{k=1}^{N} h_{N-k} \exp(j2\pi km/N), m = 0, \cdots, N - 1. \tag{10.18}$$

This property ensures that ISI can be completely removed if the inverse Fourier matrix $\mathbf{F}^\dagger$ is used as a precoder at the transmitter and $\mathbf{F}$ is used as an equalizer at the receiver. Unlike a Toeplitz equalizer at the receiver. Unlike a Toeplitz structure, which asymptotically possesses such an excellent property only as N tends to infinity, the above decomposition is true even when the orthogonal frequency-division multiplexing (OFDM) block size is *finite*. Note that the Fourier matrix $\mathbf{F}$ is totally independent of the channel matrix, implying the implementation of ISI-free transmission without the need for CSIT. A similar situation happens only in JPEG, where the discrete cosine transform is used to approximate the eigen structure of a two-dimensional Markov process of random pictures.

Thus, rather than directly transmitting the complex symbol vector s, we can precode it with inverse discrete Fourier transform (IDFT) as $\mathbf{x} = \mathbf{F}^\dagger \mathbf{s}$ and insert an appropriate prefix before transmission. The IDFT, alongside the discrete Fourier transform (DFT) taken at the receiver, jointly diagonalizes the channel matrix to implement ISI-free transmission. The time-invariant channel $\mathbf{H}$ defines multiple orthogonal subcarrier channels. Direct transmission over these subcarrier channels is the most natural way to eliminate ISI, leading to the technology of OFDM. Two of its main advantages are the avoidance of CSIT and achieving numerical efficiency by virtue of fast Fourier transform (FFT). The concept of OFDM can be traced back to the 1960s. The idea first appeared in 1961 in a conference record [13]. Its implementation with DFT was suggested by Darlington [14]. A complete implementation with efficient FFT and with a guard interval inserted to present possible ISI residual was due to Weinstein and Ebert in their seminal 1971 paper [15]. In 1985, Cinimi investigated the application of OFDM to wireless mobile communications [16].

10.6.2   OFDM: a multicarrier modulation technique

To use OFDM to diagonalize the channel matrix $\mathbf{H}_c$ in (10.17), the original symbol vector s is precoded with the IDFT matrix $\mathbf{F}^\dagger = [\mathbf{v}_0, \cdots, \mathbf{v}_{N-1}]$ at the transmitter to produce an OFDM symbol vector

$$\mathbf{x} = \mathbf{F}^\dagger \mathbf{s} = \sum_{k=0}^{N-1} s_k \mathbf{v}_k, \tag{10.19}$$

where $\mathbf{v}_{k-1}$ and s_{k-1} denote the kth column of $\mathbf{F}^\dagger$ and the kth symbol of s, respectively. The inverse Fourier transform is an efficient way to implement (10.19); this is an understanding mainly from the computational perspective. A more intuitive explanation of (10.19) is the use of different symbols in s to modulate different orthogonal discrete exponential vectors, defined by the columns of $\mathbf{F}^\dagger$. Let x_n denote the nth entry of $\mathbf{x}$. In OFDM systems, the symbol sequence $\{x_n\}$ is first pulse-shaped with $p(t)$ and then low-pass-filtered before carrier modulation to produce

$$\tilde{x}(t) = \sum_{n=-\infty}^{\infty} x_n p(t - nT_s). \tag{10.20}$$

Here and hereafter in this chapter, T_s denotes the OFDM symbol duration. OFDM is one of the multicarrier modulation techniques.

Property 10.6.2 *The OFDM signal $\tilde{x}(t)$ is equivalent to the traditional multicarrier modulation signal*

$$x_{\mathrm{MC}}(t) = \sum_{k=0}^{N-1} s_k \, e^{j2\pi \, f_k t}. \tag{10.21}$$

For ease of verification, we assume a simple shaping pulse $p(t) = \text{sinc}(t/T_s)$ of an ideal rectangular spectrum over $f \in (-\frac{1}{2T_s}, \frac{1}{2T_s})$; namely, its Fourier spectrum is $P(f) = T_s \text{rect}(fT_s)$. As such, we can rewrite (10.20), yielding

$$\tilde{x}(t) = \sum_{n=-\infty}^{\infty} x_n \, \text{sinc}[(t - nT_s)/Ts]$$

$$= \sum_{n=-\infty}^{\infty} \underbrace{\sum_{k=0}^{N-1} s_k e^{j2\pi nk/N}}_{x_n} \, \text{sinc}[(t - nT_s)/T_s], \tag{10.22}$$

which, when transformed into the frequency domain, gives

$$X(f) = \sum_{k=0}^{N-1} s_k \sum_{n=-\infty}^{\infty} e^{-j2\pi nT_s(f - \frac{k}{NT_s})}$$

$$= T_s \sum_{k=0}^{N-1} s_k \text{rect}(fT_s) \underbrace{\sum_{n=-\infty}^{\infty} e^{-j2\pi nT_s(f - \frac{k}{NT_s})}}_{\alpha(f)}. \tag{10.23}$$

We need to find a more intuitive expression for $\alpha(f)$ by inverse Fourier transforming it to a periodic delta function in time. The periodic delta function has a discrete Fourier spectrum

$$\alpha(f) = \frac{1}{T_s} \sum_{n=-\infty}^{\infty} \delta\left(f - \frac{n}{T_s} - \frac{k}{NT_s}\right), \tag{10.24}$$

among which only the term for $n = 0$ falls into the range of $\text{rect}(fT_s)$. It follows that

$$X(f) = \sum_{k=0}^{N-1} s_k \delta\left(f - \frac{k}{NT_s}\right), \tag{10.25}$$

which, when transformed back to the time domain, gives

$$x(t) = \sum_{k=0}^{N-1} s_k e^{j2\pi(k/N)\Delta ft}. \tag{10.26}$$

This is the discrete equivalent of the continuous-time multicarrier modulation.

In its continuous time form, the OFDM signal is expressible as

$$x(t) = \sum_{k=0}^{N-1} s_k p(t) e^{j2\pi(f_c + f_k)t}$$

$$= \underbrace{\left(\sum_{k=0}^{N-1} s_k \, e^{j2\pi \, f_k t}\right)}_{s(t)} p(t) e^{j2\pi \, f_c t}, \tag{10.27}$$

where $p(t)$ is the shaping pulse, and

$$f_k = \left(k - \frac{N-1}{2}\right)\Delta f \tag{10.28}$$

represents the kth subcarrier deviation from the carrier with $\Delta f = 1/T_s$. The OFDM signal has N subcarriers centered at the carrier frequency f_c. We claim that $s(t)$ in (10.27) is equivalent to the vector form **s** in (10.19), as explained below.

Sampling $s(t)$ at time $t = nT_s/N, n = 0, \cdots, N-1$ with the rate of $1/(NT_s)$ and denoting $s(n) = s(t)|_{t=nT_s/N}$, we obtain the Fourier vector in (10.19) which, when written as a time function, gives

$$\tilde{s}(t) = \sum_{n=0}^{N-1} s(n)\, \delta\left(t - \frac{nT_s}{N}\right). \tag{10.29}$$

According to the Nyquist sampling theorem, $\tilde{s}(t)$ has the same spectrum as $s(t)$ over the frequency band $(-1/T_s, 1/T_s)$. Thus, $s(t)$ is retrievable by low-pass filtering (LPF) $\tilde{s}(t)$, as shown by

$$\tilde{s}(t) \xrightarrow{\text{LPF}} s(t). \tag{10.30}$$

For exposition and without loss of the generality, we assume rectangular shaping and just focus on the complex envelope of the signal shown in (10.27) to write

$$s(t) = \sum_{k=0}^{N-1} s_k \underbrace{e^{j2\pi k \Delta f t}}_{\varphi_k(t)}. \tag{10.31}$$

Note that the orthogonality holds among $\{\varphi_k(t)\}$ for $\Delta f = 1/T_s$, as shown by

$$\int_0^{T_s} \varphi_\ell(t)\varphi_q^*(t)dt = \begin{cases} 0, & q \neq \ell \\ T_s, & q = \ell \end{cases}, \tag{10.32}$$

which is a property in parallel to the orthogonality among the Fourier vectors $\{\mathbf{v}_k\}$.

Suppose that $s(t)$ passes through a time-invariant channel $h(t)$ with Fourier transform $G(f)$. Denote $\alpha_k = G(f_k)$. Then, the received signal is given by

$$r(t) = \sum_{k=0}^{N-1} s_k \alpha_k \, e^{j2\pi \, f_k t} + n(t), \tag{10.33}$$

with $n(t)$ denoting the AWGN component. When transmitted over a dispersive channel with impulse response $h(t)$, a high-rate pulse sequence spills over, causing ISI. The strategy of OFDM is to multiplex a high-data-rate stream into N parallel low-data-rate streams, with each assigned to an orthogonal subcarrier channel for transmission. When N is sufficiently large, the bandwidth of each subchannel is far less than the channel coherent bandwidth, behaving just as a flat fading channel.

10.6.3 PAPR and statistical behavior of OFDM

Let us study the behavior of OFDM signals with a large number of subcarriers in a statistical framework. We assume that a set of complex symbols, **s**, is randomly selected from a signal constellation, and each symbol is of mean zero and variance σ_s^2. For a large block size N, the central limit theorem applies, enabling us to assert that the OFDM signal

$$\mathbf{x} = \mathbf{F}^\dagger \mathbf{s},$$

approaches a jointly Gaussian distribution with mean vector and covariance matrix given by

$$\mathbb{E}[\mathbf{x}] = \mathbf{F}^\dagger \mathbb{E}[\mathbf{s}] = \mathbf{0},$$

$$\text{Cov}(\mathbf{x}) = \mathbf{F}^\dagger \mathbb{E}[\mathbf{s}\mathbf{s}^\dagger]\mathbf{F} = \sigma_s^2 \mathbf{I}.$$

Property 10.6.3 *For a large N, the dynamic behavior of the OFDM signal can be statistically characterized by* $\mathbf{x} \sim \mathcal{CN}(\mathbf{0}, \sigma_s^2 \mathbf{I})$.

Thus, we can write the probability density function (PDF) of its ith entry x_i as

$$f(x_i) = \frac{1}{\sigma_s^2} e^{-|x_i|^2/\sigma_s^2}, \tag{10.34}$$

and its power $\xi_i = |x_i|^2 \sim \frac{\sigma_s^2}{2}\chi^2(2)$. It follows that the probability that m subcarriers out of a total N are lower than a certain level, say Ω, is given by

$$P(m|N) = \binom{N}{m}(1 - e^{-\Omega/\sigma_s^2})^m e^{-(N-m)\Omega/\sigma_s^2}. \tag{10.35}$$

The peak-value behavior of OFDM signals is of practical concern. Since OFDM and multicarrier (MC) signals are equivalent, we use the notation of the latter for ease of elucidation. For an MC signal $x(t)$, the peak-to-average power ratio (PAPR) is defined as the ratio between its maximum instantaneous power and its average counterpart:

$$\text{PAPR}[x(t)] = \frac{\max_{0 \leq t < NT}|x(t)|^2}{\mathbb{E}[|x(t)|^2]}. \tag{10.36}$$

Property 10.6.4 *For an OFDM signal* $x(t) = \sum_{k=0}^{N-1} \alpha_k e^{j2\pi kt/T}$, *the PAPR is equal to N and this happens only when all the subcarriers are equally modulated, that is, when* $\alpha_0 = \cdots = \alpha_{N-1}$.

To verify, we find that $\max[|x(t)|^2] = N^2\sigma_s^2$ and

$$\mathbb{E}[|x(t)|^2] = \sum_{k=0}^{N-1}\sum_{m=0}^{N-1} \mathbb{E}[\alpha_k \alpha_m^*] e^{j2\pi(k-m)t/T}$$

$$= \sum_{k=0}^{N-1}\sum_{m=0}^{N-1} \sigma_s^2 \delta(k-m) e^{j2\pi(k-m)t/T}$$

$$= N\sigma_s^2.$$

Taking their ratio leads to the assertion.

10.6.4   Combating PAPR

PAPR is a crucial issue to OFDM systems, since many of them employ a high-power amplifier in their transmitters to operate at or near the saturation region for maximizing their output power efficiency. High PAPR can cause various problems such as severe waveform distortions, expensive power amplifiers, and additional interference. Therefore, great efforts have been made in the last two decades for PAPR reduction, and various schemes have been proposed.

From a physical perspective, a direct approach is to clip the signal once its amplitude exceeds a certain level, resulting in the so-called signal-clipping technique. Clipping is typically done at the transmitter, and compensation for the relevant OFDM symbols needs to be done at the receiver. To this end, the clipping location and size must be estimated, but it is usually difficult to implement. The advantage of amplitude clipping lies in its simplicity. But the drawback is that it introduces in-band distortion and out-of-band spectrum leakage, which can be mitigated by using shaping windows such as Gaussian, Kaiser, Hamming, and cosine.

The second approach is of probabilistic nature. As indicated above, PAPR occurs only when all the subcarriers are equally modulated. Thus, a probability-based technique is to make the occurrence of equal symbols a very unlikely event through, for example, the method of coding. An effective scheme along this line of thought for PAPR reduction is the use of Golay complementary codes [17–19].

If a multicarrier signal is modulated by a codeword from a Golay complementary pair, the PAPR is significantly reducible to the range of (0,2) [20]. Let us consider an OFDM signal modulated by M-ary PSK symbols $\mathbf{a} = [a_0, \cdots, a_{N-1}]$, which, in continuous-time formulation, is given by

$$x(t; \mathbf{a}) = \sum_{k=0}^{N-1} e^{j2\pi a_k/M} e^{j2\pi(f_c + k\Delta f)}. \tag{10.37}$$

Without coding and assuming the independence among the symbols $\{a_k\}$, the PAPR of $x(t; \mathbf{a})$ is just equal to N, as shown in the previous subsection. However, if $\mathbf{a}$ is chosen from a Golay complementary set of size m, then

$$0 \leq \frac{|x(t; \mathbf{a})|^2}{N\sigma_s^2} \leq m. \tag{10.38}$$

The idea of coding scheme can be extended to a larger category, referred to as *signal scrambling techniques*. Other techniques can also be used to prescramble OFDM symbols by appropriate phase rotation and optimization; two schemes along this direction are partial transmission sequence (PTS) and selective mapping (SLM).

10.7   Cyclic prefix and circulant channel matrix

We have not yet answered the question of how to generate a circulant channel matrix in a practical environment.

There is another problem of relevance. In OFDM, a dispersive channel is partitioned into a number of narrowband subcarrier channels for parallel signal transmission at a much lower symbol rate. Though much smaller, residual ISI possibly remains even at a lower transmission rate. An effective measure to combat residual ISI is the insertion of a cyclic prefix into an OFDM word.

The objective of inserting a cyclic prefix (CP) in OFDM signals is twofold. First, the cyclic prefix aims to provide a guard time against interference caused by multipath propagation. The guard time is chosen to be larger than the channel impulse response length (or more accurately, the expected delay spread of a random scattering channel), so that only the prefix portion of the current block is affected by the leakage from the current and the previous block. Thus, by removing the CP at the receiver, the residual ISI is completely eliminated. Second, the insertion of CP is to create an equivalent circulant channel matrix of an OFDM vector $\mathbf{x}$, such that its eigenvectors are exactly equal to the Fourier vectors regardless of the block size of the OFDM. With a circulant structure, IDFT and DFT of finite length at the transmitter and receiver can diagonalize the ISI channel. In practical applications, the typical block size of OFDM is 512 or 1024, whereas the typical size for a cyclic prefix can be 64 or 128.

To find the equivalent channel matrix for a data vector with a cyclic prefix of length m, denote $\mathbf{P} = [\mathbf{0}_{m\times(N-m)}, \mathbf{I}_{m\times m}]$, whereby the m-symbol cyclic prefix is expressible as $\tilde{\mathbf{x}} = \mathbf{P}\mathbf{x}$. Thus, the OFDM signal with cyclic prefix is equal to

$$\mathbf{x}_a = \begin{pmatrix} \tilde{\mathbf{x}} \\ \mathbf{x} \end{pmatrix}. \tag{10.39}$$

where the subscript a is used to indicate an augmented vector. As such, the received vector is correspondingly written as $\mathbf{r}_a = \mathbf{H}\mathbf{x}_a + \mathbf{n}_a$. If we partition the augmented received vector $\mathbf{r}_a$ and noise vector $\mathbf{n}_a$ in the same way as $\mathbf{x}_a$, then we can write

$$\underbrace{\begin{pmatrix} \tilde{\mathbf{r}} \\ \mathbf{r} \end{pmatrix}}_{\mathbf{r}_a} = \mathbf{H} \begin{pmatrix} \mathbf{P} \\ \mathbf{I}_{N \times N} \end{pmatrix} \mathbf{x} + \underbrace{\begin{pmatrix} \tilde{\mathbf{n}} \\ \mathbf{n} \end{pmatrix}}_{\mathbf{n}_a}. \tag{10.40}$$

At the receiver, $\tilde{\mathbf{r}}$ is discarded and only $\mathbf{r}$ is kept. Denoting the corresponding channel matrix by $\mathbf{H}_c$, we can write

$$\mathbf{r} = \mathbf{H}_c \mathbf{x} + \mathbf{n}. \tag{10.41}$$

It can be shown that the equivalent channel matrix $\mathbf{H}_c$ is of circulant structure. The role played by CP is quite easily elucidated through an illustrating example.

📖 **Example 10.1** _____

Suppose an OFDM system employs a 4-point IDFT to generate OFDM signals $\{x_i\}$ for transmission over a time-variant channel $\mathbf{H}$ defined by three nonzero taps h_0, h_1, and h_2. The first two blocks are $\mathbf{x}_1 = [x_1; x_2; x_3; x_4]$ and $\mathbf{x}_2 = [x_5; x_6; x_7; x_8]$; the semicolon used here has the same implication as that used in MATLAB. Since the channel impulse response is of length 2, we insert two symbols as a CP for each block, so that the two blocks for transmission are $\mathbf{x}_a = [x_3; x_4; \mathbf{x}_1; x_7; x_8; \mathbf{x}_2]$. Denote $\mathbf{h} = [h_0; h_1; h_2; \mathbf{0}_{9 \times 1}]$, which is the first column that defines the channel matrix $\mathbf{H}$. Symbolically, we write $\mathbf{H} = Toeplitz(\mathbf{h})$, whereby the received signal vector

$$\mathbf{r}_a = \mathbf{H}\mathbf{x}_a + \mathbf{n}_a.$$

We are interested in the structure of the term $\mathbf{H}\mathbf{x}_a$, which, after direct calculation, leads to the following expression:

$$\mathbf{H}\mathbf{x}_a = \left[\begin{array}{cccccccc} 0 & 0 & h_0 & 0 & 0 & 0 & 0 & 0 \\ 0 & 0 & h_1 & h_0 & 0 & 0 & 0 & 0 \\ \hline h_0 & 0 & h_2 & h_1 & 0 & 0 & 0 & 0 \\ h_1 & h_0 & 0 & h_2 & 0 & 0 & 0 & 0 \\ h_2 & h_1 & h_0 & 0 & 0 & 0 & 0 & 0 \\ 0 & h_2 & h_1 & h_0 & 0 & 0 & 0 & 0 \\ \hline 0 & 0 & h_2 & h_1 & 0 & 0 & h_0 & 0 \\ 0 & 0 & 0 & h_2 & 0 & 0 & h_1 & h_0 \\ \hline 0 & 0 & 0 & 0 & h_0 & 0 & h_2 & h_1 \\ 0 & 0 & 0 & 0 & h_1 & h_0 & 0 & h_2 \\ 0 & 0 & 0 & 0 & h_2 & h_1 & h_0 & 0 \\ 0 & 0 & 0 & 0 & 0 & h_2 & h_1 & h_0 \end{array}\right] \begin{bmatrix} x_1 \\ x_2 \\ x_3 \\ x_4 \\ x_5 \\ x_6 \\ x_7 \\ x_8 \end{bmatrix}.$$

Clearly, the lines 1 and 2 in the matrix contain leakage from block 1, $\mathbf{x}_1$. On the other hand, lines 7 and 8 contain spillover from both blocks $\mathbf{x}_1$ and $\mathbf{x}_2$. By discarding lines 1, 2, 7, and 8, partitioning $\mathbf{r}_a$ and $\mathbf{n}_a$ in a manner similar to $\mathbf{x}_a$, and using similar notation, we obtain the received blocks corresponding to $\mathbf{x}_1$ and $\mathbf{x}_2$:

$$\mathbf{r}_1 = \mathbf{H}_c \mathbf{x}_1 + \mathbf{n}_1,$$

$$\mathbf{r}_2 = \mathbf{H}_c \mathbf{x}_2 + \mathbf{n}_2,$$

where the equivalent channel matrix

$$\mathbf{H}_c = \begin{bmatrix} h_0 & 0 & h_2 & h_1 \\ h_1 & h_0 & 0 & h_2 \\ h_2 & h_1 & h_0 & 0 \\ 0 & h_2 & h_1 & h_0 \end{bmatrix},$$

is indeed a circulent one.

We are now in position to summarize the procedure for generating an OFDM signal of N subcarriers. Denote the $N \times N$ DFT matrix by $\mathbf{F}$.

$$\cdots b_1 b_0 \longrightarrow \boxed{\text{Mapper}} \longrightarrow \boxed{\text{S/P}} \overset{\mathbf{s}}{\longrightarrow} \boxed{\text{IDFT}} \overset{\mathbf{x}}{\longrightarrow} \boxed{\text{Insert CP}} \overset{\mathbf{y}}{\longrightarrow}$$

The procedure to generate OFDM signals is illustrated in the above block diagram, with details as follows:

(a) The information-bearing bit sequence $\{b_k\}$ is first mapped into complex signal points in a prescribed constellation such as QAM. Partition the resulting complex symbols into blocks of size N, and take one such column block $\mathbf{s}$ as an example.
(b) Assign each symbol of $\mathbf{s}$ with a subcarrier; this corresponds to taking inverse DFT of $\mathbf{s}$ to produce $\mathbf{x} = \mathbf{F}^{-1}\mathbf{s}$. Partition $\mathbf{x}$ into the column subvector $\mathbf{x}_1$ of size $N - m$ and the one $\mathbf{x}_2$ of size m, such that $\mathbf{x} = [\mathbf{x}_1; \mathbf{x}_2]$.
(c) Cyclically prefix $\mathbf{x}_2$ to $\mathbf{x}$ to obtain an OFDM $(N + m)$-vector $\mathbf{y} = [\mathbf{x}_2; \mathbf{x}]$.
(d) Shape each OFDM symbol of $\mathbf{y}$ with the raised cosine spectrum before carrier modulation for transmission over a physical channel.

✍ Example 10.2

Consider a 5-MHz OFDM system of $N = 512$ subcarriers to operate in a channel with Doppler shift $f_D = 70$ Hz. Each subchannel occupies a bandwidth of

$$\frac{B}{N+1} = 5 \times 10^6/512 = 10^4 \text{ Hz},$$

implying that its symbol duration is equal to $T_s = 10^{-4}$ s. If we define the coherent time as the geometric mean of the one with 90% correlation and the one with 50% correlation, then it is equal to $T_c = \sqrt{\frac{9}{16\pi f_D^2}} = 0.423/f_D = 0.056$ s. Since T_c is far greater than T_s, each subchannel is a slow fading channel. ○

The OFDM system has two advantages over single-carrier systems with equalizers. A main advantage of OFDM is its capability to deal with large delay spreads with a reasonable complexity, requiring only a one-tap equalizer for phase correction of each subcarrier in the case of using a coherent receiver. A single carrier system often suffers performance degradation due to error propagation, which happens when the channel delay spread exceeds the designed value of the equalizer. The OFDM system has no such drawback. An OFDM system can even out the fading effects as long as its symbol duration is larger than the length of fades. As such, the OFDM symbols are only partly destroyed by fading. The drawbacks of OFDM are its high PAPR, sensitivity to doppler shift, timing error, and phase shift, which may destroy the orthogonality among subcarriers.

10.8 OFDM receiver

An inner-product (or correlator) receiver can be used to separate a set of received orthogonal signals, such as OFDM and multicarriers, for the retrieval of the original symbols. For multicarrier signals given in (10.33), the correlator for the kth subchannel receiver performs the inner-product operation, producing

$$\langle \varphi_k^*(t), r(t) \rangle = \alpha_k s_k + n_k, \quad k = 0, \cdots, N - 1, \tag{10.42}$$

where $\varphi(t) = (1/\sqrt{T_s}) \exp(2\pi f_k t)$ is the local reference signal for subchannel k, and n_k is the noise at the correlator output. For OFDM vectors, the operation is similar, as shown by

$$\langle \mathbf{v}_k^*, \mathbf{r} \rangle = \alpha_k s_k + n_k, \quad k = 0, \cdots, N - 1. \tag{10.43}$$

In so doing, we need N parallel correlators. The advantage of OFDM is its convenience to implement efficient computation using DFT. Upon the receipt of vector $\mathbf{r}_a$, we remove the CP and then take DFT, resulting in the decision vector

$$\mathbf{z} = \mathbf{F} \underbrace{(\mathbf{H}_c \mathbf{x} + \mathbf{n})}_{\mathbf{r}} = \begin{pmatrix} \alpha_0 s_0 \\ \vdots \\ \alpha_{N-1} s_{N-1} \end{pmatrix} + \mathbf{n}. \tag{10.44}$$

Since unitary transformation does not change the property of the AWGN vector $\mathbf{n}$, we keep the same notation after DFT. The IDFT matrix is defined by

$$\mathbf{F}^\dagger = \frac{1}{\sqrt{N}} \begin{bmatrix} 1 & 1 & \cdots & 1 \\ 1 & e^{j2\pi/N} & \cdots & e^{j2\pi(N-1)/N} \\ 1 & e^{j4\pi/N} & \cdots & e^{j4\pi(N-1)/N} \\ \vdots & \vdots & \vdots & \vdots \\ 1 & e^{j2\pi(N-1)/N} & \cdots & e^{j2\pi(N-1)(N-1)/N} \end{bmatrix}. \tag{10.45}$$

The channel gains α_k for individual subcarriers are uniquely determined from the channel frequency transfer function, via one of two different ways. If the channel impulse $h(t)$ is given in the form of a T_s-spaced tap-delayed line structure, the subchannel gains α_k can be directly calculated by using DFT just as (10.18). If, instead, the channel impulse response takes the form of the Turin model $h(t) = \sum_{i=0}^{L-1} h_i \delta(t - \tau_i)$, we calculate the Fourier transform of $h(t)$, denoted by $G(f)$. Then

$$\alpha_k = G(f)|_{f=k/N}. \tag{10.46}$$

After isolating individual subcarrier channels, we can carry out symbol-by-symbol detection for each subchannel, based on the decision vector (10.44). The symbol error rate associated with the decision can be evaluated by using the techniques previously described in Chapter 4.

10.9 Channel estimation

Channel estimation is a necessary step for the operation of an OFDM system. A succinct, yet comprehensive, overview of various channel estimation strategies is given in Ref. [21]. Early seminal work on this issue includes [22–24].

10.10   Coded OFDM

When an OFDM system operates in a frequency-selective fading channel, different subcarriers experience different fading gains. The system error performance is dominated by those subcarriers in poor channel conditions. A basic strategy to overcome this situation is to send each message symbol over multiple subcarriers, so as to even out the risk of erroneous decisions on poorly conditioned subcarriers. This can be done by cross-tone coding [25–28], or precoding before OFDM operation [29]. It is shown Ref. [29] that algebraic codes are more suitable for random errors caused by channel noise, or very fast fading as created by interleaving. Precoding is more powerful in combating structured errors introduced, for example, by ISI or frequency-selective channels. As a result of the time diversity due to coding plus inherent frequency diversity of OFDM, data loss in a subcarrier channel caused by deep fading can be recovered from coded data in other subcarriers channels. An OFDM system with coding and/or precoding is sketched below.

$$\cdots b_1 b_0 \rightarrow \boxed{\text{Coding/precoding}} \rightarrow \boxed{\text{Interleaver}} \rightarrow \boxed{\text{Mapper}} \rightarrow \boxed{\text{S/P}} \xrightarrow{\mathbf{s}} \boxed{\text{IDFT}} \rightarrow \boxed{\text{Insert CP}}$$

Channel codes used for coded OFDM include convolutional, Reed–Solomon, LDPC (low-density party-check), turbo, and complex field codes [29–31], while precoding can be implemented, for example, by complex-field coding [29].

10.11   Additional reading

Among a variety of channel decomposition techniques, OFDM and channel equalization are the two most popular ones. They have been extensively investigated and widely used in practice. Other techniques are scattered in the literature. The reader is referred to [35–51] for additional reading.

Problems

Problem 10.1   A Doppler shift can break down the orthogonality between different subcarrier vectors. Let $\mathbf{v}_p$ and $\mathbf{v}_q$ denote the length-N Fourier vectors at the discrete frequency p and q, respectively. Assume that both of them experience Doppler shift ϵ, so that the kth entry of $\mathbf{v}_p$ and $\mathbf{v}_q$ can be written as $e^{j2\pi pk(1+\epsilon)/N}$ and $e^{2\pi qk(1+\epsilon)/N}$, respectively. Show that the inner product of the two vectors has the magnitude

$$|\langle \mathbf{v}_p, \mathbf{v}_q \rangle| = \frac{\sin[\pi(1+\epsilon)(p-q)]}{\sin[\pi(1+\epsilon)(p-q)/N]}.$$

Problem 10.2   In the OFDM system, the signal vector $\mathbf{x}$ of N symbols is inverse-Fourier-transformed and then cyclic prefixed (CP) before transmission over a channel of the form $g(t) = \sum_{m=0}^{L-1} g_m \delta(t - \tau_m)$. Let $G(\cdot)$ denote the discrete Fourier transform of $g(t)$, such that $h_k = G(\frac{k}{NT_s}), k = 0, \cdots, N-1$. At the receiver, after DFT and removal of the CP, the recovered noisy signal vector becomes

$$\mathbf{y} = \mathbf{X}\mathbf{h} + \mathbf{n},$$

where $\mathbf{X} = \text{diag}(\mathbf{x})$, $\mathbf{h} = [h_0, \cdots, h_{N-1}]^T$, and $\mathbf{n} \sim \mathcal{CN}(\mathbf{0}, \sigma_n^2 \mathbf{I})$. Show that, given $\mathbf{x}$, the least-square and the MMSE estimators for the channel vector $\mathbf{h}$ are given by

$$\mathbf{h}_{ls} = [y_0/x_0, y_1/x_1, \cdots, y_{N-1}/x_{N-1}]^T,$$

$$\mathbf{h}_{\text{MMSE}} = \mathbf{R}_{hh}[\mathbf{R}_{hh} + \sigma_n^2 (\mathbf{X}\mathbf{X}^\dagger)^{-1}]^{-1}\mathbf{h}_{ls},$$

respectively, where $\mathbf{R}_{hh} = \mathbb{E}[\mathbf{h}\mathbf{h}^\dagger]$ [24].

Problem 10.3   The presence of an unknown Doppler shift in OFDM signals destroys the orthogonality of subcarriers. In this project, we investigate the use of ESPRIT, a MUSIC-like high-resolution signal processing algorithm, for the estimation of unknown Doppler shift. Run a program and verify the validity of the algorithm [32].

Problem 10.4   The OFDM signal with N subcarriers is given by

$$x(n) = \frac{1}{\sqrt{N}} \sum_{k=0}^{N-1} X(k) e^{j2\pi nk/N},$$

which, when transmitted over a multipath channel with frequency response $\{H(k)\}$, produces received signal

$$y(n) = e^{j\phi(n)} \left[\frac{1}{\sqrt{N}} \sum_{k=0}^{N-1} X(k) H(k) e^{j2\pi nk/N} \right] + w(n).$$

Besides $H(k)$, the channel also introduces AWGN $w(n) \sim \mathcal{CN}(0, \sigma_n^2)$ and phase noise $\phi(n)$. Investigate how the phase noise will destroy the orthogonality among the subcarriers and thus introduce inter-carrier interference when DFT is taken on $y(n)$ at the receiver [33].

Problem 10.5   In Chapter 9, we determined the output SNR of a communication system with ZFE based on a continuous-time model. Suppose, instead, we now consider a discrete channel model $\mathbf{H}$, so that the received signal vector is

$$\mathbf{y} = \mathbf{H}\mathbf{s} + \mathbf{n},$$

where $\mathbf{n} \sim \mathcal{CN}(0, \sigma_n^2 \mathbf{I})$ and signal $\mathbf{s}$ is of mean zero and variance σ_s^2. Denote $\gamma_0 = \sigma_s^2/\sigma_n^2$, and let α_k, $k = 1, \cdots, K$ denote all the nonzero singular values of $\mathbf{H}$.

(a) Show that, if ZFE is used, the output SINR is given by

$$\gamma_{\text{ZFE}} = \gamma_0 \left[\frac{K}{\sum_{k=1}^{K} |\alpha_k|^{-2}} \right].$$

(b) Show that this SINR is equivalent to its continuous counterpart derived in Chapter 9.
(c) If a linear MMSE equalizer is used, instead, derive the output SINR and show its equivalence to its continuous counterpart obtained in Chapter 9:

$$\gamma_{\text{MSE}} = \gamma_0 \left[\frac{K}{\sum_{k=1}^{K} \left(\gamma_0^{-1} + |\alpha_k|^2 \right)^{-1}} \right].$$

Problem 10.6   Show that the OFDM has a higher system mutual information than its counterpart with a ZFE or linear MMSE equalizer.

Problem 10.7   For $g(t) = \sum_{i=0}^{L-1} g_i \delta(t - \tau_i)$, determine an equivalent discrete channel matrix $\mathbf{H}$ for an N-point OFDM system.

Problem 10.8   An OFDM system with coherent BPSK signalling operates on a Rayleigh multipath channel $g(t) = \sum_{i=0}^{L-1} g_i \delta(t - \tau_i)$ with independent $g_i \sim \mathcal{CN}(0, \sigma_i^2)$, $i = 0, 1, \cdots, L-1$. Assume that the symbol bit energy E_b and the noise two-sided power spectral density $N_0/2$ are such that $E_b/N_0 = 1$. Determine its average bit error rate.

Problem 10.9 In Example 10.1, 4-IDFT was used to generate OFDM blocks and each block was inserted with a CP of two symbols for transmission over a three-tap channel with response $h(t) = \sum_{i=0}^{2} h_i \delta(t - iT_s)$. Here, T_s denotes the duration of an OFDM symbol. Suppose now a CP of three symbols is used, instead. Determine the channel output vectors corresponding to the first two consecutive OFDM blocks.

Problem 10.10 As indicated in (10.44), the decision variables for block ℓ at the OFDM receiver take the form

$$z_{\ell i} = \alpha_{\ell i} s_{\ell i} + n_{\ell i}, \ i = 0, 1, \cdots, N - 1, \tag{10.47}$$

where $n_{\ell i} \sim \mathcal{CN}(0, \sigma_n^2)$, $\alpha_{\ell i} \sim \mathcal{CN}(0, \sigma_\alpha^2)$, and each symbol $s_{\ell i}$ is of energy E_s. For block-static fading, the fading gain $\alpha_{\ell i}$ remains unchanged over one block but varies from one block to another independently. Suppose that M-QAM is employed and decisions are made on a symbol-by-symbol basis. Determine the average symbol error rate for each single decision.

Problem 10.11 (Timing offset in OFDM). Consider an OFDM system with N subcarriers. Due to timing synchronization error, the receiver misaligns the DFT window with the following received symbol block:

$$\mathbf{y}_k = h \begin{bmatrix} x_{k-1}[N - (\delta - l_{\text{CP}})] \\ \vdots \\ x_{k-1}[N - 1] \\ x_k[N - l_{\text{CP}}] \\ \vdots \\ x_k[N - 1] \\ x_k[0] \\ \vdots \\ x_k[N - 1 - \delta] \end{bmatrix} + \mathbf{n},$$

where $x_{k-1}[\cdot]$ is the previous OFDM symbol and $x_k[\cdot]$ is the current symbol, l_{CP} is the length of cyclic prefix, and δ is the timing offset with $\delta > l_{\text{CP}}$. In the above equation, we assume a single-tap fading channel and $\mathbb{E}[\mathbf{x}_k \mathbf{x}_k^\dagger] = \mathbf{I}$, $\mathbf{x}_k = [x_k[0] \ \cdots \ x_k[N - 1]]$, $\forall k$. Show that the average power of ISI on the nth subcarrier is

$$P_{\text{ISI}} = \mathbb{E}[|h|^2] \frac{\delta - l_{\text{CP}}}{N}.$$

[34].

References

1. G.D. Forney, "Maximum likelihood sequence estimation of digital sequences in the presence of intersymbol interference," *IEEE Trans. Inf. Theory*, vol. IT-18, pp. 363–378, 1972.
2. M. Tomlinson, "New automatic equalizer employing modulo arithmetic," *Electron. Lett.*, vol. 7, no. 5/6, pp. 138–139, 1971.
3. K. Takeda, H. Tomeba, and F. Adachi, "Joint Tomlinson-Harashima precoding and frequency-domain equalization for broadband single-carrier transmission," *IEICE Commun.*, vol. E91-B, no. 1, pp. 258–266, 2008.
4. C. Windpassinger, R.F.H. Fischer, T. Vencel, and J.B. Huber, "Precoding in multiantenna and multiuser communication," *IEEE Trans. Wireless Commun.*, vol. 3, no. 4, pp. 1305–1316, 2004.
5. H. Harashima and H. Miyakawa, "Matched-transmission technique for channels with intersymbol interference," *IEEE Trans. Commun.*, Vol. 20, no. 4, pp. 774–480, 1972.

6. Y. Jiang, W. Hager, and J. Li, "The geometric mean decomposition," *Linear Algebra Appl.*, vol. 396, pp. 373–384, 2005.

7. Y. Jiang, W.W. Hager, and J. Li, *Linear Algebra and its Applications*, North Holland: Elsevier, 2014.

8. Y. Jiang, J. Li, and W. Hager, "Joint transceiver design for MIMO communications using geometric mean decomposition," *IEEE Trans. Signal Process.*, vol. 53, no. 10, Pt. 1, pp. 3791–3803, 2005.

9. Y. Jiang, J. Li, and W. Hager, "Uniform channel decomposition for MIMO communications," *IEEE Trans. Signal Process.*, vol. 53, no. 11, pp. 4283–4294, 2005.

10. S. Lin, W.W.L. Ho, and Y.-C. Liang, "Block-diagonal geometric mean decomposition (BD-GMD) for MIMO broadcast channels," *IEEE Trans. Wireless Commun.*, Vol. 7, no. 7, p. 2–, 2008.

11. R.M. Gray, Toeplitz and Circulant Matrices, in the Website of R.M. Gray in the EE Department of Stanford University.

12. R.M. Gray, "On the asymptotic eigenvalue distribution of toeplitz matrices," *IEEE Trans. Inf. Theory*, vol. IT-18, no. 6, pp. 725–730, 1972.

13. G.A. Franco and G. Lachs, "An orthogonal coding technique for communications," *IRE Int. Conv. Rec.*, vol. 9, Pt. 8, pp. 126–133, 1961.

14. S. Darlington, "On digital single-sideband modulators," *IEEE Trans. Circuit Theory*, vol. 17, pp. 409–414, 1970.

15. S.B. Weinstein and P.M. Ebert, "Data transmission by frequency division multiplexing using the discrete Fourier transform," *IEEE Trans. Commun.*, vol. 19, no. 5, pp. 628–634, 1971.

16. L.J. Cinimi, "Analysis and simulation of a digital mobile channel using orthogonal frequency division multiplexing," *IEEE Trans. Commun.*, vol. 33, pp. 665–675, 1985.

17. J.A. Davis and J. Jedwab, "Peak-to-mean power control in OFDM: golay complementary sequences and Reed-Muller codes," *IEEE Trans. Inf. Theory*, vol. 45, no. 7, pp. 2397–2417, 1997.

18. S. Boyd, "Multitone signals with low crest factor," *IEEE Trans. Circuits Syst.*, vol. CAS-33, no. 10, pp. 1018–1022, 1986.

19. T. Jiang, G.X. Zhu, and J.B. Zheng, "A block coding scheme for reducing PAPR in OFDM systems with large number of subcarriers," *J. Electron.*, vol. 21, no. 6, pp. 482–489, 2004.

20. K.G. Paterson, "Generalized Reed-Muller codes and power control in OFDM modulation," *IEEE Trans. Inf. Theory*, vol. 46, no. 1, pp. 104–120, 2000.

21. Y. Shen and E. Martines, "Channel estimation in OFDM systems," *Application Note AN3059*, Rev.0, Freescale Semiconductor, Inc., Jan., 2006.

22. Y. Li, L.J. Cimini, and N.R. Sollenberger, "Robust channel estimation for OFDM systems with rapid dispersive fading channels," *IEEE Trans. Commun.*, vol. 46, no. 7, pp. 902–915, 1998.

23. Y. Li, "Simplified channel estimation for OFDM systems with multiple transmit antennas," *IEEE Trans. Commun.*, vol. 50, no. 1, pp. 67–75, 2002.

24. O. Edfors, M. Sandell, J.J. Van de Beek, and S.K. Wilson, "OFDM channel estimation by singular value decomposition," *IEEE Trans. Commun.*, vol. 46, no. 7, pp. 931–939, 1998.

25. B. Le Floch, M. Alard, and C. Berrou, "Coded orthogonal frequency-division multiplex," *Proc. IEEE*, vol. 83, pp. 982–996, 1995.

26. G. Ungerboeck, "Channel coding with multilevel/phase signals," *IEEE Trans. Inf. Theory*, vol. 28, pp. 56–67, 1982.

27. W.Y. Zou and Y. Wu, "COFDM: a review," *IEEE Trans. Broadcast.*, vol. 41, pp. 1–8, 1995.

28. T.N. Zoglakis, J.T. Aslanis Jr., and J.M. Cioffi, "A coded and shaped discrete multitone system," *it IEEE Trans. Commun.*, vol. 43, pp. 2941–2949, 1995.

29. Z. Wang and G.B. Giannakis, "Complex-field coding for OFDM over fading wireless channels," *IEEE Trans. Inf. Theory*, vol. 49, no. 3, pp. 707–720, 2003.

30. I.A. Chatzigeorgiou, M.R.D. Rodrigues, I.J. Wassell, and R. Carrasco, "A comparison of convolutional and Turbo coding schemes for broadband FWA systems," *IEEE Trans. Broadcast.*, vol. 53, no. 2, pp. 494–503, 2007.

31. I.B. Djordjevic, B. Vasic, and M.A. Neifeld, "LDPC coded OFDM over the atmospheric turbulence channel," *Opt. Express*, vol. 15, no. 10, pp. 6336–6350, 2007.

32. U. Tureli, H. Liu, and M.D. Zoltowski, "OFDM blind carrier offset estimation: ESPIRIT," *IEEE Trans. Commun.*, vol. 48, no. 9, pp. 1459–1461, 2000.

33. S. Wu and Y. Bar-Ness, "OFDM systems in the presence of phase noise: consequences and solutions," *IEEE Trans. Commun.*, vol. 50, no. 11, pp. 1988–1996, 2004.

34. M. Wang, L. Xiao, T. Brown, and M. Dong, "Optimal symbol timing for OFDM wireless communications," *IEEE Trans. Wireless Commun.*, vol. 8, no. 10, pp. 5328–5337, 2009.

35. T. Jiang and Y. Wu, "An overview: peak-to-average power ration reduction techniques for OFDM signals," *IEEE Trans. Broadcast.*, vol. 54, no. 2, pp. 257–268, 2008.

36. S.H. Han and J.H. Lee, "An overview of peak-to-average power ratio reduction techniques for multicarrier transmission," *IEEE Pers. Commun.*, vol. 12, no. 2, pp. 56–65, 2005.

37. IEEE Std 802.11a, Supplement to Part 11; Wireless LAN Medium Access Control (MAC) and Physical Layer (PHY) specifications: high-speed Physical Layer in the 5 GHZ Band, IEEE Std 802.11a-1999, 1999.

38. IEEE 802.11g, Part 11: Wireless LAN Medium Access Control (MAC) and Physical Layer (PHY) specifications, Amendment 4: Further Higher Data Rate Extension in the 2.4 GHz Band, 2003.

39. IEEE 802.16, Standard - Local and Metropolitan Area Networks C Part16, IEEE Std 802.16a-2003.

40. P.J. Davis, *Circulant Matrices*, New York: John Wiley & Sons, Inc., 1970.

41. R. Van Nee and R. Prasad, *OFDM for Wireless Multimedia Communications*, Norwood, MA: Artech House, 2000.

42. A. Bahai, B.R. Saltzberg, and M. Ergen, *Multi-Carrier Digital Communications: Theory and Applications of OFDM*, New York: Springer-Verlag, 2004.

43. L.J. Cimini, "Analysis and simulation of a digital mobile channel using orthogonal frequency division multiplexing," *IEEE Trans. Commun.*, vol. 33, no. 7, pp. 665–675, 1985.

44. R. Van Nee and R. Prasad, *OFDM for Wireless Multimedia Communications*, New York: Artech House, 2000.

45. R.W. Chang and J.C. Hancock, "On receiver structures for channels having memory," *IEEE Trans. Inf. Theory*, vol. IT-12, pp. 463–468, 1966.

46. K. Abend and B.D. Fritchman, "Statistical detection for communication channels with intersymbol interference," *Proc. IEEE*, vol. 58, pp. 778–785, 1970.

47. V. Aue, G.P. Fettweis, and R. Valenzuela, "Comparison of the performance of linearly equalized single carrier and coded OFDM over frequency selective fading channels using the random coding technique," *Proceedings of ICC'98*, pp. 753–757.

48. Y. Zhu and K.B. Letaief, "Single carrier frequency domain equalization with noise prediction for broadband wireless systems," *Proceedings of IEEE Global Telecommunications Conference (GLOBECOM'04)*, vol. 5, Nov. 2004, pp. 3098–3102.

49. D. Falconer and S. Ariyavisitakul, "Broadband wireless using single carrier and frequency domain equalization," *Proceedings of International Symposium on Wireless Personal Multimedia Communications, vol. 1*, pp. 27–36, Oct. 2002.

50. Y. Li, N. Seshadri, and S. Ariyavisitakul, "Channel estimation for OFDM systems with transmitter diversity in mobile wireless channels," *IEEE J. Sel. Areas Commun.*, vol. 17, no. 3, pp. 461–471, 2006.

51. T. Hwang, C. Yang, G. Wu, S. Li, and G. Y. Li, "OFDM and its wireless applications: a survey," *IEEE Trans. Veh. Technol.*, vol. 58, no. 4, pp. 1673–1694, 2009.

11

Turbo Codes and Turbo Principle

Abstract

Turbo codes are a powerful means to approach the capacity of an AWGN channel. A well-performed turbo code is usually synthesized by using two simple encoders, in serial or parallel concatenation, so that an iterative algorithm can be used for efficient decoding. A basic philosophy of turbo codes is, therefore, to create a two-device mechanism and derive an information-exchange algorithm to optimally retrieve the original message sequence. The two devices can be a pair of constituent encoders or a combination of an encoder and any other structure in a communication system. This widely applicable philosophy is known as the turbo principle. It is shown that turbo codes can effectively approach the AWGN channel capacity, leaving a gap of only $0.7\,\mathrm{dB}$ from the Shannon bound at the bit-error rate of 10^{-5}. The turbo principle is also widely used in turbo equalization and turbo CDMA.

♣

11.1 Introduction and philosophical discussion

The maximum bit rate for error-free transmission over a noisy channel is upper-bounded by its channel capacity. In his seminal paper, Shannon [1] demonstrated that the channel capacity could be theoretically approached by using an infinitely long *random* code. In practical applications, however, pure random codes are difficult to use because of the lack of appropriate structures for their decoding. On the other hand, traditional linear block codes and convolutional codes possess good algebraic structures for decoding, but lead to a considerable performance deviation from the Shannon bound. Coding to approach the Shannon limit with a tolerable complexity has been the target of the coding community for decades. The strategy is to sacrifice part of the data rate by adding certain algebraic structures to trade for a coding gain, so that the spectral efficiency approaches the Shannon limit at a lower rate than attainable by modulation. The idea is illustrated in Figure 11.1 for an AWGN channel where the uncoded 64QAM system has the bit rate of 6 bps, which deviates from the channel capacity by $7.5\,\mathrm{dB}$. One possible coding scheme is (4,3,6) 64QAM trellis-coded modulation, which reduces the data rate to 5 bps. The return of sacrificing $\Delta R_b = 1$ bps in the data rate is the coding gain of $g_{\mathrm{SNR}} = 6.4\,\mathrm{dB}$ to approach the Shannon bound at the lower level of 5 bps, shortening the gap to $3.4\,\mathrm{dB}$.

Wireless Communications: Principles, Theory and Methodology, First Edition. Keith Q.T. Zhang.
© 2016 John Wiley & Sons, Ltd. Published 2016 by John Wiley & Sons, Ltd.
Companion Website: www.wiley.com/go/zhang7749

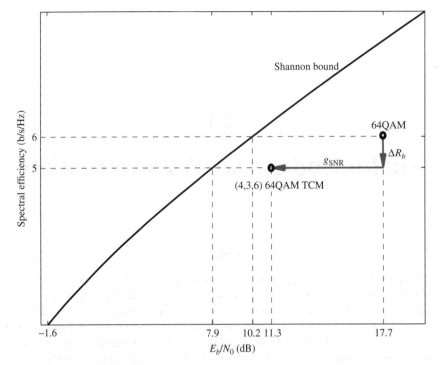

Figure 11.1   *Illustrating the idea of coding to approach capacity*

11.1.1   Generation of random-like long codes

As previously mentioned, capacity-approaching long codes must behave as random as possible on one hand while maintaining good structures for ease of decoding on the other. Code randomization can be implemented either *algebraically* by random interleaving or *geometrically* via sparse random connections among coding/checking nodes. The former leads to turbo codes [2–4], whereas the latter leads to large-density parity-check (LDPC) codes [5].

The use of long codes can make their decoding error probability drop exponentially with increased block size. Theoretically, it is not difficult to generate long codes. The difficulty is that increase in the block size leads to an exponentially increased decoding complexity. Turbo codes introduced by C. Berrou in 1993 are formed by two parallel convolutional codes that are separated by a random interleaver. The idea of turbo codes can be traced back to Forney's concatenated codes in 1962 [19]. In his original scheme, Forney cascaded two relatively simple component codes (usually block or convolutional codes) to synthesize a single concatenated code, which enjoyed exponentially decreasing probability of decoding errors while retaining polynomial-time decoding complexity. A typical example is the serial concatenation of a Reed–Solomon code (outer) with a convolutional code (inner). Inner and outer coders can be easily recognized from the system block diagram. For serial concatenation, the inner coder refers to the encoder that is closer to the physical channel while the outer encoder is the one farther from the channel.

Inserting a random interleaver into the code structure endows the resulting concatenated code with a very large block length [6]. Another family of random codes, called LDPC codes, is rooted in an idea different from concatenation. It had largely been forgotten but was recently rediscovered by researchers. Unlike algebraic codes that aim to maximize the minimum code distance to ensure the necessary error performance, random codes do not attempt to eliminate codewords of small Hamming weight from the code set, but rather minimize the probability of their occurrence. Turbo and LDPC codes have similar performance. The difference is that the former puts its main efforts on decoding, whereas the latter are more focused on encoding. A very insightful historical overview of the coding evolution to approach the Shannon limit can be found in Ref. [7] by Costello and Forney. In this chapter, we focus on turbo codes and the turbo principle.

11.1.2   The turbo principle

The success of turbo codes is not only attributed to their elegant structures but also largely relies on the turbo decoding principle, or briefly the turbo principle. Good code structures provide a potential to approach the channel capacity; but the turbo principle makes this potential implementable. Concatenated codes and the turbo principle are two aspects of the same entity, constituting the foundation of turbo codes. The core of the turbo principle is to combine two blocks or two subsystems for joint estimation in an iterative manner, so that discrete constraints are placed in the feedback/feedforward loop to convert an intractable integer optimization to a simple iteration process. However, in order for the turbo decoding algorithm to converge, these two blocks must be clearly separated so that their parameters are independent. It should be emphasized that the turbo principle is not limited to coding and decoding; it is an efficient means to solve many optimization problems with integral or discrete constraints.

Coding is a wonderland where there is no unique road to the Shannon limit. Each coding master creates masterpieces of his own style. When recalling the invention of the turbo codes, Berrou wrote [4]: "Because the foundation of digital communications relied on potent mathematical considerations, error correcting codes were believed, for a long time, to solely belong to the world of mathematics." Turbo decoding draws its inspiration from the feedback concept in electronics. In the era of algebraic codes, the design philosophy is first to identify the code structures and then search for good codes. Berrou changed the design philosophy of good codes for approaching the Shannon bound. He simply used recursive systematic convolutional (RSC) codes as constituents for synthesizing his turbo codes, based on the following justifications:

- RSC codes behave in a manner similar to a pseudo-random scrambler, producing a random-like sequence of properties required by Shannon in his proof of ideal coding to approach the channel capacity.
- They perform better at high coding rate and high noise environments.
- They have a better return-to-zero property.

11.2   Two-device mechanism for iteration

The turbo principle requires a two-device mechanism for information exchange. Similar situations happen in diverse disciplines of sciences. Before proceeding, let us recall the well-known Rayleigh-quotient-based searching technique for the maximum eigenvalue of a nonnegative definite matrix $\mathbf{A}$. A general routine to determine the eigenvalues of matrix $\mathbf{A}$ is to solve the nonlinear equation $\det(\mathbf{A} - \lambda\mathbf{I}) = 0$, which usually relies on numerical approaches. If only the maximum eigenvalue is of interest, we may use the method of Rayleigh quotient iteration for its solution, for which one algorithm is listed below.

Start from a non-zero initial vector $\mathbf{v}_1$.
for $i = 1:m$
$\quad \lambda_i = \mathbf{v}_i^\dagger \mathbf{A} \mathbf{v}_i / (\mathbf{v}_i^\dagger \mathbf{v}_i)$
$\quad \mathbf{u}_i = (\mathbf{A} - \lambda_i \mathbf{I})^{-1} \mathbf{v}_i$
$\quad \mathbf{v}_{i+1} = \mathbf{u}_i / \text{norm}(\mathbf{u}_i)$
end

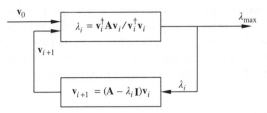

Figure 11.2   *Information-exchange principle in seeking the principal eigenvalue*

Here, λ_i and $\mathbf{u}_i$ denote, respectively, the update eigenvalue and eigenvector at the ith iteration. This algorithm can be viewed physically as a feedback loop, with its block diagram sketched in Figure 11.2.

For illustration, consider a Toeplitz matrix constructed from a vector $\mathbf{a} = [4, 2, 3, -1, 1]$. Namely, $\mathbf{A} = Toeplitz(\mathbf{a})$. It is easy to use the MATLAB function to find that its maximum eigenvalue equals 10.7278. Using the above iterative algorithm with initial vector $\mathbf{v} = [1, 1, 1, 1, 1]^T$, we find the results for the first three iterations as follows:

$$10.4000 \rightarrow 10.7274 \rightarrow 10.7278, \tag{11.1}$$

which converges to the true maximum value quickly in two steps.

It is clear that, in this algorithm, the problem of searching the maximum eigenvalue is formulated as a feedback loop to allow for information exchange between the eigenvalue and eigenvector estimates at each iteration. The reason is that both estimators carry useful information about the maximum eigenvalue since they are the two interconnected aspects of the same Toepltiz matrix. Such information exchange leads to a cubic convergence.

Let us return to communications. The above intuitive example provides three useful hints for the design of a coding/decoding scheme. First, for ease of iterative decoding, a composite code should encompass at least two components to enable information exchange between them. An extra advantage of such a composite structure, as will be shown subsequently, is that it possesses an error performance equivalent to a code of very long length. Second, an iteration mechanism must be established that is compatible with the essential features of the two constituent components. Appropriately constructing or selecting two subsystems or devices for iterative estimation is the idea central to the use of the turbo principle. Finally, in order for the turbo algorithm to converge, these two subsystems must be clearly separated, so that their parameters are independent.

An easy way to artificially create a two-device mechanism in communications is to use two simpler codes, in series or in parallel, to synthesize a code of much larger free distance. The resulting code is called a *concatenated code* or *turbo code*, the term coined by Berrou, whereas the simpler codes are called the *constituent codes*. The turbo decoding principle employs the method of log likelihood ratio (LLR) binary decisions to solve a discrete optimization problem. Suppose that a binary source vector $\mathbf{u}$ is applied to the first encoder $\mathbf{H}_1$ to produce a coded output vector $\mathbf{H}_1\mathbf{u}$, which is permuted by the interleaving matrix $\mathbf{\Pi}$ before entering the second encoder $\mathbf{H}_2$. If the composite coded output is converted to bipolar form for BPSK transmission over an AWGN channel, the received signal takes the form

$$\mathbf{y} = [2\underbrace{\mathbf{H}_2\mathbf{\Pi}(\mathbf{H}_1\mathbf{u})}_{\mathbf{x}} - 1] + \mathbf{n}, \tag{11.2}$$

where $\mathbf{n} \sim \mathcal{N}(0, N_0\mathbf{I}/2)$. Upon the receipt of the data vector $\mathbf{y}$, our task is to find the source vector $\mathbf{u}$ that maximizes the posterior probability:

$$\max_{\mathbf{u}} \Pr\{\mathbf{u}|\mathbf{y}\}, \tag{11.3}$$

$$\text{s.t.: binary } \mathbf{u}, \tag{11.4}$$

which falls into an intractable problem of integer programming. In the past, the optimization of this kind relied on time-consuming numerical search. With $\mathbf{x}$ as an intermediate variable, we can rewrite the above expression as

$$\max_{\mathbf{u}} \Pr\{\mathbf{x}|\mathbf{y}\} \Pr\{\mathbf{u}|\mathbf{x}\}, \tag{11.5}$$

which is still a complex discrete optimization problem. However, the cascade concatenation forms an effective structure to allow the two-step turbo principle to fit in. In the first step, the turbo principle uses binary decisions to solve discrete programming. To this end, letting u_i and x_i denote the ith entry of $\mathbf{u}$ and $\mathbf{x}$, respectively, we may isolate the problem as a bit-to-bit decision, with its likelihood ratio given by

$$\prod_i \frac{\Pr\{x_i = +1|\mathbf{y}\}}{\Pr\{x_i = -1|\mathbf{y}\}} \prod_k \frac{\Pr\{u_k = +1|\mathbf{x}\}}{\Pr\{u_k = -1|\mathbf{x}\}}, \tag{11.6}$$

which, in a log scale, takes the form

$$\underbrace{\sum_i \log \frac{\Pr\{\mathbf{y}|x_i = +1\} \Pr\{x_i = +1\}}{\Pr\{\mathbf{y}|x_i = -1\} \Pr\{x_i = -1\}}}_{\text{Exploit structure of } \mathbf{H}_2} + \underbrace{\sum_k \log \frac{\Pr\{\mathbf{x}|u_k = +1\} \Pr\{u_k = -1\}}{\Pr\{\mathbf{x}|u_k = -1\} \Pr\{u_k = -1\}}}_{\text{Exploit structure of } \mathbf{H}_1}. \tag{11.7}$$

Clearly, the decoding problem is more challenging than the Rayleigh quotient one, because of the discrete nature of vector $\mathbf{u}$ and the modulo-2 addition involved in the coding process. The basic strategy of the turbo principle is to convert a discrete estimation problem into binary decisions and seek its solution through feedback iteration; the details are addressed in the next three sections.

11.3 Turbo codes

Let us begin with the generation of turbo codes which, as previously mentioned, can be implemented by either serial or parallel concatenation. In the following sections, we choose parallel concatenation for illustration.

11.3.1 A turbo encoder

Figure 11.3b shows a typical turbo code [2, 3] generated by parallel concatenation of two RSC encoders. Its counterpart constructed from the two non-recursive convolutional (NRC) codes is also included for comparison. By parallel concatenation we mean that the two encoders are driven by the same set of inputs. The turbo encoder in Figure 11.3 is comprised of two binary rate-$\frac{1}{2}$ convolutional encoders connected in parallel and separated by a random interleaver, followed by an optional puncturing device. The role of puncturing is to periodically delete selected bits by following a certain pattern to reduce the coding overhead and increase the code rate. Without puncturing, the code rate would be $1/3$.

11.3.2 RSC versus NRC

Before proceeding, it is helpful to examine the relationship between RSC codes and NRC codes. Refer to Figure 11.3a. The polynomial pair for generating a rate-$\frac{1}{2}$ convolutional code traditionally takes the form of $g_n(D) = [g_1(D), g_2(D)]$, where the subscript n signifies "nonrecursive." When driven by an input sequence

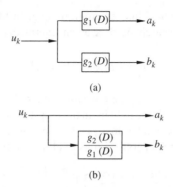

Figure 11.3   NRC encoder (a) and its RSC equivalent (b)

$\{u_k\}$, the outputs from the two generator polynomials are given, respectively, by

$$a_k = g_1(D)u_k,$$

$$b_k = g_2(D)u_k,$$ (11.8)

which are multiplexed to produce the coded output sequence $\{a_k, b_k\}$. This traditional convolutional encoder has an RSC equivalent. Rewriting the first equation as $u_k = g_1^{-1}(D)a_k$ and inserting it into the second, we obtain $b_k = [g_2(D)/g_1(D)]a_k$, which, when interlaced with a_k, leads to an equivalent recursive systematic coded sequence

$$\left\{a_k, \frac{g_2(D)}{g_1(D)}a_k\right\}.$$ (11.9)

The equivalent recursive code generator is thus equal to $g_r(D) = [1, g_2(D)/g_1(D)]$, and its block diagram is shown in Figure 11.3b. From (11.9), we have the following property:

Property 11.3.1   *The RSC encoder with input $g_1(D)u(D)$ generates the same output as the non-recursive encoder with input $u(D)$.*

Since the state of NRC codes is directly determined by the input data, this property enables the assertion that the state of the corresponding recursive code is determined by the input $u(D)g_1(D)$. A further assertion is given below.

Property 11.3.2   *The weight distribution of the output codewords is identical for the RSC and non-systematic convolutional (NSC) codes, and the two codes have the same free distance.*

✍ **Example 11.1** _____

For illustration, let us compare the transitional diagrams of the NRC and RSC encoders $[1 + D + D^2, 1 + D^2]$ and $[1, [(1 + D^2)/(1 + D + D^2)]]$.

The two codes have the same transitional and trellis diagrams, as shown in Figure 11.4. But they have different mapping between the input and output sequences, as exemplified below.

$$\left.\begin{matrix} 100(N) \\ 111(R) \end{matrix}\right\} \to 111011(wt = 5)$$

$$\left.\begin{matrix} 1100(N) \\ 1001(R) \end{matrix}\right\} \to 11010111(wt = 6)$$

$$\left.\begin{matrix} 10100(N) \\ 11011(R) \end{matrix}\right\} \to 1110001011(wt = 6)$$

$$\left.\begin{matrix} 11100(N) \\ 10101(R) \end{matrix}\right\} \to 1101100111(wt = 7)$$

$$\left.\begin{matrix} 101100(N) \\ 110001(R) \end{matrix}\right\} \to 111000010111(wt = 7)$$

$$\left.\begin{matrix} 1010101(N) \\ 1101011(R) \end{matrix}\right\} \to 11100010001011(wt = 7)$$

$$\left.\begin{matrix} 1010100(N) \\ 1101011(R) \end{matrix}\right\} \to 11100010001011(wt = 7)$$

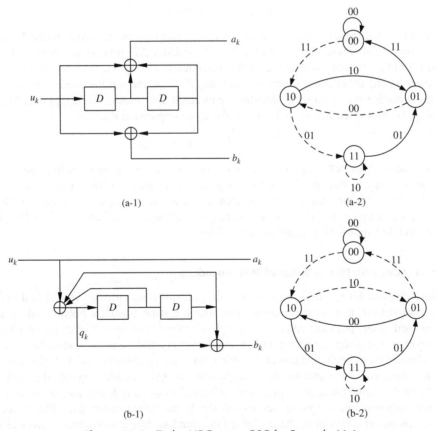

(a-1)　　　　　　　　(a-2)

(b-1)　　　　　　　　(b-2)

Figure 11.4　*Turbo NRC versus RSC for Example 11.1*

Property 11.3.3 *Regardless of their states being defined differently, the recursive and non-recursive coding systems have the same state transitional diagram and branch outputs, albeit driven by different input bits.*

To prevent degeneration, the feedback polynomial of a turbo code must be of full degree, that is, the highest order state must be included in the feedback loop so that the two transitions entering any state are driven by different bits.

There are strong reasons to choose RSC encoders for turbo codes. For each nonzero bit in its input sequence, an RSC encoder produces an infinite parity weight output and never returns to the zero state. The feedback operation has a profound impact on the code behavior. RSC encoders behave like an infinite impulse filter, with a very long equivalent memory. Referring to Figure 11.3b, we can write the output of the lower encoder as

$$b_k = \frac{g_2(D)}{g_1(D)} u_k,$$

which is expressible as a difference equation

$$g_1(D)b_k = g_2(D)u_k. \tag{11.10}$$

As example, the output of the lower encoder of Figure 11.4 is given by

$$b_k + b_{k-1} + b_{k-2} = u_k + u_{k-2}.$$

We recognize that it takes the same form as an autoregressive moving-average (ARMA) model of order $(2, 2)$ in time-series analysis, abbreviated as ARMA(2,2) . The ARMA(2,2) process is driven by a binary white sequence $\{u_k\}$. Unlike a normal ARMA process, the difference in coding is that all the additions are modulo-2 operations, hence having the effect of magnitude clipping. As such, the resulting code sequence behaves like a random process and is never divergent, while having its algebraic structures hidden in the phase in the form of zero crossing. When written in a recursive form, the above expression becomes

$$b_k = b_{k-1} + b_{k+2} + u_k + u_{k-2},$$

which is a sequence with a infinitely long memory. The encoder transforms an impulse-like input such as $\{u_k\} = 1000\cdots$ into an output of infinitely long weight. However, if the input sequence is divisible by $g_1(D) = 1 + D + D^2$, the output coded sequence has a finite weight. Such input patterns include $0 \ldots 010010 \ldots 00$ and $0 \ldots 01110 \ldots 0$, whose polynomials are $1 + D^3$ and $1 + D + D^2$. However, the events of finite weight happen with negligible probability.

11.3.3   Turbo codes with two constituent RSC encoders

A general block diagram for synthesizing a turbo code with two RSC encoders is sketched in Figure 11.5. Indeed, as compared to its non-recursive counterpart with the same generator polynomials, each recursive encoder, when used alone, does not change the weight distribution of its output codewords and minimum weights. However, its use alongside a long random interleaver in a turbo encoder can make the occurrence of low-weight codewords with negligible probability. First, for a non-recursive encoder, the lowest weight codeword is usually produced by the unit-weight input pattern. An RSC encoder converts the same unit-weight input sequence into an infinite-weight output, easily solving the difficulty facing the non-recursive encoders. Second, as previously explained, input sequences divisible by the denominator of the RSC encoder can generate finite-weight outputs. However, the use of a long, random interleaver almost surely eliminates the likelihood of the appearance of the same input pattern at the second encoder input. As such, it is nearly impossible for the output sequences at the two encoders to have low weights simultaneously.

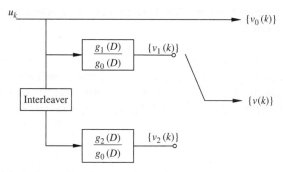

Figure 11.5   *General turbo encoder with two RSC constituent encoders*

For illustration, suppose the generator in Figure 11.5 is given by

$$g(D) = \left[1 \quad \frac{1 + D + D^3}{1 + D^2 + D^3} \quad \frac{1 + D + D^2 + D^3}{1 + D^2 + D^3}\right],$$

where the two RSC encoders are separated with an interleaver, defined by matrix

$$\Pi = \begin{pmatrix} 0 & 0 & 1 & 0 & 0 & 0 & 0 & 0 \\ 0 & 0 & 0 & 0 & 0 & 0 & 0 & 1 \\ 0 & 0 & 0 & 0 & 1 & 0 & 0 & 0 \\ 1 & 0 & 0 & 0 & 0 & 0 & 0 & 0 \\ 0 & 0 & 0 & 0 & 0 & 1 & 0 & 0 \\ 0 & 0 & 0 & 1 & 0 & 0 & 0 & 0 \\ 0 & 1 & 0 & 0 & 0 & 0 & 0 & 0 \\ 0 & 0 & 0 & 0 & 0 & 0 & 1 & 0 \end{pmatrix}.$$

If the encoder is initialized with zeros, determine the coded output in response to the input sequence $\mathbf{u} = [10110011]$ with the oldest bit on the left.

The process for the solution is as follows: The parity sequence $\mathbf{v}_1$ from the first encoder can be calculated by

$$v_1(k) = \frac{1 + D + D^3}{1 + D^2 + D^3} u_k,$$

or recursively by

$$v_1(k) = v_1(k-2) + v_1(k-3) + u_k + u_{k-1} + u_{k-3},$$

which, with input $\mathbf{u}$, produces $\mathbf{v}_1 = [11010010\cdots]$. Likewise, the input $\mathbf{u}$, after interleaving by Π, produces

$$\mathbf{u}\Pi^T = [11010101],$$

which drives the second encoder to generate the second parity vector yielding $\mathbf{v}_2 = [10101110\cdots]$. Interlace $\mathbf{u}$, $\mathbf{v}_1$, and $\mathbf{v}_2$ to obtain the rate-$\frac{1}{3}$ coded sequence:

$$\mathbf{v} = [110\ 010\ 101\ 110\ 001\ 001\ 111\ 100\ \cdots].$$

The rate-$\frac{1}{3}$ code sequence, taking the form $\{v_0(k), v_1(k), v_2(k), \cdots\}$, can be reduced to a rate-$\frac{1}{2}$ code sequence by alternatively puncturing the parity bits $v_1(k)$ and $v_2(k)$. The result is $\{v_0(k), v_1(k), v_0(k+1), v_2(k+1), \cdots\}$.

In the above illustration, we used two very short RSC encoders to synthesize a turbo code. Usually, longer RSC encoders are used in practice.

✍ Example 11.2 _____

A typical turbo code is formed by parallel concatenation of two constituent $(37, 21)$ RSC encoders that are separated by an interleaver of $65,536$ bits. Since $37 = 10011$ and $21 = 01011$, the two generators are $g_1(D) = 1 + D^4 + D^5$ and $g_2(D) = D + D^4 + D^5$. The resulting code rate is $\frac{1}{2}$, achieving the BER of 10^{-5} at $E_b/N_0 = 0.7$ dB.

○

〰〰✍

11.4   BCJR algorithm

Maximum *a posteriori* (MAP) decoding of turbo codes relies on a recursive algorithm that is capable of exploiting the inherent trellis structures of convolutional codes. Such an algorithm was derived by Bahl, Cocke, Jelinek, and Raviv, and thus has the name BCJR algorithm [8]; it is described in this section.

Serially or parallelly concatenated codes generated by a two-device mechanism can well fit to the framework of turbo decoding. Consider, again, the turbo encoder sketched in Figure 11.5, and denote the function of the interleaver that separates the two RSC encoders by matrix $\mathbf{\Pi}$. Driven by input sequence $\mathbf{u} = \{u_k\}$, the turbo encoder produces the output

$$\mathbf{v} = \{\cdots, u_k, \underbrace{[g_1(D)/g_0(D)]u_k}_{x_k}, \underbrace{[g_2(D)/g_0(D)]u'_k}_{x'_k}, \cdots\}, \tag{11.11}$$

where $\{u'_k\}$ results from $\mathbf{u} = \{u_k\}$ after permutation by $\mathbf{\Pi}$. The coded sequence, after shaping and modulation, is assumed to transmit over an AWGN channel of variance σ_n^2. Consistent with the structure of the turbo encoder, the receiver correspondingly is comprised of two decoders. In baseband form, the received signals at the two decoders at time instant k can be written, respectively, as

$$y_k = x_k(\mathbf{u}) + n_k,$$
$$y'_k = x'_k(\mathbf{u}) + n'_k, \tag{11.12}$$

where $k = 0, 1, \cdots$ and n_k and n'_k are independent zero-mean AWGN components. The two decoders iteratively process $\{y_k\}$ and $\{y'_k\}$, respectively, to estimate message bits and exchange their information to accelerate the convergence. Owing to the RSC code structure, the receiver also receives samples solely containing u_k plus noise. This feature greatly facilitates the implementation of the BCJR algorithm.

The derivation of the BCJR algorithm is based on a single RSC code. It suffices to focus on the first decoder since the operation of the second decoder is similar. In what follows, the first decoder is simply termed the *decoder* for brevity. To proceed, denote the received data $\mathbf{y} = [y_1, y_2, \cdots, y_N]$, and denote its subvector containing entries y_i to y_m as $\mathbf{y}_i^m = [y_i, y_{i+1}, \cdots, y_m]^T$. With this notation, $\mathbf{y} = \mathbf{y}_1^N$. Given the received data $\mathbf{y}$, the decoder needs to estimate message bit u_k at time k, which can be done in the MAP framework for a binary decision, based on the ratio of posterior probabilities:

$$L(u_k) = \log\left[\frac{f(u_k = +1|\mathbf{y})}{f(u_k = -1|\mathbf{y})}\right], \tag{11.13}$$

with $f(\cdot)$ denoting the PDF of u_k. Decoding is performed by confining to feasible paths in the trellis diagram. The above expression is too general, and we need to make it more manageable by exploiting the trellis structures of RSC codes. Note that all the information about u_k is contained in the states of the decoder at time instant k, which are termed hereafter the *current states* for brevity. Each current state is evolved from the previous ones along relevant feasible paths in the trellis diagram of the RSC encoder. The entering transitions to the current states can be driven either by $u_k = 1$ or by $u_k = -1$. We may therefore partition these transitions into two subsets. Suppose that the current information bit u_k causes the state transition from $s_{k-1} = s'$ to $s_k = s$. Let S^+ denote the subset of transitions between $t = (k-1)T_b$ and $t = kT_b$ with input bit 1; that is

$$S^+ = \{(s', s) : s_{k-1} \xrightarrow{u_k = 1} s_k\}. \tag{11.14}$$

Likewise, we define

$$S^- = \{(s', s) : s_{k-1} \xrightarrow{u_k = -1} s_k\}. \tag{11.15}$$

Note that the decoding process is based the *entire* trellis structure of the encoder over the time span defined by $\mathbf{y}$, but not just confined to a single time segment from time $k-1$ to time k. The problem of estimating u_k, in practice, is to test a binary hypothesis. Namely, given $\mathbf{y}$, which subset—the subset of all feasible paths that go through S^+ or the subset of all feasible paths that go through S^-—is more likely to have generated the current bit u_k?

The decision can be made in a framework based on the LLR of the *a posteriori* probabilities, as defined by

$$L(u_k) = \log\left[\frac{\sum_{S^+} f(s_{k-1} = s'; s_k = s; \mathbf{y})/f(\mathbf{y})}{\sum_{S^-} f(s_{k-1} = s'; s_k = s; \mathbf{y})/f(\mathbf{y})}\right]$$

$$= \log\left[\frac{\sum_{S^+} f(s_{k-1} = s'; s_k = s; \mathbf{y})}{\sum_{S^-} f(s_{k-1} = s'; s_k = s; \mathbf{y})}\right]. \tag{11.16}$$

We need to link the current state transition to its evolutional history and to its future trajectory, so as to fully exploit all the useful information contained in the trellis structure. By simple probabilistic operations, it yields

$$f(s', s, \mathbf{y}) = f(s', s, \mathbf{y}_1^{k-1}, y_k, \mathbf{y}_k^N)$$

$$= f(s', \mathbf{y}_1^{k-1}) f(s, y_k, \mathbf{y}_{k+1}^N | s', \mathbf{y}_1^{k-1})$$

$$= f(s', \mathbf{y}_1^{k-1}) f(s, y_k | s', \mathbf{y}_1^{k-1}) f(\mathbf{y}_{k+1}^N | s', s, y_k, \mathbf{y}_1^{k-1}). \tag{11.17}$$

Denote parameters

$$\alpha_k(s) = f(s_k = s, \mathbf{y}_1^k) : \quad \text{past to current,}$$

$$\beta_k(s) = f(\mathbf{y}_{k+1}^N | s_k = s) : \quad \text{current to future,}$$

$$\gamma_k(s', s) = f(s_k = s, y_k | s_{k-1} = s') : \quad \text{transition,} \tag{11.18}$$

where $\gamma_k(s', s)$ can be considered as a transition probability. Thus, the above three parameters represent the forward, backward, and branch transition probabilities of the MAP decoder, respectively. With the above notations, we can rewrite $L(u_k)$ as

$$L(u_k) = \frac{\sum_{S^+} \alpha_{k-1}(s') \gamma_k(s', s) \beta_k(s)}{\sum_{S^-} \alpha_{k-1}(s') \gamma_k(s', s) \beta_k(s)}. \tag{11.19}$$

The parameter $\alpha_k(s)$ represents the total probability arriving at $s_k = s$ from all possible previous states $s' \in S$. Such a path evolution enables us to write it in a recursive form, yielding

$$\alpha_k(s) = \sum_{s' \in S} \alpha_{k-1}(s')\gamma_k(s', s). \tag{11.20}$$

Similarly, $\beta_{k-1}(s') = f(\mathbf{y}_k^N | s_{k-1} = s')$ represents the probability of future data $\mathbf{y}_k^N$ stemming from the current state $s_{k-1} = s'$, and can be recursively expressed as

$$\beta_{k-1}(s') = \sum_{s \in S} f(\mathbf{y}_{k+1}^N | s_k = s) f(s_k = s, y_k | s_{k-1} = s')$$

$$= \sum_{s \in S} \beta_k(s)\gamma_k(s', s). \tag{11.21}$$

Owing to the discrete property of feasible paths, all information in $\mathbf{y}_1^{k-1}$ about s' is condensed in $s_{k-1} = s'$, and thus we can write

$$f(s, y_k | s', \mathbf{y}_1^{k-1}) = f(s, y_k | s') = \gamma(s', s). \tag{11.22}$$

Using the chain rule to represent $f(s, y_k | s')$ and noting that the transition from the status s' to s is caused by input u_k, we obtain

$$\gamma_k(s', s) = f(s|s')f(y_k|s', s) = f(u_k)f(y_k|u_k). \tag{11.23}$$

We further need to exploit the RSC structure of the turbo code by incorporating the systematic sample, which consists only of u_k plus noise. To this end, we slightly modify the received model in vector form to include the systematic sample, such that

$$\mathbf{y} = \mathbf{x} + \mathbf{n}, \tag{11.24}$$

where $\mathbf{n} \sim \mathcal{N}(\mathbf{0}, \sigma_n^2 \mathbf{I})$, and $\mathbf{x} = [\mathbf{x}_1, \cdots, \mathbf{x}_N]$ is the coded binary sequence of ± 1 entries, for which $\mathbf{x}_k = [u_k, x_k]$. Here, the signal has been normalized to have unit energy, so that the noise variance is equal to the reciprocal of the SNR ρ; that is, $\rho^{-1} = \sigma_n^2$. The noisy versions of the source bit u_k and the parity bit x_k are denoted by y_k^s and y_k^p, respectively. With these symbols, we have

$$f(y_k|u_k) = \frac{1}{\sqrt{2\pi\sigma_n^2}} \exp\left[-\frac{(y_k^s - u_k)^2 + (y_k^p - x_k^p)^2}{2\sigma_n^2}\right]. \tag{11.25}$$

For calculation of the current bit u_k, the transition $\gamma_k(s', s)$ is a constant for all the possible path segments under S^+ or S^-, as shown by

$$\gamma_k(s', s|S^\pm) = f(u_k = \pm 1) \times \frac{1}{\sqrt{2\pi\sigma_n^2}} \exp\left[-\frac{(y_k^s \mp 1)^2 + (y_k^p - x_k^p)^2}{2\sigma_n^2}\right]. \tag{11.26}$$

As such, we can take it out of the summation sign in the numerator and the denominator of $L(u_k)$ to form a ratio

$$\frac{\gamma_k(s', s|S^+)}{\gamma_k(s', s|S^-)} = \frac{f(u_k = +1)}{f(u_k = -1)} e^{\rho(2y_k^s + 2y_k^p x_k^p)}, \tag{11.27}$$

which, after taking logs, leads to

$$\ln \frac{\gamma_k(s', s|S^+)}{\gamma_k(s', s|S^-)} = L_a(u_k) + 2\rho \, y_k^s + 2\rho \, y_k^p x_k^p. \tag{11.28}$$

Unlike $\gamma_k(s', s)$ for the current bit decision, which assumes only those branches driven by $u_k = 1$ or driven by $u = -1$ and is thus relatively simple to determine, the transition probability $\gamma_i(s', s)$ for $i \neq k$ implicitly involved in $\alpha's$ and $\beta's$ includes both branches driven by $u_i = \pm 1$. Such $\gamma_i(s', s)$ appears in both the numerator and the denominator of $L(u_k)$; the only difference is that one leads to the current branch driven by $u_k = 1$, whereas the other leads to that driven by $u_k = -1$. Thus, it is more convenient to consider the ratio

$$\frac{\gamma_i(s', s | S, u_k = +1)}{\gamma_i(s', s | S, u_k = -1)} = \frac{[f(u_i = 1)f(y_i|u_i = 1) + f(u_i = -1)f(y_i|u_i = -1)]_{u_k=+1}}{[f(u_i = 1)f(y_i|u_i = 1) + f(u_i = -1)f(y_i|u_i = -1)]_{u_k=-1}}$$

$$= \frac{[\Lambda(u_i)\frac{f(y_i|u_i=+1)}{f(y_i|u_i=-1)} + 1]_{u_k=+1}}{[\Lambda(u_i)\frac{f(y_i|u_i=+1)}{f(y_i|u_i=-1)} + 1]_{u_k=-1}}$$

$$= \frac{[\Lambda(u_i)e^{\rho(2y_k^s+2y_k^P x_k^P)} + 1]_{u_k=+1}}{[\Lambda(u_i)e^{\rho(2y_k^s+2y_k^P x_k^P)} + 1]_{u_k=-1}}, \tag{11.29}$$

where $\Lambda(u_k) = f(u_k = 1)/f(u_k = -1)$ is the likelihood ratio that is related to the *a priori* LLR of u_k, that is, $L_a(u_k)$, by $\Lambda(u_k) = e^{L_a(u_k)}$. Dividing both the numerator and denominator of (11.29) by $\sqrt{\Lambda(u_i)}\ e^{\rho(y_k^s+y_k^P x_k^P)}$ leads to

$$\frac{\gamma_i(s', s | S, u_k = +1)}{\gamma_i(s', s | S, u_k = -1)} = \frac{\left[\sum_{u_i=\pm 1} e^{\frac{1}{2}u_i L_a(u_i)}e^{u_k(y_k^s+y_k^P x_k^P)\rho}\right]_{u_k=+1}}{\left[\sum_{u_i=\pm 1} e^{\frac{1}{2}u_i L_a(u_i)}e^{u_i(y_i^s+y_i^P x_i^P)\rho}\right]_{u_k=-1}}, \tag{11.30}$$

where, for a unified general expression, u_i is inserted, wherever necessary, to indicate the signs of some exponents and LLRs. Based on this argument, we may redefine

$$\gamma_i(s', s) = e^{u_i L_a(u_i)/2}e^{u_i(y_i^s+y_i^P x_i^P)\rho}, \tag{11.31}$$

which is applied to both the cases of $u_i = 1$ and $u_i = -1$. Inserting (11.31) into (11.19) and taking logarithm, we get

$$L(u_k) = \ln \frac{\sum_{S^+}\alpha_{k-1}(s')\gamma_k(s', s)\beta_k(s)}{\sum_{S^-}\alpha_{k-1}(s')\gamma_k(s', s)\beta_k(s)}. \tag{11.32}$$

Using (11.28) and denoting

$$\gamma_k(s', s) = e^{u_k\rho\ y_k^P x_k^P}, \tag{11.33}$$

we can further separate $L(u_k)$ into three terms, yielding

$$L(u_k) = L_a(u_k) + \underbrace{2\rho\ y_k^s}_{L_c(u_k)} + \underbrace{\ln \frac{\sum_{S^+}\alpha_{k-1}(s')\tilde{\gamma}_k(s', s)\beta_k(s)}{\sum_{S^-}\alpha_{k-1}(s')\tilde{\gamma}_k(s', s)\beta_k(s)}}_{L_e(u_k)}. \tag{11.34}$$

The first term $L_a(u_k)$ is the prior information about u_k usually supplied by the other device, the second term represents the channel value, and the third term $L_e(u_k)$ is the extrinsic information for passing to the other device. The other device can be the second decoder or another subsystem such as a channel or multiuser detector, depending upon the application. From the above derivations, it is clear that the *a posteriori* likelihood ratio $L(u_k)$ exploits information from both past and future received samples through forward and backward iterations.

11.5 Turbo decoding

To decode a turbo code formed by parallel concatenation of two RSC codes, two constituent decoders perform the BCJR algorithm individually, but exchanging their estimation information to expedite the iteration convergence. The first decoder processes the input data $\mathbf{y}$ to produce an extrinsic output $\{L_e^{(1)}(u_i)\}$, which, after interleaving, is fed forward to the second decoder to assist its decoding of u_k. Collectedly denote the extrinsic LLRs, $\{L_e^{(i)}(u_\ell), \forall\, \ell\}$, as a column vector $\{L_e^{(i)}(\mathbf{u})\}$ for $i = 1, 2$, where the superscripts (i) signifies the ith decoder. Define

$$\mathbf{f}_1 = \mathbf{\Pi}\, L_e^{(1)}(\mathbf{u}),$$

$$\mathbf{f}_2 = \mathbf{\Pi}^{-1}\, L_e^{(2)}(\mathbf{u}). \tag{11.35}$$

Decoder 1 feeds its permuted extrinsic LLRs $\mathbf{f}_1$ to Decoder 2 as the prior information $L_a^{(2)}(\mathbf{u})$. Likewise, Decoder 2 feeds its deinterleaved extrinsic LLRs to Decoder 1 as prior information $L_a^{(1)}(\mathbf{u})$. Symbolically, we can write

$$\mathbf{f}_1 \rightarrow L_a^{(2)}(\mathbf{u}),$$

$$\mathbf{f}_2 \rightarrow L_a^{(1)}(\mathbf{u}). \tag{11.36}$$

Note that Decoder 1 must exclude the kth entry of $\mathbf{f}_1$ from updating its LLR, $L_e^{(1)}(u_k)$, of u_k since, by definition, u_k involved therein must be treated as a deterministic variable, as detailed in (11.31). A similar treatment happens when Decoder 1 updates its $L_e^{(2)}(u_k)$. More intuitively, we have

$$\xrightarrow{\ \mathbf{f}_2\backslash k\ } \boxed{\text{Decoder 1}} \xrightarrow{\ \{\gamma_i^{(1)}(s',s),\, i\neq k\}\ } L_e^{(1)}(u_k)$$

$$\xrightarrow{\ \mathbf{f}_1\backslash k\ } \boxed{\text{Decoder 2}} \xrightarrow{\ \{\gamma_i^{(2)}(s',s),\, i\neq k\}\ } L_e^{(2)}(u_k) \tag{11.37}$$

where the set minus sign $\backslash k$ indicates the exclusion of the kth entry from a vector. Excluding $L_a^{(1)}(u_k)$ from updating the current extrinsic information $L_e^{(1)}(u_k)$ is to avoid repetitive use of the same information. Otherwise, it can cause the failure of the iteration to converge. A similar treatment is done in Decoder 2.

The turbo receiver is essentially a feedback loop formed by two decoders, with information exchange between them through an interleaver or a deinterleaver. The soft output from one decoder, usually in the form of LLR and called *extrinsic* information, is fed into the input of another decoder.

11.6 Illustration of turbo-code performance

The superior error performance of turbo codes is exemplified in Figure 11.6 through two rate-$\frac{1}{3}$ turbo codes. The two turbo codes are generated by using two identical RSC encoders given by $(1 + D^2)/(1 + D + D^2)$ and $(1 + D^4)/(1 + D + D + D^2 + D^3)$, respectively. It is observed that both codes substantially outperform their uncoded counterparts. We also observe that the code built on $(7, 5)$ is superior to the one with $(15, 17)$. The inferior bit error performance of the latter code is attributed to the fact that it excludes the connection to the highest degree from its feedback loop. In turbo design, the component codes are chosen to maximize the free distance of the code. This is equivalent to maximizing the minimum code weight, which usually guarantees good performance in the region of large SNRs. On the other hand, optimizing the weight distribution of codewords can improve the error performance of the code at lower SNRs.

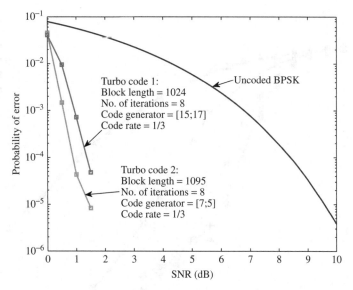

Figure 11.6 *Error performance of two rate-$\frac{1}{3}$ turbo codes*

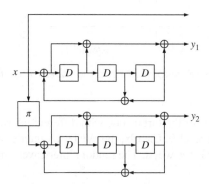

Figure 11.7 *Encoder structure of a turbo code with two constituents*

To further demonstrate the error performance of turbo codes, we consider a turbo code generated by a turbo encoder shown in Figure 11.7. Both constituent encoders have the same recursive generator

$$g_i(D) = \frac{1 + D + D^3}{1 + D^2 + D^3}, i = 1, 2$$

and are separated by a simple 3×3 interleaver. Three operational modes of the interleaver are considered; they are (1) randomized interleaving, (2) row read-in and column readout, and (3) optimized interleaver used in WCDMA. Log-MAP decoders are assumed to be used alongside the turbo code.

A practical concern is the speed at which a turbo iteration converges. The block-error rate (BLER) performance as a function of SNR is plotted in Figure 11.8, with parameter I indicating that the performance curve is obtained at the end of the Ith iteration. It is observed that the BLER falls off very fast, reaching the level of 10^{-7} just in a few steps. We next investigate the influence of the interleaver on the BLER performance of the

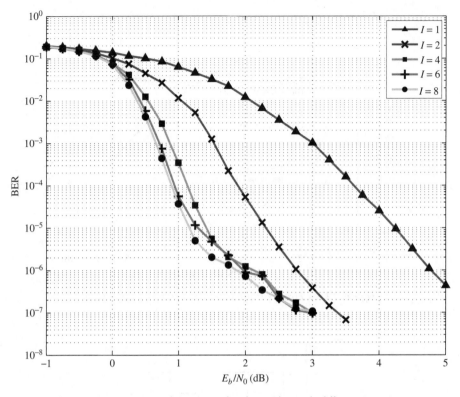

Figure 11.8   *BLER performance of turbo codes with different iterations*

turbo code. The BLER performance of the turbo code with different interleavers achievable at the end of the eighth iteration is depicted in Figure 11.9. Clearly, the WCDMA interleaver performs the best, whereas the row-in and column-out interleaver is the worst. The random interleaver runs between the two.

11.7   Extrinsic information transfer (EXIT) charts

We can imagine that it is the behavior of the information exchange between the two decoders that determines the convergence of the turbo decoding algorithm. We, therefore, need a metric to quantify the information exchange involved therein. An effective metric to this end is the mutual information between the information bit u_k and its soft estimate $L^{(i)}(u_k)$ from the ith decoder, as advocated by ten Brink [9]. Since the treatment for decoders 1 and 2 are the same, we drop the superscript of the decoder and the iteration step k in the subsequent analysis for simplicity. Then, by definition, the mutual information between u_k and $L_e(u_k)$ is given by

$$I(u, L_e) = \frac{1}{2} \sum_{u=\pm 1} \int_{-\infty}^{\infty} f(L_e|u) \log_2 \frac{f(L_e|u)}{\frac{1}{2}[f(L_e|u = +1) + f(L_e|u = -1)]} \, dL_e, \qquad (11.38)$$

where the denominator inside the logarithm represents the PDF of u_k. The argument of the logarithm can be simplified by dividing both the numerator and denominator by $f(L_e|u)$, using the symmetry property of the

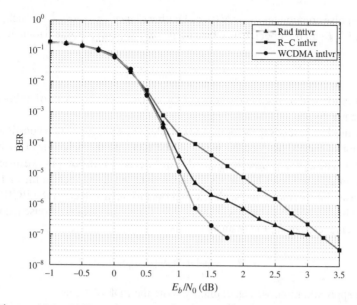

Figure 11.9 *BLER performance of turbo codes with different interleavers*

$L-$value distribution, namely $f(L_e|u = -1) = f(-L_e|u = 1)$, and noting that $f(L_e|u)/f(-L_e|u) = e^{uL_e}$. The result turns out to be

$$I(u, L_e) = 1 - \frac{1}{2}\sum_{u=\pm 1}\int_{-\infty}^{\infty} f(L_e|u)\log_2(1 + e^{-uL_e})\, dL_e$$

$$= 1 - \int_{-\infty}^{\infty} f(L_e|u = 1)\log_2(1 + e^{-L_e})\, dL_e$$

$$= 1 - \mathbb{E}_{L_e}[\log_2(1 + e^{-L_e})]. \tag{11.39}$$

To work out the expectation, we need the PDF of the L-values, which, however, is usually unknown and non-Gaussian due to the high nonlinearity involved in the BCJR algorithm.

The PDF $f(L_e|u)$ can be determined by virtue of a semianalytic and semi-simulation method. The output extrinsic LLR L_e is a very complicated function of the input coded bits, so we need to establish a simple statistical model for it so that simulation can be conducted. To get the inspiration, we start with the LLR of binary transmission over an AWGN channel, for which the received signal is given by

$$z = x + n, \tag{11.40}$$

where the symbol $x = \pm 1$ is transmitted with unit energy and corrupted by noise $n \sim \mathcal{N}(0, \sigma_n^2)$. Denoting the conditional PDF of z by $p(z|x)$ and denoting $\mu_\ell = 2/\sigma_n^2$, the LLR of z is expressible as

$$\ell = \ln\frac{p(z|x = +1)}{p(z|x = -1)} = \frac{2z}{\sigma_n^2}$$

$$= \mu_\ell(x + n)$$

$$= \mu_\ell\, x + \underbrace{\mu_\ell\, n}_{n_\ell}. \tag{11.41}$$

It is easy to assert the noise component $n_\ell \sim \mathcal{N}(0, \sigma_\ell^2)$ with variance $\sigma_\ell^2 = 4\sigma_n^{-2}$. As such, we have $\mu_\ell = \frac{1}{2}\sigma_\ell^2$. In other words, the LLR ℓ can be modeled as a Gaussian variable with a single parameter, as shown by

$$\ell = \frac{\sigma_\ell^2}{2}x + \mathcal{N}(0, \sigma_\ell^2). \tag{11.42}$$

This important observation will be used in the study of the convergence behavior of turbo decoding. The SNR associated with ℓ is equal to $\rho = \sigma_\ell^2/2$.

The use of a large interleaver tends to decouple the correlation between the two constituent decoders. Thus, to predict the behavior of the BCJR decoder, we may first focus on the behavior of each single constituent decoder, say decoder 1. The *a priori* value L_a applied to this decoder is the output extrinsic LLR $L_e^{(2)}$ from the second decoder. The large interleaver also tends to decorrelate the L_a sequence, thus allowing us to handle L_a on a bit-by-bit basis. Furthermore, the sophisticated operations involved in the BCJR algorithm justifies the use of the central limit theorem to approximate L_a as a Gaussian variable. Consequently, we may invoke the model (11.42) to characterize L_a to obtain

$$L_a = \frac{\sigma_a^2}{2}u + \mathcal{N}(0, \sigma_a^2). \tag{11.43}$$

Using this Gaussian approximation, we can explicitly write the PDF of L_a as

$$f_{L_a}(z|u) = \frac{1}{\sqrt{2\pi\sigma_a^2}} \exp\left(-\frac{(z - x\sigma_a^2/2)^2}{2\sigma_a^2}\right), \tag{11.44}$$

which, when used to replace $f(L_e|u)$ on the second line of (11.39), enables us to determine the mutual information I_a associated with the *a priori* LLR L_a, though not in a closed form. The resulting I_a is a statistical average, being a function of the only parameter σ_a^2. For notational convenience, we write $I_a = \Psi(\sigma_a^2)$. It can be shown that I_a is a monotonically increasing function of σ_a^2 ranging between

$$\lim_{\sigma_a^2 \to 0} \Psi(\sigma_a^2) = 0 \ \ and \ \ \lim_{\sigma_a^2 \to \infty} \Psi(\sigma_a^2) = 1. \tag{11.45}$$

The function $\Psi(\cdot)$ is reversible, allowing us to obtain σ_a^2 for any given $I_a \in (0, 1)$, symbolically expressible as

$$\sigma_a^2 = \Psi^{-1}(I_a). \tag{11.46}$$

But we need to have a more convenient way for calculation; the inverse operator of Ψ can be approximated, leading to [10]

$$\sigma_a^2 = \begin{cases} 1.09542I_a^2 + 0.214217I_a + 2.33727\sqrt{I_a}, & 0 \le I_a \le 0.3646 \\ -0.706692 \log_2[0.386013(1 - I_a)] + 1.75017I_a, & 0.3646 < I_a \le 1 \end{cases}. \tag{11.47}$$

We next turn to the calculation of the mutual information I_e associated with the output extrinsic LLR of the constituent decoder using, again, the formula in (11.39). But this time we need the PDF of L_e to fully characterize the statistical behavior defined by the BCJR decoder. To this end, we employ a semi-simulation approach by tracing the following steps:

(a) For a given value of I_a, determine σ_a^2 using (11.47).
(b) Generate a corresponding *a priori* LLR sequence $\{L_a\}$ using the model (11.43), typically with a length of 10^4 systematic bits, which can be set to an all-1 sequence.
(c) Run the BCJR algorithm to produce the output extrinsic sequence whereby to obtain a histogram as an approximate PDF of $f(L_e|u)$ for the calculation of I_e in (11.39).

The I_e so obtained is a function of $\sigma_a^2 = \Psi^{-1}(I_a)$; hence, we may formally write $I_e = T(I_a)$, with $T(\cdot)$ representing the information transfer between I_a and I_e.

In practice, one can use an alternative simulation method, other than the histogram based one, for calculating I_e. From (11.39), we need to take the expectation over L_e. Note that L_e includes the sign information $b = \operatorname{sgn}(L_e)$ and the magnitude information $\varphi = |L_e|$, such that $L_e = \operatorname{sgn}(L_e)|L_e| = b\varphi$, with b and φ representing the hard binary decision based on L_e and the corresponding reliability, respectively. With these results, it follows that

$$
\begin{aligned}
I(u, L_e) &= 1 - \mathbb{E}_{L_e}[\log_2(1 + e^{-L_e})] \\
&= 1 - \mathbb{E}_{\varphi}[\mathbb{E}_b[\log_2(1 + e^{-b\varphi})]] \\
&= \int_{-\infty}^{\infty} \sum_{b=\pm 1} P(b|u)\log_2(1 + e^{-b\varphi})f(\varphi)d\varphi.
\end{aligned}
\tag{11.48}
$$

Similar to the previous arguments, we can argue that $P(b|u = 1) = (1 + e^{-b\varphi})^{-1}$, which is essentially the binary decision error probability P_b based on L_e. As such, we can simplify the above expression as

$$
\begin{aligned}
I(u, I_e) &= 1 - \int_{-\infty}^{\infty} \underbrace{[P_b \log P_b - (1 - P_b)\log_2(1 - P_b)]}_{H_b(P_b)} f(\varphi)d\varphi \\
&= 1 - \mathbb{E}_{\varphi}[H_b(P_b)],
\end{aligned}
\tag{11.49}
$$

where H_b denotes the entropy associated with the binary decision. The last line can be calculated using the large sample average, according to the ergodic theory. As such, suppose for given a given SNR ρ, we observe a large number N of samples of LLR L_e from a decode output $\{b_n\varphi_n, n = 1, 2, \cdots, N\}$. Then, we have

$$
I(u, L_e) \approx 1 - \frac{1}{N}\sum_{n=1}^{N} \log_2(1 + e^{-b_n\varphi_n}),
\tag{11.50}
$$

which enables us to numerically estimate the mutual information directly based on the sign and magnitude of the output extrinsic LLRs without the need for a histogram.

Having known how to determine the information transfer in *a single* constituent decoder, we are in position to determine the EXIT chart for the *two* constituent decoders. To proceed, we replace I_a with the extrinsic LLR fed back from the other decoder, and add subscript $i = 1, 2$ to the relevant symbols to indicate their association with the ith decoder. The EXIT chart is plotted on a two-dimensional plane, with its horizontal and vertical axes representing I_{2e} and I_{1e}, respectively. We first plot the curve for each single constituent decoder, namely plotting I_{1e} as the function of the variable I_{2e} over $0 \le I_{2e} < 1$ for the first decoder, and plot I_{2e} as the function of variable I_{1e} over $0 \le I_{1e} < 1$ for the second decoder.

We next characterize the information iterative evolution between the decoders, as follows:

$$
\cdots I_{2e}^{(k-1)} \xrightarrow{(k)} \boxed{I_{1e}^{(n)} = T_1(\Psi_2^{-1}(I_{2e}^{(k-1)}))} \xrightarrow{(k+1)} \boxed{I_{2e}^{(k+1)} = T_2(\Psi_1^{-1}(I_{1e}^{(k)}))} \xrightarrow{(k+2)} \cdots
$$

where the number (k) on the arrows is used to indicate step k. In step 1, Decoder 1 starts with the case without information fed back from Decoder 2, that is, $L_{2e}^{(0)} = 0$. Namely, the decoder works on the channel input alone, producing output extrinsic LLR $L_{1e}^{(1)}$, which is applied to Decoder 2 as *a priori* LLR in the second step, generating output extrinsic LLR $L_{2e}^{(2)}$. The iteration continues until the LLR reaches its maximum. The resulting

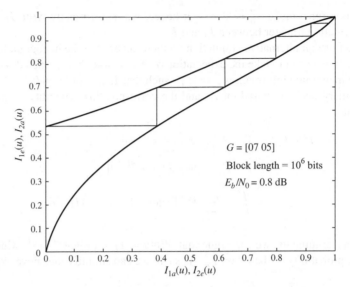

Figure 11.10   *EXIT chart for illustrating the convergence behavior*

graphs present the EXIT chart, originated by ten Brink [9], providing an intuitive way to visualize the convergence behavior of a turbo decoding process. A typical EXIT chart is shown in Figure 11.10.

✍ Example 11.3

An RSC code generator used by Brink in his illustration, in octal, is given by $(G_r, G) = (023, 037)$, with G_r defining the recursive coefficients. Now, we have

$$023 = 2^4 + 2^2 + 1; \quad 037 = 2^4 + 2^3 + 2^2 + 2 + 1,$$

which, in polynomial form, is expressible as

$$\left(1, \frac{1 + D + D^2 + D^3 + D^4}{1 + D^2 + D^4} \right).$$

Other RSC encoders used are $(023, 011)$, and $(023, 035)$; the former is good at the beginning, while the latter behaves the other way around. The encoder $(037, 021)$ is good for low to moderate L_a. ○

11.8   Convergence and fixed points

The EXIT chart intuitively describes the convergence behavior of extrinsic information exchange between the two constituent decoders, thereby providing a guideline for turbo design. Yet, it is not a rigorous proof of the convergence of the iterative turbo decoding.

The convergence issue is, in essence, a fixed-point problem of the operator defined by the BCJR algorithm. The convergence of the iterative BCJR algorithm for turbo decoding is proved in Refs [11–13] by showing that the operator defined by the algorithm has a fixed point regardless of the constituent codes. The proofs follow different methodologies. Reference [12] takes an algebraic approach, while Ref. [11] reaches the same conclusion geometrically.

11.9 Statistics of LLRs

The two-device mechanism used in turbo codes is artificially designed. In communication, a *natural* two-device mechanism exists in many scenarios. Very often, FEC codes are chosen as one device, but there are many ways to choose the other. Up to a particular application, the other device can be an ISI channel, a multiuser channel, or a MIMO system. In turbo codes, information exchange between the two decoders is conducted on the basis of the LLR-type of soft information. In many other applications such as turbo equalization, turbo MIMO, and turbo CDMA, however, the criterion of minimum mean-squared error (MMSE) may be used, where the average statistics of LLRs is required instead. Such average statistics is derived in this section.

11.9.1 Mean and variance of LLRs

Given a received vector $\mathbf{y}$, which carries information about a binary variable u_k, the soft information about u_k is contained in its LLR $L(u_k)$. The important characteristics of $L(u_k)$ are its first two moments. Let us first consider the mean. By definition, we can write

$$
\begin{aligned}
\mathbb{E}[u_k|\mathbf{y}] &= 1 \times \Pr\{u_k = 1|\mathbf{y}\} - 1 \times \Pr\{u_k = -1|\mathbf{y}\} \\
&= \frac{\Pr\{u_k = 1|\mathbf{y}\} - \Pr\{u_k = -1|\mathbf{y}\}}{\Pr\{u_k = 1|\mathbf{y}\} + \Pr\{u_k = -1|\mathbf{y}\}} \\
&= \tanh[L(u_k)/2] \in (-1, 1),
\end{aligned}
\tag{11.51}
$$

where we have used the fact that the denominator of the second line is unity. We proceed to determine the variance of $L(u_k)$. By dividing both the numerator and the denominator on the second line by $\Pr\{u_k = -1|\mathbf{y}\}$ and using the definition of the LLR, the third line is obtained. Since the power of u_k is unity, we further obtain the conditional variance of u_k as

$$
\sigma_{u_k|\mathbf{y}}^2 = 1 - \mathbb{E}^2[u_k|\mathbf{y}] = 1 - \tanh^2[L(u_k)/2].
\tag{11.52}
$$

The parameters $\mathbb{E}[u_k|\mathbf{y}]$ and $\sigma_{u_k|\mathbf{y}}^2$ provide soft information about u_k, and are often required in iteration algorithms for turbo equalization and turbo CDMA.

11.9.2 Mean and variance of hard decision

For comparison, it is interesting to see the average statistics of the corresponding hard-decision estimator, which is defined by

$$
\hat{u}_k = \text{sgn}[L(u_k)]
\tag{11.53}
$$

and is obtainable from the model

$$
y = u_k + n,
\tag{11.54}
$$

where $u_k = \pm 1$ and $n \sim \mathcal{N}(0, \gamma^{-1})$, with SNR $\gamma = 2E_b/N_0$. Hard decision is simple. However, it depends only the sign of $L(u_k)$ while dropping the magnitude information. The hard decision estimator has the following statistics:

$$
\mathbb{E}[\hat{u}_k|u_k] = \begin{cases} -Q(\sqrt{\gamma}) + [1 - Q(\sqrt{\gamma})] = 1 - 2Q(\sqrt{\gamma}), & u_k = 1 \\ Q(\sqrt{\gamma}) - [1 - Q(\sqrt{\gamma})] = 2Q(\sqrt{\gamma}) - 1, & u_k = -1 \end{cases}
\tag{11.55}
$$

$$
\text{MSE} = (1 - 1)^2[1 - Q(\sqrt{\gamma})] + (1 + 1)^2 Q(\sqrt{\gamma}) = 4Q(\sqrt{\gamma}).
\tag{11.56}
$$

Since $\mathbb{E}[\hat{u}_k] = (1/2)\mathbb{E}[\hat{u}_k|u_k = 1] + (1/2)\mathbb{E}[\hat{u}_k|u_k = -1] = 0$, it follows that

$$\text{bias} = 2Q(\sqrt{\gamma}),\tag{11.57}$$

$$\text{var}(\hat{x}) = \text{MSE} - \text{bias}^2 = 4Q(\sqrt{\gamma}).\tag{11.58}$$

The bias can be unity for low SNR.

11.10   Turbo equalization

In this section we investigate turbo equalization, which results from the combination of FEC codes and the structures introduced by an ISI channel.

Suppose that a length-N information bit vector $\mathbf{b}$ is coded with the operator $\mathbf{C}$, converted to bipolar pulses $\mathbf{c} = 2\mathbf{C}(\mathbf{b}) - 1$, and permuted by interleaver $\boldsymbol{\Pi}$ before transmission over a time-invariant dispersive channel characterized by the matrix $\mathbf{H}$. In baseband form, the received signal can be written as

$$\mathbf{z} = \mathbf{H}\boldsymbol{\Pi}\underbrace{[2\mathbf{C}(\mathbf{b}) - 1]}_{\mathbf{x}} + \mathbf{n} = \mathbf{H}\boldsymbol{\Pi}\mathbf{c} + \mathbf{n} = \sum_{i=1}^{N} x_i \mathbf{h}_i + \mathbf{n},\tag{11.59}$$

where $\mathbf{x} = [x_1, \cdots, x_N]^T$ is the permuted BPSK vector, $\mathbf{n} \sim \mathcal{N}(\mathbf{0}, \frac{N_0}{2}\mathbf{I})$ is AWGN, $\mathbf{h}_i$ denotes the ith column vector of $\mathbf{H}$, and all the matrices and vectors are assumed compatible so that their multiplications and additions are meaningful. The channel state information is assumed to be known at the receiver. Upon the reception of $\mathbf{z}$, the turbo equalizer iteratively solves for the information vector $\mathbf{b}$ through information exchange between its channel detector and BCJR decoder, as illustrated in Figure 11.11.

In the figure, the vector $L_E(\mathbf{x})$ denotes the *posterior* LLR of $\mathbf{x}$, while $L_a(\mathbf{x})$ denotes its *prior* LLR provided by the BCJR decoder. Their difference

$$L_e(\mathbf{x}) = L_E(\mathbf{x}) - L_a(\mathbf{x})\tag{11.60}$$

represents the extrinsic LLR of $\mathbf{x}$ produced by the detector and fed to the BCJR decoder as prior information for use in decoding. Similarly, we can define the posterior and prior LLRs of $\mathbf{c}$, denoted by $L_E(\mathbf{c})$ and $L_a(\mathbf{c})$, respectively, such that

$$L_e(\mathbf{c}) = L_E(\mathbf{c}) - L_a(\mathbf{c})\tag{11.61}$$

represents the extrinsic information of $\mathbf{c}$ produced by the decoder and fed back to the detector. The extrinsic LLR from one device is used as prior information by the other in the next iteration, as shown by

$$L_a(\mathbf{x}) = \boldsymbol{\Pi}[L_e(\mathbf{c})]$$

$$L_a(\mathbf{c}) = \boldsymbol{\Pi}^{-1}[L_e(\mathbf{x})]\tag{11.62}$$

with $\boldsymbol{\Pi}^{-1}$ denoting the operation of deinterleaving.

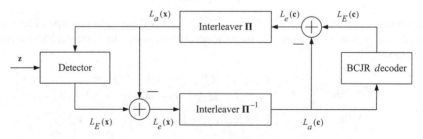

Figure 11.11   *Information exchange in turbo equalization*

Now let us show how to generate $L_e(\mathbf{x})$ in the detector. Because of the use of random interleaver and deinterleaver, adjacent symbols in $\mathbf{x}$ can be treated as an independent sequence. Thus, calculating the LLR of $\mathbf{x}$ reduces to the calculation of its counterpart for individual x_k, for which we have

$$L_E(x_k) = \ln \frac{P(x_k = +1|\mathbf{z})}{P(x_k = -1|\mathbf{z})}$$

$$= \underbrace{\ln \frac{P(\mathbf{z}|x_k = +1)}{P(\mathbf{z}|x_k = -1)}}_{L_e(x_k)} + \underbrace{\ln \frac{P(x_k = +1)}{P(x_k = -1)}}_{L_a(x_k)}, \tag{11.63}$$

where $P(\mathbf{z}|x_k)$ denotes the distribution of $\mathbf{z}$ conditioned on x_k.

To calculate $L_e(x_k)$ on the last line of (11.63), the key is to determine $P(\mathbf{z}|x_k)$. Given x_k and the feedback information $L_a(\mathbf{x})$ from the decoder, the received vector in (11.59) can be approximately treated as jointly Gaussian distributed variates, which, after direct manipulation, can be shown to be

$$\mathbf{z}|_{x_k, L_a(\mathbf{x})} \sim \mathcal{CN}(x_k \mathbf{h}_k + \mathbf{m}_k, \mathbf{R}_k) \tag{11.64}$$

with vector $\mathbf{m}_k$ and covariance matrix $\mathbf{R}_k$ given by

$$\mathbf{m}_k = \sum_{i=1, i \neq k}^{N} \bar{x}_i \mathbf{h}_i,$$

$$\mathbf{R}_k = \sigma_n^2 \mathbf{I} + \sum_{i=1, i \neq k}^{N} v_i \mathbf{h}_i \mathbf{h}_i^\dagger. \tag{11.65}$$

Note that here the feedback information $L_a(\mathbf{x})$ from the decoder has been incorporated into the conditional distribution calculation via the relationship

$$\bar{x}_i = \tanh\left(\frac{1}{2}L_a(x_i)\right), \quad v_i = 1 - \bar{x}_i^2$$

for $i = 1, \cdots, N$. It is also noted that, given x_k, x_k is a *deterministic* variable and, thus, $L_a(x_k)$ must be *excluded* from the the LLR calculation for x_k. It follows from (11.63) and (11.64) that

$$L_e(x_k) = \ln \frac{\exp[-(\mathbf{z} - \mathbf{m}_k - \mathbf{h}_k)^\dagger \mathbf{R}_k^{-1}(\mathbf{z} - \mathbf{m}_k - \mathbf{h}_k)]}{\exp[-(\mathbf{z} - \mathbf{m}_k + \mathbf{h}_k)^\dagger \mathbf{R}_k^{-1}(\mathbf{z} - \mathbf{m}_k + \mathbf{h}_k)]}$$

$$= 2\Re\{\mathbf{h}_k^\dagger \mathbf{R}_k^{-1}(\mathbf{z} - \mathbf{m}_k)\}. \tag{11.66}$$

The last line has a clear physical significance. The detector employs the prior information to remove the ISI caused by the other symbols, producing $\hat{\mathbf{z}}_k = \mathbf{z} - \mathbf{m}_k$ which is then processed by the MMSE filter defined by $\mathbf{w}_k = \mathbf{R}_k^{-1}\mathbf{h}_k$. The MMSE filter points its main beam to the desired signal vector $\mathbf{h}_k$ while posing its nulls again in the directions defined by the interfering vectors $\{\mathbf{h}_i, i \neq k\}$ through the inverse matrix $\mathbf{R}_k^{-1}$. Clearly, the role played by the prior information fed back from the BCJR decoder is twofold; it is used for the cancelation of ISI and for the construction of the MMSE filter.

The computational complexity of $L_e(x_k)$ is on the order of magnitude $O(N^3)$. To reduce the computation, we may use the matrix inversion lemma to simplify (11.66) and denote

$$\mathbf{R} = \sigma_n^2 \mathbf{I} + \sum_{i=1}^{N} v_i \mathbf{h}_i \mathbf{h}_i^\dagger,$$

obtaining

$$L_e(x_k) = \frac{2\Re\{\mathbf{h}_k^\dagger \mathbf{R}^{-1}(\mathbf{z} - \mathbf{m}_k)\}}{1 - v_k \mathbf{h}_k^\dagger \mathbf{R}^{-1} \mathbf{h}_k}, \tag{11.67}$$

whose computational complexity is $O(N^2)$.

The LLRs $L_e(\mathbf{x}) = \{L_e(x_k), k = 1, \cdots, N\}$ are used as the prior information to the BCJR algorithm described in Section 11.4 for decoding, as illustrated below.

$$\xrightarrow{L_e(\mathbf{X})} \boxed{\Pi^{-1}} \xrightarrow{L_a(\mathbf{c})} \boxed{\text{BCJR decoder}} \xrightarrow{L_e(\mathbf{c})} \tag{11.68}$$

The BCJR decoder produces soft LLR $L_e(\mathbf{c})$ for $\mathbf{c}$. By the same argument, the prior information $L_a(c_k)$ is excluded from the determination of $L_e(c_k)$. As $L_e(\mathbf{c})$ passes through the deinterleaver and is fed back to the detector as prior information, another iteration starts.

In this section, we have shown how to derive the output extrinsic information $L_e(x_k)$ from the channel detector in the framework of log-likelihood ratio test. In fact, by a similar token, we can derive the extrinsic information by using the MMSE criterion [27].

✍ Example 11.4

This is a numerical example. A binary sequence is coded by a rate-$\frac{1}{2}$ RSC encoder and BPSK-modulated and permuted by a random interleaver before transmission over a length-3 Proakis-B channel with impulse response $\mathbf{h} = [0.407, 0.815, 0.407]$. The RSC encoder, defined by $(1 + D^2)/(1 + D + D^2)$, each time processes a block of $10,240$ information bits with zero tail bits appended for states clearance at the end of each block. The receiver consists of a BCJR-based MAP decoder and an equalizer implemented under either the MMSE or the MAP criterion. By information exchange between the two devices, the algorithm converges quickly in 10 iteration, as clearly indicated in Figure 11.12. It is also clear that after 10 iterations, the bit error performances of the MAP and MMSE equalizers are nearly identical, both approaching the BER lower bound of the AWGN channel.

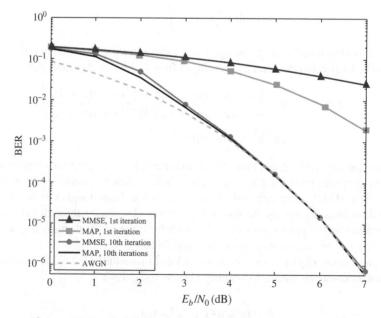

Figure 11.12 *Turbo equalizers: error performance comparison*

11.11   Turbo CDMA

The second possible application of the turbo principle is turbo CDMA. Consider a multiuser system for uplink transmission, where each user, say k, sends his coded and permuted a message vector $\mathbf{x}_k$ over a flat fading channel characterized by a single tap weight gain. Vector $\mathbf{x}_k$ results from passing a message vector $\mathbf{b}_k$ through an encoder defined by matrix $\mathbf{C}_k$ and an interleaver defined by matrix $\mathbf{\Pi}_k$, and then mapping to the BPSK constellation, such that

$$\mathbf{x}_k = \mathbf{\Pi}_k[2\mathbf{C}_k\mathbf{b}_k - 1]. \tag{11.69}$$

The BPSK modulation is considered here just for simplicity. The channel gain for user k is constant over each coded symbol period, say i, but varying from one symbol to the next, and is thus denoted by $h_k(i)$. Each entry (i.e., each coded and permuted symbol) of user k is spread by his spreading vector $\mathbf{s}_k$ before mapping to constellation for transmission. The BPSK modulation is a simple linear operation, just converting a unipolar coded sequence into bipolar, and is thus omitted hereafter for brevity. Assume that the length of $\mathbf{x}_k$ is N, let $\mathbf{H}_k = \mathrm{diag}\{\mathbf{h}_k\}$, where $\mathbf{h}_k = [h_k(1), \cdots, h_k(N)]^T$ denote the channel vector of user k over N consecutive symbol periods, and let $\otimes$ denote the Kronecker product. The coded and BPSK-modulated vectors $\mathbf{x}_i$ are then spread by their respective spreading codes $\mathbf{s}_i$ before passing through the flat fading channels. As such, the received signal from the K users can be represented as

$$\mathbf{z}_L = \sum_{i=1}^{K}(\mathbf{H}_i \otimes \mathbf{s}_i)\mathbf{x}_i + \mathbf{n}, \tag{11.70}$$

where $\mathbf{n} \sim \mathcal{CN}(\mathbf{0}, \sigma_n^2\mathbf{I})$, and the dimensions of the matrices and vectors are compatible, so that the operations are meaningful.

Given a received vector $\mathbf{z}_L$, the task of multiuser detection is to exploit the channel state information (CSI), users, spreading vectors, and code structures available at the receiver to reliably retrieve the users' message information. For illustration, let us confine to user k. We need to find $\mathbf{b}_k$ such that

$$\hat{\mathbf{b}}_k = \arg\max_{\mathbf{b}_k} \Pr\{\mathbf{b}_k|\mathbf{z}_L\}. \tag{11.71}$$

Just as the case for turbo equalization, we partition the entire process of estimation into two devices:

$$\mathbf{z}_L = \sum_{i=1}^{K}\underbrace{(\mathbf{H}_i \otimes \mathbf{s}_i)}_{\mathbf{K}_i}\underbrace{\mathbf{\Pi}_i(2\mathbf{C}_i\mathbf{b}_i - 1)}_{\mathbf{x}_i} + \mathbf{n}, \tag{11.72}$$

so that the two-step turbo procedure is applied, as sketched below.

$$\boxed{\mathbf{z}_L = \sum_k \mathbf{K}_k\mathbf{x}_k + \mathbf{n}} \xrightarrow{\hat{\mathbf{x}}_k} \boxed{\mathbf{\Pi}_k^{-1}} \longrightarrow \boxed{\mathbf{y}_k = \mathbf{C}_k(\mathbf{b}_k) + \mathbf{e}_k} \longrightarrow \hat{\mathbf{b}}_k. \tag{11.73}$$

where $\mathbf{y}_k = \mathbf{\Pi}_k^{-1}\hat{\mathbf{x}}_k$. The first device is the multiuser detector that exploits the combined structure of the channel matrices and spreading vectors to estimate vector $\mathbf{x}_k$ from $\mathbf{z}_L$. The estimate is then fed to the second device comprised of K BCJR decoders, whose operating principle has been addressed in the previous sections. The kth BCJR decoder exploits the code structure $\mathbf{C}_k$ to decode the original message vector $\mathbf{b}_k$. The results from all the K decoders are fed back to the multiuser detector for refining the estimation of $\mathbf{x}_k$, hence forming an iterative mechanism for information exchange.

The independence among symbols and randomization due to long interleaving allow the channel detection to be done symbol by symbol. Thus, it suffices to focus on the detection of a single symbol inside $\mathbf{x}_k$. Denote the corresponding received signal block by $\mathbf{z}$, which is a subset of $\mathbf{z}_L$. For simplicity, we only keep the user indices while dropping the time indices for the channel gains and for the symbol location inside $\mathbf{x}$'s, so that $\mathbf{h}_i$ reduces to h_i and $\mathbf{x}_i$ reduces to x_i. With this notation, we can write

$$\mathbf{z} = \sum_{i=1}^{K} x_i h_i \mathbf{s}_i + \mathbf{n} = x_k h_k \mathbf{s}_k + \underbrace{\sum_{i=1,i\neq k}^{K} x_i h_i \mathbf{s}_i + \mathbf{n}}_{\mathbf{v}_k}. \tag{11.74}$$

Given $\mathbf{z}$, we want to estimate x_k. The prior information on x's available at the receiver is fed back from the decoders, in the form of $L_a[x_i] = \gamma_i$, with which the mean and variance of x_i are given by

$$\mu_i = \mathbb{E}[x_i] = \tanh(\gamma_i/2)$$
$$v_i = \mathbb{E}[|x_i - \mu_i|^2] = 1 - \mu_i^2 \tag{11.75}$$

for $i = 1, \cdots, K$.

To find the LLR of x_k for given $\mathbf{z}$, we need to determine the distributions of $\mathbf{z}$ under the two hypotheses on x_k. With the results shown above, we have

$$\mathbb{E}[\mathbf{z}|x_k = \pm 1] = \pm h_k \mathbf{s}_k + \underbrace{\sum_{i=1,i\neq k}^{K} \mu_i h_i \mathbf{s}_i}_{\mathbf{m}_k},$$

$$\text{Cov}(\mathbf{z}|x_k = \pm 1) = \sigma_n^2 \mathbf{I} + \underbrace{\sum_{i=1,i\neq k}^{K} |h_i|^2 (1 - \mu_i^2) \mathbf{s}_i \mathbf{s}_i^\dagger}_{\mathbf{\Omega}_k}. \tag{11.76}$$

It follows that $\mathbf{z}$ for $x_k = \pm 1$ is distributed, respectively, as

$$\mathbf{z}|_{(x_k=\pm 1)} \sim \mathcal{CN}(\mathbf{m}_k \pm h_k \mathbf{s}_k, \sigma_n^2 \mathbf{I} + \mathbf{\Omega}_k), \tag{11.77}$$

whereby the LLR for x_k can be determined, yielding

$$L_e(x_k) = \frac{p(x_k = 1|\mathbf{z}, \{L_a(x_i), i \neq k\})}{p(x_k = -1|\mathbf{z}, \{L_a(x_i), i \neq k\})}$$
$$= 4\Re\{h_k(\mathbf{z} - \mathbf{m}_k)(\sigma_n^2 \mathbf{I} + \mathbf{\Omega}_k)^{-1} \mathbf{s}_k\}. \tag{11.78}$$

This is the extrinsic information to be forwarded to the decoders.

It is clear that the LLR method incorporates the average prior information $\{v_i, \mu_i, i \neq k\}$ into the update estimation of $L_e(x_k)$. The extrinsic LLR in (11.78) has a form similar to that of turbo equalization. The factor $(\mathbf{z} - \mathbf{m}_k)$ represents removing, from the received vector $\mathbf{z}$, the multiuser interfering component that is predictable on the basis of the prior information provided by the decoders. This signal-enhanced vector is then projected onto an MMSE filter for an improved soft information $L_e(x_k)$ about x_k. Note, again, that the *a priori* information about x_k must be discarded from its current update estimation since, at this time, x_k is treated as deterministic according to the definition of conditional distributions.

The above LLR treatment allows for nonorthogonal spreading codes among multiple users. When orthogonal spreading codes are used for all the users, the above LLR reduces to a parallel despreading done individually for each user, followed by a soft decision.

11.12   Turbo IDMA

As the last example of applying the turbo principle, we consider another turbo multiuser detector called turbo IDMA, which is based on the interleaver-division multiple-access (IDMA) scheme. In IDMA, users are distinguished by their interleavers, denoted by $\mathbf{\Pi}_i$. In an AWGN channel, the received signal vector can be written as

$$y = \sum_{i=1}^{K} \mathbf{\Pi}_i \mathbf{C}_i \mathbf{s}_i + \mathbf{n}, \tag{11.79}$$

where $\mathbf{s}_k = [s_{k1}, s_{k2}, \cdots, s_{kL}]^T$ denotes the signal vector for user k. Here, L is the appropriate block size chosen for iterative turbo detection; a typical value is $L = 30$. The symbol vector from user i is coded by $\mathbf{C}_i$ before interleaving by $\mathbf{\Pi}_i$. Since each user transmits L symbols, there are totally KL symbols to be retrieved. In order to make the problem resolvable, the number of equations in $\mathbf{y}$ must, at least, be equal to KL. To this end, the code rate of $\mathbf{C}_i$ cannot exceed $1/K$. In its simplest form, $\mathbf{C}_i$ can be an m-times repetitive code, which, in mathematics, is expressible as $\mathbf{C}_i = (\mathbf{I} \otimes 1_m)$ where $\otimes$ denotes the Kronecker product and 1_m denotes the m-dimensional all-1 column vector. For the order-3 repetitive code, $\mathbf{C}_i$ takes the form

$$\mathbf{C}_i = \begin{bmatrix} 1 & 0 & 0 & \cdots & \cdots & 0 \\ 1 & 0 & 0 & \cdots & \cdots & 0 \\ 1 & 0 & 0 & \cdots & \cdots & 0 \\ 0 & 1 & 0 & \cdots & \cdots & 0 \\ 0 & 1 & 0 & \cdots & \cdots & 0 \\ 0 & 1 & 0 & \cdots & \cdots & 0 \\ \vdots & & & \vdots & & \vdots \\ 0 & 0 & \cdots & & 0 & 1 \\ 0 & 0 & \cdots & & 0 & 1 \\ 0 & 0 & \cdots & & 0 & 1 \end{bmatrix}. \tag{11.80}$$

Coding schemes other than simple repetition, such as convolutional codes, can offer better performance. Random interleaver matrices play a central role in the IDMA. They tend to eliminate the correlation between different users, thus having the effect of suppressing multiuser interference. The signal model of (11.79), in its mathematical form, is similar to that of CDMA. Thus, the procedure described for information retrieval therein is applicable and not repeated here.

There is a huge literature addressing different aspects of turbo codes and turbo decoding. Reader can consult [24] for finding good constituent convolutional codes to construct a turbo code, consult [23, 25] for the minimum distance of turbo codes and their performance, consult [6, 13, 29] for the convergence of turbo iteration, consult [20] for soft decoding, consult [21] for a more in-depth understanding of EXIT charts, consult [30, 31] for multiuser detection, and consult [26] for IDMA.

11.13   Summary

On the method of coding to approach the Shannon limit, turbo codes and turbo principle play an important role. By concatenating two or multiple RSC codes, it possible for a turbo encoder to generate a long coded sequence of sufficiently random behavior. On the other hand, the turbo principle ensures the tractability of the resulting decoding problem with a tolerable computational complexity by converting a sophisticated integer programming to a feedback iteration process. The general convergence of the algorithm, though unsolved mathematically, is not a problem to most practical applications. The drawback of turbo codes is time delay. For example, the use of a turbo code of 2318 bits implies the latency of 2318 bits.

The turbo principle, originally designed for two concatenated component codes, is widely applicable to many other application scenarios, as long as there exists a two-device mechanism. Such a mechanism can result from combining a code with a time dispersive channel, with a CDMA system, or with a MIMO channel. The key consideration is that one device must contain unknown variables subject to integer or discrete constraints, so that binary decisions are applied.

As pointed out in the chapters Introduction and Signal Processing Strategies, the ultimate goal of the receiver in a digital wireless system is to recover the original discrete message symbols under the MAP criterion or in an equivalent framework. The objective function involves multiple devices, including, for example, digital modulation, codec, ISI channels, multiple access, and multiple antennas. The turbo principle makes a good move to the global solution; yet, its extension to the general case of multiple devices still needs great efforts. Probably, a new principle is needed, a direction that is worth further investigation.

Problems

Problem 11.1   This is a small project. Study the application of Viterbi equalizer in the GSM system [14].

Problem 11.2   Study RSC codes in a different framework of autoregressive modeling.

Problem 11.3   A RSC code can be derived from an NSC code by adding a feedback loop and setting one of the two outputs equals to the input bit. Assume an NSC encoder, as given in Figure 11.13. (1) Derive an RSC encoder from it. (2) Determine the state transition diagrams of the two codes.

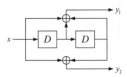

Figure 11.13   *Nonsystematic convolutional code*

Problem 11.4   For the NSC given in Figure 11.13 and the corresponding RSC you have derived, determine their output sequences if the input sequence is $1000\cdots$. What happens if the input sequence is changed to $1110\cdots$? Explain the phenomenon you observe.

Problem 11.5   Draw a turbo encoder with an interleaver of length $K = 8$ using an interleaver with two RSC encoders, as shown in Figure 11.14. The interleaving function is as follows: $\pi(i) = i + 1$ for $i = 1, 2, \cdots, 7$ and $\pi(1) = 8$, that is, the output sequence of the interleaver is a circular right shift of the input. Determine a termination scheme (the encoder states are zeros at both the beginning and the end, so tail bits may be needed). Write down the output sequence when the input is 10100011.

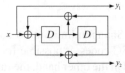

Figure 11.14   *Recursive systematic convolutional code*

Problem 11.6   Let v_t denote a binary sequence of ± 1-bits from a convolutional encoder. Suppose that $\{v_t\}$ goes through a channel of memory length $L = 2$, so that the received signal is given by

$$r_t = h_0 v_t + h_1 v_{t-1} + h_2 v_{t-2}.$$

(a) Draw the trellis diagram of states transition for the ISI channel.

(b) Derive the γ metric expression of the MAP algorithm for the ISI equalizer.

Problem 11.7   Both the Viterbi and BCJR algorithms exploit the Markovian properties of convolutional codes. What is the difference in methodology between the two? Give a detailed description and make your comments.

Problem 11.8   In this chapter, we only considered turbo equalizers with binary modulation. It is more difficult to produce soft output information for a turbo equalizer with higher level modulation, such as QAM. Nevertheless, due to the orthogonality of the I and Q channels of QAM, bit estimation on the I/Q channels can be done separately and, thus, we can just focus on a single branch, say the I channel. The QPSK corresponds to two parallel 2PAM signals, while 16QAM corresponds to two parallel 4PAM signals, as shown in Figures 11.15 and 11.16, respectively. Denote the information bit in 2PAM or information dibit in 4PAM by s, which is mapped into signal points through mapping $\psi(\cdot)$, so that the received signal in a single branch can be written as $r = a\psi(s) + n$. Here, $n \sim \mathcal{N}(0, \sigma^2)$ is the AWGN, and a denotes the physical quantization level in signal magnitude. Suppose that the LLR method is used for the soft demodulation of signal level s.

(a) Show that for QPSK, the soft LLR output from a single branch is given by

$$\text{LLR}_b^{2\text{PAM}} = \frac{2ar}{\sigma^2};$$

(b) Further, show that for the decision of each 4PAM of the 16QAM, the soft LLR output for the first bit is given by

$$\text{LLR}_{b_1}^{4\text{PAM}} \approx \begin{cases} 4a(r-a)/\sigma^2 & \text{if } r > 2a \\ 4a(r+a)/\sigma^2 & \text{if } r < -2a \\ 2ar/\sigma^2 & \text{else,} \end{cases}$$

while its counterpart for the second bit is given by

$$\text{LLR}_{b_2}^{4\text{PAM}}(r) = -\text{LLR}_b^{2\text{PAM}}(|r| - 2a).$$

In your analysis, you may assume that Gray mapping is used in QPSK and 16QAM.

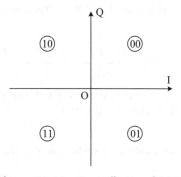

Figure 11.15   *Constellation of QPSK*

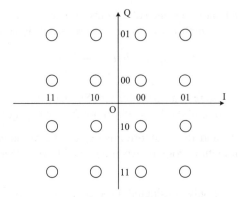

Figure 11.16   *Constellation of 16QAM*

Problem 11.9   Suppose that K users transmit their coded bipolar binary signals $x_k, k = 1, \cdots, K$ through their respective flat fading channels h_k and arrive at the receiver, producing channel output

$$y = \sum_{k=1}^{K} h_k x_k + n, \qquad (11.81)$$

where $x_k = \pm 1$ and $n \sim \mathcal{N}(0, \sigma_n^2)$ is the AWGN. The *a priori* soft LLR information on x_k is $L_a[x_k] = \gamma_k$. Denoting $\overline{x}_k = \tanh(\gamma_k/2)$, show that the LLR of x_k based on y is

$$\lambda_k = L[x_k] = \frac{2h_k(y - a_k)}{v_k},$$

where

$$a_k = \sum_{i=1, i \neq k}^{K} h_k \overline{x}_k,$$

$$v_k = \sigma_n^2 + \sum_{i=1, i \neq k}^{K} |h_k|^2 (1 - \overline{x}_k^2).$$

Problem 11.10   In turbo CDMA and turbo equalization, what is the role played by the feedback information from the BCJR decoders in the multiuser or channel detectors?

Problem 11.11   For a synchronous IDMA system in a block flat fading channel, the receiver signal vector is expressible as

$$z = \sum_{k=1}^{K} \underbrace{h_k \mathbf{\Pi}_k (\mathbf{I} \otimes \mathbf{e}_k)}_{\mathbf{A}_k} \underbrace{\mathbf{C}_k(\mathbf{b}_k)}_{\mathbf{x}_k} + \mathbf{n},$$

where $\mathbf{e}_k$ is an all-1 column vector representing repetitive transmission of each coded symbol for user k. Suppose the feedback LLR information on x_i from the decoders is $\lambda_i = L_a[x_i]$. Derive the multiuser detector's soft output LLR for x_k.

Problem 11.12   Consider the BCJR decoding of a systematic recursive convolutional code on the AWGN channel with generator matrix

$$G(D) = [1 \ \ 1/(1 + D)].$$

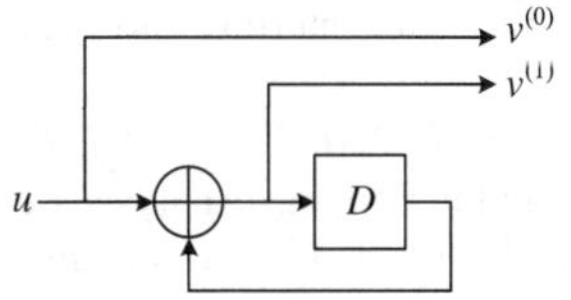

Figure 11.17   *Systematic recursive convolutional code*

Let $u = (u_0, u_1, u_2, u_3)$ denote the input vector of length 4 and $v = (v_0, v_1, v_2, v_3)$ denote the codeword of length 8. We assume $E_s/N_0 = 1/4 = -6\,\mathrm{dB}$. The received vector $r = (+0.8, +0.1; +1.0, -0.5; -1.8,$ $+1.1; 1.6, -1.6)$. The encoder uses the terminated code. Assume that the information bits are equally likely with $L_a(u) = 0$. Use the log-MAP algorithm to compute the APP L-value of input bits. The systematic recursive convolutional code is illustrated in Figure 11.17

Problem 11.13   Compose a MATLAB program to implement the BCJR algorithm. Test the program by using it to decode a turbo-encoded and AWGN-corrupted sequence. (Hint: In the computation of log likelihood ratios, the immediate values of forward recursion, backward recursion, and transition probabilities can be normalized to prevent a possible vanishing outcome.)

Problem 11.14   Consider a systematic turbo encoder. Let **b** be an $N \times 1$ source vector of independent random variables. The encoded output is denoted as $c = (\mathbf{b}, \mathbf{c}_p)$, where $\mathbf{c}_p$ is the parity part of the codeword. The codeword is transmitted over a memoryless noisy channel characterized by the transitional probability $p(\mathbf{r}|\mathbf{b})$ of the received vector **r**. Similarly, $\mathbf{r} = (\mathbf{r}_s, \mathbf{r}_p)$ is composed of the systematic and the parity parts. Show that the posterior probability of b_i is expressible as

$$q(b_i = a) \stackrel{\mathrm{def}}{=} p(b_i = a|\mathbf{r}) = \mu \psi_i(a) \pi_i(a) \sum_{b:b_i=a} p(\mathbf{r}_p|\mathbf{c}_p) \prod_{j=1, j \neq i}^{N} \psi_j(b_j) \pi_j(b_j),$$

where $\mu = 1/p(\mathbf{r})$, $\psi_j(b_j) = p(r_{sj}|b_j)$, and $\pi_j(b_j) = p(b_j)$.

Problem 11.15   A systematic turbo code is composed of two parallel RSC encoders with the same generating polynomial $[15, 13]$. The two encoders are connected by an interleaver with permutation

$$\Pi = \begin{pmatrix} 1 & 2 & 3 & 4 & 5 & 6 & 7 & 8 \\ 3 & 8 & 1 & 2 & 7 & 5 & 4 & 6 \end{pmatrix}.$$

Find out the encoder output if the input is given by $\mathbf{b} = \begin{bmatrix} 1 & 1 & 0 & 0 & 1 & 0 & 1 & 0 \end{bmatrix}$.

Problem 11.16   Because of space limitation, LDPC codes are not addressed in the book. Design of LDPC codes for a MIMO system is investigated in Ref. [15]. As a small project of relevance to LDPC codes, write a program to examine the idea and technique described therein [15].

References

1. C.E. Shannon, "A mathematical theory of communication," *Bell Syst. Tech. J.*, vol. 27, pp. 379–423 (Part I), pp. 623–656 (Part II), July 1948.
2. C. Berrou, A. Glavieux, and P. Thitimajshima, "Near Shannon limit error correcting coding and decoding: turbo codes," *Proceedings of ICC'93*, vol. 2, pp. 1064–1070, 1993.
3. C. Berrou and A. Glavieux, "Near optimum error correcting coding and decoding: turbo-codes," *IEEE Trans. Commun.*, vol. 44, no. 10, pp. 1261–1271, 1996.

4. C. Berrou, "The ten-year-old Turbo codes are entering into service," *IEEE Commun. Mag.*, vol. 41, no. 8, pp. 110–116, 2003.
5. R.G. Gallager, "Low-density parity-check codes," *IRE Trans. Inf. Theory*, vol. IT-8, pp. 21–28, 1962.
6. S. Benedetto, D. Divsalar, G. Montorsi, and F. Pollara, "Analysis, design, and iterative decoding of double serially concatenated codes with interleavers," *IEEE J. Sel. Areas Commun.*, vol. 16, no. 2, pp. 231–244, 1998.
7. D.J. Costello and G.D. Forney, "Channel coding: the road to channel capacity," *Proc. IEEE*, vol. 95, no. 6, pp. 1150–1177, 2007.
8. L. Bahl, J. Cocke, F. Jelinek, and J. Raviv, "Optimal decoding of linear codes for minimizing symbol error rate," *IEEE Trans. Inf. Theory*, vol. 20, no. 2, pp. 284–287, 1974.
9. S. ten Brink, "Convergence behaviour of iteratively decoded parallel concatenated codes," *IEEE Trans. Commun.*, vol. 49, no. 10, pp. 1727–1737, 2001.
10. S. ten Brink, "Design of concatenated coding schemes based on iterative decoding convergence," Ph.D. dissertation, University at Stuttgart, Shaker Verlag, Aachen, 2002.
11. T.J. Richardson, "The geometry of turbo decoding dynamics," *IEEE Trans. Inf. Theory*, vol. 46, pp. 9–23, 2000.
12. L. Duan and B. Rimoldi, "The iterative-turbo decoding algorithm has fixed points," *IEEE Trans. Inf. Theory*, vol. 47, no. 7, pp. 2993–2995, 2001.
13. M. Ferrari and S. Bellini, "Existence and uniqueness of the solution for Turbo decoding of parallel concatenated single parity check codes," *IEEE Trans. Inf. Theory*, vol. 49, no. 3, pp. 722–726, 2003.
14. B. Sklar, "Rayleigh fading channels in mobile digital communication systems-Part II: mitigation," *IEEE Commun. Mag.*, vol. 35, no. 7, pp. 102–109, 1997.
15. J. Hou, P.H. Siegel, and L.B. Milstein, "Design of multi-input multi-output systems based on low-density parity-check codes," *IEEE Trans. Commun.*, vol. 53, no. 4, pp. 601–611, 2005.
16. R.W. Hamming, "Error detecting and error correcting codes," *Bell Syst. Tech. J.*, vol. 29, pp. 147–160, 1950.
17. P. Elias, "Coding for noisy channels," *IRE Conv. Rec.*, Part 4, pp. 37–47, 1955.
18. R.M. Tanner, "A recursive approach to low complexity codes," *IEEE Trans. Inf. Theory*, vol. IT-27, pp. 533–547, 1981.
19. G.D. Forney Jr., *Concatenated Codes*, Cambridge, MA: MIT Press, 1966.
20. J. Hagenauer, "Soft is better than hard," *Communications, Coding and Cryptology*, Kluwer Publication, 1994.
21. J. Hagenauer, "The EXIT chart-introduction to extrinsic information transfer in iterative processing," *EUSIPCO*, pp. 1541–1548, Sep. 2004.
22. S. Benedetto and G. Montorsi, "Unveiling turbo codes: some results on parallel concatenated coding schemes," *IEEE Trans. Inf. Theory*, vol. 42, pp. 409–428, 1996.
23. S. Benedetto, D. Divsalar, G. Montorsi, and F. Pollara, "Serial concatenation of interleaved codes: performance analysis, design, and iterative decoding," *IEEE Trans. Inf. Theory*, vol. 44, no. 5, pp. 909–926, 1998.
24. S. Benedetto, R. Garello, and G. Montorsi, "A search for good convolutional codes to be used in the construction of Turbo codes," *IEEE Trans. Commun.*, vol. 46, no. 9, pp. 1101–1105, 1998.
25. A. Perotti and S. Benedetto, "A new upper bound on the minimum distance of turbo codes," *IEEE Trans. Inf. Theory*, vol. 50, no. 12, pp. 2985–2997, 2004.
26. H. Herzberg, "Multilevel turbo coding with short interleavers," *IEEE J. Sel. Areas Commun.*, vol. 16, no. 2, pp. 303–308, 1998.
27. R. Koetter, A.C. Singer, and M. Tuechler, "Turbo equalization," *IEEE Signal Process. Mag.*, vol. 21, no. 1, pp. 67–80, 2004.
28. L. Ping, L. Liu, K. Wu, and W.K. Leung, "Interleave-division multiple-access," *IEEE Trans. Wireless Commun.*, vol. 5, no. 4, pp. 938–947, 2006.
29. P.A. Regalia, "Contractility in turbo iterations," *ICASSP'2004*, no. 4, pp. 637–640, 2004.
30. X. Wang and H.V. Poor, "Iterative multiuser decoder for near-capacity communication," *IEEE Trans. Commun.*, vol. 47, no. 7, pp. 1046–1061, 1999.
31. V. Poor, "Iterative multiuser detection," *IEEE Signal Process. Mag.*, vol. 21, pp. 81–88, 2004.

12

Multiple-Access Channels

Abstract

The primary thought behind various multiple-access (MA) schemes in current use is rooted in the paradigm of orthogonality. Collision of two arbitrary symbol waveforms is unavoidable if they have components in the same one-dimensional space. To allow multiple users to share the same channel without interference with each other, the dimension of the space defined by the channel must be extended, so that each user exclusively occupies a mutually *orthogonal* subspace. The principle behind various MA schemes is to seek appropriate basis functions for multiple users, and this can be done in the time domain, the frequency domain, or their combination. Mathematically, the space defined by a channel remains the same, irrespective of the choice of its basis functions. Physically, however, different choices have different system implications.

12.1 Introduction

Modern communication systems often support multiple users, thereby requiring a multiple-access (MA) technique to enable the users to share the same physical channel. MA technology is at the core of modern wireless communications. The choice of appropriate MA schemes is often the issue of debates for each generation of cellular systems.

When multiple users share the same channel resource for transmission, signals from different users can interfere with each other, resulting in inter-user interference (IUI), or more often called multiple-access interference (MAI). This interference is a dominating factor that dictates the system error performance. In search of inspiration for combating IUI, consider K fishes in an aquarium where competition for the use of space is unavoidable. Note that competition never happens if we partition the aquarium into smaller containers and put one fish in one container. A natural question is how to create an exclusive "container", formally referred to as *user space*, for each user. This simple thought constitutes the idea of various MA techniques.

A physical channel possesses two basic physical elements, namely time and frequency. Thus, a possible scheme is to partition the physical channel into frequency or temporal slices and appropriately allocate these slotted subchannels to different users. More generally, we can partition the two-dimensional time–frequency space in different ways into orthogonal or roughly orthogonal subspaces, so that each

Wireless Communications: Principles, Theory and Methodology, First Edition. Keith Q.T. Zhang.
© 2016 John Wiley & Sons, Ltd. Published 2016 by John Wiley & Sons, Ltd.
Companion Website: www.wiley.com/go/zhang7749

user has his own subspace for IUI-free transmission. This is, in fact, the application of the paradigm of orthogonality in MA systems.

We have come across a similar situation in QPSK signaling, where two parallel binary data streams are transmitted, separately, along the in-phase and quadrature carriers without interference with each other. The same principle can be used for transmission of signals from two users. More generally, multiple-user signals are retrievable if they are transmitted with orthogonal or near-orthogonal carriers. Such orthogonal carriers can be sine/cosine waveforms with appropriate frequency separation, translated rectangular pulses, or their combinations. From the geometric point of view, each user should have his or her own subspace. Finding orthogonal waveforms is, in essence, the process of constructing a set of basis functions to span a higher dimensional signal space to accommodate all multiple users. Theoretically, the dimension of user space must exceed the number of users. Otherwise, there must be at least one dimension containing two users' signals, unavoidably causing signal collision. Constructing orthogonal basis functions, no matter how done in the temporal domain, the frequency domain, or their combination, always causes an expansion in the system bandwidth. Thus, spectrum spreading is a *natural consequence* of signal-space expansion, but is *not the nature* of the MA problem.

More generally, the distinction of multiple users' signals can be implemented by endowing each user with an exclusive *algebraic* structure, for which orthogonality is only one choice. However, allocation with orthogonal subspaces makes it easy for signal retrieval by simply using linear correlation receivers.

When operating in a cellular environment, an MA scheme is subject to many requirements other than orthogonality, which, for example, include the following:

- intra-cell interference suppression;
- cross-cell interference suppression;
- flexibility in variable bandwidth allocation among users; and
- capability to cope with frequency-selective fading.

In multiuser environments, the system performance is eventually limited by MAI. Today, the data rate of LTE-A systems reaches the order of 100 Mb/s, producing bandwidth much wider than the channel coherence bandwidth. In other words, most systems operate in a frequency-selective channel. A good MA scheme should consider MAI on one hand, and provide a simple and effective means to tackle frequency-selective channels on the other.

The following notations will be used throughout this chapter.

$\mathbf{c}_k$	spreading vector for user k
$c_k(i)$	the ith chip of $\mathbf{u}_k$
Δf	frequency separation between adjacent subcarriers
K	number of users
k	subscript of user k
N_c	number of chips
N_f	number of sub-carriers
E_s	symbol energy
$\mathbf{e}_K(k)$	K-by-1 vector with all-zero entries except the kth entry of unity that is,
	$\mathbf{e}_K(k) = [\mathbf{0}_{1\times(k-1)}, 1, \mathbf{0}_{1\times(K-k)}]^T$
$\text{rect}(t\|T)$	$= 1$ for $t \in (0, T)$, and $= 0$ otherwise
$p(t\|T)$	unit-magnitude rectangular pulse defined over $0 \le t < T$, which is used interchangeably with $\text{rect}(t\|T)$
$p_c(t)$	the short form of $p(t\|T_c)$.
T_c	chip duration

T_F	frame duration
T_s	symbol duration for a single-user system
T_{sam}	sampling interval
$\otimes$	Kronecher product
$s(t) \star h(t)$	convolution

12.2   Typical MA schemes

Let T_s denote the symbol duration of a single-user system before multiplexing with other users. We will use two notations to denote the same rectangular pulse:

$$\text{rect}(t|T) = \begin{cases} 1, & 0 \le t < T \\ 0, & \text{elsewhere} \end{cases}$$

$$p(t|T) = \text{rect}(t|T) : \quad \text{alternative notation of } \text{rect}(t|T).$$

The two symbols are used interchangeably in what follows. When $T = T_c$ and $T = T_s$, $p(t|T)$ is further abbreviated as $p_c(t)$ and $p_s(t)$, respectively, for brevity. That is

$$p_c(t) = p(t|T_c)$$

$$p_s(t) = p(t|T_s),$$

which are graphically sketched in Figure 12.1.

To accommodate K users, we must extend the dimension of the signal space, at least, to K. The extension can be done in time, frequency, or their combination, resulting in the MA schemes called time-division MA (TDMA), code-division MA (CDMA), multi-carrier CDMA (MC-CDMA), multi-carrier direct-sequence CDMA (MC-DS-CDMA), and so on. Multi-access data streams are organized in frames, and the frame for each scheme has a unique structure.

Let us begin with the TDMA scheme, in which each user transmits m symbols over a disjoint time slot. As such, the frame duration is $T_F = mT_s$, and the actual time duration for each user to transmit a symbol within each frame is

$$t_s = T_s/K. \tag{12.1}$$

For notational simplicity, let $m = 1$. The corresponding TDMA signal can be written as

$$x_{\text{TDMA}}^\infty(t) = \sum_{i=-\infty}^{\infty} \sum_{k=1}^{K} s_k(i)p(t - iT_F - kt_s|t_s), \tag{12.2}$$

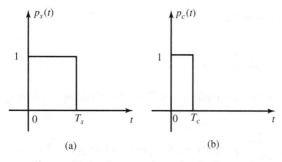

(a)   (b)

Figure 12.1   *Rectangular pulses to be used*

where $T_F = T_s$ and $s_k(i)$ denotes the user k's symbol transmitted over the ith frame. The extension to a general m is straightforward, as shown subsequently in the vector form of TDMA signal. In TDMA, each user is allocated an exclusive time slot over each frame, thereby avoiding IUI.

In CDMA, each user transmits one symbol over a frame and, thus, $T_F = T_s = N_c T_c$. Though named as code-division multiple-access, the CDMA scheme partitions its channel space also in the *time* domain, just as done in TDMA. The only difference is that they choose different coordinate systems. In CDMA, the temporal coordinate axes are defined by the spreading code vectors. A comparison is illustrated in Figure 12.1. The CDMA signal is given by

$$x_{\text{CDMA}}^{\infty}(t) = \sum_{i=-\infty}^{\infty} \sum_{k=1}^{K} s_k(i) \sum_{m=0}^{N_c-1} c_k(m) p(t - iT_s - mT_c|T_c) \cos \omega_0 t, \tag{12.3}$$

where $\mathbf{c}_k = [c_k(m), m = 0, \cdots, N_c - 1]^T$ is the spreading code vector for user k, and each entry is called a *chip*.

In fact, partitioning the channel resource can be done in the *frequency* domain, as is the case of MC-CDMA. In MC-CDMA, the spreading chips are distributed over different carriers, resulting in a signal of the form

$$x_{\text{MC}}^{\infty}(t) = \sum_{i=-\infty}^{\infty} p(t - iT_s|T_s) \sum_{k=1}^{K} \sum_{j=0}^{N_c-1} s_k(i) c_k(j) \cos 2\pi(f_0 + j\Delta f)t \tag{12.4}$$

for which $T_F = T_s$. Note that CDMA employs the temporal chips $p_c(t - jT_c)$, whereas MC-CDMA employs the frequency chips $p_s(t) \cos \omega_j t$. MC-CDMA is a combination of CDMA and OFDM. Its advantages are (1) efficient implementation by using fast Fourier transform (FFT), and (2) robustness to frequency-selective fading. Since its introduction in 1993 [1, 2], MC-CDMA has captured a great deal of attention.

The combination of CDMA and multicarrier techniques leads to the MC-DS-CDMA scheme, as shown by

$$x_{\text{MD}}^{\infty}(t) = \sum_{i=-\infty}^{\infty} \sum_{k=1}^{K} \sum_{\ell=1}^{N_f} s_k(iN_f + \ell) \sum_{j=0}^{N_c-1} c_k(j) p(t - iT_s' - jT_c'|T_c') \cos 2\pi(f_0 + \ell\Delta f')t, \tag{12.5}$$

where T_s' is the symbol duration used in the MC-DS-CMDA. In the MC-DS-CDMA scheme, each user splits its data sequence into N_f parallel subsequences for transmission over N_f carriers. Symbols on each carrier are then spread by the same user's spreading code, in exactly the same way as done by the standard DS-CMDA technique. In doing so, we can achieve frequency diversity of order N_f. By the use of multiple carriers for parallel transmission of data from a single user of symbol rate $1/T_s$, the MC-DS-CDMA can take a lower transmission rate or a long symbol duration equal to $T_s' = N_f T_s$. As such, the resulting chip duration and carrier spacing are given by

$$T_c' = T_s'/N_c = (N_f/N_c)T_s \quad \text{and} \quad \Delta f' = (1/T_c') = N_c/(N_f T_s), \tag{12.6}$$

respectively. The frame duration is $T_F = T_s'$.

12.3   User space of multiple-access

Let us focus on a single symbol, for example, $i = 0$. Then, by denoting $s_k(i)$ as s_k, we can simplify the above expressions to yield

$$x_{\text{TDMA}}(t) = \sum_{k=0}^{K-1} s_k \, \text{rect}(t - kt_s|t_s) \cos \omega_0 t, 0 \le t < T_s, \tag{12.7}$$

$$x_{\text{CDMA}}(t) = \sum_{k=1}^{K} \sum_{j=0}^{N_c-1} s_k c_k(j) \underbrace{\text{rect}(t - jT_c|T_c)}_{p_c(t-jT_c)} \cos \omega_0 t, \ 0 \le l < T_s, \tag{12.8}$$

$$x_{\text{MC}}(t) = \sum_{k=0}^{K-1} \sum_{j=0}^{N_c-1} s_k c_k(j) \underbrace{\text{rect}(t|T_s)}_{p_s(t)} \cos 2\pi(f_0 + j \, \Delta f)t, \ 0 \le t < T_s, \tag{12.9}$$

$$x_{\text{MD}}(t) = \sum_{k=0}^{K-1} \sum_{\ell=1}^{N_f} \sum_{j=0}^{N_c-1} s_k(\ell) c_k(j) \, \text{rect}(t - jT_c'|T_c') \cos 2\pi(f_0 + \ell \, \Delta f')t,$$

$$0 \le t < N_f T_s. \tag{12.10}$$

The physical nature of multiple access is to allow multiple users to share the same physical channel. In this and the subsequent sections, we show how the idea of channel sharing is mathematically implemented in various MA schemes by appropriately choosing a set of orthogonal basis functions.

12.3.1 User spaces for TDMA

To begin with, we assume, for simplicity, that each TDMA user transmits only one symbol during each time frame. In baseband form, we may rewrite $x_{\text{TDMA}}(t)$ to yield

$$x_{\text{TDMA}}(t) = \sum_{k=1}^{K} s_k \sqrt{t_s} \underbrace{p_s(t - kt_s|t_s)/\sqrt{t_s}}_{\phi_k(t)}, \tag{12.11}$$

which, when mapped onto the Euclidean space, leads to an equivalent vector expression:

$$\mathbf{x}_{\text{TDMA}} = \sum_{k=1}^{K} s_k \sqrt{t_s} \mathbf{e}_k(k). \tag{12.12}$$

This implies that the space of TDMA is spanned by $\{e_K(1), \dots, e_K(k)\}$. It is easy to extend the TDMA to a more general case, in which each user, say user k, is allowed, to transmit multiple (say m) symbols, denoted by vector $\mathbf{s}_k = [s_k(1), \dots, s_k(m)]^T$, over each frame. Then in baseband form, the TDMA signal can be written as

$$x_{\text{TDMA}}(t) = \sum_{k=1}^{K} \sum_{i=1}^{m} \sqrt{t_s} s_k(i) \underbrace{\sqrt{1/t_s} \text{rect}(t - ((k-1)m + i)T_s|t_s)}_{\phi_i(t)}, \ 0 \le t < T_F, \tag{12.13}$$

where the frame duration $T_F = mT_s$. Mapping the above expression into the Euclidean space enables us to obtain the generalized TDMA signal in vector form, as

$$\mathbf{x}_{\text{TDMA}} = \sum_{k=1}^{K} \sqrt{t_s} \mathbf{e}_K(k) \otimes \mathbf{s}_k. \tag{12.14}$$

It follows that, in TDMA, the subspace for user k is mK-dimensional, defined by

$$\mathbf{V}_{\text{TDMA}}^{(k)} = \text{span}\{\mathbf{e}_K(k) \otimes \mathbf{e}_m(1), \mathbf{e}_K(k) \otimes \mathbf{e}_m(2), \cdots, \mathbf{e}_K(k) \otimes \mathbf{e}_m(m)\}. \tag{12.15}$$

12.3.2   User space for CDMA

For CDMA signals, $T_s = N_c T_c$. We can write

$$x_{\text{CDMA}}(t) = \sum_{k=1}^{K} s_k \sqrt{T_c} \sum_{i=0}^{N_c-1} \underbrace{c_k(i) p_c(t - iT_c)/\sqrt{T_c}}_{\phi_k(t)}, \tag{12.16}$$

which, when mapped onto the N_c-dimensional Euclidean space, leads to an equivalent vector expression

$$\mathbf{x}_{\text{CDMA}} = \sum_{k=1}^{K} s_k \sqrt{T_c}\, \mathbf{c}_k, \tag{12.17}$$

where $\mathbf{c}_k = [c_k(0), \cdots, c_k(N_c - 1)]^T$ with $c_k(i) = \pm 1$ is the spreading vector for user k. In CDMA, the separation among users relies on the orthogonality of $\mathbf{c}'s$. Denote

$$\mathbf{C} = [\mathbf{c}_1, \cdots, \mathbf{c}_K]/\sqrt{N_c}$$

$$\mathbf{s} = [s_1, \cdots, s_K], \tag{12.18}$$

whereby (12.17) can be re-expressed as

$$\mathbf{x}_{\text{CDMA}} = \sqrt{T_s}\mathbf{Cs}. \tag{12.19}$$

Clearly, each user is allocated a one-dimensional space, as shown by

$$\mathbf{V}_{\text{CDMA}}^{(k)} = \text{span}\{\mathbf{c}_k\}. \tag{12.20}$$

For orthogonal CDMA, we have $\mathbf{CC}^\dagger = \mathbf{C}^\dagger\mathbf{C} = \mathbf{I}_K$, and there is a much simpler way to determine its user space. To this end, we redefine the basis functions

$$\phi_k(t) = \frac{1}{\sqrt{T_s}} \sum_{i=0}^{N_c-1} c_k(i) p_c(t - iT_c) \tag{12.21}$$

and rewrite (12.16) as

$$x_{\text{CDMA}}(t) = \sum_{k=1}^{K} s_k \sqrt{T_s}\, \phi_k(t). \tag{12.22}$$

It is easy to examine

$$\int_{-\infty}^{\infty} \phi_i(t)\phi_j(t)dt = \begin{cases} 1, & i = j \\ 0, & \text{otherwise} \end{cases}. \tag{12.23}$$

Thus, another vector representation of orthogonal CDMA signals is

$$\mathbf{x}_{\text{CDMA}} = \sum_{k=1}^{K} \sqrt{T_s} s_k \mathbf{e}_K(k) = \mathbf{s}\sqrt{T_s}, \tag{12.24}$$

which differs from (12.19) only by a unitary transformation.

12.3.3   User space for MC-CDMA

In DS-CDMA, spreading chips take the form of serial rectangular pulses in the time domain. MC-CDMA implements its spectrum spreading by allocating its chips over multiple carriers, with the signal model shown

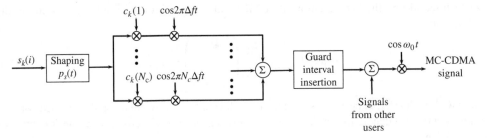

Figure 12.2   *Generation of MC-CDMA signals*

in (12.9). Mapping (12.9) into the K-dimensional Euclidean space yields the vector form of the MC-CDMA signal, as

$$\mathbf{x}_{\mathrm{MC}} = \sum_{k=1}^{K} \sqrt{T_s} s_k \mathbf{e}_K(k) \tag{12.25}$$

which, mathematically, has the same form as (12.24) for DS-CDMA. The subspace for user k is then

$$\mathbf{V}_{\mathrm{MC}}^{(k)} = \mathrm{span}\{\mathbf{e}_K(k)\}. \tag{12.26}$$

However, the two schemes have different physical implications. DS-CDMA achieves the effect of temporal diversity, while MC-CDMA achieves the advantage of frequency diversity.

A more insightful understanding can be gained if we directly discretize the carriers in (12.9) as a Fourier matrix $\mathbf{F}$ and rewrite the MC-CDMA, in vector form, as

$$\mathbf{x}_{\mathrm{MC}} = \sum_{k=1}^{K} s_k \mathbf{F} \mathbf{c}_k = \mathbf{F}\mathbf{C}\mathbf{s}, \tag{12.27}$$

where $\mathbf{s} = [s_1, \cdots, s_K]^T$ and $\mathbf{C} = [\mathbf{c}_1, \cdots, \mathbf{c}_K]$. Since $\mathbf{C}\mathbf{s}$ represents the DS-CDMA signal, the above expression implies that the MC-CDMA signal is the Fourier transform of its CDMA counterpart. The generation of MC-CDMA signals is shown in Figure 12.2.

12.3.4   MC-DS-CDMA

The MC-DS-CDMA signal is given by

$$x_{\mathrm{MD}} = \sum_{k=0}^{K-1} \sum_{\ell=1}^{N_f} \sum_{m=0}^{N_c-1} s_k(\ell) c_k(m) \, \mathrm{rect}(t - mT_c'|T_c') \cos 2\pi(f_0 + \ell\Delta f')t, \ 0 \leq t < N_f T_s. \tag{12.28}$$

Unlike CDMA and MC-CDMA, the MC-DS-CDMA signal experiences spreading in both the frequency and time domains, with basis functions defined by

$$\phi_{\ell m}(t) = \sqrt{\frac{2}{T_c'}} \, p(t - mT_c'|T_c') \cos 2\pi(f_0 + \ell\Delta f')t, \ 0 \leq t < N_f T_s, \tag{12.29}$$

which, when inserted into the last equation of (12.10) and mapped into the Euclidean space, produces an $N_f N_c \times 1$ vector

$$\mathbf{x}_{\mathrm{MD}} = \sqrt{\frac{T_c'}{2}} \sum_{k=1}^{K} \begin{pmatrix} s_k(1) \\ \vdots \\ s_k(N_f) \end{pmatrix} \otimes \mathbf{c}_k. \tag{12.30}$$

Clearly, in MC-DS-CDMA, each user is exclusively allocated an N_f-dimensional subspace for his signal transmission. Let $\mathbf{e}_{N_f}(i)$ denote the $N_f \times 1$ vector whose entries are all zeros except the ith entry, which is equal to 1, and denote

$$\mathbf{v}_{ki} = \sqrt{2/T_c'}\, \mathbf{e}_{N_f}(i) \otimes \mathbf{c}_k. \tag{12.31}$$

Then, the subspace for user k is explicitly expressible as

$$\mathbf{V}_{\mathrm{MD}}^{(k)} = \mathrm{span}\{\mathbf{v}_{k1}, \cdots, \mathbf{v}_{kN_f}\} \tag{12.32}$$

and his signal is given by $\mathbf{x}_k = \sum_{i=1}^{N_f} s_i \mathbf{v}_{ki}$. The orthogonality (or approximate orthogonality) among users is ensured by the users' spreading code vectors (namely the users' signature vectors), while the capability against ISI among symbols of each user is guaranteed by the orthogonality of the subcarriers.

12.3.5   User space for OFDMA

OFDMA follows the same principle as OFDM, with the difference that the latter is used for simultaneous transmission of parallel data streams, while the former is used for multiple user transmission. In OFDMA, each user is uniquely allocated one or several subcarriers for transmission, and the subcarrier channel allocation may change over time slots to match the channel conditions and the users' traffic load. Multiple-carrier allocation in OFDMA is virtually implemented via FFT and, thus, the OFDMA is physically a single-carrier system. It is easier to elucidate by considering an example. Consider an OFDMA system with six subcarriers denoted by the column vectors $\mathbf{w}_i, i = 1, 2, 3, 4, 5, 6$ to support four users. Denote the baseband symbol blocks from the *four users* by the row vectors $\mathbf{a}$, $\mathbf{b}$, $\mathbf{c}$, and $\mathbf{d}$, respectively. The OFDMA provides a lot of flexibility for subcarrier channels allocation. A TDMA-type of allocation over a frame of 8 time slots can be written as

$$\mathbf{X} = \underbrace{[\mathbf{w}_1, \mathbf{w}_2, \mathbf{w}_3, \mathbf{w}_4, \mathbf{w}_5, \mathbf{w}_6]}_{\text{Sub-carriers } 1 \to 6} \underbrace{\begin{bmatrix} a & b & c & d & a & b & c & d \\ a & b & c & d & a & b & c & d \\ a & b & c & d & a & b & c & d \\ a & b & c & d & a & b & c & d \\ a & b & c & d & a & b & c & d \\ a & b & c & d & a & b & c & d \end{bmatrix}}_{\text{Time slots } 1 \to 8}. \tag{12.33}$$

The matrix on the right stands for the users' symbol matrix and is denoted by $\mathbf{S}$ hereafter. Its first row is transmitted by carrier $\mathbf{w}_1$, and the second row by $\mathbf{w}_2$, and so on. The arrangement of matrix $\mathbf{S}$ provides flexibility for scheduling the time–frequency resources among different users, according to their need and their channel conditions. In the meantime, the traditional idea of TDMA, FDMA, and frequency hopping can be easily incorporated into resource allocation. The users' symbol matrix shown above is an example of OFDM–TDMA combination. All the subcarriers in the first time slot are dedicated to user 1, the second time slot is allocated to user 2, and so on. Note that here and hereafter, the same notation **a** appearing in different time slots or on different locations of the same slot is used only to indicate data blocks from user 1. In general, **a**'s at different locations and different time slots have *different* contents except for a special design to achieve diversity gains. The same remarks apply to data blocks for other users.

The OFDM–FDMA allocation represents another extreme, as exemplified by the allocation

$$
\mathbf{S} = \begin{bmatrix}
b & b & b & b & b & b & b & b \\
b & b & b & b & b & b & b & b \\
d & d & d & d & d & d & d & d \\
a & a & a & a & a & a & a & a \\
c & c & c & c & c & c & c & c \\
c & c & c & c & c & c & c & c
\end{bmatrix}.
\qquad (12.34)
$$

$$
\underbrace{\qquad\qquad\qquad\qquad}_{\text{Time slots } 1\to 8}
$$

In this example, users 1, 2, 3, and 4 use fixed frequency allocation. They exclusively employ subcarrier channels $\mathbf{w}_4$, $[\mathbf{w}_1, \mathbf{w}_2]$, $[\mathbf{w}_5, \mathbf{w}_6]$, and $\mathbf{w}_3$, respectively, for their transmission over the eight time slots.

In many MA environments, different users may experience different fading channel conditions. The fading channel gains for different users can vary from time to time and from subcarrier to subcarrier. The optimal allocation is to schedule each user with its best channels at each time slot, so that the system can exploit the advantage of user diversity in multipath fading. An example is shown below.

$$
\mathbf{S} = \begin{bmatrix}
a & b & c & b & b & d & d & d \\
d & a & b & c & c & c & a & a \\
d & d & a & a & a & b & b & d \\
c & c & d & d & a & a & c & c \\
a & a & c & c & d & c & b & b \\
c & d & d & a & a & b & b & b
\end{bmatrix}.
\qquad (12.35)
$$

$$
\underbrace{\qquad\qquad\qquad\qquad}_{\text{Time slots } 1\to 8}
$$

The user data matrix given in (12.35) is used to illustrate a dynamic carrier-channel allocation among the four users. In this example, the subcarrier channels allocated to user 1 over the eight time slots are found to be $\{\mathbf{w}_1, \mathbf{w}_5\}$, $\{\mathbf{w}_2, \mathbf{w}_5\}$, $\mathbf{w}_3$, $\{\mathbf{w}_3, \mathbf{w}_6\}$, $\{\mathbf{w}_3, \mathbf{w}_4, \mathbf{w}_6\}$, $\mathbf{w}_4$, $\mathbf{w}_2$, and $\mathbf{w}_2$, respectively. In this setting, the subspace for each user has a different dimension from slot to slot.

In general, an OFDMA system employs N subcarriers, $\mathbf{w}_1, \cdots, \mathbf{w}_N$, to support $K < N$ users. Note that here $\mathbf{F} = [\mathbf{w}_1, \cdots, \mathbf{w}_N]$ constitutes the $N \times N$ Fourier matrix. If we pick up one column from the data matrix, denoted as $\tilde{\mathbf{s}}$, then the corresponding OFDMA signal can be written as

$$
\mathbf{x}_{\text{OFDMA}} = \mathbf{F}\tilde{\mathbf{s}},
\qquad (12.36)
$$

where $\tilde{\mathbf{s}}$ is an $N \times 1$ vector with its entries from different users, in the order determined by the subcarrier allocation pattern.

12.3.6   Unified framework for orthogonal multiaccess schemes

In the previous subsections, we have shown how various MA schemes are defined by their unique set of basis functions. Different sets of basis functions define different user subspaces for different MA schemes. It would, therefore, be more insightful to compare these user subspaces in the same framework.

By inspecting (12.7)–(12.10), we can find the basis functions for various MA schemes as follows:

$$\phi_k(t) = \begin{cases} p(t - kt_s|t_s)/\sqrt{t_s}, & \text{TDMA} \\[2ex] \frac{1}{\sqrt{N_c T_c}} \sum_{j=1}^{N_c} c_k(j) p_c(t - jT_c), & \text{CDMA} \\[2ex] \sqrt{\frac{2}{T_s}}\, p_s(t) \sum_{j=0}^{N_c-1} c_k(j) \cos \omega_j t, & \text{MC-CDMA} \\[2ex] \text{sinc}[2Bt] \cos 2\pi(f_0 + k\Delta f)t, & \text{FDMA} \end{cases} \tag{12.37}$$

for $k = 1, \cdots, K$ and $0 \le t < T_F$. Here, for completeness, also included in the list are the basis functions for FDMA, where we have used the fact that $\text{rect}(f/2B) \to 2B\text{sinc}(2Bt)$. The basis functions for MC-DS-CDMA are defined over the time and frequency domains and, thus, have two subscript indexes, given by

$$\phi_{k\ell}(t) = \sqrt{\frac{2}{T_s'}}\, p(t|T_s') \sum_{j=0}^{N_c-1} c_k(j) p(t - jT_c'|T_c') \cos 2\pi(f_0 + \ell\Delta f')t,$$

$$\ell = 1, \cdots, N_f, \qquad \text{MC-DS-CDMA}. \tag{12.38}$$

For *orthogonal* MA schemes, it is easy to examine the orthogonality of the above basis function sets. In particular, for TDMA, FDMA, CDMA, and MC-CDMA,

$$\int_{-\infty}^{\infty} \phi_i(t)\phi_j(t)dt = \begin{cases} 1, & i = j, \\ 0, & i \ne j; \end{cases} \tag{12.39}$$

and for MC-DS-CDMA,

$$\int_{-\infty}^{\infty} \phi_{i\ell}(t)\phi_{jq}(t)dt = \begin{cases} 1, & i = j \text{ and } \ell = q \\ 0, & \text{otherwise} \end{cases}. \tag{12.40}$$

With these basis functions, we can directly obtain vector representation(12.14) for TDMA, (12.24) for CDMA, and (12.25) for MC-CDMA.

From (12.37), it is observed that both the TDMA ad CDMA schemes, indeed, employ a set of *temporal* basis functions, albeit defined differently. This situation is shown in Figure 12.3, where we intentionally assume that each user of TDMA occupies a time slot of T_c and transmits only one symbol in each frame. This special setting is made only for an intuitive comparison with the CDMA scheme. The basis functions for OFDMA are also included in the figure.

For ease of comparison, the geometric and bandwidth features of various commonly used multiple access schemes are summarized in Table 12.1.

12.4   Capacity of multiple-access channels

Without loss of generality, we focus on uplink MA channels, which, in turn, can be classified into two cases of flat fading and frequency-selective fading.

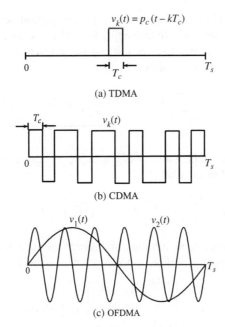

$$v_k(t) = p_c(t - kT_c)$$

(a) TDMA

$v_k(t)$

(b) CDMA

$v_1(t)$ $v_2(t)$

(c) OFDMA

Figure 12.3 *Basis functions for different access schemes where the slotted channel for TDMA users is intentionally assumed of duration T_c for intuitive comparison, and $v_k(t)$ has the same waveform as $\phi_k(t)$ except for a scale factor*

Table 12.1 *Characteristic comparison of various multiple-access schemes*

Scheme	Basis functions $(k = 1, \cdots, K)$	Signal representation in Euclidean space	Original BW	Spreading BW
TDMA	$p_c(t - kT_c)p_s(t)/\sqrt{T_c}$	$\sqrt{T_c}\, s_k \mathbf{e}_k$	$1/T_s$	$1/T_c$
CDMA	$p_c(t - kT_c)p_s(t)/\sqrt{T_c}$	$\sqrt{T_s}\, s_k \mathbf{c}_k$	$1/T_s$	$1/T_c$
OFDMA	$\sqrt{2/T_s}\, p_s(t) \cos \omega_k t$	$\sqrt{T_s/2}\, s_k \mathbf{e}_k$	$1/T_s$	K/T_s
MC-CDMA	$\sqrt{2/T_s}\, p_s(t) \cos \omega_k t$	$\sqrt{T_s/2}\, s_k \mathbf{c}_k$	$1/T_s$	K/T_s

12.4.1 Flat fading

Consider an uplink where K users transmit their zero-mean independent Gaussian random signals $s_k(t), k = 1, \cdots, K$ through a flat-fading channel with gain g_k to the base station, so that the received signal can be written as

$$r(t) = \sum_{k=1}^{K} g_k s_k(t) + n(t), \tag{12.41}$$

where $n(t)$ is a zero-mean AWGN process of two-sided power spectral density $N_0/2$. Without SCIT, each user employs equal power for transmission. In the setting for capacity evaluation, we assume that each signal $s_k(t)$ is of bandwidth B Hz and variance σ_k^2. Thus, it can be represented in terms of the sinc function and Nyquist sampling interval $\mathcal{T}_s = 1/(2B)$, as shown by

$$s_k(t) = \sum_{n=-\infty}^{\infty} s_k(n)\sqrt{\mathcal{T}_s}\underbrace{\frac{1}{\sqrt{\mathcal{T}_s}}\mathrm{sinc}2B(t - n\mathcal{T}_s)}_{\phi_n(t)}, \tag{12.42}$$

where $s_k(n)$ is the value of $s_k(t)$ sampled at $t = n\mathcal{T}_s$, and $\{\phi_n(t)\}$ are a complete set of orthonormal basis functions over $-\infty < t < \infty$. It follows that

$$r(t) = \sum_{n=-\infty}^{\infty} \left(\sum_{k=1}^{K} g_k s_k(n)\sqrt{\mathcal{T}_s}\right)\phi_n(t) + n(t), \tag{12.43}$$

which can be isometrically mapped into an infinite-dimensional vector in the Euclidean space. For the evaluation of spectral efficiency, however, it suffices to focus a finite vector, for example, of N samples

$$\mathbf{r} = \begin{bmatrix} \sum_{k=1}^{K} g_k s_k(1) \\ \vdots \\ \sum_{k=1}^{K} g_k s_k(N) \end{bmatrix}\sqrt{\mathcal{T}_s} + \mathbf{n}, \tag{12.44}$$

where $\mathbf{n} \sim \mathcal{N}(\mathbf{0}, N_0\mathbf{I}/2)$. Note that the random samples of $r(t)$ spaced by $\mathcal{T}_s$ are mutually independent. Since there are $2B$ samples over each second, the spectral efficiency of the MA uplink channel can be determined as

$$C = \frac{1}{B}\log\left(1 + \frac{\sum_{k=1}^{K} |g_k|^2\sigma_k^2\mathcal{T}_s}{N_0/2}\right)^{2B/2} = \log\left(1 + \sum_{k=1}^{K} \rho_k|g_k|^2\right) \text{ bits/s/Hz}, \tag{12.45}$$

where $\rho_k = \sigma_k^2/(BN_0)$ is the transmit SNR of user k.

The capacity obtained here is consistent with the result of Cover and Thomas [3]:

$$C = \frac{1}{2}\log\left(1 + \sum_{k=1}^{K} \rho_k|g_k|^2\right) \text{ bits/s/Hz}, \tag{12.46}$$

which is directly derived from the discrete-time model $y = \sum_{k=1}^{K} g_k s_k + n$. The difference in the factor of 2 comes from the Nyquist sampling rate to convert the continuous signal to a discrete one.

12.4.2   Frequency-selective fading

With a very high transmission rate, a large number of multiple paths become resolvable, and the transmitted signal will suffer frequency-selective fading.

Consider a frequency-selective MA uplink in which K mobile users transmit their Gaussian random signals $s_k(t), k = 1, \cdots, K$, respectively, through a corresponding channel $h_k(t)$ to the base station, so that received signal at the base station can be written as

$$r(t) = \underbrace{\sum_{k=1}^{K} s_k(t) \star h_k(t)}_{x(t)} + n(t),$$

where $*$ denotes convolution and $n(t)$ is the AWGN process with two-sided spectral density $N_0/2$ W/Hz. It can be shown that, without CSIT, the channel capacity is given by

$$C = \int_{-\infty}^{\infty} \log_2 \left(1 + \frac{1}{N_0} \sum_{k=1}^{K} S_k(f)|H_k(f)|^2 \right) df, \qquad (12.47)$$

where $S_k(f)$ denotes the power spectral density of $s_k(t)$, and $H_k(f)$ denotes the Fourier transform of $h_k(t)$. The proof of this result is left in Problem (12.17) as an exercise to the reader.

DS-CDMA and OFDM–CDMA tackle frequency-selective fading by following different philosophies. The former transmit short pulses so that paths are resolvable allowing the use of path diversity combining to enhance error performance, whereas the latter implements frequency diversity by sending users' signals over orthogonal subcarriers.

12.5   Achievable MI by various MA schemes

Having investigated the capacity of MA channels, we are interested in mutual information (MI) achievable by various practical MA schemes that are rooted in different design philosophies. Random signals are assumed in the capacity analysis of MA channels, so that each user can evenly distribute its signal energy over *all* the orthogonal basis functions for channel rate maximization. For practical MA schemes, however, we need to insert certain algebraic structures into the signature waveforms for user separation at the receiver. A commonly used structure is orthogonality, accurate or approximate, among users, as described in the preceding sections. The constraint of mutual orthogonality forces the signal energy of each user to be conveyed only through a limited number (typically one) of basis functions. It is such a reduction in the degrees of freedom that decreases the system mutual information. Certainly, random signals are sometimes also used in the analysis of practical MA schemes such as TDMA and CDMA, so that the Karhunen–Loeve theorem can be employed for orthogonal expansion of the received signal [4].

In our analysis, we adopt deterministic waveforms to reflect the reality of MA operation. For exposition, let us concentrate on synchronous uplink. We investigate the mutual information of various MA schemes in two cases: AWGN and flat-fading channels.

12.5.1   AWGN channel

Let us consider the MA signal over the AWGN channel, for which the received signal is given by

$$\mathbf{y} = \begin{cases} \sum_{k=1}^{K} \sqrt{t_s}\, \mathbf{e}_K(k) \otimes \mathbf{s}_k + \mathbf{n}, & \text{TDMA} \\ \sqrt{T_c}\, \mathbf{Cs} + \mathbf{n}, & \text{CDMA} \\ \sqrt{T_c}\, \mathbf{FCs} + \mathbf{n}, & \text{MC-CDMA} \\ \sqrt{T_s}\, \mathbf{F\tilde{s}} + \mathbf{n}, & \text{OFDMA} \\ \sqrt{T_c'}\, \mathbf{Av} + \mathbf{n}, & \text{MC-DS-CDMA} \end{cases} \qquad (12.48)$$

where, again, $\mathbf{s} = [s_1, \cdots, s_K]^T$ with s_k denoting a symbol from user k, $\mathbf{C} = [\mathbf{c}_1, \cdots, \mathbf{c}_K]$ with $\mathbf{c}_k$ denoting the spreading code vector for user k, and $\tilde{\mathbf{s}}$ was defined in (12.36). For MC-DS-CDMA, matrix $\mathbf{A}$ is the $mN_c \times mN_c$ block diagonal matrix formed by m repetitions of $\mathbf{C}$, and $\mathbf{v}$ is an augmented symbol vector defined by

$$A = \begin{pmatrix} C & & & \\ & C & & \\ & & \ddots & \\ & & & C \end{pmatrix}, \quad v = \begin{pmatrix} \tilde{s}_1 \\ \tilde{s}_2 \\ \vdots \\ \tilde{s}_m \end{pmatrix}. \tag{12.49}$$

In deriving the mutual information for various MA schemes, we assume independent complex Gaussian signals $s_k \sim \mathcal{CN}(0, \sigma_k^2 I_m)$, and note that $n \sim \mathcal{CN}(0, N_0 I)$. We calculate the entropy of y by using $H(y) = \log_2 \det[\text{cov}(y)]$ and the entropy of y the given signals by using $H(y|\text{signals}) = \log_2 \det[\text{cov}(y|\text{signals})]$. Their difference is the mutual information, that is, $I = H(y) - H(y|\text{signals})$. The derivation is very straightforward. The frame duration and the system bandwidth of various MA schemes are needed in the determination of their mutual information, and they are listed below.

$$(T_F, B) = \begin{cases} (mT_s, K/T_s), & \text{TDMA} \\ (T_s, N_c/T_s), & \text{CMDA} \\ (T_c, N_c/T_c), & \text{MC-CDMA} \\ (T_s, N/T_s), & \text{OFDMA} \\ (T_s', N_f N_c/T_s'). & \text{MC-DS-CDMA} \end{cases} \tag{12.50}$$

These, alongside the signal models given in (12.48), enable us to determine the MI (in bits/s/Hz) for various MA schemes:

$$I = \begin{cases} \frac{1}{mK} \log_2 \prod_{k=1}^{K} \det\left(I_m + \frac{\sigma_k^2 t_s}{N_0} I_m\right), & \text{TDMA} \\[2ex] \frac{1}{N_c} \log_2 \det\left(I_K + \frac{T_c}{N_0} C\, \mathbb{E}[ss^\dagger] C^\dagger\right) & \text{CDMA} \\[2ex] \frac{1}{N_c} \log_2 \det\left(I_{N_c} + \frac{T_c}{N_0} FC\, \mathbb{E}[ss^\dagger] C^\dagger F^\dagger\right), & \text{MC-CDMA} \\[2ex] \frac{1}{N} \log_2 \det\left(I_{N_c} + \frac{T_s}{N_0} F\, \mathbb{E}[\tilde{s}\tilde{s}^\dagger] F^\dagger\right), & \text{OFDMA} \\[2ex] \frac{1}{N_f N_c} \log_2 \det\left(I_{N_f N_c} + \frac{T_c'}{N_0} A\, \mathbb{E}[vv^\dagger] A^\dagger\right), & \text{MC-DS-CDMA} \end{cases} \tag{12.51}$$

Assuming all the users transmit independent zero-mean complex Gaussian signals each of variance σ_s^2, and noting that $F^\dagger F = I_{N_c}$, the above expressions can be simplified to

$$I = \begin{cases} \log_2(1 + \rho_s), & \text{TDMA} \\ \frac{1}{N_c} \log_2 \det\left(I_K + \rho_s N_c^{-1} CC^\dagger\right), & \text{CDMA} \\ \frac{1}{N_c} \log_2 \det(I_{N_c} + \rho_s N_c^{-1} CC^\dagger), & \text{MC-CDMA} \\ \log_2(1 + \rho_s), & \text{OFDMA} \\ \frac{1}{N_c} \log_2 \det(I_{N_c} + \rho_s N_c^{-1} CC^\dagger), & \text{MC-DS-CDMA} \end{cases} \tag{12.52}$$

where the average SNRs are defined as

$$\rho_s = \begin{cases} \sigma_s^2 t_s/N_0, & \text{TDMA} \\ \sigma_s^2 (N_c T_c)/N_0, & \text{CDMA} \\ \sigma_s^2 (N_c T_c)/N_0, & \text{MC-CDMA} \\ \sigma_s^2 T_s/N_0, & \text{OFDMA} \\ \sigma_s^2 (N_c T_c')/N_0. & \text{MC-DS-CDMA} \end{cases} \tag{12.53}$$

Two properties can be drawn from (12.52).

Property 12.5.1 *For CDMA, MC-CDMA, and MC-DS-CDMA systems operating in AWGN channels with i.i.d. Gaussian symbols, their mutual information is maximized if orthonormal spreading code vectors are used. That is*

$$I = \frac{1}{N_c}\log_2 \det(\mathbf{I}_{N_c} + \rho_s N_c^{-1}\mathbf{CC}^\dagger) \leq \log_2(1 + \rho_s), \tag{12.54}$$

for which, the maximum is attained when $N_c^{-1}\mathbf{CC}^\dagger = \mathbf{I}$.

The assertion can be easily proved by invoking Hadamard's inequality, which states [3] that for a symmetric nonnegative definite matrix $\mathbf{K}$, its determinant is less than the product of its diagonal elements. This property, when stated in another way, reveals that correlation among spreading code vectors reduces system mutual information.

The column (or row) vectors of a Walsh–Hadamard matrix are commonly used as orthogonal spreading code vectors for the forward link of CDMA systems, such as IS-95. The m-sequences are pseudo-noise (PN) sequences. Though not exactly orthogonal, they are widely used as alternative candidates for the same purpose.

✍ **Example 12.1** ───

Suppose a CDMA system supports 15 users in AWGN by assigning each user a 15-bit m-sequence. Let $\mathbf{C}$ denote the resulting 15×15 user-code matrix. Then, we have

$$\mathbf{R}_c = (1/15)\mathbf{C}^\dagger\mathbf{C} = \text{toeplitz}(\mathbf{r})$$

where $\mathbf{r} = [1, -1/15, \cdots, -1/15]$ is the first row of $\mathbf{R}_c$. Assume the SNR $\rho_s = 10\,\text{dB}$. Determine the spectral efficiency loss caused by the correlated spreading codes.

Solution

With this correlation, the mutual information is given by

$$I_{\text{CDMA}} = \frac{1}{15}\log_2 \det(\mathbf{I}_{15} + 10\mathbf{R}_c) = 3.3572 \text{ b/s/Hz}$$

as compared to the ideal orthogonal case for which $I_{\text{CDMA}} = \log_2(1 + 10) = 3.4594$ b/s/Hz. Thus, there is a small spectral efficiency loss of 0.1022 b/s/Hz. ○

〰〰✍

Property 12.5.2 *Suppose that orthogonal MA schemes are used to share an AWGN channel by transmitting independent complex Gaussian symbols with mean zero but user-dependent variance. We assert that various MA schemes have an identical spectral efficiency, given by*

$$I = \frac{1}{K}\sum_{k=1}^{K}\log_2(1 + \rho_k),$$

where ρ_k denotes the average SNR of user k, defined in a way similar to (12.53) except that σ_s^2 is replaced by the power σ_k^2 of user k.

This property is easy to understand if we take a close look at (12.52), where one MA scheme can be obtained from another by a unitary transform. Thus, we may say that for AWGN channels, the mutual information of an orthogonal MA scheme is invariant under unitary transformation.

12.5.2 Flat-fading MA channels

To investigate the MI on flat-fading channels, we can assume, without loss of generality, that all the users transmit i.i.d. complex Gaussian symbols with mean zero and variance σ_s^2, since unequal user power can be absorbed into the corresponding channel fading gains. As such, we can write the received signals for various MA channels as

$$
\mathbf{y} = \begin{cases}
\sum_{k=1}^{K} \sqrt{t_s}\, g_k \mathbf{e}_K(k) \otimes \mathbf{s}_k + \mathbf{n}, & \text{TDMA} \\[2mm]
\sqrt{T_c}\, \mathbf{CGs} + \mathbf{n}, & \text{CDMA} \\[2mm]
\sqrt{T_c}\, \mathbf{F} \sum_{k=1}^{K} \mathbf{H}_k \mathbf{c}_k s_k + \mathbf{n}, & \text{MC-CDMA} \\[2mm]
\sqrt{T_s}\, \mathbf{F}\tilde{\mathbf{H}}\tilde{\mathbf{s}} + \mathbf{n}, & \text{OFDMA} \\[2mm]
\sqrt{T_c'}\, \mathbf{Av} + \mathbf{n}, & \text{MC-DS-CDMA}
\end{cases}
\tag{12.55}
$$

where, again, $\mathbf{s} = [s_1, \cdots, s_K]^T$ with s_k denoting a symbol from user k, $\mathbf{C} = [\mathbf{c}_1, \cdots, \mathbf{c}_K]$ with $\mathbf{c}_k$ denoting the spreading code vector for user k, and $\tilde{\mathbf{s}}$ was defined in (12.36). For MC-DS-CDMA, matrix $\mathbf{A}$ is the $mN_c \times mN_c$ block diagonal matrix formed by m repetitions of $\mathbf{C}$, and $\mathbf{v}$ is an augmented symbol vector defined by

$$
\mathbf{A} = \begin{pmatrix}
\mathbf{CG}_1 & & & \\
& \mathbf{CG}_2 & & \\
& & \ddots & \\
& & & \mathbf{CG}_m
\end{pmatrix}, \quad
\mathbf{v} = \begin{pmatrix}
\tilde{\mathbf{s}}_1 \\
\tilde{\mathbf{s}}_2 \\
\vdots \\
\tilde{\mathbf{s}}_m
\end{pmatrix},
\tag{12.56}
$$

where $\tilde{\mathbf{s}}_i$ signifies the symbol vector sent through the ith subcarrier whose kth entry comes from user k. The channel matrices are defined as

$$
\begin{aligned}
\mathbf{G} &= \mathrm{diag}(g_1, \cdots, g_K), \\
\mathbf{G}_i &= \mathrm{diag}(g_{i1}, \cdots, g_{iK}), \\
\mathbf{H}_k &= \mathrm{diag}(h_{k1}, \cdots, h_{kN_c}),
\end{aligned}
\tag{12.57}
$$

where g_k denotes the flat-fading channel gain for user k in a TDMA or CDMA channel, h_{ki} denotes the flat-fading channel gain of user k on the ith subcarrier channel in MC-CDMA, and g_{ik} denotes the flat-fading channel gain of user k on subcarrier i. The diagonal matrix $\tilde{\mathbf{H}}$ is defined similar to $\mathbf{H}_k$ except that its diagonal entries stand for the channel gains for different users, compatible with the subcarrier allocation specified in $\tilde{\mathbf{s}}$. It should be pointed out that, for ease of differentiation, we use two types of symbols, $\mathbf{G}$ and $\mathbf{H}$, to signify channel gains for TDMA and CDMA channels, and for a user (or users) to operate on different subcarriers, respectively. Their physical significance and implications must be understood in their corresponding system context.

It is observed from (12.55) that the channel gain matrix in MC-CDMA destroys the orthogonality among the spreading code vectors. If we use a zero-forcing filter to remove the influence, the consequence is the amplification of the white noise component. The best we can do is to employ an optimal filter to match the wanted signal vector while placing nulls against MAI. However, the presence of residual MAI worsens the system error performance. It is evident that, though equivalent in AWGN, various orthogonal MA schemes suffer differently in fading channels.

Just as for AWGN channels, the mutual information for various MA schemes are obtained as follows:

$$I = \begin{cases} \frac{1}{K} \sum_{k=1}^{K} \log_2(1 + \rho_s |g_k|^2), & \text{TDMA} \\ \frac{1}{N_c} \log_2 \det \left(\mathbf{I} + \rho_s N_c^{-1} \mathbf{CGG}^\dagger \mathbf{C}^\dagger\right), & \text{CDMA} \\ \frac{1}{N_c} \det \left(\mathbf{I}_{N_c} + \rho_s N_c^{-1} \sum_{k=1}^{K} \mathbf{H}_k \mathbf{c}_k \mathbf{c}_k^\dagger \mathbf{H}_k^\dagger\right), & \text{MC-CDMA} \\ \frac{1}{N} \det(\mathbf{I}_{N_c} + \rho_s \tilde{\mathbf{H}} \tilde{\mathbf{H}}^\dagger), & \text{OFDMA} \\ \frac{1}{N_f N_c} \det(\mathbf{I}_{N_f N_c} + \rho_s N_c^{-1} \mathbf{AA}^\dagger), & \text{MC-DS-CDMA.} \end{cases} \quad (12.58)$$

Here, we have used the property of the Fourier matrix, namely $\mathbf{F}^\dagger \mathbf{F} = \mathbf{I}_{N_c}$, to simplify the results for MC-CDMA and OFDMA. Consider orthogonal spreading code vectors for which $N_c^{-1} \mathbf{CC}^\dagger = \mathbf{I}_{N_c}$ and $N_c^{-1} \mathbf{C}_i \mathbf{C}_i^\dagger = \mathbf{I}_{N_c}$, $i = 1, \cdots, N_f$. Then, the above expressions can be simplified to give

$$I = \begin{cases} \frac{1}{K} \sum_{k=1}^{K} \log_2(1 + \rho_s |g_k|^2), & \text{TDMA} \\ \frac{1}{N_c} \sum_{k=1}^{K} \log_2(1 + \rho_s |g_k|^2), & \text{CDMA} \\ \frac{1}{N} \sum_{k=1}^{K} \sum_{i=1}^{N_k} \log_2(1 + \rho_s |h_{ki}|^2), & \text{OFDMA} \\ \frac{1}{N_f N_c} \sum_{i=1}^{N_f} \sum_{j=1}^{N_c} \log_2(1 + \rho_s |g_{ij}|^2), & \text{MC-DS-CDMA} \end{cases} \quad (12.59)$$

where N_k in OFDMA denotes the number of subcarriers allocated to user k, such that $\sum_{k=1}^{K} N_k = N$. In OFDMA, user-data symbols are directly allocated to frequency subchannels and therefore there is no frequency diversity. On the contrary, the MC-CDMA offers frequency diversity.

Using the inequality for geometric and arithmetic means to (12.59) leads to

$$I \le \begin{cases} \log_2\left(1 + \frac{1}{K} \sum_{k=1}^{K} \rho_s |g_k|^2\right), & \text{TDMA} \\ \log_2\left(1 + \frac{1}{K} \sum_{k=1}^{K} \rho_s |g_k|^2\right), & \text{CDMA} \\ \log_2\left(1 + \frac{1}{N_c} \sum_{k=1}^{K} \rho_s^2 |h_i|^2\right), & \text{MC-CDMA} \\ \log_2\left(1 + \frac{1}{N} \sum_{k=1}^{K} \rho_s |h_i|^2\right), & \text{OFDMA} \\ \log_2\left(1 + \frac{1}{N_f N_c} \sum_{i=1}^{N_f} \sum_{j=1}^{N_c} \rho_s |g_{ij}|^2\right), & \text{MC-DS-CDMA.} \end{cases} \quad (12.60)$$

These upper bounds are much smaller than the channel capacity given in (12.45). The reasons behind this big gap are twofold. First, the derivation of the MA channel capacity is based on random signals, while practical MA schemes employ deterministic waveforms subject to the constraint of exact or approximate orthogonality for user separation. The cost for the orthogonality among users is the reduction in the degrees of freedom (DFs), thus lowering the system mutual information. As comparison, a random signal of bandwidth B creates $2B$ independent random samples per second, or 2 DF/s/Hz. The CDMA system has N_c orthogonal waveforms subject to $N_c - 1$ constraints of mutual orthogonality, leaving only one degree of freedom. To obtain this degree of freedom, the system spends bandwidth $B = 1/T_c$ and time span of $T_s = N_c T_c$, resulting in the normalized DF of $1/N_c$ DF/s/Hz. Second, with a random transmitted signal, each user fully exploits all the dimensions of the entire time–frequency space that defines the channel. The consequence is the superposition of the SNRs of all the users. With orthogonal user waveforms in practical MA schemes, on the contrary, each user is confined to one or a few dimensions of the time–frequency space, and superposition is applied only to the mutual information. Certainly, the penalty for using random signals is that users are linearly unresolvable.

From the discussion in the previous section, various orthogonal MA schemes are mathematically equivalent when operating on an AWGN channel, in the sense that they lead to the same mutual information. However, they exhibit different physical features in multipath fading, as clearly indicated in (12.59). The MC-CDMA system spreads each symbol of its users over multiple subcarriers for transmission, thus inherently possessing the advantage of frequency diversity. The MC-DS-CDMA system allocates its symbols to different subcarriers to avoid frequency fading in a particular subchannel. The OFDMA system allows for a flexible subcarrier channel allocation among users to obtain user diversity gains. The CDMA, on the other hand, is a wideband system, making it multiple paths resolvable and thus easy to implement path diversity over scattering channels. OFDMA has been adopted by both the WiMAX and LTE-4G standards. The former standard well exploits its frequency diversity, while the latter nicely exploits its user diversity.

For illustrating the practical application of MA schemes, we consider IS-95A CDMA and LTE systems in the next five sections.

12.6   CDMA-IS-95

The IS-95A CMDA system consists of the forward link (the downlink) from the base station to a user and reverse link (the uplink) from users to the base station, with each operating on a different frequency band. The forward and reverse links support a number of functional channels, called *logical* channels, as listed below.

$$
\text{Forward link} \begin{cases} \text{Pilot channel} \\ \text{Sync channel} \\ \text{Paging channel} \\ \text{Traffic channels} \end{cases}
$$

and

$$
\text{Revese link} \begin{cases} \text{Access channel} \\ \text{Traffic channels} \end{cases}
$$

The users' traffic data in IS-95A is organized in frames, each of 20 ms. Traffic channels are used to carry user information (speech or data) between the base and the mobile station. The access channel in the reverse link is used to send control information for initiating a call or responding to page when a mobile unit places a call. In CDMA cellular, each user has a *unique* long PN sequence, and each cell is allocated a *unique* pair of short PN sequences. These PN sequences play an important role in the operation of CDMA systems, as detailed in the subsequent sections.

12.6.1   Forward link

The forward link is a broadcast channel and operates in a synchronous CDMA mode and, thus, requires a pilot channel for channel estimation and a sync channel for synchronization so as to maintain strict orthogonality among all the users in the same cell.

At the base station, a variable-rate speech coder is used, based on the Qualcomm's QCELP (Qualcomm code-excited LPC) vocoder method, to produce output data rates at 9.6, 4.8, 2.4, or 1.2 kb/s. The basic rate is 9.6 kb/s, since any lower rates can easily meet this standard rate simply by repetition.

Let us take a close look and see how the data frames from a particular user are formatted in the base station for downlink transmission. Each 20 ms-data frame contains 172 message bits. It is cyclic redundancy check (CRC)-coded by the voice coder by appending 12 bits to the information block. The use of CRC coding helps the receiver checking the frame quality of its received signal. The CRC-coded data is further encoded for forward error control by using a rate-$\frac{1}{2}$ constraint length-9 convolutional code with generator polynomials

$$g_1(x) = 1 + x + x^2 + x^3 + x^5 + x^7 + x^8$$
$$g_2(x) = 1 + x^2 + x^3 + x^4 + x^8,$$

which, when written in octal, are given by $(753, 561)$ or equivalently in the reverse order by $(657, 435)$. The resulting code possesses a free distance of 12, having 11 paths of Hamming weight 12, fifty paths of Hamming weight 14, and none of Hamming weight 13. To facilitate the decoding operation at the mobile, the encoder state is forced to zero at the end of each block by appending eight zeros to the end of each block, called tailed bits. Thus, the data rate after convolutional coding equals

$$2(172 + 12 + 8)/(20 \times 10^{-3}) = 19.2 \text{ kb/s.}$$

The convolutionally coded output is then permuted with a 24×16 block interleaver, operating on a block of 384 symbols in each 20 ms. The purpose of interleaving is to mitigate the influence of burst errors caused by multipath fading.

The forward link is a broadcast channel, and data streams for multiple users are simultaneously transmitted. The diagram of a typical down-link transmitter for IS-95 is sketched in Figure 12.4. Thus, the CDMA system needs to differentiate between multiple users in downlink transmission. The base station achieves this goal by allocating to each user a particular 64-bit Walsh vector, a specific 42-bit PN code sequence, and a pair of 15-bit short PN sequences which are shared by all the users in the given cell. The Walsh vectors are strictly orthogonal, thus making it possible to completely eliminate intra-cell IUI. To maintain rigorous signal orthogonality at a forward-link mobile receiver, a pilot channel is inserted for channel estimation and a sync channel is inserted for timing alignment. The cell-specific short PN sequence pair provides protection against *cross-cell* interference to ensure that all the user traffic channels are 100% reused over the entire service

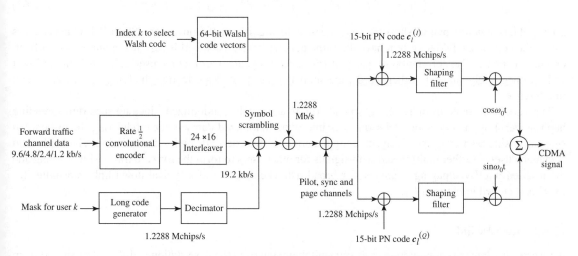

Figure 12.4 *IS-95 CDMA downlink transmitter for cell l*

area. As the third measure, the data frames of each user are masked (scrambled) by a unique user-specific 42-bit-long PN sequence, providing additional protection against intra-cell or cross-cell interference. Again, all these user-specific long sequences are derived from the same 42-bit long PN sequence by allocating to each user a particular time offset or phase, usually in form of an *electronic serial number* (ESN). A 32-bit ESN for phase initializing to the 42-bit-long sequence generator is embedded by the manufacturer, which defines the unique offset in the long PN sequence to identify its user. The PN generator is synchronized to the mobile during initialization.

Masking the data frames of each user by its specific long PN sequence requires a prior treatment because of the difference in data rate among users. The long sequence used in IS-95A is generated by the polynomial [4]

$$g(x) = 1 + x^7 + x^9 + x^{11} + x^{15} + x^{16} + x^{17} + x^{20} + x^{21} + x^{23} + x^{24} + x^{25}$$
$$+ x^{26} + x^{32} + x^{35} + x^{36} + x^{37} + x^{39} + x^{40} + x^{41} + x^{42} \tag{12.61}$$

or by Ziemer and Peterson [6]

$$g(x) = 1 + x + x^2 + x^3 + x^5 + x^6 + x^7 + x^{10} + x^{16} + x^{17} + x^{18} + x^{19} + x^{21}$$
$$+ x^{22} + x^{25} + x^{26} + x^{27} + x^{31} + x^{33} + x^{35} + x^{42} \tag{12.62}$$

at the clock rate of 1.2288 megachips per second (Mcps), which is 64 times the data rate of users' coded sequences. Therefore, the long PN sequence must be decimated before its use to scramble a coded sequence, and this is done by sampling itself once every 64 chips. The coded and scrambled frames of each user are then spread by its uniquely allocated Walsh vector to provide orthogonal protection against *intra-cell* MAI. In IS-95, there are a total of 64 Walsh vectors. Among them, Walsh vectors #0, #1, and #32 are used for pilot, paging, and sync channels, respectively. The remaining 61 vectors are allocated to users. The resulting 1.2288 Mcps spread data is applied to a quadratic modulator with two orthogonal branches (I/Q). The orthogonality of the branches is defined by two 15-bit short PN sequences and sine/cosine carriers. The two PN sequences for the I and Q branch are created, respectively, by the two m-sequence generators:

$$g_I(x) = 1 + x^2 + x^6 + x^7 + x^8 + x^{10} + x^{15}$$
$$g_Q(x) = 1 + x^3 + x^4 + x^5 + x^9 + x^{10} + x^{11} + x^{12} + x^{15}. \tag{12.63}$$

Each cell has a *unique* pair of short PN sequences to distinguish itself from others. Short PN sequence pairs for all the cells in a CDMA system have the same generators, but each cell is assigned a unique time offset, or a unique phase. The correlation property of m-sequences guarantees that the user's signals from different cells are negligibly small. In other words, the short PN pair provides effective shielding against cross-cell interference.

The same data frame from a user is "spread" (more accurately, randomized, since no spectrum spreading happens here) by the two short PN sequences into two randomized datasets for modulating the orthogonal sine and cosine carriers, respectively, creating the effect of an order-2 diversity transmission. The modulated signal of a user is further multiplexed with signals for other users to form the forward traffic signal, alongside the inserted pilot/sync/paging channels, for broadcast. The diagram of a typical down-link transmitter for IS-95 is sketched in Figure 12.4.

12.6.2 Reverse link

In reverse-link transmission, mobile units transmit their signals to the base station at different times and from different locations. It is too expensive and too inefficient to dedicate a specific pilot channel for each uplink

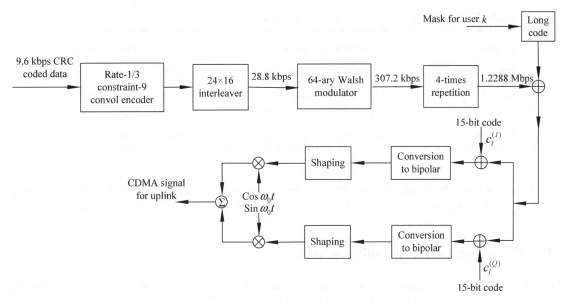

Figure 12.5 *IS-95 CDMA uplink transmitter for cell l*

user to implement synchronous transmission. Thus, the reverse link operates in an asynchronous CDMA mode. However, to prevent intra-cell and cross-cell MAI, *weak* correlation among users that share the same frequency resource must be maintained. This goal is achieved in the reverse link by allocating each user a user-specific long PN sequence and a pair of cell-specific short PN sequence. The long PN sequence and the short PN sequence pair are the same as described for use in the forward link. When a mobile unit is adopted to the traffic, it is allocated an ESN for the use of its specific long code mask. As a result of masking by their corresponding long PN sequences and randomization by the cell-specific short sequence pair, all the co-channel users have very weak correlation no matter they are located intra-cell or across cells. A block diagram for an uplink transmitter is depicted in Figure 12.5.

The reverse link employs the same 64×64 Walsh matrix as that used in the forward link, but with a totally different purpose. In the reverse link, the Walsh matrix is used for orthogonal modulation. Let us see the formation of users' traffic data frames. The CRC-coded data rates in the reverse link are the same as those in the forward link, taking the values of 1.2, 2.4, 4.8, and 9.6 kbps. Such data frames are forward error correction (FEC)-coded by a convolutional encoder of rate $\frac{1}{3}$ and constraint length 9, and the resulting data rates are then tripled. The coded data are repeated, whenever necessary, so as to reach the same standard rate of 28.8 kbps before they are applied to a 24×16 block interleaver. Each coded and interleaved data frame is partitioned into 6-bits symbols, and each of them is mapped into a 64-chip orthogonal modulated symbol defined by the 64×64 Walsh matrix. The resulting chip rate is 307.2 kcps. To comply with the chip rate (1.2288 Mcps) of the scrambling long PN sequence, each Walsh chip is spanned over *four* chips of the PN sequence. This corresponds to a spreading gain of 4. The scrambled data frame is multiplied by the two cell-specific short PN sequences before the offset quaternary carrier modulation. Using the same set of data, with an offset of half a chip, to modulate two orthogonal carriers (sine and cosine), provides diversity gain and makes it possible to implement noncoherent demodulation of the 64-bit orthogonal symbols. Introducing the offset of half a chip between the two carriers helps to reduce the peak amplitude of the modulated signal.

Table 12.2   *Functions of various sequences in IS-95*

Component	Forward traffic channel	Reverse traffic channel
Walsh codes	Intra-cell user separation	Orthogonal modulation
Short PN pair	(1) Cell identification (2) Against cross-cell IUI	(1) Cell identification (2) Against cross-cell IUI
Long PN	Scrambling for extra protection	User distinction

The roles of long PN sequences, short PN sequences, and Walsh matrix in the forward and reverse links are briefly summarized in Table 12.2, where the abbreviation IUI means inter-user interference.

12.7   Processing gain of spreading spectrum

In CDMA, a direct consequence of spanning each symbol of a user over its signature sequence is the spreading of its power spectrum. For a synchronous CDMA system with strictly synchronous signature sequences, IUI is completely eliminated at each receiver. For asynchronous CDMA systems or systems with nonorthogonal signature sequences (such as m-sequences), the effect of spectrum spreading is traded for a processing gain.

Processing gain associated with spreading spectrum in a CDMA system has an intuitive explanation. For simplicity, let us elucidate it with a two-user system, as sketched in Figure 12.6 showing the variation of the power spectrum density (PSD) at different stages of signal processing. The symbols of both Users 1 and 2 are spread by their N_c-chip signature vectors $\mathbf{c}_1$ and $\mathbf{c}_2$, respectively. The symbol rate is R_s and the chip rate is $R_c = N_c R_s$, with N_c times spectrum spreading, as shown in line 2 of the figure. At the receiver for User 1, $\mathbf{c}_1$ is used for despreading and the spectrum of User 1's signal is back to the original. However, User 2's signal turns out to be a noise-like component due to the correlation operation between $\mathbf{c}_1$ and $\mathbf{c}_2$. Only (R_s/R_c) of this noise-like component can pass through the subsequent low-pass filter, implying a processing gain of N_c, as shown on the last line of the figure.

12.8   IS-95 downlink receiver and performance

The receiver of any user should have the capability to reject interference from other users. This is not difficult for intra-cell users in the downlink since we can allocate to them a set of perfectly orthogonal spreading codes generated, for example, from a Walsh matrix. Given that the same Walsh codes are reused in all adjacent cells, it would cause a serious co-channel interference. Without appropriate measure, a user using a Walsh code in the home cell can be hit with probability of $1/64$ by its counterpart from any other cell. Clearly, we need to maintain the orthogonality among users in the cell on one hand while randomizing spreading codes between different cells on the other. A good strategy is to assign a unique pair of PN sequences to each cell.

Consider the forward link transmission to user k in cell l, which is denoted hereafter as user lk for simplicity. Let cell 0 be the home cell, and we quantitatively study interference suppression at a downlink receiver by focusing on the signal for user 0κ.

The information bit sequence of user lk is CRC- and convolutional-coded, interleaved, and masked by a long-code-producing data stream, producing a new sequence $\{d_{lk}(i), i = \cdots, -1, 0, 1, \cdots\}$. The resulting sequence $d_{lk}(i)$ is then spread by the Walsh code vector $\mathbf{w}_k = [w_k(1), \cdots, w_k(N_c)]^T$, pulse-shaped by $p_c(t)$,

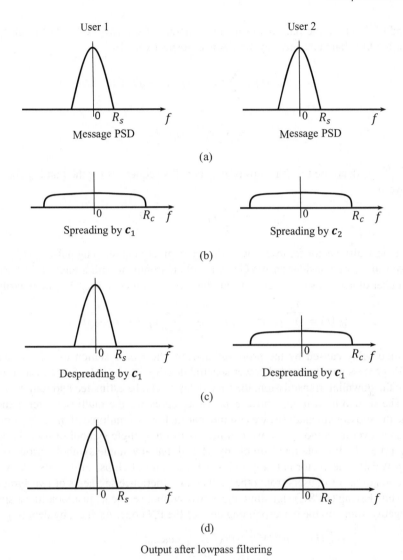

Figure 12.6 *Processing gain of spread spectrum systems: a heuristic explanation*

and I/Q modulated by $C_l^{(I)}(t)\cos\omega_0(t)$ and $C_l^{(Q)}(t)\sin\omega_0(t)$, respectively. The resulting transmitted waveform from the lth base station to user lk is then expressible as

$$g_{lk}(t) = \sum_{i=-\infty}^{\infty} d_{lk}(i) \underbrace{\sum_{j=1}^{N_c} w_k(j) p_c[t - iT_s - (j-1)T_c]}_{\text{Walsh waveform: } w_k(t-iT_s)}$$

$$\times [C_l^{(I)}(t)\cos\omega_0(t) + C_l^{(Q)}(t)\sin\omega_0(t)], \qquad (12.64)$$

where $C_l^{(I)}(t)$ and $C_l^{(Q)}(t)$ are the pulsed waveforms generated from two $(2^{15} - 1)$-bit, independent, short PN sequences for the I/Q channels, and they are defined, respectively, by

$$C_l^{(I)}(t) = \sum_{j=1}^{N_c} c_l^{(I)}(j)p_c(t - jT_c)$$

$$C_l^{(Q)}(t) = \sum_{j=1}^{N_c} c_l^{(Q)}(j)p_c(t - jT_c) \tag{12.65}$$

with $c_l^{(I)}(j)$ and $c_l^{(Q)}(j)$ denoting the jth chips of the short PN sequences for the I and Q channel of the cell l. The pulsed waveform

$$w_k(t) = \sum_{j=1}^{N_c} w_k(j)p_c(t - jT_c) \tag{12.66}$$

is generated from the Walsh vector for user k with the rectangular chip-shaping pulse $p_c(t)$.

For ease of exposition, we consider an AWGN channel environment with a total of M base stations. Let K_l denote the number of active users in cell l. Then, the received signal at user 0κ can be written as

$$r_{0\kappa}(t) = \sum_{k=1}^{K_0} g_{0k}(t - \tau_0) + \sum_{l=1}^{M-1} \sum_{k=1}^{K_l} g_{lk}(t - \tau_l) + n(t), \tag{12.67}$$

where τ_l is the time delay caused by the propagation from the l base station to user 0κ and $n(t)$ is the zero-mean AWGN process with two-sided power spectral density $N_0/2$ W/Hz. Since synchronous transmission is adopted in the downlink transmission, the time delay τ_0 can be estimated and removed. Thus, we can assume $\tau_0 = 0$. The second term in the above expression represents the multiuser interference from other cells. To facilitate the subsequent analysis, we assume that each τ_l is a multiple of the chip duration T_c. Since the detection is done over a symbol basis, we can just focus on a single symbol of user 0κ, say $i = 0$, and briefly denote $d_{lk}(i)$ as d_{lk} for notational simplicity. Recall that any single symbol period corresponds to a complete 64-chip Walsh code vector but only to a small portion (64 chips) of the bits' PN sequence. Over each symbol duration of user 0κ, users from other cells may experience segments of two symbols. However, for binary phase shift-keying (BPSK) signaling, these two symbol segments just look like a single symbol if we absorb the negative sign into the corresponding chip of the PN code. As such, by denoting

$$\phi_{lk}^{(I)}(t) = w_k(t)C_l^{(I)}(t)\sqrt{2/T_s} \, \cos\omega_0 t$$

$$\phi_{lk}^{(Q)}(t) = w_k(t)C_l^{(Q)}(t)\sqrt{2/T_s} \, \sin\omega_0 t, \; 0 \le t \le T_s, \tag{12.68}$$

and focusing on the 0th symbol of user 0κ, the received signal given in (12.67) can be rewritten as

$$z_{0\kappa}(t) = \sum_{k=0}^{K-1} d_{0k}[\phi_{0k}^{(I)}(t) + \phi_{0k}^{(Q)}(t)] + \sum_{l=1}^{M-1} \sum_{k=0}^{K_l - 1} d_{lk}[\phi_{lk}^{(I)}(t - \tau_l) + \phi_{lk}^{(Q)}(t - \tau_l)] + n(t). \tag{12.69}$$

The receiver of mobile user κ performs cross correlation with $z_{0\kappa}(t)$ with local references $\phi_{l\kappa}^{(I)}(t)$ and $\phi_{l\kappa}^{(Q)}(t)$ at its upper and lower branches to retrieve the data $d_{l\kappa}(0)$, producing two output variables

$$\xi_I = \int_0^{T_s} z_{0\kappa}(t)\phi_{l\kappa}^{(I)}(t)dt$$

$$\xi_Q = \int_0^{T_s} z_{0\kappa}(t)\phi_{l\kappa}^{(Q)}(t)dt. \tag{12.70}$$

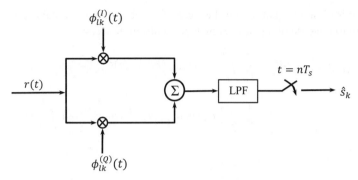

Figure 12.7 *IS-95 CDMA downlink receiver*

The functional structure of a down-link IS-95 receiver is shown in Figure 12.7. To proceed with the evaluation of ξ_1 and ξ_2, we need to investigate the properties of $\phi_{l\kappa}^{(I)}(t)$ and $\phi_{l\kappa}^{(Q)}(t)$.

Note that each symbol is masked only a small portion (only $N_c = 64$ bits) of the short PN sequence ($2^{15} - 1$ bits) which is, thus, not expected to possess the autocorrelation or cross-correlation property of the original PN sequence. As such, in what follows, we will treat chips in $C_l(t)$ as being randomly chosen from the original sequence.

Property 12.8.1 Orthogonality in the same cell.
Any pair of I and Q channel signals from the same cells are always mutually orthogonal, no matter they belong to the same or different users Such orthogonality, up to a constant symbol product, is expressible as.

$$\int_0^{T_s} \phi_{lk}^{(I)}(t)\phi_{l\kappa}^{(Q)}(t)dt = 0. \tag{12.71}$$

A pair of I channel signals from any two users in the same cell are orthogonal, and the same is true for Q channel signals; namely,

$$\int_0^{T_s} \phi_{lk}^{(I)}(t)\phi_{l\kappa}^{(I)}(t)dt = \delta(\kappa - k)$$

$$\int_0^{T_s} \phi_{lk}^{(Q)}(t)\phi_{l\kappa}^{(Q)}(t)dt = \delta(\kappa - k), \tag{12.72}$$

where $\delta(k)$ is the Dirac delta function, which equals 1 if $k = 0$ and zero otherwise.

The cross-branch orthogonality (12.71) comes from the orthogonality of the sine and cosine functions, whereas the same-branch orthogonality (12.72) stems from the orthogonality of the Walsh vectors. For a proof, let us begin with (12.71). In particular, we have

$$\int_0^{T_s} \phi_{lk}^{(I)}(t)\phi_{l\kappa}^{(Q)}(t)dt = \frac{2}{T_s} \int_0^{T_s} w_k(t)w_\kappa(t)C_l^{(I)}(t)C_l(Q)(t)dt$$

$$= \frac{1}{T_s} \sum_{m=1}^{N_c} w_k(m)w_\kappa(m)c_l^{(I)}(m)c_l^{(Q)}(m) \int_0^{T_c} \sin 2\omega_0 t \, dt$$

$$= 0. \tag{12.73}$$

The integral on the third line is equal to zero because of the fact that T_c contains a multiple of the carrier periods. We then turn to the case of a pair of signals from different users:

$$\int_0^{T_s} \phi_{lk}^{(I)}(t)\phi_{l\kappa}^{(I)}(t)dt = \frac{2}{T_s} \int_0^{T_s} w_k(t)w_\kappa(t)C_l^{(I)}(t)C_l(I)(t)dt$$

$$= \frac{1}{T_s} \sum_{m=1}^{N_c} w_k(m)w_\kappa(m)c_l^{(I)}(m)c_l^{(I)}(m) \int_0^{T_c} \cos^2\omega_0 t \, dt$$

$$= \frac{T_c}{T_s} \sum_{m=1}^{N_c} w_k(m)w_\kappa(m)$$

$$= \delta(\kappa - k). \tag{12.74}$$

To obtain line 3, we have used the fact that the integral on line 2 equals T_c and $[c_l^{(I)}(m)]^2 = 1$. We further used the orthogonality property of Walsh codes to obtain the final result. The derivation for the Q channel signals is similar. For ease of presentation, a pair of signals is referred to as *same-phased* signals or *same-phased* channels if both of them come from I channels or from the Q channels, and is referred to as *quadric-phased* signals or *quadric-phased* channels if one of them comes from the I channel while another comes from the Q channel. Clearly, for signals in the same cell, the intra-cell interference caused by quadrature-phased signals is completely eliminated as a result of the orthogonality of sine and cosine carriers, and its counterpart caused by same-phased signals is completely removed by the orthogonality of the Walsh codes. Both kinds of orthogonality are maintained owing to synchronous transmission in the downlink.

Property 12.8.2   Orthogonality between different cells.
The cross-correlation between users from different cells are orthogonal if the two waveforms come from two quadratic-phased branches. More specifically, $\forall l \geq 1$, we have

$$\int_0^{T_s} \phi_{0\kappa}^{(I)}(t)\phi_{lk}^{(Q)}(t - \tau_l)dt = 0,$$

$$\int_0^{T_s} \phi_{0\kappa}^{(Q)}(t)\phi_{lk}^{(I)}(t - \tau_l)dt = 0. \tag{12.75}$$

This property addresses the cross-cell interference from other cells. Let us show this property by focusing on the first expression. For simplicity, assume $\tau_l = jT_c$. In the same way as used in (12.73), we obtain

$$\int_0^{T_s} \phi_{0\kappa}^{(I)}(t)\phi_{lk}^{(Q)}(t)dt = \frac{1}{T_s} \sum_{m=1}^{N_c} w_\kappa(m)[w_k(m+j)]c_0(m)[c_l(m+j)] \underbrace{\int_0^{T_c} \sin 2\omega_c t \, dt}_{=0}$$

$$= 0, \tag{12.76}$$

where $[w_k(m+j)] = w_k(m')$ is a cyclic-shifted version of $w_k(m)$ with m' defined by $m' = (m+j)_{\mathrm{mod}-N_c}$. From the above result, it is clear that the rejection of cross-cell interference from quadratic channels in other cells relies on the orthogonality between sine and cosine functions.

Having obtained the properties of the composite basis functions $\phi_{lk}^{(\alpha)}(t), \alpha = I, Q$, we are in position to simplify the correlator output ξ's in (12.70) yielding

$$\xi_I = d_{0\kappa} + \sum_{l=1}^{M-1} \sum_{k=0}^{K_l-1} d_{lk} \eta_{lk}^{(I)} + n_I,$$

$$\xi_Q = d_{0\kappa} + \sum_{l=1}^{M-1} \sum_{k=0}^{K_l-1} d_{lk} \eta_{lk}^{(Q)} + n_Q, \tag{12.77}$$

where

$$\eta_{lk}^{(I)} = \int_0^{T_s} \phi_{lk}^{(I)}(t - t_l)\phi_{0\kappa}^{(I)}(t)dt,$$

$$\eta_{lk}^{(Q)} = \int_0^{T_s} \phi_{lk}^{(Q)}(t - t_l)\phi_{0\kappa}^{(Q)}(t)dt. \tag{12.78}$$

The uncorrelated noise components $n_I \sim \mathcal{CN}(0, N_0/2)$ and $n_Q \sim \mathcal{CN}(N_0/2)$ have the same distribution as $n(t)$. We need to work out $\eta_{lk}^{(I)}$ and $\eta_{lk}^{(Q)}$. To proceed, let N_c denote the number of chips in each Walsh vector, and take $\eta_{lk}^{(I)}$ as an example. The composite chips of $\phi_{0\kappa}^{(I)}(t)$ can be hit by $\phi_{lk}^{(I)}(t - t_l), l \geq 1$ with different patterns. In both $\phi_{0\kappa}^{(I)}(t)$ and $\phi_{lk}^{(I)}(t - t_l), l \geq 1$, the corresponding Walsh vectors are masked by a random segment of their short PN sequences, thus allowing us to assume that individual composite chips of user 0κ are hit independently, each with a probability of $1/2$. Thus, the probability of q chips being hit is given by

$$P(q) = \binom{N_c}{q} (1/2)^q (1/2)^{N_c-q} = 2^{-N_c} \binom{N_c}{q}. \tag{12.79}$$

The two corresponding vectors have q places with the same values and the remaining $(N_c - q)$ places with opposite values. Thus, after cancellation, the cross product of the two vectors reduces to a summation of $(N_c - q)(-1) + q \times 1 = 2q - N_c$ terms with each of magnitude $\sqrt{E_s/N_c}$. As such, the interfering power from user lk of cell l can be evaluated as

$$\mathbb{E}[|\eta_{lk}^{(I)}|^2] = \sum_{q=0}^{N_c} P(q) \left(\frac{2q - N_c}{N_c}\right)^2 E_s$$

$$= \frac{E_s}{2^{N_c}} \sum_{q=0}^{N_c} \binom{N_c}{q} \left(1 - \frac{2q}{N_c}\right)^2$$

$$= \frac{E_s}{2^{N_c}} \left[2^{N_c} - \frac{4}{N_c} \sum_{q=0}^{N_c} \binom{N_c}{q} q + \frac{4}{N_c^2} \sum_{q=0}^{N_c} \binom{N_c}{q} q^2\right]. \tag{12.80}$$

The second and third terms inside the square brackets can be determined by taking the first two-order derivatives of both sides of $(a + b)^{N_c} = \sum_{q=0}^{N_c} \binom{N_c}{q} a^q b^{N_c-q}$ and setting $a = b = 1$. The results are

$$\sum_{q=0}^{N_c} q \binom{N_c}{q} = N_c 2^{N_c-1}$$

$$\sum_{q=0}^{N_c} q^2 \binom{N_c}{q} = N_c(N_c - 1)2^{N_c-2} + N_c 2^{N_c-1} \tag{12.81}$$

which, when inserted into (12.80), leads to

$$\mathbb{E}[|\eta_{lk}^{(I)}|^2] = E_s/N_c \tag{12.82}$$

where N_c, as defined above, is the number of chips in each Walsh vector. Following the same token, we can obtain the same variance for $\eta_{lk}^{(Q)}$.

It is interesting to note that this rigorous analysis leads to the same result as obtained by the intuitive argument of processing gain. The multiplication of a Walsh code with a randomized segment of the short PN code makes the resulting sequence behave like m-sequences in resisting cross-cell interference from the same-phased channels. Thus, the interference from other cells reduces by the factor of the processing gain N_c.

Hence, we can model the second terms in ξ_I and ξ_Q of (12.77), denoted by η_I and η_Q, respectively, as zero-mean Gaussian variables

$$\eta_i \sim \mathcal{CN}\left(0, (E_s/N_c)\sum_{l=1}^{M-1} K_l\right), i = I, Q \tag{12.83}$$

and rewrite (12.77) as

$$\xi_I = d_{0\kappa} + \eta_I + n_I,$$
$$\xi_Q = d_{0\kappa} + \eta_Q + n_Q. \tag{12.84}$$

By coherently combining these two branches, we obtain a single decision variable for symbol $d_{0\kappa}$, as shown by

$$\xi = \xi_I + \xi_Q = 2d_{0\kappa} + \eta_I + \eta_Q + n_I + n_Q, \tag{12.85}$$

for which $(\eta_I + \eta_Q) \sim \mathcal{CN}(0, 2(E_s/N_c)\sum_{l=1}^{M-1} K_l)$ and $(n_I + n_Q) \sim \mathcal{CN}(0, N_0)$. As such, the SINR in ξ is equal to

$$\gamma = \frac{4}{\rho^{-1} + (2/N_c)\sum_{l=1}^{M-1} K_l} \overset{\rho \to \infty}{\longrightarrow} \frac{2N_c}{\sum_{l=1}^{M-1} K_l}, \tag{12.86}$$

where $\rho = E_s/N_0$.

✍ Example 12.2 ───────────────────────────────────

Consider the use of a four-finger Rake receiver for coherent detection of IS-95A downlink signals on an AWGN channel with a total number of $K = \sum_{l=1}^{M-1} K_l$ cross-cell interfering users. Suppose that hard-decision output is used for convolutional decoding. Assuming that the SNR ρ is large so that its ρ^{-1} is negligible, determine the decoding bit error rate for $K = 30$.

Solution

The spreading gain is $N_c = 64$, the I/Q channel diversity is 2, and the path diversity gain contributed by the coherent Rake receiver is 4. Thus, the SINR without coding is $\gamma_0 = 8N_c/K$. Since 1/2-convolutional code is employed, the SINR for coded bit is equal to

$$\gamma = \frac{1}{2}\gamma_0 = \frac{256}{K}.$$

Thus, the error rate for each hard-decision is

$$p_c = Q(\sqrt{256/K}) = 0.0017.$$

The convolutional code has a free distance of 12, which corresponds to a 5-bit correction capability. As such, the decoding bit error is

$$P_b = 1 - \sum_{q=0}^{5} \binom{384}{q} p_c^q (1 - p_c)^{384-q} = 5.975 \times 10^{-5}.$$

In this example, we have not exploited the fact that the distance from the mobile receiver to cross-cell interfering base stations is typically 2 times that to the home basestation. Taking the propagation loss into consideration, the value of K can be tripled or quadrupled.

12.9 IS-95 uplink receiver and performance

The uplink receivers usually employ noncoherent detection, and we briefly derive its structure in this section. From the description of the reverse link in Section 12.6, it is straightforward to write down the expression for the received signal at the base station. Let us concentrate on user k in cell l whose 6-bit word is mapped to one of 64 possible Walsh vectors. Denote this message-bearing Walsh vector by $\mathbf{w}_{lk}$, and denote $1_4 = [1; 1; 1; 1]$. Then, the transmitted vector $\mathbf{s}_{lk}$ is a four-time repetition of $\mathbf{w}_{lk}$ expressible in terms of the Kronecker product, as shown by $\mathbf{s}_{lk} = \mathbf{w}_{lk} \otimes 1_4$. Let $s_{lk}(i)$ denote the ith entry of $\mathbf{s}_{lk}$, and let $c_l^{(I)}(i)$ and $c_l^{(Q)}(i)$ denote the corresponding short-code chips for the I and Q branches of cell l, respectively. For exposition, it suffices to focus on the time span $0 \le t < T_{ws}$, where $T_{ws} = 256T_c$. Then, the received signal at the home base station can be written as

$$r_l(t) = \sum_{i=0}^{255} s_{lk}(i) c_{lk}^{(I)}(i) z_{lk}(i) p_c(t - iT_c) \cos(\omega_c t + \theta_0)$$

$$+ \sum_{i=0}^{255} s_{lk}(i) c_{lk}^{(Q)}(i) z_{lk}(i) p_c(t - iT_c - 0.5T_c) \sin(\omega_c t + \theta_0)$$

$$+ \underbrace{(\text{cross-cell MAI terms})}_{\zeta_{lk}} + \underbrace{(\text{intra-cell MAI terms})}_{\eta_{lk}} + n(t), \qquad (12.87)$$

where $n(t)$ is the AWGN and θ_0 is a random phase. Both intra-cell and cross-cell MAI terms, ζ_{lk} and η_{lk}, have similar expressions to the first term with cosine function and the second term with sine function except the changes in their indexes. In uplink reception, timing synchronization is assumed, but no attempt is made to remove the phase ambiguity due to θ_0.

The detection of the Walsh symbol $\mathbf{w}_{lk}$ can be done in much the same way as noncoherent detection of MFSK signals, though they are different in correlator functions. Let us show how to accumulate the symbol energy in the I channel. Note that $\mathbf{s}_{lk}$ can be any one derived from 64 possible Walsh symbols $\{\mathbf{w}_q, q = 0, \cdots, 63\}$. We, therefore, construct a bank of 64 filters with each matched to the waveform

$$\phi_q(t) = \sum_{i=0}^{255} s_q(i) c_{lk}^{(I)}(i) z_{lk}(i) p_c(t - iT_c) \cos \omega_c t \qquad (12.88)$$

for a particular Walsh vector $\mathbf{w}_q$. These matched filters rely on cell-specific short codes to eliminate most of cross-cell MAI while rejecting intra-cell MAI via a user-specific long PN sequence. We sample the matched filter's output at $t = jT_{ws}$, and for $j = 0$ we obtain

$$y_c(\mathbf{w}_q) = (r(t), \phi_q(t)), \tag{12.89}$$

where $(\cdot)$ denotes the inner product or correlation operation. There is another bank of 64 filters matched, respectively, to the waveform

$$\psi_q(t) = \sum_{i=0}^{255} s_q(i) c_{lk}^{(Q)}(i) z_{lk}(i) p_c(t - iT_c - 0.5T_c) \sin \omega_c t \tag{12.90}$$

for Walsh vectors $\mathbf{w}_q, q = 0, \cdots, 63$. Sample the matched filter's output at $t = jT_{ws}$ and for $j = 0$, we obtain

$$y_s(\mathbf{w}_q) = (r(t), \psi_q(t)). \tag{12.91}$$

The total symbol energy on the I branch can be accumulated to yield

$$\xi_I(\mathbf{w}_q) = y_c^2 + y_s^2, \tag{12.92}$$

where the squared-sum operation removes the random phase through the fact that $\cos^2\theta_0 + \sin^2\theta_0 = 1$. The aforementioned matched filter banks can also be used to collect the symbol energy on the Q branch by sampling their outputs at the time instants $t = jT_{ws} + 0.5T_c$. Denote the corresponding output samples by y_c' and y_s', respectively, whereby the accumulated energy equals $\xi_Q(\mathbf{w}_q) = (y_c')^2 + (y_s')^2$. Finally, combine ξ_I and ξ_Q to implement branch diversity reception, with the resulting decision variable given by

$$\xi(\mathbf{w}_q) = \xi_I(\mathbf{w}_q) + \xi_Q(\mathbf{w}_q). \tag{12.93}$$

It follows that the 6-bit transmitted symbol is estimated as

$$\hat{\mathbf{w}}_{lk} = \arg \max_{q=0,\cdots,63} \xi(\mathbf{w}_q). \tag{12.94}$$

The performance analysis for the noncoherent detection of uplink orthogonal Walsh signals is similar, in principle, to its counterpart for orthogonal MFSK signals, and is thus omitted here for brevity.

12.10   3GPP-LTE uplink

The MA scheme for 3GPP-LTE must be flexible enough to support variable-bandwidth requirements by users, and enable 100% frequency reuse in each cell, leaving room to support MIMO.

To support wireless broadband services for fast-growing mobile users, the International Mobile Telecommunication-Advanced (IMT-Advanced) entered the phase in the International Telecommunication Union (ITU) to tackle this issue, leading to the IMT-Advanced system with its technical specifications and performance detailed in the recommendation delivered in November 2008 [7]. The Third Generation Partnership Project (3GPP) working group then put its efforts on establishing a Long Term Evolution Advanced (LTE-A) standard Release-10 (Rel-10) for the implementation of IMT-Advanced. The LTE-A is an enhanced version of the previously delivered LTE Rel-8, allowing for a total data rate of up to 100 Mb/s at high mobility, as opposed to 20 Mb/s of the latter. In addition, the LTE-A also needs a backward compatibility with LTE Rel-8 systems, as well as the flexibility to support users of unequal frequency bandwidths. To meet the requirement of higher data rate and backward compatibility simultaneously, the technique of carrier aggregation (CA) becomes a natural choice. With CA, the LTE Rel-8 system can be used as a building block

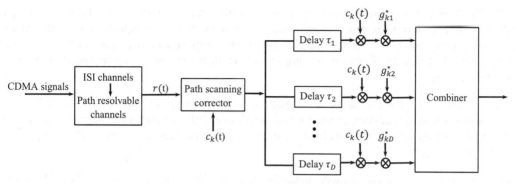

Figure 12.8   *Rake receiver for CDMA*

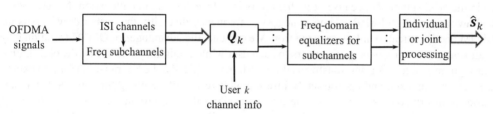

Figure 12.9   *Idea of OFDMA to combat dispersive channels*

for LTE-A. However, the challenge lies in the design of MA schemes. A good MA scheme should have the capability

- to eliminate intra-cell and inter-cell multiuser interference as much as possible;
- to appropriately handle frequency-selective fading channels as confronted by LTE systems that support high data-rate transmission;
- to maintain an acceptable PAPR, which is particularly important to uplink transmission;
- to be backward compatible with the existing systems in operation.

ISI is not a big issue to all commonly used MA schemes as long as users are allocated an orthogonal or near-orthogonal channels.

During the standardization of 3GPP LTE-A, there were serious debates regarding the use of CDMA or the OFDM-based access scheme OFDMA. Each scheme has its own advantages and weaknesses. As illustration, the ideas used in the two schemes for combatting dispersive channels are compared in Figure 12.8 and Figure 12.9. The advantages of OFDMA are its higher spectral efficiency and lower implementation complexity by exploiting the property of FFT. However, the weakness is its poor inter-cell interference performance, which is rooted in the poor autocorrelation property of Fourier vectors. On the contrary, the capability of resisting inter-cell interference is the strength of the CDMA scheme. Because of the consideration of monopoly CDMA techniques, OFDMA was eventually adopted by 3GPP-LTE for downlink transmission. Given the use of OFDM-based techniques for LTE-A, there are still debates between two schemes for uplink transmission, that is, whether to use OFDMA or SC-OFDMA. The scheme of SC-OFDMA was eventually chosen for 3GPP-LTE downlink because of its relatively low peak-to-average power ratio (PARR).

3GPP-LTE-A needs to support the maximum user data bandwidth up to 100 MHz and backward-compatible with the LTE Release-8 systems with user bandwidth of 20 MHz. A key technique for this bandwidth expansion is *carrier aggregation* (CA), which aggregates multiple component carriers, each of the same structure as adopted in the LTE Release-8 systems, to support a higher data rate. However, adopting the CA technique poses some challenges.

- In uplink, the adoption of CA needs an enhanced MA scheme to maintain an acceptable PAPR level.
- The use of multiple carrier components for transmission on different carrier frequencies requires that the LTE-A be able to handle channels with different propagation loss, fading conditions, channel bandwidths, and frequency offsets.

Each user can be allocated one or more subcarriers in an OFDM system, and the resulting MA scheme is called orthogonal frequency division multiple-access (OFDMA). In contrast, the multicarrier CDMA scheme transmits, in parallel, the chips of a spread-spectrum system on different subcarriers. The OFDMA has been adopted in the B3G standards because of its flexibility in subcarrier channel allocation. Nevertheless, it suffers the drawbacks of high PARR and possible loss of frequency diversity; the former drawback is inherited from OFDM and is critical to uplink transmission. We can take the advantages of OFDM technique, by using DFT to allocate a set of frequency subcarriers to each user while retaining the lower PAPR of a single-carrier signal through Inverse fast Fourier transform (IFFT). More specifically, the Fourier-transformed data for multiple users are multiplexed and appropriately inserted with zeros before going through the IFFT operation. The Fourier-transformed data with zeros inserted, after taking IFFT, reproduce a more *densely interpolated* temporal waveforms of the relevant users, thus behaving similar to a single-carrier signal. To see the effect of a single carrier, we consider the case of two users sharing a chunk. Their transmitted data vectors $\mathbf{s}_1$ and $\mathbf{s}_2$ are spread by DFTs $\mathbf{F}_1$ and $\mathbf{F}_2$, respectively, before IFFT by $\mathbf{F}^\dagger$. Thus, we can write the transmit vector as

$$\mathbf{x} = \mathbf{F}^\dagger \begin{pmatrix} \mathbf{0} \\ \mathbf{F}_1\mathbf{s}_1 \\ \mathbf{F}_2\mathbf{s}_2 \\ \mathbf{0} \end{pmatrix} = \mathbf{F}^\dagger \underbrace{\begin{bmatrix} \mathbf{0} & & & \\ & \mathbf{F}_1 & & \\ & & \mathbf{F}_2 & \\ & & & \mathbf{0} \end{bmatrix}}_{\mathrm{diag}\{\mathbf{0},\, e^{-j\alpha}\mathbf{I},\, e^{-j\beta}\mathbf{I},\, \mathbf{0}\}} \begin{pmatrix} \mathbf{0} \\ \mathbf{s}_1 \\ \mathbf{s}_2 \\ \mathbf{0} \end{pmatrix} = \begin{pmatrix} \mathbf{0} \\ e^{-j\alpha}\mathbf{s}_1 \\ e^{-j\beta}\mathbf{s}_2 \\ \mathbf{0}, \end{pmatrix}$$

which, up to two phase factors, is the same as the original data vectors.

This technique is usually known as the *DFT-spread OFDM*, or simply DFT-S-OFDM. The baseband transmitted vector of a general MC system is given by

$$\mathbf{x} = \mathbf{F}^\dagger_{N\times N}\mathbf{Q}_{N\times M}\mathbf{P}_{M\times M}\mathbf{s}_{M\times 1}. \tag{12.95}$$

The complex $M \times 1$ modulated vector $\mathbf{s}$ is precoded by $\mathbf{P}$ before loading to a higher dimensional $\mathbf{F}^\dagger$ through a switching matrix $\mathbf{Q}$. The switching can be localized (L) or distributed (D), with their $\mathbf{Q}$ matrix defined, respectively, by Ciochina et al. [8]

$$\mathbf{Q}_{N\times M} = \begin{pmatrix} \mathbf{0}_{q\times M} \\ \mathbf{I}_M \\ \mathbf{0}_{(N-q-M)\times 1} \end{pmatrix} : \text{DFT-S-OFDM-L} \tag{12.96}$$

$$\mathbf{Q}_{N\times M} = \mathbf{I}_m \otimes \begin{pmatrix} \mathbf{0}_{n\times 1} \\ 1 \\ \mathbf{0}_{(K-n-1)\times 1} \end{pmatrix} : \text{DFT-S-OFDM-D}. \tag{12.97}$$

The commonly used precoder $\mathbf{P}$ can be a Walsh matrix or a DFT matrix. The Walsh–DFT combination results in a precoded OFDMA called SS-MC-MA. Its advantages include cell range extension and spectrum spreading to ensure its robustness against cell interference. The DFT–IDFT combination leads to SC-OFDMA, one advantage of which is PARR reduction.

In the previous study, we saw that multiuser wireless systems required various orthogonal sequences and pseudo-random (PN) sequences for users/cells distinction, for user scrambling, and even for orthogonal modulation. These sequences must have very good autocorrelation properties to avoid interference caused by multi-path propagation, and very good cross-correlation properties to prevent IUI and adjacent-cell interference. Commonly used PN sequences include m-sequences, Walsh sequences, Gold sequences, and CAZAC sequences; among them, Walsh sequences are strictly orthogonal. We, therefore, dedicate the next three sections, that is, Sections 12.11–12.13, to the study of these sequences. A Gold sequence is generated by combining two properly chosen m-sequences and is, thus, not included in the subsequent discussion because of space limitation.

12.11 m-Sequences

Given the important role played by orthogonal sequences in cellular systems, from this section on, we turn to their generation and properties. We first address ml-sequences. A pseudo-noise (PN) sequence is defined as a coded sequence of 1's and 0's with certain autocorrelation properties. As indicated by its name, a PN sequence is not purely random, but rather has a period in the sense that a sequence of 1's and 0's repeats itself exactly with a known period. The most commonly used periodic PN sequences are *maximum-length* sequences, or briefly m-sequences.

An m-sequence is usually generated by a *primitive* polynomial. Primitive generating polynomials have their foundation in abstract algebra. A primitive polynomial of degree q defined over GF(2) constitutes an extended Galois field $GF(2^q)$, which consists of a zero and $(2^q - 1)$ nonzero elements. Consider the Galois field defined by a third degree primitive polynomial

$$f(x) = x^3 + x + 1. \tag{12.98}$$

Let us find its $(2^3 - 1) = 7$ nonzero elements. To this end, let α be one of the roots of $f(x)$. Then, its nonzero elements can be written as $\alpha^i, i = 0, 1, \cdots, 6$; all of them are expressible in terms of polynomials in α of degree $\leq (n - 1)$. Specifically, α^i can be represented as the remainder by dividing α^i by $f(\alpha)$. For example,

$$\frac{\alpha^5}{\alpha^3 + \alpha + 1} = \alpha^2 + \alpha + 1, \tag{12.99}$$

which can be alternatively represented as a vector of 110. Note that, in this division, the polynomial $f(\alpha)$ is organized in *descending* order. We tabulate the results in Table 12.3. It is observed that the vector set repeats

Table 12.3   Elements of GF(8) defined by $f(x) = x^3 + x + 1$

Power of α	Polynomial	Vectors
0	0	000
α^0	1	001
α^1	α	010
α^2	α^2	100
α^3	$\alpha + 1$	011
α^4	$\alpha^2 + \alpha$	110
α^5	$\alpha^2 + \alpha + 1$	111
α^6	$\alpha^2 + 1$	101

itself after α^7 and, thus, the period is equal to $N = 2^3 - 1$. In general, for a polynomial of degree q, the maximum possible period is $N = 2^q - 1$. The polynomials satisfying this condition are called *m-sequence* generators.

12.11.1   PN sequences of a shorter period

Note that the maximum-periodic property applies only to m-sequence generators but *not* to an arbitrary polynomial.

✎ **Example 12.3** _____

Determine whether the polynomial $f(x) = x^4 + x^2 + 1$ is an m-sequence generator.

Solution

Let us initialize the generator with different values, say 0001 and 1111, and examine the period of the corresponding output sequences. We find that, for an initial value of 0001, the output sequence is 100010 with period 6, whereas for initial value of 1111 the output sequence is 101 with period 3. Thus, we assert that $f(x) = x^4 + x^2 + 1$ is *not* an m-sequence generator, since the period of the resulting sequence depends not only on the feedback logic but also on the initial setting.   ◯

~~~✎

12.11.2   Conditions for m-sequence generators

In order for an mth-degree polynomial $f(x)$ to be an m-sequence generator, it must satisfy all the following conditions:

- $f(x)$ must be irreducible in the sense that it has no factor polynomials other than a constant or itself.
- $f(x)$ can divide $x^{2^m - 1} + 1$.
- $f(x)$ cannot divide $(x^q + 1)$ for any integer $0 < q < 2^m - 1$.

We skip the proof of these conditions.

Some examples of m-sequence generator polynomials are listed in Table 12.4. A detailed table covering m-sequence generators of degree up to 40 is given in Ref. [6]. A much more complete list is available in Refs [9, 10].

Table 12.4   *Examples of m-sequence generators*

Degree m	Sequence length	Coefficients in octal
3	7	13
4	15	23
5	31	45
6	63	103
7	127	203
8	255	435
9	511	1021
10	1023	2011
15	32767	see Equation (12.63)
42	$2^{42} - 1$	see Equation (12.61)–(12.62)

✍ Example 12.4

Find the fifth-degree generator polynomial that corresponds to $(45)_8$ and the sixth-degree polynomial that corresponds to $(103)_8$, respectively.

Solution

It is straightforward to determine

$$(45)_8 = 4 \times 8 + 5 = 2^2 \times 2^3 + 2^2 + 1 = 2^5 + 2^2 + 1,$$

which corresponds to the generator polynomial $f(x) = x^5 + x^2 + 1$, the same as that shown in Table 12.3. We next calculate

$$(103)_8 = 8^2 + 3 = 2^6 + 2 + 1,$$

which leads to $f(x) = x^6 + x + 1$. ○

〰〰✍

12.11.3 Properties of m-sequence

Maximum-length sequences have many of the properties possessed by a truly random binary sequence. A random binary sequence is a sequence in which the presence of a binary symbol 1 or 0 is equally probable. Some properties of m-sequences are listed below:

Property 12.11.1 *In each period of an m-sequence, the number of 1's is always one more than the number of 0's. This property is called the balance property.*

Property 12.11.2 *There are 2^{m-1} runs in any m-sequence produced by a shift register of m stages. Among 2^{m-1} runs, one-half are of length 1, one-fourth are of length 2, one-eighth are of length 3, and so on, as long as these fractions represent meaningful numbers of runs. This property is called the run property.*

Property 12.11.3 *A cyclic shift of an m-sequence is still an m-sequence.*

Property 12.11.4 *The autocorrelation function of an m-sequence is periodic. If we convert bit 0 to bit −1, then we have*

$$R_c(k) = \frac{1}{N} \sum_{n=1}^{N} c_n c_{n-k}$$

$$= \begin{cases} 1 & k = \ell N \\ -\frac{1}{N} & k \neq \ell N \end{cases}.$$
(12.100)

A plot is given in Figure 12.10.

Property 12.11.5 *The total number of primitive polynomials of degree m is equal to*

$$N_{pp} = \frac{n}{m} \prod_{p_i \mid n} \left(1 - \frac{1}{p_i} \right),$$
(12.101)

where the product is performed over all distinct prime numbers of $n = 2^m - 1$.

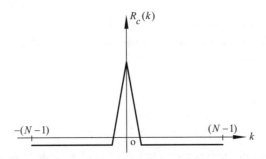

Figure 12.10   *Autocorrelation of m-sequences*

As illustration, consider an m-sequence generated by $f(x) = x^4 + x^3 + 1$, as given by 100010011011110. Now $m = 4$, hence $2^{m-1} = 8$. The distribution of runs are as follows:

Run length	Number of runs
1	4
2	2
3	1
4	1

Total number of runs is 8.

✍ Example 12.5

Find all primitive polynomials of degree 4 with which to examine Property 12.11.5.

Solution

For $m = 4$, we have $n = 2^m - 1 = 15$, which has two primitive factors, 3 and 5. Thus, according to Property 12.11.5, the number of degree-4 primitive polynomials is given by

$$N_{pp} = \frac{15}{4}\left(1 - \frac{1}{3}\right)\left(1 - \frac{1}{5}\right) = 2.$$

Next we turn to the factorization of $f(x) = x^{15} + 1$, yielding

$$x^{15} + 1 = (x^4 + x + 1)(x^4 + x^3 + 1)(x^4 + x^3 + x^2 + x + 1)$$
$$\times (x^2 + x + 1)(x + 1).$$

Among the total of five factor polynomials, three are of degree 4. However, $(x^4 + x^3 + x^2 + x + 1)$ is not an m-sequence generator since it can divide $x^5 + 1$. The remaining two satisfy all the conditions of an m-sequence, agreeing with the theoretical assertion. ○

〜〜✍

12.11.4   Ways to generate PN sequences

A generator polynomial of degree m is in general expressible as

$$f(x) = x^m + a_{m-1}x^{m-1} + \cdots + a_1 x + 1, \tag{12.102}$$

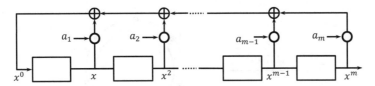

Figure 12.11   *Shift register for PN sequence generation*

where the coefficients a's can take on the value of 1 or zero except that a_m and a_0 must be 1. It can be implemented by using a linear feedback shift register, as shown in Figure 12.11. A feedback connection exists if the corresponding $a = 1$ and vanishes if $a = 0$. All additions are modulo-2, and a negative sign has the same effect as a positive sign. In this book, we adopt the convention that we organize the shift registers in ascending order so that the highest degree is rightmost. In so doing, the long division method will generate the same PN sequence as a shift register. If the shift register organized in descending order is also feasible, but it generates the same PN sequence in a reverse order.

A PN sequence can be generated by using different methods, three of which are introduced below. Let us illustrate these methods through a simple generator $f(x) = 1 + x + x^3$ initialized with state $[011]$.

12.11.4.1   *Method 1: based on linear feedback shift register*

Refer to the shift register with linear feedback shown in Figure 12.12. According to the structure, we need to calculate $(M_1 + M_3)_{\mathrm{mod}\ 2}$ from the current line, and put the result on the position M_1 of the next line, with the results tabulated below.

Clock pulse	M_1	M_2	M_3
0	0	1	1
1	1	0	1
2	0	1	0
3	0	0	1
4	1	0	0
5	1	1	0
6	1	1	1
7	0	1	1

(12.103)

Since the shift register shifts to the right by one bit, the resulting data matrix has a Toeplitz structure. We observe that, from the seventh line on, the state contents start to repeat, implying that the period of resulting sequences is equal to $L = 2^3 - 1 = 7$. We observed that, during one period, the stage $[M_1, M_2, M_3]$ experiences all possible three-tuples except all-zeros. One period of the last column forms a PN sequence 1101001. The two other columns are simply its cyclically shifted version. This assertion becomes obvious if we plot the output sequence from the last column on a cycle, as shown in Figure 12.13.

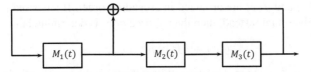

Figure 12.12   *Example to illustrate PN sequence generation*

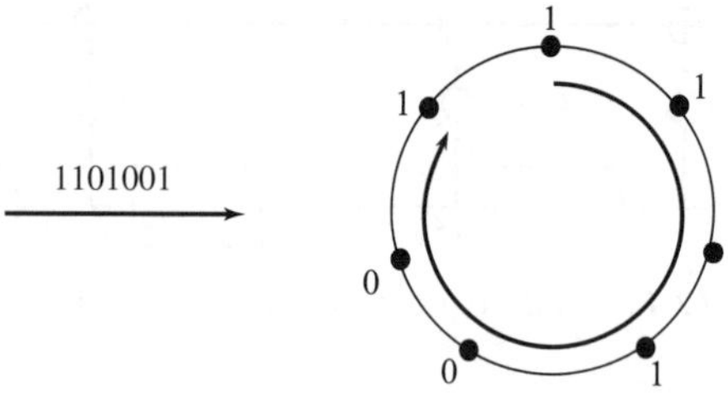

Figure 12.13   *Way to count the number of runs*

From the above exposition, for engineering implementation, the simplest way to allocate PN sequences to users is to assign each user *a set of initial values*.

12.11.4.2   *Method 2: transition matrix*

Alternatively, the circuit connection of the above shift register defines a transition matrix $\mathbf{T}$, given by

$$\underbrace{\begin{bmatrix} M_1(t+1) \\ M_2(t+1) \\ M_3(t+1) \end{bmatrix}}_{\mathbf{M}(t+1)} = \underbrace{\begin{bmatrix} 1 & 0 & 1 \\ 1 & 0 & 0 \\ 0 & 1 & 0 \end{bmatrix}}_{\mathbf{T}} \underbrace{\begin{bmatrix} M_1(t) \\ M_2(t) \\ M_3(t) \end{bmatrix}}_{\mathbf{M}(t)}. \tag{12.104}$$

whereby the behavior of the shift register can be represented by the following transition equation. The output PN sequences are listed as a matrix as below.

$$[\mathbf{M}(0), \mathbf{M}(1), \cdots, \mathbf{M}(7)] = \begin{bmatrix} 0 & 1 & 0 & 0 & 1 & 1 & 1 & 0 \\ 1 & 0 & 1 & 0 & 0 & 1 & 1 & 1 \\ 1 & 1 & 0 & 1 & 0 & 0 & 1 & 1 \end{bmatrix}. \tag{12.105}$$

To obtain the above output matrix, we first put the initial vector $\mathbf{M}(0) = [0; 1; 1]$ on the first column of the matrix. Then we determine the second column by using $\mathbf{M}(1) = \mathbf{TM}(0) = [1; 0; 1]$. By the same token, we calculate the columns 2–7 and find that, after the seventh column, the matrix repeats its previous pattern again; thus, stop there.

Comparing (12.103) and (12.105), we see that the methods lead to the same results.

12.11.4.3   *Method 3: Long division*

Given an mth-degree generator polynomial $f(x) = a_0 + a_1 x + \cdots + a_m x^m$ and a set of initial values $[q_0, q_1, \cdots, q_{m-1}]$, we determine the output polynomial by using long division, as shown by

$$g(x) = \frac{q(x)}{f(x)}, \tag{12.106}$$

where $q(x) = q_0 + q_1 x + \cdots + q_{m-1} x^{m-1}$ is the polynomial representation of the given initial values, and both the polynomials $q(x)$ and $f(x)$ are arranged in ascending order. If a desired sequence is given and we need to find an initial polynomial instead, then the required $q(x)$ is determined as

$$q(x) = \{f(x)g(x)\}_{\text{degree}<m}, \tag{12.107}$$

where the subscript degree $< m$ means that $q(x)$ only collects all the terms with degree less than m.

✍ Example 12.6

As illustration, let us show how to use the long division method to generate the sequence listed on the last column of (12.103).

Solution

Since $f(x) = 1 + x + x^3$ and $\mathbf{g} = [1; 1; 0; 1; 0; 0; 1] \to g(x) = 1 + x + x^3 + x^6$, we have

$$q(x) = \{f(x)g(x)\}_{\text{degree} < m} = \{1 + x^2 + x^7 + x^9\}_{\text{degree} < m} = 1 + x^2.$$

It should be pointed out that in calculating the polynomial product, it suffices to just work out the terms with degree less than n. It is easy to check that

$$\frac{1 + x^2}{1 + x + x^3} = 1 + x + x^3 + x^6 \to [1101001]$$

with remainder $x^7 + x^9$, which is a repeated pattern of $1 + x^2$. In mapping an output polynomial to a PN sequence, we must include all the terms in ascending order, even those with zero coefficients.   ○

12.12   Walsh sequences

Walsh sequences are also called Walsh–Hadamard sequences or simply Walsh codes. Unlike PN codes, Walsh codes are perfectly orthogonal and, thus, are widely used in CDMA downlink transmission. The Walsh codes can be easily generated by using the recurrence rule as shown by the recurrence relation

$$\mathbf{W}_N = \begin{pmatrix} \mathbf{W}_{N/2} & \mathbf{W}_{N/2} \\ \mathbf{W}_{N/2} & -\mathbf{W}_{N/2} \end{pmatrix}_{N \times N} \quad \text{with} \quad \mathbf{W}_1 = 1. \tag{12.108}$$

The orthogonality of the codes so obtained can be verified by using the method of mathematical induction. Let us first examine the simplest case of $N = 2$, for which

$$\mathbf{W}_2 = \begin{pmatrix} 1 & 1 \\ 1 & -1 \end{pmatrix}.$$

It is easy to check the orthogonality that $\mathbf{W}_2 \mathbf{W}_2^T = 2\mathbf{I}_2$. Next, we assume that the assertion holds true for N, that is, $qW_N \mathbf{W}_N^T = 2^N \mathbf{I}_N$. Then, we have

$$\mathbf{W}_{2N} = \begin{pmatrix} \mathbf{W}_N & \mathbf{W}_N \\ \mathbf{W}_N & -\mathbf{W}_N \end{pmatrix} \begin{pmatrix} \mathbf{W}_N & \mathbf{W}_N \\ \mathbf{W}_N & -\mathbf{W}_N \end{pmatrix}^T = 2^{2N} \mathbf{I}_{2N},$$

which completes the proof.

12.13   CAZAC sequences for LTE-A

Constant amplitude zero autocorrelation (CAZAC) sequences have been adopted by 3GPP LTE as preamble signatures for the random access channel (RACH) in the uplink. The RACH is a common uplink channel that enables a mobile user for an initial contact to the base station and for short data packet transportation. CAZAC sequences can be also used for channel estimation and time synchronization.

CAZAC sequences were developed by Zadoff [11] and Chu [12], defined by

$$c_k(n) = \begin{cases} \exp\left[j\pi\ n(n+3)k/L\right], & L \text{ is odd} \\ \exp[j\pi\ n(n+2)k/L), & L \text{ is even} \end{cases} \tag{12.109}$$

for $n = 0, 1, \cdots, L - 1$. Any positive integer k that is prime to L corresponds to a Zadoff–Chu CAZAC sequence. CAZAC sequences have very good cross-correlation and autocorrelation properties. The cross-correlation between two CAZAC sequences is defined by

$$R_{k,\kappa}(m) = \frac{1}{\sqrt{L}} \sum_{n=0}^{L-1} c_k(n) c_\kappa^*(n - m)_{\text{mod } L}. \tag{12.110}$$

The normalization by $\sqrt{L}$ will become apparent shortly.

Property 12.13.1 *For a prime number L, the cross-correlation between the k- and κ-sequences is equal to*

$$R_{k,\kappa}(m) = \frac{1}{\sqrt{L}} \sum_{n=0}^{L-1} \exp(j\pi\ n(n+3)k/L) \exp[-j\pi\ (n-m)(n-m+3)\kappa/L]. \tag{12.111}$$

The autocorrelation can be obtained if we set $\kappa = k$ in the above expression, leading to

$$R_k(m) = \frac{1}{\sqrt{L}} \exp[j\pi\ k(3m - m^2)/L] \sum_{n=0}^{L-1} [\exp(j2\pi\ k\ m/L)]^n, \tag{12.112}$$

for $m = 0, 1, \cdots, L - 1$. In particular, $R_k(0) = \sqrt{L}$; that is why we use the normalization factor $\sqrt{L}$ in the definition of the cross-correlation.

It is straightforward to prove this property by using the definition. To remove the mod-L operation, we write the cross-correlation into two parts, as shown by

$$R_{k,\kappa}(m) = \frac{1}{\sqrt{L}} \sum_{n=0}^{m-1} \exp(j\pi\ n(n+1)k/L) \underbrace{\exp[-j\pi\ (n-m+L)(n-m+L+3)\kappa/L]}_{\xi}$$

$$+ \frac{1}{\sqrt{L}} \sum_{n=m}^{L-1} \exp(j\pi\ n(n+1)k/L) \exp[-j\pi\ (n-m)(n-m+3)\kappa/L]. \tag{12.113}$$

To simplify the factor ξ, we expand its exponent and note that both $2(n - m)$ and $(L + 3)$ are even numbers, resulting in

$$\xi = \exp\{-j\pi[(n-m)(n-m+3) + 2L(n-m) + L(L+3)]\kappa/L\}$$

$$= \exp[-j\pi(n-m)(n-m+3)\kappa/L], \tag{12.114}$$

which, when inserted into (12.113), leads to the results in (12.111):

$$|R_k(m)| = \frac{1}{\sqrt{L}} \left| \frac{1 - \exp(j2\pi\ m)}{1 - \exp(j2\pi\ m/L)} \right|$$

$$= \frac{1}{\sqrt{L}} \frac{\sin(\pi\ km)}{\sin(\pi\ km/L)}$$

$$= \begin{cases} \sqrt{L}, & m = 0 \\ 0, & m = 1, \cdots, L - 1. \end{cases} \tag{12.115}$$

Property 12.13.2   *The magnitude of $R_k(m)$ equals $\sqrt{L}$ for $m = 0$ and equals zero for $m = 1, \cdots, L - 1$.*

12.14   Nonorthogonal MA schemes

As an example of nonorthogonal MA schemes, let us consider the one known as interleave division multiple-access (IDMA) in which each user, say user k, is allocated a particular permute matrix (i.e., interleaver) $\mathbf{\Pi}_k$ [13]. The signal vector $\mathbf{s}_k$ of user k is m-time repetitively coded before interleaving by $\mathbf{\Pi}_k$, producing the output

$$\mathbf{x}_k = \mathbf{\Pi}_k(\mathbf{s}_k \otimes 1_m), \ k = 1, \cdots, K. \tag{12.116}$$

In more detail, we can write $\mathbf{s}_k = [s_{k1}, s_{k2}, \cdots, s_{kL}]^T$, where L is the block size for iterative multiuser detection, and its typical value is $L = 30$. An intuitive understanding of $\mathbf{\Pi}_k$ is the whitening of $\mathbf{s}_k$. When signals from all users pass through an AWGN channel, the received signal is given by

$$\mathbf{y} = \sum_{k=1}^{K} \mathbf{\Pi}_k(\mathbf{s}_k \otimes 1_m) + \mathbf{n}, \tag{12.117}$$

where $\mathbf{n} \sim \mathcal{CN}_{mL}(\mathbf{0}, 0.5 N_0 \mathbf{I})$.

The question is: given $\mathbf{x}$ and $\mathbf{B}_k$, is $\mathbf{s}_k$ uniquely resolvable? To this end, denote $\mathbf{C} = \mathbf{I}_L \otimes 1_m$, and

$$\mathbf{\Pi} = \begin{pmatrix} \mathbf{\Pi}_1 & & & \\ & \mathbf{\Pi}_2 & & \\ & & \ddots & \\ & & & \mathbf{\Pi}_K \end{pmatrix}, \ \mathbf{s} = \begin{pmatrix} \mathbf{s}_1 \\ \mathbf{s}_2 \\ \vdots \\ \mathbf{s}_K \end{pmatrix}, \tag{12.118}$$

whereby we can write (12.117) more compactly as $\mathbf{y} = \mathbf{\Pi C s} + \mathbf{n}$. For a synchronous uplink system, the signal vector for all users can be estimated as

$$\hat{\mathbf{s}} = (\mathbf{C}^\dagger \mathbf{\Pi}^\dagger \mathbf{\Pi C})^{-1} (\mathbf{\Pi C})^\dagger \mathbf{y}$$
$$= (\mathbf{C}^\dagger \mathbf{C})^{-1} (\mathbf{\Pi C})^\dagger \mathbf{y}$$
$$= \mathbf{s} + (\mathbf{C}^\dagger \mathbf{C})^{-1} (\mathbf{\Pi C})^\dagger \mathbf{n}, \tag{12.119}$$

where, on the second line, we have used the fact that $\mathbf{\Pi}^\dagger \mathbf{\Pi} = \mathbf{I}$. The last line shows that the noise component is correlated and amplified.

In general, each user's signal $\mathbf{s}_k$ is independently coded as $\mathbf{c}_k$, for which the received signal is expressible as

$$\mathbf{y} = \sum_{k=1}^{K} \mathbf{\Pi}_k \mathbf{c}_k + \mathbf{n}. \tag{12.120}$$

Users' signals are resolved by virtue of their code structures and interleavers. The code rate of each user should be less than $1/K$ in order for users to be resolvable. However, the receivers no longer take the form of a simple correlator, but must rely on the turbo iteration.

Orthogonal MA schemes are used in practice simply because of their implementation simplicity. However, as mentioned before, the restriction of orthogonality reduces mutual information of the system. Nonorthogonal MA schemes certainly have the advantage of offering a higher MI. The availability of the turbo principle opens a new horizon with other possible MA schemes. Sometimes you are asked to pick up a passenger in an airport you have never seen except for some descriptive features such as thin, tall, middle aged, in grey suit, and with a pair of gold-rimmed glasses. These features are not "orthogonal" to those of the other passengers; yet, it enables you to recognize the right person from the crowd with a very high probability. This example can be considered as the application of an intelligent MA scheme, which is worth further investigation.

12.15   Summary

MA is a successful scenario to apply the paradigm of orthogonality. IUI stems from a simple fact that a κ-dimensional signal space is competed by $K > \kappa$ users. According to the paradigm of orthogonality, the key is to partition the signal space to ensure that each user can operate on an exclusive user space. The partition can be done in time, frequency, or their combination, leading to various multi-access schemes such as TDMA, CDMA, MC-CDMA, OFDMA, MC-DS-CDMA, and so on. Various orthogonal multi-access schemes are equivalent on AWGN channels, in the sense that they have the same mutual information. Yet, they exhibit different features when operating on flat and frequency-selective fading channels, thus providing abundant choices to a wireless system designer.

The predominance of orthogonal (exact or appropriate) MA schemes in current wireless communications is attributed to their simplicity, allowing for the use of linear receivers for user separation. The cost of this simplicity is the reduction in the degrees of freedom, leading to much lower system mutual information as compared to the channel capacity.

Pursuing the channel capacity and orthogonal separation among users should meet two conflicting requirements in the design of an MA scheme. With the invention of the turbo principle, the restriction of orthogonality among users can be relaxed. Use of nonlinear algebraic features such as codes for user separation opens a new horizon to tackle this issue and is, thus, worth further investigation.

A huge literature is available for multiple access schemes and relevant orthogonal sequences. The reader is referred to [18]–[46] for additional reading.

Problems

Problem 12.1   Determine the number of primitive polynomials of degree 5 based on the following factorization:

$$x^{31} + 1 = (1 + x^2 + x^5)(1 + x^3 + x^5)(1 + x + x^3 + x^4 + x^5)(1 + x + x^2 + x^4 + x^5)$$
$$\times (1 + x + x^2 + x^3 + x^5)(1 + x^2 + x^3 + x^4 + x^5)(1 + x).$$

Problem 12.2   Is $g(x) = x^{111} + x^{11} + x + 1$ is an m-sequence generator polynomial? Justify your assertion.

Problem 12.3   Give an m-sequence generated by $g(x) = x^4 + x^3 + 1$.

Problem 12.4   Suppose that a base station transmits BPSK signals $\{s_k\}$ to $K = 50$ user terminals. The K users share the same channel by using m-sequences as their spreading signature (column) vectors $\{\mathbf{c}_k, k = 1, 2, \cdots, K\}$. These m-sequence vectors are generated by the polynomial $x^{10} + x^3 + 1$ with different initial states, and are converted into bipolar format, namely, with their entries taking values of 1 or -1. In vector form, the received signal at user terminal k over one symbol period is expressible as

$$\mathbf{x} = \sum_{k=1}^{K} s_k \mathbf{c}_k + \mathbf{n}, \tag{12.121}$$

where s_k is the symbol for user k with symbol energy E_s, and vector $\mathbf{n}$ denotes additive white Gaussian noise. The entries of $\mathbf{n}$ are independent, each having mean zero and variance σ_n^2. Further, denote $\rho = E_s/\sigma_n^2$. Usually, despreading is done by using a correlator type of receiver.

(a) Determine the number of 1's in an m-sequence generated by the generator polynomial.
(b) Derive an expression for the signal-to-interference plus noise ratio (SINR) at the despreader's output.
(c) Using the SINR expression obtained, determine the SINR for $\rho = 10$ and $20\,\mathrm{dB}$, respectively.
(d) Under what conditions is the noise component negligible compared to the interference?
(e) Sketch the block diagram of a correlator receiver structure for user k.
(f) Suggest a decision variable and decision rule for the detection of BPSK signals for user.

Hint: In part (b), the problem setting is not the same as the processing gain we previously learnt; you may use the cross-correlation property of m-sequences.

Problem 12.5 Describe the idea of user space, and its application to various MA schemes.

Problem 12.6 Suppose a CDMA system supports K users in an AWGN channel with SNR of $\rho = 12\,\mathrm{dB}$ by assigning each user a K-bit m-sequence. Determine system spectral efficiency for $K = 10$ and $K = 50$, respectively.

Problem 12.7 Various multiple-access (MA) schemes are usually analyzed in the context of multiple users who share an AWGN channel. Suppose now that a MA system is to operate on a time-invariant frequency-selective channel instead. Which MA scheme will you choose? Justify your choice.

Problem 12.8 Determine the number of all the m-sequence generators of degree 5, and find their corresponding generator polynomials.

Problem 12.9 Determine all the Walsh code vectors of length 32.

Problem 12.10 Let $\mathbf{H}$ denote the $2^m \times 2^m$ Hadamard matrix where m is a positive integer.

(a) Determine $\log_2 \det(\mathbf{H}^T\mathbf{H})$.
(b) Determine $\mathrm{tr}(\mathbf{H}^T\mathbf{H})$.
(c) Write a MATLAB program for the generation of $\mathbf{H}$ with arbitrary $m \geq 1$.

Problem 12.11 In deriving the SINR for coherent detection of a downlink signal of IS-95, we assume for simplicity that signals of both the home-cell and cross-cell users experience the same propagation loss of unity. Suppose that the 7-cell (i.e., $N = 7$) reuse pattern is used and that the home-cell users have a propagation gain of unity while all the cross-cell users experience propagation power loss equal to $q = (3N)^{n/2}$ with n denoting the power loss exponent.

(a) By assuming that the cross-cell interference is dominated by the six adjacent cells, show that the SINR at the coherent receiver output is given by

$$\gamma = \frac{4}{\rho^{-1} + \frac{2}{qN_c}\sum_{l=1}^{N-1} K_l}.$$

Here, N_c is the length of the Walsh codes, $\rho = E_s/N_0$ is the average transmit SNR, and K_l is the user number in the lth cell.
(b) Determine γ for the case of $n = 4$, $\rho = 10\,\mathrm{dB}$, and $K_1 = \cdots = K_6 = 50$.

Problem 12.12 In an MC-CDMA system with K users, the received signal, in vector form, is given by

$$\mathbf{y} = \sum_{k=1}^{K} \mathbf{F}\mathbf{H}_k \mathbf{c}_k s_k + \mathbf{n},$$

where s_k denotes the symbol from user k which is spread by the spreading vector $\mathbf{c}_k$, $\mathbf{H}_k$ being a diagonal matrix with its (i, i)th entry denoting the channel gain for user k on the ith subcarrier, and AWGN $\mathbf{n} \sim \mathcal{CN}(\mathbf{0}, N_0)$. Assuming CSIR, determine the optimal receiver

(a) for $K = 1$;

(b) for the detection of user k's symbol s_k with a general K.

Problem 12.13 An OFDMA system with N subcarriers to support K users ($K < N$) suffers from inter-carrier interference due to carrier frequency offset. The N subcarriers are assigned to the K users, so that each user has a different set of subcarriers. By collectively writing the data symbols of all the users as an $N \times 1$ vector $\mathbf{s}$, the symbols of user k can be represented by

$$\mathbf{s}^{(k)} = \mathbf{F}^{\dagger}\boldsymbol{\Phi}^{(k)}\mathbf{s},$$

where $\mathbf{F}$ is the N-point DFT matrix, and $\boldsymbol{\Phi}^{(k)}$ is the subcarrier-allocation diagonal matrix for user k, $k = 1, \cdots, K$, such that its n-th diagonal element equals 1 if the nth subcarrier is allocated to user k, and equals 0 otherwise. Due to frequency synchronization errors, the signal $\mathbf{s}^{(k)}$ passes through its channel $\mathbf{H}^{(k)}$ to the base station, arriving at the receiver with a normalized carrier frequency offset $\xi^{(k)}$.

(a) Denote $\boldsymbol{\Omega} = \sum_{k=1}^{K} \mathbf{F}\boldsymbol{\Xi}^{(k)}\mathbf{F}^{\dagger}\boldsymbol{\Phi}^{(k)}$ and $\mathbf{x} = \sum_{k=1}^{K} \mathbf{H}^{(k)}\mathbf{s}^{(k)}$. Show that after DFT, the received symbol is expressible as

$$\mathbf{y} = \boldsymbol{\Omega}\mathbf{x} + \mathbf{F}\mathbf{z},$$

where $\boldsymbol{\Xi}^{(k)} = \mathrm{diag}(1, e^{j2\pi\xi^{(k)}/N}, \cdots, e^{j2\pi\xi^{(k)}(N-1)/N})$ and $\mathbf{z}$ is the white Gaussian noise vector.

(b) Show that the power of the (m, n) entry of $\boldsymbol{\Omega}$ is equal to

$$|\boldsymbol{\Omega}|_{m,n}^{2} = \frac{\sin^{2}(\pi(n - m + \xi^{(k)}))}{N^{2}\sin^{2}(\pi(n - m + \xi^{(k)})/N)}, \tag{12.122}$$

assuming that the channel coefficients and data symbols are zero-mean Gaussian variables with unit variance.

(c) In (a), the effect of carrier frequency offset is incorporated into the interference matrix $\boldsymbol{\Omega}$. Can we multiply $\mathbf{y}$ with $\boldsymbol{\Omega}^{-1}$ to mitigate the inter-carrier interference? If so, what is the side effect with this solution? Are there other solutions?

(Z. Cao et.al, 2007, [14]).

Problem 12.14 Suppose that an OFDMA system employs four subcarriers to support four users, producing an output signal

$$\mathbf{X} = [\mathbf{w}_1, \mathbf{w}_2, \mathbf{w}_3, \mathbf{w}_4]\mathbf{S}.$$

Design the channel-allocation matrix $\mathbf{S}$ if the channel gains of the four users on the four subcarrier channels over time slot 1 are given by

$$\mathbf{G}_1 = \begin{pmatrix} 0.2 + j0.8 & -0.5 + j0.4 & 0.8 - j0.4 & -0.4 - j0.5 \\ -0.3 + j0.3 & 0.8 + j0.8 & -0.1 - j0.2 & 0.2 + j0.1 \\ 0.1 + j0.2 & 0.1 + j0.1 & 0.6 + j0.4 & -0.9 + j0.3 \\ 0.7 + j0.7 & 0.2 + j0.5 & 0.4 + j0.3 & 0.4 + j0.5 \end{pmatrix},$$

and over time slot 2 is given by

$$\mathbf{G}_2 = \begin{pmatrix} 0.3 + j0.9 & -0.5 + j0.4 & 0.5 - j0.6 & -0.6 - j0.5 \\ -0.8 + j0.5 & 0.8 + j0.7 & -0.2 - j0.2 & 0.2 + j0.1 \\ 0.2 + j0.2 & 0.1 + j0.3 & 0.7 + j0.5 & -0.7 + j0.5 \\ 0.3 + j0.3 & 0.5 + j0.6 & 0.6 + j0.8 & 0.5 + j0.7 \end{pmatrix},$$

where the four entries of row i represent the channel gains of user i on $\mathbf{w}_1, \mathbf{w}_2, \mathbf{w}_3$, and $\mathbf{w}_4$, respectively.

Problem 12.15 Consider an OFDMA system with $N = 512$ subcarriers. The OFDM symbol interval is $T_s = 512\,\mu s$ and the rolloff factor of the pulse shape is $\beta = 1.5$. The symbols are transmitted over a channel with coherence bandwidth $B_c = 100\,\text{kHz}$ (the delay spread of the channel is assumed to be $T_c = 1/B_c$). To mitigate ISI between OFDM symbols, a cyclic prefix of length $L = 10$ is used. What is the bandwidth of each subchannel, and the total bandwidth of the system? Calculate the data rate when 16QAM modulation is employed.

Problem 12.16 Assume an OFDMA system operating on a discrete-time FIR channel with response $h(n) = 0.75 + 0.52\delta(n-1) + 0.33\delta(n-2)$. The system has $N = 6$ subcarriers and adopts cyclic prefix of length 2. Find the circulant convolution matrix $\mathbf{H}$ such that the received symbols $\mathbf{y} = [y_5, \cdots, y_0]$ can be represented as $\mathbf{y} = \mathbf{Hx} + \mathbf{n}$, where $\mathbf{x} = [x_5, \cdots, x_0]$. Calculate the singular values of $\mathbf{H}$ and its eigenvectors.

Problem 12.17 Consider a frequency-selective multiple-access uplink in which K mobile users transmit their Gaussian random signals $s_k(t), k = 1, \cdots, K$, respectively, through a corresponding channel $h_k(t)$ to the base station, so that the received signal at the base station can be written as

$$r(t) = \underbrace{\sum_{k=1}^{K} s_k(t) \star h_k(t)}_{x(t)} + n(t),$$

where $\star$ denotes convolution and $n(t)$ is an AWGN process with two-sided spectral density $N_0/2\,\text{W/Hz}$. Show that without CSIT, the channel capacity is given by

$$C = \int_{-\infty}^{\infty} \log_2 \left(1 + \frac{1}{N_0} \sum_{k=1}^{K} S_k(f)|H_k(f)|^2 \right) df,$$

where $S_k(f)$ denotes the power spectral density of $s_k(t)$ and $H_k(f)$ denotes the Fourier transform of $h_k(t)$. Hint: Apply the Karhunen–Loeve expansion to $s(t)$ to represent $r(t)$ in vector form [15].

Problem 12.18 Can we use the formula (12.47) to derive the mutual information of CDMA in frequency-selective fading? Justify your answer.

Problem 12.19 Suggest a method to derive the mutual information of the CDMA system operating in frequency-selective fading.

Problem 12.20 For a prime number P, let $x_u(n)\triangleq \exp(-j\pi un(n+1)/P), 0 \le n < P$, be the uth root Zadoff–Chu sequence of length P, and let $X_u(k)$ denote its DFT. Show that

$$X_u(k) = x_u^\dagger(u^{-1}k)X_u(0), \quad 0 \le k < P, \tag{12.123}$$

where u^{-1} is the multiplicative inverse of u modulo P [16].

Problem 12.21 Let N be an even positive number, and let M be an integer prime to N. Construct an $N \times N$ matrix $\mathbf{A}$ by the product

$$\mathbf{A} = \mathbf{FZ},$$

where $\mathbf{F}$ is an IDFT matrix with its (m, n)th entry specified by $\exp(j2\pi nm/N)$, and $\mathbf{Z}$ is a diagonal matrix formed by a Zadoff–Chu sequence, that is

$$\mathbf{Z} = \text{diag}\left(1, \cdots, \exp\left(\frac{-j\pi Mn(n+2g)}{N}\right), \cdots, \exp\left(\frac{-j\pi M(N-1)(N-1+2g)}{N}\right)\right)$$

with g an integer. Show that the cross-correlation between the k-th and the l-th rows of $\mathbf{A}$ is given by

$$R_{kl}(\tau) = \begin{cases} N \exp\left(\frac{j\pi(M\tau^2 - 2(l-Mg))\tau}{N}\right), & \mathrm{mod}(l - k - M\tau, N) = 0 \\ 0, & \mathrm{mod}(l - k - M\tau, N) \neq 0. \end{cases}$$

(Li and Huang, 2007 [17]).

Problem 12.22 Conduct a literature search to find the application of CAZAC sequences in wireless communications.

Problem 12.23 Compare the advantages and disadvantages of OFDMA and CDMA, and offer your comments.

Problem 12.24 Pinpoint the technical and nontechnical reasons that lead to the choice of OFDMA as the multiple-access scheme in 4G cellular, and describe the measures used in 4G to prevent or mitigate cross-cell interference.

Problem 12.25 With the advent of the turbo principle, investigate the possible application of nonorthogonal spreading codes in future multiuser systems.

References

1. N. Yee, J.-P. Linnardtz, and G. Fettweis, "Multi-carrier CDMA for indoor wireless radion networks," *Proceedings of PIMRC'93*, pp. 109–113, Sep. 1993.
2. A. Chpoly, A. Brajal, and S. Jourdan, "Orthogonal multi-carrier techniques applied to direct-sequence spread spectrum CDMA systems," *Globecom'93*, Conference Records vol. 3, pp. 1723–1728, Nov.-Dec. 1993, Houston, TX.
3. T.M. Cover and J.A. Thomas, *Elements of Information Theory*, 2nd ed., New York: John Wiley & Sons, Inc., 2006.
4. S. Shamai and A.D. Wyner, "Information-theoretic consideration for symmetric, cellular, multiple-access fading channels-Part I," *IEEE Trans. Inf. Theory*, vol. 43, no. 6, pp. 1877–1894, 1997.
5. S. Haykin, *Digital Communications*, 4th ed., New York: John Wiley & Sons, Inc., 2012.
6. R.E. Ziemer and R.L. Peterson, *Introduction to Digital Communication*, 2nd ed., Upper Saddle River, NJ: Prentice Hall, 2001.
7. ITU-R M.2134 Rep., "Requirements Related to Technical Performance for IMT-Advanced Radio Interface(s)," Nov. 2008.
8. C. Ciochina, D. Mortier, and H. Sari, "An analysis of three multiple access techniques for the uplink of future cellular mobile systems," *Eur. Trans. Telecommun.*, vol. 19, no. 5, pp. 581–588, 2008.
9. M. Zivkovic, "A table of primitive binary polynomials," *Math. Comput.*, vol. 62, no. 205, pp. 385–386, 1994.
10. M. Zivkovic, "Table of primitive binary polynomials: II," *Math. Comput.*, vol. 63, no. 207, pp. 301–306, 1994.
11. R.L. Frank and S.A. Zadoff, "Phase shift pulse codes with good periodic correlation properties," *IRE Trans. Inf. Theory*, vol. 8, pp. 381–382, 1962.
12. D. Chu, "Polyphase codes with good periodic correlation properties," *IEEE Trans. Inf. Theory*, vol. 18, pp. 531–532, 1972.
13. P. Li, L. Liu, K.Y. Wu, and W.K. Leung, "Interleave division multiple-access," *IEEE Trans. Wireless Commun.*, vol. 5, no. 4, pp. 938–947, 2006.
14. Z. Cao, U. Tureli, and Y.-D. Yao, "Low-complexity orthogonal spectral signal construction for generalized OFDMA uplink with frequency synchronization errors," *IEEE Trans. Veh. Technol.*, vol. 56, no. 3, pp. 1143–1154, 2007.
15. L.H. Ozarow, S. Shamai, and A.D. Wyner, "Information-theoretic considerations for mobile radio," *IEEE Trsans. Veh. Technol.*, vol. 43, no. 2, pp. 359–378, 1994.
16. D. Sarwate, "Bounds on crosscorrelation and autocorrelation of sequences (corresp.)," *IEEE Trans. Inf. Theory*, vol. 25, no. 6, pp. 720–724, 1979.

17. C.-P. Li and W.-C. Huang, "A constructive representation for the fourier dual of the zadoff-chu sequences," *IEEE Trans. Inf. Theory*, vol. 53, no. 11, pp. 4221–4224, 2007.

18. 3GPP R1-060908, "On the performance of LTE RACH," Mar. 27–31, 2006, Athens, Greece.

19. 3GPP TSG RAN1#53bis, R1-082609, "Uplink Multiple Access for LTE-Advanced," Warsaw, Poland, June 30-July 4, 2008.

20. 3GPP TR 36.913, "Requirements for Further Advancements for Evolved Universal Terrestrial Radio Access (E-UTRA)," v.8.0.1, Mar. 2009.

21. 3GPP TR 36.814 V9.0.0, "Further Advancements for E-UTRA Physical Layer Aspects," Mar. 2010.

22. S. Hara and R. Prasad, "Overview of multicarrier CDMA," *IEEE Commun. Mag.*, vol. 35, pp. 126–133, 1997.

23. L.L. Yang and L. Hanzo, "Multicarrier DS-CDMA: a multiple access scheme for ubiquitous broadband wireless communications," *IEEE Commun. Mag.*, vol. 41, no. 10, pp. 116–124, 2003.

24. R.H. Mahadevappa and J.G. Proakis, "Multigating multiple access interference and intersymbol interference in uncoded CDMA systems with chip-level interleaving," *IEEE Trans. Wireless Commun.*, vol. 1, no. 4, pp. 781–792, 2002.

25. A. Jamalipour, T. Wada, and T. Yamazato, "A tutorial on multiple access technologies for beyond 3G mobile networks," *IEEE Commun. Mag.*, vol. 43, no. 2, pp. 110–117, 2005.

26. R.L. Pickholtz, L.B. Milstein, and O.L. Schilling, "Spread spectrum for mobile communications," *IEEE Trans. Veh. Technol.*, vol. 40, no. 2, pp. 313–322, 1991.

27. S. Zhou, M. Zhao, J. Wang, and Y. Yao, "Distributed wireless communications systems: a new architecture for future public wireless access," *IEEE Commun. Mag.*, vol. 41, no. 3, pp. 108–113, 2003.

28. B.M. Popovic, "Efficient despreaders for multi-code CDMA systems," *Proceedings of IEEE ICUPC'97*, San Diego, CA, USA, Oct. 12-16, 1997, pp. 516–520.

29. X. Wang and H.V. Poor, "Iterative (turbo) soft interference cancellation and decoding for coded CDMA," *IEEE Trans. Commun.*, vol. 47, no. 7, pp. 1046–1061, 1999.

30. E.H. Dinan and B. Jabbari, "Spreading codes for direct sequence CDMA and wideband CDMA cellular networks," *IEEE Commun. Mag.*, vol. 36, pp. 48–54, 1998.

31. B.J. Wysocki and T.A. Wysocki, "Modified Walsh-Hadamard sequences for DS-CDMA wireless systems," *Int. J. Adapt. Control Signal Process.*, vol. 16, pp. 589–602, 2002.

32. F. Adachi, "Effects of orthogonal spreading and Rake combining on DS-CDMA forward link in mobile radio," *IEICE Trans. Commun.*, vol. E80-B, pp. 1703–1712, 1997.

33. E.K. Hong, K.J. Kim, and K.C. Whang, "Performance evaluation of DS-CDMA system with M-ary orthogonal signaling," *IEEE Trans. Veh. Technol.*, vol. 45, pp. 57–63, 1996.

34. V.M. DaSilva and E.S. Sousa, "Performance of orthogonal CDMA codes for quasi-synchronous communication systems," *Proceedings of IEEE UPC'1993*, Ottawa, Canada, Oct. 12-15, 1993, pp. 995–999.

35. J.-W. Choi, Y.-H. Lee, and Y.-H. Kim, "Performance analysis of forward link DS-CDMA systems using random and orthogonal spreading sequences," *Proceedings of IEEE ICC'2001*, St.-Petersburg, Russia, pp. 1446–1450, June 11-15, 2001.

36. G. Xiang and T.S. Ng, "Performance of asynchronous orthogonal multicarrier CDMA system in frequency selective fading channel," *IEEE Trans. Commun.*, vol. 47, no. 7, pp. 1084–1091, 1999.

37. K.H.A. Karkkainen and P.A. Leppanen, "Comparison of the performance of some linear spreading code families for asynchronous DS/SSMA systems," *Proceedings of IEEE MILCOM'91*, 4–7, pp. 784–790, Nov. 1991.

38. Y. Li and S. Yoshida, "Near capacity multiple access scheme for interference channel using complex-valued signals," *IEEE Electron. Lett.*, vol. 34, no. 22, pp. 2096–2097, 1998.

39. D.V. Sarwate and M.B. Pursley, "Cross-correlation properties of pseudorandom and related sequences," *IEEE Proc.*, vol. 68, pp. 593–619, 1980.

40. P. Fan and M. Darnell, *Sequence Design for Communication Applications*, UK: Research Studies Press, 1996.

41. J. Oppermann and B.S. Vucetic, "Complex spreading sequences with a wide range of correlation properties," *IEEE Trans. Commun.*, vol. 45, pp. 365–375, 1997.

42. S. Rahardja, W. Ser, and Z. Lin, "UCHT-based complex sequences for asynchronous CDMA system," *IEEE Trans. Commun.*, vol. 51, pp. 618–626, 2003.

43. R.G. Gallager, "*An inequality on the capacity region of multiaccess multipath channels,*" in *Communication and Cryptography: Two Sides of Tapestry*, R.E. Blahut, etc., Eds. Boston, MA: Kluwer Academic Publishers, 1994.

44. D.N.C. Tse and S.V. Hanly, "Multiaccess fading channels-Part I: polymatroid structure, optimal resource allocation and throughput capacities," *IEEE Trans. Inf. Theory*, vol. 44, no. 7, pp. 2796–2815, 1998.

45. G.R. Cooper, *Modern Communications and Spread Spectrum*, McGraw-Hill, 1996, Chapter 11 (PN sequence and m-sequence generation).

46. B.M. Popovic, "Generalized chip-like polyphase sequence with optimum correlation properties," *IEEE Trans. Inf. Theory*, vol. 38, pp. 1406–1409, 1992.

13

Wireless MIMO Systems

Abstract

The abnormal capacity inherent in multi-input multi-output (MIMO) fading channels makes it possible to implement high-speed wireless transmission, as demonstrated by Telatar and Foschini. One major challenge is the inter-antenna (or cross-antenna) interference (IAI) caused by spatial correlation among antennas. The idea used to combat inter-symbol interference arising in temporally dispersive channels is also applied to MIMO systems. Various possible transceivers that tackle IAI are derived and analyzed in this chapter under different criteria such as maximal mutual information, zero forcing, and minimum mean-square error (MMSE). The difference is that spatial correlation usually does not possess a Toeplitz structure.

A more aggressive strategy to combat IAI, however, is to eliminate its source of generation. Alamouti achieves this goal by introducing temporal orthogonality at the transmitter, which is virtually converted into equivalent independent channels at the receiver whereby the IAI is completely eliminated. More generally, the IAI problem automatically vanishes if we adopt a different philosophy of treating a set of parallel-transmitted data streams as a single matrix symbol. A MIMO channel is of multiple degrees of freedom (DF), which can be converted into higher data rate or into transmission reliability. The tradeoff between the two is also discussed.

♣

13.1 Introduction

As mobile units move around a ground surface, scatterers such as buildings, towers, and foliage rebound and reflect the transmitted electromagnetic waves, thus forming a two-dimensional (2-D) random field. The consequence is fading in the received signal magnitudes, a phenomenon that leads to a significant drop in transmission reliability. This harmful effect is only one aspect of a 2-D random field. The random field also possesses abnormal channel capacity. Such a potential, when fully exploited, will radically change our vision of wireless communications. The technique to be used is MIMO.

We need to address two issues: a theoretic issue of potential capacity in MIMO Channels, and a practical issue of transceiver structures for its exploitation. A theoretical breakthrough began with the seminal paper

Wireless Communications: Principles, Theory and Methodology, First Edition. Keith Q.T. Zhang.
© 2016 John Wiley & Sons, Ltd. Published 2016 by John Wiley & Sons, Ltd.
Companion Website: www.wiley.com/go/zhang7749

by Telatar, which kicked off the study of MIMO channels and systems. Telatar's discovery totally changed our notion: the randomness of scattering fields is a resource. By placing antennas in a collocated manner or distributively over a random field, we can exploit its abnormal capacity. The huge capacity of MIMO channels was shortly verified by an experimental system called vertical Bell Labs layered space-time (VBLAST), which further stimulated the widespread study of MIMO systems. Three sections of this chapter are dedicated to the study of MIMO channel capacity, followed by four sections dedicated to possible transceivers of practical interest.

Today, MIMO techniques are recommended for use in various wireless standards. Typical examples include the adoption of Alamouti space-time codes in 3GPP LTE standards, MIMO technology in the 802.11n Wireless LAN study group, and spatial multiplexing techniques in 802.16 WiMAX systems. In the standardization for the coming 5G, massive MIMO is proposed as a basic structure.

We use the following symbols throughout this chapter.

N_t : the number of transmit antennas
N_r : the number of receive antennas
P_s : the total transmit power
σ_n^2 : the noise variance at each receive antenna
$M = \max\{N_t, N_r\}$
$m = \min\{N_t, N_r\}$
$\sigma_s^2 = P_s/N_t$: Power at each transmit antenna
$\rho = \sigma_s^2/\sigma_n^2$: SNR at each transmit antenna

13.2   Signal model and mutual information

The maximum data rate achievable by a MIMO system depends on the available channel state information at the transmitter and receiver. In this chapter, we assume that channel state information (CSI) is always available at the receiver side (CSIR) but not necessarily at the transmitter side (CSIT). Thus, at the transmitter side, we can classify two cases: with and without CSIT. Suppose that an $N_t \times 1$ random symbol vector s of total power P_s transmits over an $N_r \times N_t$ MIMO channel **H**, producing an $N_r \times 1$ output vector

$$\mathbf{y} = \mathbf{HQs} + \mathbf{n},\tag{13.1}$$

where $\mathbf{n} \sim \mathcal{CN}_{N_r}(\mathbf{0}, \sigma_n^2\mathbf{I})$, $\mathbf{s} \sim \mathcal{CN}_{N_t}(\mathbf{0}, \sigma_s^2\mathbf{I})$, with σ_s^2 denoting the average power on each transmit antenna, and **Q** denotes the precoding matrix at the transmitter. In what follows, the subscript in a complex Gaussian distribution to signify its dimension will be dropped for notational simplicity, as long as no confusion is introduced.

A MIMO channel **H** can have correlation at the transmit and receive sides because of electromagnetic coupling between antennas and the impinging angle of the wavefronts. For tractability, we often assume that the entries of **H** are jointly complex Gaussian-distributed, represented as

$$\mathbf{H} \sim \mathcal{CN}(\mathbf{M}, \mathbf{R}_r \otimes \mathbf{R}_t).\tag{13.2}$$

This expression means that **H** has mean matrix **M**, and its stack vector vec(**H**) has the covariance structure of $\mathbf{R}_r \otimes \mathbf{R}_t$. The vector vec(**H**) is formed by stacking the columns of **H** one over another. The Kronecker type of covariance structure is assumed partly for ease of treatment and partly reflects the practical situations, implying that the MIMO channel has correlation matrix $\mathbf{R}_t$ and $\mathbf{R}_r$ at the transmitter and receiver sides, respectively. Define matrix

$$\mathbf{W} = \begin{cases} \mathbf{H}^\dagger\mathbf{H}, & N_t \le N_r \\ \mathbf{HH}^\dagger, & N_r < N_t, \end{cases}\tag{13.3}$$

which will be used in the subsequent analysis.

The purpose of this chapter is twofold: investigating the channel capacity of various MIMO channels, and designing possible transceivers for its exploitation. Note that the signal model in (13.1) looks very similar to that for ISI channels. However, unlike a time-invariant ISI channel, $\mathbf{H}$ has no Toeplitz structure. This difference determines that the OFDM signaling technique is not applicable to MIMO channels. Keeping these similarity and dissimilarity in mind is helpful in understanding various transceiver structures for MIMO channels. As a special case, a MIMO system can install a single antenna at the transmitter or the receiver end, reducing to the classical diversity reception or transmission, usually referred to as SIMO and MISO, respectively.

A MIMO system can experience both spatial and temporal correlations, forming the so-called frequency-selective fading MIMO channel. The temporal correlation can be resolved by using the OFDM technique so that a frequency-selective channel is reducible to a bank of frequency flat fading MIMO channels. This chapter is, therefore, just focused on frequency flat fading MIMO channels.

The determination of the mutual information of the MIMO channel is similar to the determination of that for a scalar flat fading channel, that is, calculate the difference between the entropies given and without given s. The only difference is that the variance is now replaced by the *generalized* variances given and without given s, which are equal to $\det(\sigma_n^2\mathbf{I} + \sigma_s^2\mathbf{Q}^\dagger\mathbf{H}^\dagger\mathbf{H}\mathbf{Q})$ and $\det(\sigma_n^2\mathbf{I})$, respectively. The logarithm of these generalized variances represents the corresponding entropies. As such, conditioned on $\mathbf{H}$, the channel mutual information is given by

$$C = \log \det(\mathbf{I} + \rho\mathbf{Q}^\dagger\mathbf{H}^\dagger\mathbf{H}\mathbf{Q}), \qquad (13.4)$$

where $\rho = \sigma_s^2/\sigma_n^2$. The mutual information C is measured in bits/s/Hz if the base of the logarithm is 2. In this chapter, the default base is 2 unless otherwise stated. We need to find the semidefinite positive matrix $\mathbf{Q}$ (symbolically denoted as $\mathbf{Q} \geq 0$) for which the *instantaneous* mutual information or its ensemble average is maximized to achieve the corresponding channel capacity. The treatment for the two cases with and without CSIT is different, and is therefore addressed separately in the subsequent sections.

13.3 Capacity with CSIT

With CSIT, one can optimize the precoder $\mathbf{Q}$ to maximize the mutual information in (13.4) to reach the instantaneous channel capacity

$$C = \max_{\mathbf{Q}} \log \det(\mathbf{I} + \rho\mathbf{Q}^\dagger\mathbf{H}^\dagger\mathbf{H}\mathbf{Q}). \qquad (13.5)$$

According to Hadamard's inequality described in Chapter 2, $\mathbf{Q}$ should be chosen such that $\mathbf{Q}^\dagger\mathbf{H}^\dagger\mathbf{H}\mathbf{Q}$, the matrix argument of the determinant, is diagonalized. The desired $\mathbf{Q}$ takes the form

$$\mathbf{Q} = \mathbf{UP},$$

where $\mathbf{P} = \mathrm{diag}(p_1, \cdots, p_m)$ is the power-weighting allocation matrix subject to a constant constraint $\mathrm{tr}(\mathbf{P}^\dagger\mathbf{P}) = \Omega_t$, $\mathbf{U}$ is the unitary matrix formed by the eigenvectors of $\mathbf{H}^\dagger\mathbf{H}$ such that

$$\mathbf{H}^\dagger\mathbf{H} = \mathbf{U\Lambda U}^\dagger,$$

and $\mathbf{\Lambda} = \mathrm{diag}(\lambda_1, \cdots, \lambda_m)$ is formed by its eigenvalues. As such, we can simplify (13.5) to yield

$$C = \max_{\mathrm{tr}(\mathbf{P}^\dagger\mathbf{P})=\Omega_t} \sum_{i=1}^{m} \log\left(1 + |p_i|^2\lambda_i\rho\right). \qquad (13.6)$$

The optimal power-weighting allocation can be determined by using the method of Lagrange multiplier, leading to the so-called water-filling technique defined by

$$|p_i|^2 = \left[\frac{1}{\alpha} - \frac{1}{\rho\lambda_i}\right]^+, \qquad (13.7)$$

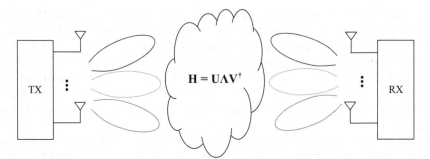

Figure 13.1   *Parallel eigenbeams transmission*

where $[a]^+ = \max\{0, a\}$, and the multiplier α (i.e., the water level) is chosen to meet the constraint

$$\sum_{i=1}^{m} \left[\frac{1}{\alpha} - \frac{1}{\rho\lambda_i} \right]^+ = \Omega_t. \tag{13.8}$$

With this result, the received signal reduces to

$$\mathbf{y} = \sum_{i=1}^{m} p_i \sqrt{\lambda_i} \, \rho \mathbf{v}_i + \mathbf{n},$$

which has a clear physical implication. Namely, the best transmission strategy with CSIT is to transmit symbol signals in parallel along the eigenbeams defined by the channel matrix, with power allocated according to the rule of water filling. The situation of eigenbeam transmission is sketched in Figure 13.1 Good channels with higher instantaneous SNRs are allocated more power, hoping to achieve a higher efficiency in power-to-data rate conversion, while poorly conditioned channels are likely to be switched off.

13.4   Ergodic capacity without CSIT

In many applications, channel information is not available at the transmitter. The best strategy is to uniformly distribute power over the spatial directions defined by the channel matrix. As such, the precoder $\mathbf{Q}$ takes the form $\mathbf{Q} = \sigma_s^2 \mathbf{I}$ for which the mutual information of the MIMO channel reduces to

$$C = \log \det(\mathbf{I} + \rho\mathbf{W}) \tag{13.9}$$

with $\mathbf{W}$ defined in (13.3). As a function of random matrix $\mathbf{W}$, the mutual information C is a random variable. We are interested in its mean and higher order moments. Directly working on its moments is not easy, given the nonlinearity of the $\log \det$ operator. All the information of interest is embodied in the characteristic function (CHF) of C, defined by $\mathbb{E}[\exp(zC)]$. The CHF provides a natural way of removing the nonlinear logarithm operation in C, leading to a relatively simple expression

$$\phi_C(z) = \mathbb{E}[\det(\mathbf{I} + \rho\mathbf{W})^{z/\ln 2}]$$

$$= \mathbb{E}\left[\prod_{i=1}^{M} (1 + \rho\lambda_i)^{z/\ln 2} \right]. \tag{13.10}$$

The second line represents the determinant using the eigenvalues of $\mathbf{W}$, denoted by $\{\lambda_i\}$. The kth moment of C can be determined by taking the derivative of C, as shown by

$$\mathbb{E}[C^k] = \frac{d^k \phi_C(z)}{dz^k}\Big|_{z=0}, \ k = 1, 2, \cdots. \tag{13.11}$$

The mean value $\mathbb{E}[C]$ is called the *ergodic* channel capacity, assuming that a long temporal average of C is equal to its ensemble average.

Given the adoption of the CHF approach, there are, yet, two different techniques [1, 2, 3]. The first technique starts from the second line of (13.10) by taking the average over the marginal PDF of the sample eigenvalues of $\mathbf{W}$. Following the second technique, one can directly work on the Wishart-distributed matrix $\mathbf{W}$, as shown on the first line of (13.10), by representing the determinant as a hypergeometric function $_0F_0$ with matrix arguments. The strategy for handling such hypergeometric functions is to expand them as a zonal series, a powerful tool developed by A.T. James in 1964 [4], whereby expectation can be evaluated term by term. Surprisingly, the final series so obtained is often expressible in terms of incomplete gamma functions [1, 3, 5]. In the subsequent analysis, we follow the first approach.

13.4.1 i.i.d. MIMO Rayleigh channels

For i.i.d. Gaussian MIMO channels, the sample eigenvalues of $\mathbf{W}$ follow the joint identical distribution, which is well known in the literature. Starting from the joint PDF of the unordered sample eigenvalues of $\mathbf{W}$, Telatar obtained the ergodic capacity of MIMO Rayleigh channel under power constraint ρ, as shown by Telatar [6]

$$\mathbb{E}[C] = \int_0^\infty \log(1 + \rho z) \sum_{k=0}^{m-1} \frac{k!}{(k+M-m)!} [L_k^{M-m}(z)]^2 z^{M-m} e^{-z}, dz \tag{13.12}$$

where L_j^i are the associated Laguerre polynomials defined by

$$L_j^i(z) = \frac{1}{j!} e^z z^{-i} \frac{d^j}{dz^j}(e^{-z} z^{i+j}). \tag{13.13}$$

13.4.2 Ergodic capacity for correlated MIMO channels

There are two different methods used for correlated MIMO. Here, we present the results due to Kang and Alouini [1].

The fundamental tool for MIMO capacity analysis is the theory of Wishart distribution developed in multivariate statistical analysis. Theoretically, we can handle general MIMO channels with full correlation in the form of a Kronecker product, resulting in an intractable capacity expression in terms of a hypergeometric function of three matrix arguments. To obtain a numerically computable result, we usually confine ourselves to the cases with semi-correlation. Two cases are of practical interest. (1) Correlation occurs at the side with fewer antennas; (2) correlation occurs at the side with more antennas. The first case directly fits into the framework of Wishart distribution; the second case needs some extension. These two cases need different mathematical treatments and lead to different results, as described below.

13.4.2.1   *Correlation occurs at the least-antenna side*

Suppose the correlation occurs at the least antenna side with an $m \times m$ correlation matrix Θ, the m eigenvalues of which, in the descending order, are given by $\theta_1 \geq \cdots \geq \theta_m > 0$. Then, by denoting

$$\beta_1 = \frac{1}{\det(\mathbf{V}_1) \prod_{i=1}^{m} \Gamma(M - i + 1)}, \tag{13.14}$$

the CHF of the mutual information is given by

$$\phi_C(z) = (\beta_1 / \ln 2) \det(\mathbf{\Psi}_1), \tag{13.15}$$

where $\mathbf{\Psi}_1$ is an $m \times m$ matrix with its (i,j)th entry given by

$$\{\mathbf{\Psi}_1\}_{i,j} = \int_0^\infty (1 + \rho x)^{\frac{z}{\ln 2}} x^{M-i} e^{-\frac{x}{\theta_j}} \, dx \tag{13.16}$$

and $\mathbf{V}_1$ is an $m \times m$ matrix whose determinant is given by

$$\det(\mathbf{V}_1) = \det((-1)^{m-j} \, \theta_i^{M-m+j})$$

$$= \left(\prod_{i=1}^{m} \theta_i^M \right) \prod_{1 \leq \ell < k \leq m} \left(\frac{1}{\theta_k} - \frac{1}{\theta_\ell} \right). \tag{13.17}$$

On the first line, the matrix $\mathbf{V}_1$ is simply shown in terms of its (i,j)th entry for brevity. The (i,j)th entry in (13.16) can be simplified to

$$\{\mathbf{\Psi}_1\}_{ij} = \rho^{-M+i-1} \Gamma(M - i + 1) U \left(M - i + 1, M - i + 2 + \frac{z}{\ln 2}, \frac{1}{\rho \theta_j} \right), \tag{13.18}$$

where $U(\cdot, \cdot, \cdot)$ is the confluent function of the second kind and can be easily calculated by directly calling the Mathematica function.

The CHF given in (13.15) enables us to determine the MIMO ergodic capacity with correlation at the least antenna side, yielding

$$\mathbb{E}[C] = (\beta_1 / \ln 2) \sum_{k=1}^{m} \det(\mathbf{\Psi}_1(k)), \tag{13.19}$$

where $\mathbf{\Psi}_1(k)$ is an $m \times m$ matrix with its (i,j)th entry given by

$$\{\mathbf{\Psi}_1(k)\}_{i,j} = \begin{cases} \int_0^\infty \ln(1 + \rho x) x^{M-i} e^{-x/\theta_j} \, dx, & i = k \\ \theta_j^{M-i+1} \Gamma(M - i + 1), & i \neq k \end{cases}. \tag{13.20}$$

The second moment of C is given by

$$\mathbb{E}[C^2] = (\beta_1 / \ln^2 2) \sum_{k=1}^{m} \sum_{\ell=1}^{m} \det(\mathbf{\Omega}(k, \ell)), \tag{13.21}$$

where $\mathbf{\Omega}_1(k, \ell), k, \ell = 1, \cdots, m$, are $m \times m$ matrices with entries given by

$$\{\mathbf{\Omega}_1(k, \ell)\}_{i,j} = \begin{cases} \int_0^\infty \ln^2(1 + \rho x) x^{M-i} e^{-x/\theta_j} \, dx, & i = k = \ell \\ \int_0^\infty \ln(1 + \rho x) x^{M-i} e^{-x/\theta_j} \, dx, & i = k \text{ or } i = \ell; k \neq \ell. \\ \theta_j^{M-i+1} \Gamma(M - i + 1), & i \neq k, i \neq \ell \end{cases} \tag{13.22}$$

13.4.2.2 Correlation occurs at the most-antenna side

Suppose that the correlation occurs in the most-antenna side with an $M \times M$ correlation matrix $\mathbf{R}$, whose eigenvalues in the descending order are given by $r_1 \geq \cdots \geq r_M > 0$. Then, by denoting

$$\beta_2 = \frac{(-1)^{m(M-m)}}{\det(\mathbf{V}_2) \prod_{i=1}^{m} \Gamma(m-i+1)}, \tag{13.23}$$

the CHF of the mutual information is expressible as

$$\phi_C(z) = \beta_2 \det\left(\begin{bmatrix} \boldsymbol{\Psi}_{2A} \\ \boldsymbol{\Psi}_{2B} \end{bmatrix}\right), \tag{13.24}$$

where $\Gamma(\cdot)$ is the Gamma function, the $(M-m) \times M$ matrix $\boldsymbol{\Psi}_{2A}$ is defined by

$$\boldsymbol{\Psi}_{2A} = [(-r_j)^{m-M+i}]_{i=1,\cdots,M-m; j=1,\cdots,M}, \tag{13.25}$$

and $\boldsymbol{\Psi}_{2B}$ is the $m \times M$ matrix with entries given by

$$\{\boldsymbol{\Psi}_{2B}\}_{i,j} = \int_0^\infty (1+\rho x)^{\frac{z}{\ln 2}} x^{m-i} e^{-\frac{x}{r_j}} dx, i = 1\cdots, m; j = 1, \cdots, M; \tag{13.26}$$

and $\mathbf{V}_2$ is an $m \times m$ matrix whose determinant is given by

$$\det \mathbf{V}_2) = \det[(-1)^{M-j} r_i^{m-M+j}]_{i,j=1,\cdots,M}$$

$$= \left(\prod_{i=1}^{M} r_i^m\right) \prod_{1 \leq \ell < k \leq M} \left(\frac{1}{r_k} - \frac{1}{r_\ell}\right). \tag{13.27}$$

In the above expression, the matrix $\mathbf{V}_2$ is simply represented by its (i, j)th entries. It follows that

$$\mathbb{E}[C] = (\beta_2/\ln 2) \sum_{k=1}^{m} \det\left(\begin{bmatrix} \boldsymbol{\Psi}_{2A} \\ \boldsymbol{\Phi}_{2B}(k) \end{bmatrix}\right), \tag{13.28}$$

where the lower $m \times M$ matrix $\boldsymbol{\Phi}_{2B}(k)$ has its (i, j)th entry defined by

$$\{\boldsymbol{\Phi}_{2B}(k)\}_{i,j} = \begin{cases} \int_0^\infty \ln(1+\rho x) x^{m-i} e^{-x/r_j} dx & i = k \\ r_j^{m-i+1} \Gamma(m-i+1), & i \neq k \end{cases} \tag{13.29}$$

for $1 \leq i \leq m$ and $1 \leq j \leq M$.

It also follows that

$$\mathbb{E}[C^2] = (\beta_2/\ln^2 2) \sum_{k=1}^{m} \sum_{\ell=1}^{m} \det\left(\begin{bmatrix} \boldsymbol{\Psi}_{2A} \\ \boldsymbol{\Sigma}_{2B}(k, \ell) \end{bmatrix}\right) \tag{13.30}$$

with the $m \times M$ matrices $\boldsymbol{\Sigma}_{2B}(k, \ell), k, \ell = 1, \cdots, m$, have their entries defined by

$$\{\boldsymbol{\Sigma}_{2B}(k,\ell)\}_{i,j} = \begin{cases} \int_0^\infty \ln^2(1+\rho x) x^{m-i} e^{-x/r_j} dx, & i = k = \ell \\ \int_0^\infty \ln(1+\rho x) x^{m-i} e^{-x/r_j} dx, & i = k \text{ or } i = \ell; k \neq \ell \\ r_j^{m-i+1} \Gamma(m-i+1), & i \neq k, i \neq \ell \end{cases} \tag{13.31}$$

for $i = 1, \cdots, m$ and $j = 1, \cdots, M$.

The ergodic capacity is an important characteristic and metric of a MIMO channel. It is desirable to have a high ergodic capacity. However, two MIMO channels with the same ergodic capacity may have different engineering implications. The one with a larger second moment means that the instantaneous mutual information has a larger fluctuation around the ergodic capacity. A consequence is that a system operating in such an environment experiences more frequent capacity outage or unreliable communications.

Having obtained the characteristic functions of the mutual information, we can calculate the cumulative distribution function (CDF) of the capacity using

$$F(x) = \Pr\{C < x\} = \frac{1}{2} - \frac{1}{\pi} \int_0^\infty \frac{\Im\{\Phi_C(j\omega)e^{-j\omega x}\}}{\omega} \, d\omega, \tag{13.32}$$

which, with an appropriate threshold setting, is also interpreted as capacity outage.

13.5   Capacity: asymptotic results

The exact expression for MIMO channel capacity derived in the previous sections is very useful. However, it is sometime helpful to have simple yet insightful asymptotic results. Results presented in this section represent the efforts along this direction.

13.5.1   Asymptotic capacity with large MIMO

We focus on the i.i.d. MIMO channel matrix without CSIT and without assuming a particular distribution on the channel matrix. According to (13.3), we have $(1/M)\mathbf{W} \to \mathbf{I}$ as $M \to \infty$. By setting $\mathbf{Q} = \mathbf{I}$ and denoting $\rho_s = \rho M$, the ergodic capacity reduces to

$$\mathbb{E}[\log \det(\mathbf{I} + (\rho_s/M)\mathbf{W})] \to m \, \log(1 + \rho_s), \tag{13.33}$$

which is linearly proportional to $m \triangleq \min\{N_t, N_r\}$. This is the first asymptotic result.

A more accurate asymptotic expression for the ergodic capacity can be derived. Denote $\mathbf{A} = (1/M)\mathbf{W}$ and $\tau = M/m \geq 1$. For large m and M, and $M/m \to \tau \geq 1$, it can be shown that [7]

$$\mathbb{E}[C] = \int_0^\infty \log\left(1 + \frac{\rho_s m}{N_t}v\right) dF^{\mathbf{A}}(v), \tag{13.34}$$

where $F^{\mathbf{A}}(v)$ is the empirical cumulative distribution that the number of the eigenvalues of the $m \times m$ Hermitian matrix $\mathbf{A}$ is less than v.

Denoting $v_\pm = (\sqrt{\tau} \pm 1)^2$, we have

$$\frac{dF^{\mathbf{A}}(v)}{dv} = \begin{cases} \frac{1}{2\pi}\sqrt{(\frac{v_+}{v} - 1)(1 - \frac{v_-}{v})}, & \text{for } v \in [v_-, v_+]; \\ 0, & \text{otherwise.} \end{cases}$$

Thus, for large N_t and N_r, we have

$$C = \frac{m}{2\pi} \int_{v_-}^{v_+} \log\left(1 + \frac{\rho_s m}{N_t}v\right)\sqrt{\left(\frac{v_+}{v} - 1\right)\left(1 - \frac{v_-}{v}\right)} \, dv. \tag{13.35}$$

Both (13.33) and (13.35) uncover that the ergodic capacity is linear in m. The findings above enable the following assertion:

Property 13.5.1 *As an approximate rule, the ergodic capacity of MIMO channels is proportional to $m = \min\{N_t, N_r\}$.*

This property provides a strong justification to support the use of MIMO systems for wireless communications. Modern wireless networks face a serious challenge of ever-increasing data rates, which, theoretically, can be solved by exploiting new frequency bands. However, spectrum resources are very limited and precious. The above finding indicates that we can explore, instead, the potential in random fields by using MIMO of increased number of antennas. This is the thought of massive MIMO recommended for use in 5G cellular.

13.5.2 Large SNR approximation

We focus on a MIMO channel with a correlated complex Gaussian distributed channel matrix $\mathbf{H}$, and investigate its asymptotic capacity as the average SNR $\rho \to \infty$.

Specifically, for large ρ, a widely used approximation for the mutual information is given by

$$C = \log \det(\mathbf{I} + \rho \mathbf{W}) \overset{\rho \to \infty}{\to} \log \det(\rho \mathbf{W}) = \log \left[\rho^m \det(\mathbf{W}) \right] \tag{13.36}$$

with $\mathbf{W}$ defined in (13.3). To see the accuracy of this approximation, we assume that $\mathbf{H}$ is a semi-correlated channel with spatial correlation occurring at the least-antenna side with m antennas. Correspondingly, the other side is called the with M antennas is called the most-antenna side. Let $\mathbf{h}_i$ denote the channel vector linking all the antennas at the m-side to antenna i at the M-side. Assume that $\mathbf{h}_i, i = 1, \cdots, m$ are mutually independent, each of distribution $\mathbf{h}_i \sim \mathcal{CN}_m(\mathbf{0}, \mathbf{R}_h)$. Then, $\mathbf{W}$ follows the complex Wishart distribution $\mathbf{W} \sim \mathcal{CW}_m(M, \mathbf{R}_h)$. Note that $\mathbf{W}$ is a random matrix having very spread eigenvalues $\{\ell_1 \geq \cdots \geq \ell_m\}$ even for a moderately correlated matrix $\mathbf{R}_h$. The smallest sample eigenvalue ℓ_m is likely to be negligible, though with a small probability. Thus, using $\rho \ell_m$ to approximate the factor $(1 + \rho \ell_m)$ can produce a huge discrepancy from the true value of $\det(\mathbf{I} + \rho \mathbf{W})$. We are therefore interested in the accuracy provided by the aforementioned approximation.

One way to see the discrepancy is to compare the CDFs of C given in Subsection 13.4.2 and the approximation $\log \det(\rho \mathbf{W})$. The determinant of $\mathbf{W}$ is found to be distributed as the product of m independent chi-square variables, as shown by

$$\det(\mathbf{W}) \sim 2^{-m} \det \mathbf{R}_h \prod_{i=0}^{m-1} \chi^2(2M - 2i), \tag{13.37}$$

a lemma due to C.R. Rao [8]. By using an argument based on the Gram–Schmidt procedure and using the theorem to handle the loss in degrees of freedom caused by the constraints on the random variables, we can show that, given the threshold $\eta = \rho^r$, the capacity outage equals

$$P_{\text{out}} = \Pr\{C < r \log \rho\}$$

$$= \Pr \left\{ \prod_{i=1}^{m-1} \frac{1}{2} \chi^2(2M - 2i) < \Lambda \right\}, \tag{13.38}$$

where

$$\Lambda = \frac{2^\eta}{\rho^m \det \mathbf{R}_r}. \tag{13.39}$$

According to M.D. Springer, the PDF of the variable product $x = \prod_{i=1}^{n-1} \frac{1}{2} \chi^2(2m - 2i)$ can be represented in terms of the Meijer G function, as shown by

$$f(x) = \frac{G_{0,m}^{m,0}(x | M - 1, M - 2, \cdots, M - m)}{\sum_{i=1}^{m} \Gamma(M - i + 1)} \tag{13.40}$$

whereby the capacity outage probability is determined as

$$P_{\text{out}} = \int_0^{\Lambda} \frac{G_{0,m}^{m,0}(x|M-1, M-2, \cdots, M-m)}{\sum_{i=1}^{m} \Gamma(M-i+1)} dx. \tag{13.41}$$

This integral can be easily calculated by using the function "Integrate" in Mathematica, where there is a built-in program "MeijerG" for the Meijer function.

✍ Example 13.1

The approximation accuracy using (13.36) is illustrated in Figures 13.2 and 13.3, where the capacity outage probability of a 3×3 MIMO channel is shown as a function of the spectral efficiency for different ρ's. Figure 13.2 shows the case for a weakly correlated MIMO channel with eigenvalues equal to 0.9, 1.0, and 1.1, while Figure 13.3 shows the case for a highly correlated MIMO channel with eigenvalues 0.4, 0.8, and 1.8. For both cases, a large discrepancy is observed for small and moderate SNR of $\rho \leq 10\,\text{dB}$. A good approximation is found for $\rho \geq 30\,\text{dB}$.

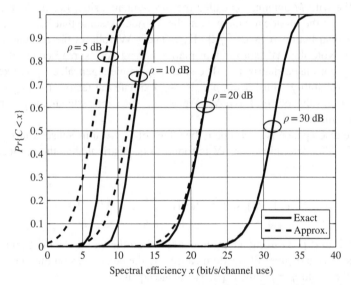

Figure 13.2 *Exact versus approximate capacity outage probability of a weakly correlated MIMO channel*

Thus, we can conclude that the use of $\log \det(\rho \mathbf{W})$ seriously underestimates the capacity of MIMO channels. It provides an accurate approximation for large SNRs, say, $\rho \geq 30\,\text{dB}$.

13.6   Optimal transceivers with CSIT

The optimal transceiver for MIMO channels with CSIT/CSIR corresponds to the OFDM transceiver for temporally dispersive channels, in the sense that both transmit signals along the eigenbeams of the channel matrix. The difference is that MIMO channels lack a Teoplitz structure, forcing their eigenbeams to significantly differ from the Fourier vectors. As such, the OFDM technique is not applicable to MIMO channels and, hence, CSIT is required for optimal beamforming.

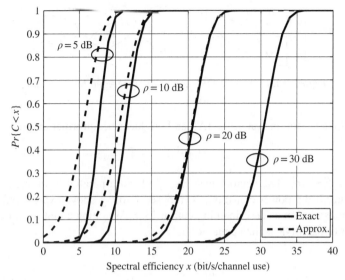

Figure 13.3 *Exact versus approximate capacity outage probability of a highly correlated MIMO channel*

Consider the parallel transmission (spatial multiplexing) of an $N_t \times 1$ signal vector $\mathbf{s}$ through an $N_r \times N_t$ MIMO channel $\mathbf{H} = [\mathbf{h}_1, \cdots, \mathbf{h}_{N_t}]$ with $N_r \geq N_t$, producing an $N_r \times 1$ channel output $\mathbf{y}$, as shown in (13.1). Let $\mathbf{w}_t$ and $\mathbf{w}_r$ denote the transmit and receive beamformers, respectively, such that $\|\mathbf{w}_t\| = \|\mathbf{w}_r\| = 1$. Denote the precoder matrix $\mathbf{Q} = [\mathbf{q}_1, \cdots, \mathbf{q}_{N_t}]$ in terms of its column vectors such that $\|\mathbf{q}_i\| = 1$. The N_t column vectors of $\mathbf{Q}$ can be considered as N_t parallel beamformers with each carrying its corresponding symbol for transmission over the MIMO channel. Further, denote the singular decomposition of $\mathbf{H}$ as

$$\mathbf{H} = \sum_{i=1}^{N_t} \omega_i \mathbf{u}_i \mathbf{v}_i^\dagger, \tag{13.42}$$

where $\mathbf{U} = [\mathbf{u}_1, \cdots, \mathbf{u}_{N_t}]$, $\mathbf{V} = [\mathbf{v}_1, \cdots, \mathbf{v}_{N_t}]$ are the eigenvectors, and $\omega's$ are the singular values such that $|\omega_1| \geq \cdots \geq |\omega_{N_t}|$. Note that among N_t singular values, only m are nonzero. Then, the received signal takes the form

$$\mathbf{y} = \mathbf{H} \sum_{k=1}^{N_t} s_k \mathbf{q}_k + \mathbf{n}, \tag{13.43}$$

which is correlated with the receive beamformer $\mathbf{w}_i$ to produce a symbol estimate of s_i:

$$\hat{s}_i = \mathbf{w}_i^\dagger \mathbf{y} = \mathbf{w}_i^\dagger \left(\sum_{\ell=1}^{N_t} \omega_\ell \mathbf{u}_\ell \mathbf{v}_\ell^\dagger \right) \left(\sum_{k=1}^{N_t} s_k \mathbf{q}_k \right) + \mathbf{w}_i^\dagger \mathbf{n}. \tag{13.44}$$

The question is how to design the transmit and receive beamformers $\mathbf{q}'s$ and $\mathbf{w}'s$, so as to meet a certain criterion.

13.6.1 Optimal eigenbeam transceiver

Two criteria are considered here. The first is to maximize the transmission data rate. As shown in Section 13.3, to achieve the MIMO channel capacity, we must set

$$\mathbf{w}_i = \mathbf{u}_i, \mathbf{q}_i = \mathbf{v}_i, \tag{13.45}$$

Figure 13.4 *MIMO-MRC with principal eigen beamforming*

which ensures the complete elimination of the IAI, and produces beamformer output

$$\hat{s}_i = \omega_i s_i + \mathbf{w}_i^\dagger \mathbf{n}, \ i = 1, \cdots, m \tag{13.46}$$

with SNR equal to $|\omega_i|^2 \sigma_s^2 / \sigma_n^2$. The vectors $\mathbf{w}'s$ and $\mathbf{q}'s$ form a parallel beamformer network to achieve the maximum data rate. If the power constraint is imposed into the above expression, we have the case of maximum data rate transmission as described in Section 13.3. Assuming a unit-power constraint and noting that $\|\omega_i\|^2 = \lambda_i$, the average SNR is then equal to

$$\gamma = \frac{1}{m}(\lambda_1 + \cdots + \lambda_m)\rho. \tag{13.47}$$

With a power constraint, if our goal is to maximize the transmission reliability instead, we can put all the power into the principal eigenbeams, by setting $\mathbf{q}_1 = \mathbf{u}_1$ and $\mathbf{w}_1 = \mathbf{v}_1$, to transmit a *single* symbol s_1 at a time. With this setting, the output SNR is equal to

$$\gamma = \lambda_1 \rho, \tag{13.48}$$

which reaches the maximum possible output SNR given a power constraint of unity. Such a system is also called MIMO-MRC, and its idea is intuitively sketched in Figure 13.4.

13.6.2 Distributions of the largest eigenvalue

Techniques used for deriving the distribution of the largest sample eigenvalue of MIMO channels vary with the distribution assumption on the channel covariance matrix. Here, we only consider the case of Ralyleigh fading MIMO channels. To begin with, we introduce results due to McKay et al. [9, 10].

Lemma 13.6.1 *Let* $\mathbf{X} \sim \mathcal{CN}_{p,q}(\mathbf{0}, \boldsymbol{\Sigma} \otimes \boldsymbol{\Omega})$*, with* $q \leq p$*, and* $\boldsymbol{\Sigma} \in \mathcal{C}^{p \times p}$ *and* $\boldsymbol{\Omega} \in \mathcal{C}^{q \times q}$ *are Hermitian positive-definite matrices with eigenvalues* $\sigma_1 < \cdots < \sigma_p$ *and* $\omega_1 < \cdots < \omega_q$*, respectively. The CDF of the maximum eigenvalue of the* $q \times q$ *matrix* $\mathbf{X}^\dagger \mathbf{X}$ *is*

$$F_{\lambda_{\max}}(x) = \frac{(-1)^q (\prod_{i=1}^{q-1} i!) \det(\boldsymbol{\Sigma})^{p-1} \det(\boldsymbol{\Omega})^{q-1} \det(\boldsymbol{\Psi}(x))}{(-x)^{q(q-1)/2} \prod_{i<j}^{p}(\sigma_j - \sigma_i) \prod_{i<j}^{q}(\omega_j - \omega_i)}, \tag{13.49}$$

where $\boldsymbol{\Psi}(x)$ *is a* $p \times p$ *matrix with its* (i,j)*th component defined by*

$$(\boldsymbol{\Psi}(x))_{i,j} = \begin{cases} \sigma_j^{i-p}, & i \leq p - q \\ e^{g_{i,j}(x)} - \sum_{k=0}^{p-1}[g_{i,j}(x)]^k / k!, & i > p - q \end{cases} \tag{13.50}$$

and $g_{i,j}(x)$ *defined as*

$$g_{i,j}(x) = -\frac{x}{\omega_{i-p+q}\,\sigma_j}.$$

For single-side correlated MIMO channels, the general result given in (13.49) can be simplified. In particular, assuming that the correlation occurs at the least-antenna side with covariance matrix $\mathbf{R}$ with distinct eigenvalues, the matrix $\mathbf{W}$ defined in (13.3) has its maximum eigenvalue distributed as follows:

Lemma 13.6.2 *For a single-side correlated MIMO Rayleigh channel with correlation matrix $\mathbf{R}$ at the least-antenna side, the $m \times m$ matrix $\mathbf{W}$ follows the complex Wishart distribution $\mathbf{W} \sim \mathcal{CW}_m(M, \mathbf{R})$ with $M > m$. The eigenvalues of $\mathbf{R}$, in descending order, are denoted by $r_1 > \cdots > r_m > 0$. Then, the CDF of the largest eigenvalue (denoted here by λ_1) of $\mathbf{W}$ is given by Ordonez et al. [11] as*

$$F_{\lambda_1}(x) = \left(\prod_{i=1}^{m} \frac{1}{r_i^M (M-i)!} \prod_{i<j}^{m} \frac{r_i r_j}{r_j - r_i} \right) \det[\mathbf{Y}(x)], \tag{13.51}$$

where the $m \times m$ matrix $\mathbf{Y}$ is defined as

$$[\mathbf{Y}(x)]_{u,v} = r_u^{M-m+v} \gamma(M - m + v, x/r_u), \ u, v = 1, \cdots, m, \tag{13.52}$$

in terms of the lower incomplete gamma function defined by

$$\gamma(n, z) = \int_0^z e^{-t} t^{n-1} dt = (n-1)! \left(1 - e^{-z} \sum_{k=0}^{n-1} \frac{z^k}{k!} \right). \tag{13.53}$$

Note that the series expansion on the last line holds only for integer argument n [12].

For the case of i.i.d. Gaussian channels, we have the following results:

Lemma 13.6.3 *For an i.i.d. MIMO channel $\mathbf{H} \sim \mathcal{CN}_{q,p}(\mathbf{0}, \mathbf{I} \otimes \mathbf{I})$, denote $m = \min(p, q)$ and $M = \max(p, q)$. The CDF of the maximum eigenvalue of the $m \times m$ matrix*

$$\mathbf{W} = \begin{cases} \mathbf{H}\mathbf{H}^\dagger, & q \leq p \\ \mathbf{H}^\dagger\mathbf{H}, & p \leq q \end{cases} \tag{13.54}$$

is given by Khatri [13] as

$$F_{\lambda_1}^{i.i.d}(x) = \Pr\{\lambda_{\max} < x\} = \frac{\det[\mathbf{\Psi}_c(x)]}{\prod_{k=1}^{m} \Gamma(M - k + 1)\Gamma(m - k + 1)}, \tag{13.55}$$

where $\mathbf{\Psi}_c(x)$ is an $m \times m$ Hankel matrix with its (i, j)th entries given by

$$\{\mathbf{\Psi}_c(x)\}_{ij} = \gamma(M - m + i + j - 1, x), i, j = 1, \cdots, m. \tag{13.56}$$

Correspondingly, the PDF of λ_1 is given by Kang and Alouini [14, 15] as

$$f_{\lambda_{\max}}^{i.i.d}(x) = \frac{\det[\mathbf{\Psi}_c(x)]}{\prod_{k=1}^{m} \Gamma(M - k + 1)\Gamma(m - k + 1)} \times tr(\mathbf{\Psi}_c^{-1}(x)\mathbf{\Phi}_c(x)), \ x > 0, \tag{13.57}$$

where $\mathbf{\Phi}_c(x)$ is an $m \times m$ matrix with its (i, j)th entries defined by

$$\{\mathbf{\Phi}_c(x)\}_{ij} = x^{M-m-i-j-2} e^{-x}, \ i, j = 1, \cdots, m.$$

13.6.3   Average symbol-error probability

Having obtained the marginal distributions for the largest sample eigenvalue of the channel matrix in Subsection 13.6.2, it is theoretically simple to determine the outage and symbol error performance of the MIMO-MRC.

We would like to determine the capacity outage performance of the optimal eigenbeam transceiver. The capacity outage probability of correlated MIMO Rayleigh channels with a minimum data rate $R_{\min}$ can be easily calculated from the above lemma. The result is

$$
\begin{aligned}
P_{\text{out}} &\triangleq \Pr\{\log(1 + \rho\lambda_1) < R_{\min}\} \\
&= \Pr\{\lambda_1 < (2^{R_{\min}} - 1)/\rho\}.
\end{aligned}
\tag{13.58}
$$

For the case of i.i.d. Rayleigh MIMO channels, the treatment is totally different due to the multiplicity of eigenvalues, and can be found in Ref. [14].

13.6.4   Average mutual information of MIMO-MRC

Because of the theoretical difficulty, we focus here only on the case of i.i.d. MIMO Raylegh fading channels with covariance matrix $\boldsymbol{\Sigma} = \sigma\mathbf{I}$. To determine the ergodic capacity $\mathbb{E}[\log(1 + \rho\lambda_1)]$ associated with the MIMO-MRC, we require the PDF of λ_1, the maximum eigenvalue of $\mathbf{HH}^\dagger$. Certainly, we can start from the CDF in (13.51). It is more straightforward to directly invoke the results from Ref. [16], which is restated below. Denote $x = \rho\lambda_1$. The PDF of the received SNR, ρx, can be obtained by differentiating the corresponding CDF available in Ref. [13], and is given by

$$
f_{\rho x}(\rho x) = \sum_{k=1}^{m} \sum_{\ell=n-m}^{(n-m-2k)k} \frac{b_{k\ell}}{\ell!} \left(\frac{k}{\rho}\right)^{\ell-1} x^\ell e^{kx/\rho},
\tag{13.59}
$$

where

$$
b_{k\ell} = a_{k\ell} K_{mn} \ell! / k^{\ell-1},
$$
$$
K_{mn} = \left[\prod_{i=1}^{m}(m-i)!\right]^{-1} \left[\prod_{i=1}^{m}(n-i)!\right]^{-1}.
\tag{13.60}
$$

From this expression, it is observed that the PDF of the received SNR is a linear summation of chi-square PDFs. The coefficients are parameters resulting from fitting the derivative curve of the CDF of x [15]. A more accurate algorithm for their determination is provided in Ref. [16]. It turns out that

$$
\mathbb{E}[\log(1 + \rho\lambda_1)] = \log(e) \sum_{k=1}^{m} e^{k/\rho} \sum_{\ell=n-m}^{(n-m-2k)k} b_{k\ell} \sum_{j=1}^{\ell-1} E_{\ell-2-j}(k/\rho),
\tag{13.61}
$$

where the exponential integral function $E_\ell(\mu)$ is related to the complementary incomplete Gamma function $\Gamma(\ell, \mu)$ by Abramowitz and Stegun [12], (6.5.9):

$$
E_\ell(\mu) = \int_1^\infty \frac{\exp(-\mu\gamma)}{\gamma^\ell} d\gamma = \mu^{\ell-1}\Gamma(1-\ell, \mu).
\tag{13.62}
$$

13.6.5 Average symbol-error probability

Given the PDFs and CDFs of the largest sample eigenvalues of $\mathbf{H}^\dagger\mathbf{H}$, we can determine the symbol error probability of the MIMO-MRC transceiver. The conditional probability of symbol error for various coherent modulation schemes is usually expressible as $P_e(\gamma) = \alpha\, Q(\sqrt{b\gamma}) - \beta\, Q^2(\sqrt{b\gamma})$. Now, the instantaneous SNR is $\gamma = \rho\lambda_1$, where ρ denotes the total average SNR at transmit antennas. By using the rule for differentiating an integral, the average probability of symbol error can be determined as

$$P_e = \int_0^\infty P_e(\rho x)\, dF_{\lambda_1}(x)$$

$$= \underbrace{F_{\lambda_1}(x)P_e(\rho x)\,\big|_{x=0}^\infty}_{0} - \int_0^\infty F_{\lambda_1}(x)\, dP_e(\rho x)$$

$$= \int_0^\infty \frac{F_{\lambda_1}(x)\sqrt{b\rho}\, e^{-b\rho x/2}}{\sqrt{2\pi x}} \left[\frac{\alpha}{2} - \beta Q(\sqrt{b\rho x})\right] dx. \tag{13.63}$$

For MIMO channels of i.i.d. entries, $F_{\lambda_1}(x)$ should be replaced by $F_{\lambda_1}^{i.i.d.}(x)$. If the SNR is large, we can ignore the influence of the term Q^2 by setting $\beta = 0$.

✍ **Example 13.2** ──

Consider a MIMO-MRC system operating on an 8×4 Rayleigh MIMO channel with semi-correlation at the receive end of exponential type as shown by $\mathbf{R} = [|r|^{i-j}e^{2\pi(j-i)\theta}]$. Its probability of symbol error as a function of the average SNR per transmit antenna is depicted in Figure 13.5. ○

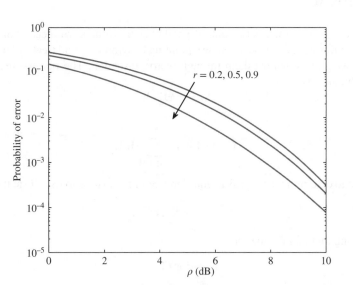

Figure 13.5 *Error performance of 8 × 4 MIMO-MRC versus the average SNR per transmit antenna*

13.7   Receivers without CSIT

In the previous discussions, we described the transceiver structures for the case with CSIT. For most applications, it is more realistic to acquire channel state information at the receiver (CSIR) but not at the transmitter. In the remaining sections, we only assume CSIR and study the structures and properties of the corresponding receivers. In particular, in Sections 13.13–13.14.1, we focus on single-snapshot processing and investigate three possible receiver structures. These receivers are of the zero forcing (ZF), minimum mean-square error (MMSE), and vertical Bell Labs layered space-time architectures, respectively. By a snapshot, we mean N_t parallel data symbols that are transmitted at a time instant by N_t antennas. When focusing on a single $N_t \times 1$ snapshot $\mathbf{x}$, the $N_r \times 1$ received signal takes the form

$$\mathbf{y} = \mathbf{H}\mathbf{x} + \mathbf{n} \tag{13.64}$$

$$= \sum_{i=1}^{N_t} x_i \mathbf{h}_i + \mathbf{n}, \tag{13.65}$$

where the column channel vector $\mathbf{h}_i$ is obtained by linking transmit antenna i to all the N_r receive antennas. A reader familiar with statistics immediately recognizes that this is a *regressive model*. Given that $\mathbf{h}'s$ are not mutually orthogonal, cross-antenna interference is unavoidable. Given the received vector $\mathbf{y}$, the estimation problem is to estimate $\mathbf{x}$ as accurately as possible but with a reasonable complexity. For the solvability of (13.65), we need $N_r \geq N_t$. In the subsequent analysis, we assume $\mathbf{h}_i \sim \mathcal{CN}(\mathbf{0}, \mathbf{R}_h)$.

Four possible receiver structures for MIMO channels with only CSIR are optimal beamformer-based receiver, ZF receiver, MMSE receiver, and VBLST, which are studied in the subsequent Sections 13.8–13.11, respectively.

13.8   Optimal receiver

Given CSIR, the receiver can construct a bank of parallel beamformers, with each pointing to a transmitted symbol. The kth beamformer has its beam pattern pointing to symbol s_k and treats all the remaining symbols as interferers by posing nulls against them through matrix inversion. The optimal receive beamformer for symbol k is thus obtained as

$$\mathbf{w}_k = \mathbf{\Theta}_k^{-1} \mathbf{h}_k, \tag{13.66}$$

where

$$\mathbf{\Theta}_k = \mathbf{I} + \rho \sum_{i=1, i \neq k}^{N_t} \mathbf{h}_i \mathbf{h}_i^\dagger. \tag{13.67}$$

It is straightforward to determine the signal-to-interference-plus-noise ratio (SINR) at the optimal combiner's output as

$$\gamma_k = \rho \mathbf{h}_k^\dagger \mathbf{\Theta}_k^{-1} \mathbf{h}_k \tag{13.68}$$

and the corresponding mutual information

$$I_k = \log\left(1 + \gamma_k\right), \tag{13.69}$$

whereby the total mutual information of the N_t beams is equal to

$$C = \sum_{k=1}^{N_t} \log(1 + \rho \mathbf{h}_k^\dagger \mathbf{\Theta}_k^{-1} \mathbf{h}_k). \tag{13.70}$$

The PDF of $\gamma_k = \rho \mathbf{h}_k \Theta_k^{-1} \mathbf{h}_k$ can be found in Ref. [17]. To determine $[I_k]$, we apply the property of a determinant that $\det(\mathbf{A} + \mathbf{u}\mathbf{v}^\dagger) = \det(\mathbf{A})(1 + \mathbf{v}^\dagger \mathbf{A}^{-1}\mathbf{u})$ to our case, yielding

$$
1 + \gamma_k = \frac{\det(\mathbf{I} + \rho \sum_{i=1}^{N_t} \mathbf{h}_i \mathbf{h}_i^\dagger)}{\det(\mathbf{I} + \rho \sum_{i=1,i\neq k}^{N_t} \mathbf{h}_i \mathbf{h}_i^\dagger)}. \tag{13.71}
$$

It follows that

$$
\mathbb{E}[\log(1 + \gamma_k)] = \mathbb{E}\log\det\left(\mathbf{I} + \rho \sum_{i=1}^{N_t} \mathbf{h}_i \mathbf{h}_i^\dagger\right) - \mathbb{E}\log\det\left(\mathbf{I} + \rho \sum_{i=1,i\neq k}^{N_t} \mathbf{h}_i \mathbf{h}_i^\dagger\right), \tag{13.72}
$$

which can be simplified to an explicit expression by using the formulas by Telatar, Kiessling, or Chiani [6, 5, 3].

13.9 Zero-forcing MIMO receiver

In this section, we assume $N_r \geq N_t$, as before, and $\mathbf{H} \sim \mathcal{CN}_{N_t,N_r}(\mathbf{0}, \mathbf{R}_t \otimes \mathbf{I})$, namely, with antenna correlation at the transmitter side.

From (13.65), the simplest way to eliminate the cross-antenna interference is to remove the $\mathbf{H}$ matrix in front of the desired vector $\mathbf{x}$, implementing what we call *zero forcing* (ZF). ZF is implemented by dividing both sides of (13.65) with the pseudo-inverse of $\mathbf{H}$, denoted by $\mathbf{H}^\#$. The output vector turns out to be

$$
\mathbf{z} = \mathbf{H}^\# \mathbf{y} = (\mathbf{H}^\dagger \mathbf{H})^{-1} \mathbf{H}^\dagger \mathbf{y}
$$
$$
= \mathbf{x} + \underbrace{(\mathbf{H}^\dagger \mathbf{H})^{-1} \mathbf{H}^\dagger \mathbf{n}}_{\mathbf{v}}. \tag{13.73}
$$

The ZF filter decorrelates the signal components while colorizing the noise. It is easy to find the mean and covariance matrix of $\mathbf{z}$ to be $\mathbb{E}[\mathbf{z}] = \mathbf{x}$ and $\mathbf{R}_v = \mathbb{E}[\mathbf{v}\mathbf{v}^2] = \sigma_n^2 (\mathbf{H}\mathbf{H})^{-1}$, enabling us to write

$$
\mathbf{z} \sim \mathcal{CN}(\mathbf{x}, \sigma_n^2 (\mathbf{H}\mathbf{H})^{-1}). \tag{13.74}
$$

Having decorrelated the signal component in $\mathbf{y}$, the receiver can estimate each symbol independently. Let us focus on the detection of symbol x_i, based on entry i of $\mathbf{z}$:

$$
z_i = x_i + v_i, \tag{13.75}
$$

where v_i is the ith entry of noise vector $\mathbf{v}$, which becomes correlated due to the pseudo-inverse operation. Clearly, the variance of v_i is the (i,i)th entry of $\mathbf{R}_v$ and is denoted as σ_i^2. To more explicitly represent σ_i^2, denote $\mathbf{e}_i$ as the all-zero $N_t \times 1$ vector except its ith entry of unity. Thus, we can write

$$
\sigma_i^2 = \sigma_n^2 \mathbf{e}_i^\dagger (\mathbf{H}\mathbf{H}^\dagger)^{-1} \mathbf{e}_i \tag{13.76}
$$

Hence, the SINR is given by

$$
\gamma_i = \frac{\mathbb{E}|x_i|^2}{\sigma_i^2} = \frac{\rho}{\mathbf{e}_i^\dagger (\mathbf{H}^\dagger \mathbf{H})^{-1} \mathbf{e}_i}. \tag{13.77}
$$

We need the PDF of γ_i to determine the ergodic capacity, capacity outage probability, and detection performance of the ZF receiver. The result was first obtained by Winters et al. [18] and independently by Gore et al. [19]. The former employed an indirect CHF-based approach, while the latter directly invoked

Table 13.1   *Influence of antenna correlation on η_3*

$|c|$	0.1	0.2	0.3	0.4	0.5	0.6	0.7	0.8	0.9
η_3	0.980	0.923	0.835	0.724	0.6	0.471	0.342	0.220	0.105

a lemma from multivariate statistical analysis. We basically follow the approach of Ref. [19]. Note that $\mathbf{H}^\dagger\mathbf{H} \sim \mathcal{CW}_{N_t}(N_r, \mathbf{R}_t)$. An important tool that helps us simplify the analysis is a lemma extended from R.J. Muirhead [20].

Lemma 13.9.4 *If $\mathbf{A} \sim \mathcal{CW}_{N_t}(N_r, \mathbf{R}_t)$ and $\mathbf{e}_i$ is as defined above, then*

$$\xi = (\mathbf{e}_i^\dagger\mathbf{A}^{-1}\mathbf{e}_i)^{-1} \sim \mathcal{CW}_1(N_r - N_t + 1, \underbrace{(\mathbf{e}_i^\dagger\mathbf{R}_t^{-1}\mathbf{e}_i)^{-1}}_{\alpha_i}), \tag{13.78}$$

where α_i signifies the (i, i)th entry of $\mathbf{R}_t^{-1}$.

Recall that $\mathcal{CW}_1(\cdot, \cdot)$ degenerates to a gamma distribution. Specifically, by denoting $p = N_r - N_t$ and $\eta_i = 1/(\mathbf{e}_i^\dagger\mathbf{R}_t^{-1}\mathbf{e}_i)$, we have

$$f_\xi(u) = \frac{\alpha_i^{p+1}}{p!}u^p e^{-\alpha_i u} \sim \frac{\eta_i}{2}\chi^2(2p+2). \tag{13.79}$$

Note that η_i can be alternatively represented as $\eta_i = \det(\mathbf{R}_t)/\det(\mathbf{R}_{t\backslash i,i})$, where $\mathbf{R}_{t\backslash i,i}$ results from $\mathbf{R}_t$ by simply excluding the column i and row i of the latter. By inserting (13.79) into (13.77), we get

$$\gamma_i \sim \frac{\rho}{2\alpha_i}\chi^2(2p+2) = \frac{\rho\eta_i}{2}\chi^2(2p+2). \tag{13.80}$$

From (13.80), it is clear that each symbol decision in the ZF receiver enjoys the diversity of $2(N_r - N_t + 1)$ degrees of freedom. The opposite effect is the factor of $\eta_i = \det(\mathbf{R}_t)/\det(\mathbf{R}_{t\backslash i,i})$, representing the drop in average SNR, due to the noise amplification. For insightful understanding, we assume $\mathbf{R}_t$ of an exponential-type Toeplitz correlation structure with the correlation between two adjacent antennas given by $c = |c|\exp(j\theta)$, such that the first row of $\mathbf{R}_t$ is equal to $[|c|^m\exp(jm\theta), m = 0, \ldots, N_t - 1]$ and its first column is equal to $[|c|^m\exp(-jm\theta), m = 0, \ldots, N_t - 1]$. From its determinant ratio representation, η_i is expected to behave similar to the power of an order-$(N_t - 1)$ linear predictor. For illustration, set $N_t = 5$ and calculate η_3; the results are given in Table 13.1.

Note that the phase of c has no impact on the value η_i. As pointed out in the previous study of channel equalization, the ZF equalizer always tries to retrieve symbol information from the subspace totally free of corruption from its adjacent ones. The ZF MIMO receiver adopts the same philosophy. As the antenna correlation increases, the uncorrupted signal component in the pristine subspace becomes small, thus leading to a poor estimation, as indicated by a rapidly dropping η_i. Thus, when the antenna correlation exceeds 0.6, it is not a good idea to use the ZF MIMO receiver.

✐ Example 13.3

Assume an i.i.d. MIMO channel $\mathbf{H}$, and determine the total average mutual information of the MIMO ZF receiver.

Solution

For an i.i.d. channel, $\mathbf{R}_t = \mathbf{I}$ for which $\eta_i = 1$ and, hence,

$$\mathbb{E}[C] = N_t \, \mathbb{E}\log\left(1 + \gamma_i\right) = N_t \, \mathbb{E}\log\left[1 + \frac{\rho}{2\alpha}\chi^2(2p+2)\right]$$

$$= N_t[\log\rho + \log\left(2p+2\right) - \frac{1}{2}(p+1)^{-1}].$$

13.10 MMSE receiver

The MMSE receiver is represented by a tap-weight matrix $\mathbf{W}$ such that the symbol estimate is as close to $\mathbf{s}$ as possible, in the MMSE sense that $\mathbb{E}\|\mathbf{s} - \mathbf{W}\mathbf{y}\|^2$ is minimized. One may directly work on this minimization to obtain the optimal $\mathbf{W}$ and thus the estimate of $\mathbf{s}$. Alternatively, a more intuitive method is to find $\mathbf{W}$, such that the estimation error vector is orthogonal to the space spanned by the data $\mathbf{y}$; namely

$$(\mathbf{s} - \mathbf{W}\mathbf{y})\perp\mathbf{y}, \tag{13.81}$$

which, when translated into mathematics, gives

$$\mathbb{E}[(\mathbf{s} - \mathbf{W}\mathbf{y})\mathbf{y}^{\dagger}] = \mathbb{E}\{[\mathbf{s} - \mathbf{W}(\mathbf{H}\mathbf{s} + \mathbf{n})](\mathbf{s}^{\dagger}\mathbf{H}^{\dagger} + \mathbf{n}^{\dagger})\} = 0 \tag{13.82}$$

with the expectation taken over $\mathbf{s}$. After simple calculation and noting $\mathbb{E}[\mathbf{s}\mathbf{s}^{\dagger}] = \sigma_s^2\mathbf{I}$ and $\mathbb{E}[\mathbf{n}\mathbf{n}^{\dagger}] = \sigma_n^2\mathbf{I}$, it yields

$$\mathbf{W} = \sigma_s^2\mathbf{H}^{\dagger}(\sigma_n^2\mathbf{I} + \sigma_s^2\mathbf{H}\mathbf{H}^{\dagger})^{-1}$$

$$= \rho\mathbf{H}^{\dagger}\underbrace{(\mathbf{I} + \rho\mathbf{H}\mathbf{H}^{\dagger})^{-1}}_{\mathbf{O}^{-1}}. \tag{13.83}$$

Its ith row defines the MMSE filter for symbol s_i, as shown by

$$\mathbf{w}_i = \rho\mathbf{h}_i^{\dagger}\mathbf{O}^{-1}. \tag{13.84}$$

To proceed, denote

$$\mathbf{O}_i = \mathbf{I} + \rho\sum_{\ell=1,\ell\neq i}^{N_t} \mathbf{h}_{\ell}\mathbf{h}_{\ell}^{\dagger},$$

and invoke the matrix inversion lemma to calculate $\mathbf{O}^{-1}$ yielding

$$\mathbf{O}^{-1} = (\mathbf{O}_i + \rho\mathbf{h}_i\mathbf{h}_i^{\dagger})^{-1}$$

$$= \mathbf{O}_i^{-1} - \rho\mathbf{O}_i^{-1}\mathbf{h}_i(1 + \rho\mathbf{h}_i^{\dagger}\mathbf{O}_i^{-1}\mathbf{h}_i)^{-1}\mathbf{h}_i^{\dagger}\mathbf{O}_i^{-1} \tag{13.85}$$

which, when inserted into (13.84), produces

$$\mathbf{w}_i = \rho\mathbf{h}_i^{\dagger}\mathbf{O}_i^{-1} - \frac{\rho^2(\mathbf{h}_i^{\dagger}\mathbf{O}_i^{-1}\mathbf{h}_i)}{1 + \rho\mathbf{h}_i^{\dagger}\mathbf{O}_i^{-1}\mathbf{h}_i}\mathbf{h}_i^{\dagger}\mathbf{O}_i^{-1}$$

$$= \rho\underbrace{(1 + \rho\mathbf{h}_i^{\dagger}\mathbf{O}_i^{-1}\mathbf{h}_i)^{-1}}_{\alpha_i}\underbrace{\mathbf{h}_i^{\dagger}\mathbf{O}_i^{-1}}_{\mathbf{w}_{\text{opt}}}. \tag{13.86}$$

Clearly, the MMSE filter $\mathbf{w}_i$ has the same form as the optimal beamformer given in (13.66), except for a row-dependent scaling factor α_i. After filtering the received signal (13.1) by the MMSE filter defined in (13.84), it is easy to determine the output SINR as

$$\gamma_i = \rho \mathbf{h}_i^\dagger \mathbf{O}_i^{-1} \mathbf{h}_i. \tag{13.87}$$

Property 13.10.1 *The MMSE filter is an optimal beamformer with a specially chosen scaling factor α_i, enabling it to simultaneously take the advantages of Wiener filter and matched filter. The spatially matched filtering enables to achieve its maximized output SINR given by (13.87). The characteristic of Wiener filter allows it to achieve the minimum MSE.*

$$\begin{aligned}
\mathrm{MSE}_i &= \mathbb{E}[|s_i - \hat{s}_i|^2] \\
&= \mathbb{E}[|s_i - \rho \mathbf{h}_i^\dagger \mathbf{O}^{-1}(\mathbf{Hs} + \mathbf{n})|^2] \\
&= \sigma_s^2 - \rho \mathbf{h}_i^\dagger (\sigma_n^2 \mathbf{I} + \sigma_s^2 \mathbf{HH}^\dagger)^{-1} \mathbf{h}_i \\
&= \sigma_s^2 (1 - \rho \mathbf{h}_i^\dagger \mathbf{O}^{-1} \mathbf{h}_i) \\
&= \sigma_s^2 (1 + \rho \mathbf{h}_i^\dagger \mathbf{O}_i^{-1} \mathbf{h}_i)^{-1}.
\end{aligned} \tag{13.88}$$

It reveals a relationship between the SINR and MSE, given by

Property 13.10.2 *The MSE and output SINR of the MMSE filter $\mathbf{w}_i$ are closely related by*

$$\mathrm{MSE}_i = \sigma_s^2 (1 + \gamma_i)^{-1}, \ \text{ or } \ \gamma_i = \frac{\sigma_s^2}{MSE_i} - 1. \tag{13.89}$$

Clearly, the criterion of minimizing MSE is consistent with that of maximizing SINR.

The PDF of γ_i is required for many theoretical evaluations of MMSE-type receivers, and has been investigated by a number of researchers. The result due to Gao et al. [17] is simple and easy to use. Quoted below is a summarized version due to Hedayat and Nosratinia [21].

Property 13.10.3 *If all $N_t N_r$ channels in $\mathbf{H}$ are i.i.d. complex Gaussian distributed, the CDF of γ_i is given by*

$$F_{\gamma_i}(y) = 1 - e^{-y/\rho} \sum_{i=1}^{N} \frac{A_i(y)}{(i-1)!} \left(\frac{y}{\rho}\right)^{i-1}, \tag{13.90}$$

where

$$A_i(y) = \begin{cases} 1, & i \le N_r - N_t + 1 \\ \left(1 + \sum_{k=1}^{N_r-i} c_k y^k\right) / (1+y)^{N_t-1}, & i > N_r - N_t + 1 \end{cases} \tag{13.91}$$

and c_k is the coefficient of y^k in the expansion of $(1+y)^{N_t-1}$.

It should be noted that $\{\gamma_i, i = 1, \cdots, N_t\}$ at the output of both MMSE and ZF receivers are generally correlated.

✍ **Example 13.4** ────────────────────────────────────

Consider a simple case studied in Ref. [22], where m antennas are used for reception of a Gaussian signal $b_0 \sim \mathcal{CN}(0, 1)$ corrupted by a single Gaussian interferer $b_1 \sim \mathcal{CN}(0, \sigma_c^2)$, such that the received vector can be written as $\mathbf{y} = b_0\mathbf{s} + b_1\mathbf{c} + \mathbf{n}$, where the m−vectors $\mathbf{s} \sim \mathcal{CN}_m(\mathbf{0}, \sigma_s^2\mathbf{I})$ and $\mathbf{c}$ represent the channel gains for signal and interferer, respectively. The SINR is then given by

$$\gamma = \mathbf{s}^\dagger(\sigma_n^2\mathbf{I} + \sigma_c^2\mathbf{c}\mathbf{c}^\dagger)^{-1}\mathbf{s}. \tag{13.92}$$

Determine the CHF of γ conditioned on $\mathbf{c}$.

Solution

Denote $\rho_s = \sigma_s^2/\sigma_n^2$ and $\rho_c = \sigma_c^2/\sigma_n^2$. Given $\mathbf{c}$, we can directly obtain the CHF of γ

$$\phi_\gamma(z|\mathbf{c}) = \det(\mathbf{I} - z\sigma_s^2\mathbf{I}(\sigma_n^2\mathbf{I} + \sigma_c^2\mathbf{c}\mathbf{c}^\dagger)^{-1})^{-1}$$
$$= \det(\mathbf{I} - z\rho_s(\mathbf{I} + \rho_c\mathbf{c}\mathbf{c}^\dagger)^{-1})^{-1}. \tag{13.93}$$

The matrix $(\mathbf{I} + \rho_c\mathbf{c}\mathbf{c}^\dagger)$ has an eigenvalue 1 of multiplicity $(m - 1)$ and a simple eigenvalue $1 + \rho_c\mathbf{c}^\dagger\mathbf{c}$. Thus, we can factor

$$\phi_\gamma(z|\mathbf{c}) = (1 - z\rho_s)^{-(m-1)}(1 - \varrho z)^{-1}, \tag{13.94}$$

where $\varrho = \rho_s/(1 + \rho_c\mathbf{c}^\dagger\mathbf{c})$. Thus, γ can be decomposed into two terms as

$$\gamma \sim \frac{\rho_s}{2}\chi^2(2m - 2) + \frac{\varrho}{2}\chi^2(2). \tag{13.95}$$

The optimal receiver forms a null against the interferer while pointing its main beam to the signal source. The resulting γ has two terms. The second term of two degrees of freedom represents the SINR leaking from the null, whereas the first term signifies the SINR gained from the main beam. When the interferer disappears, the second term reduces to $(\rho_s/2)\chi^2(2)$. Hence, γ becomes $(\rho_s/2)\chi^2(2m)$, in the same form as the output SNR of the MRC in i.i.d. Rayleigh fading. In the other extreme with $\rho_c \to \infty$, the second term vanishes. ◯

13.11   VBLAST

In the early development of transceiver structures for MIMO channels, a well-known receiver structure was proposed by the Bell Labs, called VBLAST, which is rooted in interference cancelation. In this scheme, data streams are parallel-coded, interleaved, and modulated on each antenna, generating a T-symbol codeword $[x_k[i], i = 1, 2, \cdots, T]$ at antenna k. Each antenna, say antenna k, transmits one symbol at a time, which reaches the receiver through the $N_r \times 1$ receive-antenna vector, producing the received vector for the ith interval

$$\mathbf{y}[i] = \sum_{k=1}^{N_t} x_k[i]\mathbf{h}_k + \mathbf{n}, \tag{13.96}$$

where $\mathbf{h}'s$ constitute the MIMO channel matrix such that $\mathbf{H} = [\mathbf{h}_1, \cdots, \mathbf{h}_{N_t}]$. In what follows, the time index i will be dropped for brevity. The VBLAST adopts a symbol-by-symbol detection strategy, and starts the detection from the symbol with the largest SNR while treating the remaining symbols as interference. As such, the symbol is correctly detected with a relatively high probability. The VBLAST then moves to the next symbol of highest SNR, and subtracts the estimated symbol from the interference to enhance the signal detection for a better performance. This process repeats until the detection of all the N_t transmitted symbols is completed.

The rationale of the VBLAST is clearly based on a heuristic argument. However, it has a more profound information-theoretic justification. To this end, let us invoke a lemma due to Kelly [23].

Lemma 13.11.5 *For $M \times m$ matrix* $\mathbf{H} = [\mathbf{h}_1, \cdots, \mathbf{h}_m]$, *denote* $\mathbf{S}_k = \mathbf{I} + \rho \sum_{i=1}^{k} \mathbf{h}_i \mathbf{h}_i^\dagger$. *We have*

$$\det(\mathbf{I} + \rho \mathbf{H}\mathbf{H}^\dagger) = (1 + \rho \mathbf{h}_1^\dagger \mathbf{h}_1)(1 + \rho \mathbf{h}_2^\dagger \mathbf{S}_1^{-1} \mathbf{h}_2) \cdots (1 + \rho \mathbf{h}_m^\dagger \mathbf{S}_{m-1}^{-1} \mathbf{h}_m). \tag{13.97}$$

In the context of MIMO channels, the physical significance of this lemma is very clear. It simply asserts that the channel capacity can be rewritten as

$$C = \sum_{k=1}^{m} \log(1 + \rho \mathbf{h}_k^\dagger \mathbf{S}_{k-1}^{-1} \mathbf{h}_k) \tag{13.98}$$

which, in turn, implies that the optimal beamforming is the best strategy to achieve the channel capacity. From the information-theoretic prospective, the symbol-by-symbol detection strategy is justified. The best strategy is to suppress the interference by using the inverse operation $\mathbf{S}_k^{-1}$ and removing the influence of the estimated symbols. The VBLAST practises the same philosophy in a slightly different way by noise cancelation. The PDF of $\gamma_k = \rho \mathbf{h}_k \mathbf{S}_{k-1}^{-1} \mathbf{h}_k$ can be found in Ref. [17].

In the implementation of VBLAST, the symbol s_κ of the highest SNR can be found by

$$\kappa = \arg \min_k \det \left(\mathbf{I} + \rho \sum_{i=1, i \neq k}^{N_t} \mathbf{h}_i \mathbf{h}_i^\dagger \right). \tag{13.99}$$

Namely, delete a transmit antenna such that the volume of the remaining manifold is minimized. The SINR for this symbol is determined as

$$\xi_\kappa = \mathbf{h}_\kappa^\dagger (\rho^{-1} \mathbf{I} + \sum_{i=1, i \neq \kappa}^{N_t} \mathbf{h}_i \mathbf{h}_i^\dagger)^{-1} \mathbf{h}_\kappa. \tag{13.100}$$

13.11.1   Alternative VBLAST based on QR decomposition

The VBLAST algorithm can be alternatively implemented by using the QR decomposition [24]

$$\mathbf{H} = \mathbf{QR}, \tag{13.101}$$

where the column vectors of $\mathbf{Q}$ are orthonormal bases obtained via the Gram–Schmidt type of orthogonalization procedure with

$$\mathbf{R} = \begin{pmatrix} r_{11} & r_{12} & \cdots & r_{1N_t} \\ 0 & r_{22} & \cdots & r_{2N_t} \\ \vdots & \ddots & \ddots & \vdots \\ 0 & \cdots & 0 & r_{N_r N_t} \end{pmatrix}. \tag{13.102}$$

Here, r_{11} represents the length of $\mathbf{h}_1$, column 1 of $\mathbf{H}$, and thus is distributed as $\chi(2N_r)$. The values r_{12} and r_{22} denote the projection of $\mathbf{h}_2$ onto the first and second columns of $\mathbf{Q}$, respectively, and, hence, we can assert $\mathbf{r}_{22} \sim \chi(2N_r - 2)$. The reduction of two degrees of freedom is due to the requirement of orthogonality to the first base. Other columns of $\mathbf{R}$ have a similar explanation. With (13.101), we can write

$$\mathbf{y} = \mathbf{QRx} + \mathbf{n}. \tag{13.103}$$

Multiplying both sides with the orthonormal matrix $\mathbf{Q}^{-1}$ does not change the property of the noise vector, leading to

$$\hat{\mathbf{y}} = \mathbf{Rx} + \hat{\mathbf{n}}, \tag{13.104}$$

whereby the retrieval of $\mathbf{x}$ can be done easily. Decoding starts from x_{N_r} with diversity order $(2N_r - 2N_t + 2)$.

13.12   Space–time block codes

Various receiver structures studied in previous sections, including the ZF, MMSE, VBLSAT, and optimal beamformer, adopt the same *defensive* philosophy of transmitting and processing information symbols on a snapshot-by-snapshot basis. By a snapshot we mean a set of symbols that are transmitted in parallel at a time from N_t antennas; an example is vector **s** described in (13.1). With this snapshot processing scheme, spatially parallel symbols are transmitted through nonorthogonal channels, and are corruptly superposed at each receive antenna forming a strong inter-antenna interference (IAI). Such superposed IAI components, once formed, are difficult to remove, since both the signal and IAI components coexist in the same space. The receiver strategies described in the previous sections were simply to minimize the influence of the IAI in one way or another, representing a philosophy of remedy in combatting IAI.

A proactive strategy is to create a set of orthogonal space–time channels for parallel symbol transmission. To introduce orthogonal channels, we must increase the temporal dimension by transmitting a *temporal* symbol vector, instead of a single symbol, at each transmit antenna. It implies that we adopt a multi-snapshot transmission and processing scheme. This innovative idea, originally due to Alamouti [25], aims to eliminate the source that generates spatial interference.

More generally, one can combine the temporal and spatial dimensions to constitute space–time block codes (STBCs) (or simply space–time codes), with certain embedded algebraic structures. Each space–time code is treated as a matrix symbol, thereby leaving no room for IAI.

The two space–time coding techniques, Alamouti's codes and general STBCs, are expounded in Sections 13.13 and 13.15.

13.13   Alamouti codes

The central idea of Alamouti's space-time codes is to construct a set of *temporally* orthogonal signal vectors at the transmitter, thereby converting an i.i.d. MIMO channel into a set of *virtual* orthogonal channels when viewed from the receiver. The Alamouti scheme is motivated by a simple idea: with multiple signals transmitted over a MISO channel, is it possible to implement the effect of maximum ratio combining with a *single-antenna* receiver. The challenge is that when multiple transmitted symbols arrive at the one-dimensional receive space formed by a single receive antenna, collision is unavoidable. This assertion is correct because we are confined to a single time instant. The story is totally different if we extend our vision over multiple dimensions in the temporal space.

Alamouti achieves his goal by transmitting the same set of symbols over different transmit antennas, in an orthogonal matrix pattern, and by properly performing a conjugation operation at the receiver. The resulting space–time codes are known as Alamouti codes. The nature of Alamouti codes is to convert the *temporal* orthogonality of the code matrix to a set of equivalent *orthogonal channels*, thereby eliminating the IAI and implementing maximum ratio combining.

13.13.1   One receive antenna

For illustration, let us begin with the simplest Alamouti's code for a 2×1 MISO channel, with a 2×2 transmit matrix given by

$$\mathbf{X} = \frac{1}{\sqrt{2}} \begin{bmatrix} x_1 & -x_2^* \\ x_2 & x_1^* \end{bmatrix}, \tag{13.105}$$

where $1/\sqrt{2}$ is the power-normalized factor, such that $\mathbf{X}\mathbf{X}^\dagger = [(|x_1|^2 + |x_2|^2)/2]\mathbf{I}$. The first row represents the consecutive symbols transmitted by antenna 1, while the second row represents the corresponding symbols for antenna 2. The two consecutively received samples, in vector form, are given by

$$\begin{bmatrix} y_1 \\ y_2 \end{bmatrix} = h_1 \underbrace{\begin{bmatrix} x_1 \\ -x_2^* \end{bmatrix}}_{TX\ 1} + h_2 \underbrace{\begin{bmatrix} x_2 \\ -x_1^* \end{bmatrix}}_{TX\ 2} + \mathbf{n} \tag{13.106}$$

with $h_i, i = 1, 2$ representing the channel gain that links the receiver to the transmit antenna i, and $\mathbf{n}$ denoting the AWGN vector at the receiver. Taking the conjugate of the second equation and defining equivalent channel vectors

$$\mathbf{g}_1 = \begin{bmatrix} h_1 \\ h_2^* \end{bmatrix}, \quad \mathbf{g}_2 = \begin{bmatrix} h_2 \\ -h_1^* \end{bmatrix}, \quad \mathbf{G} = [\mathbf{g}_1 \ \mathbf{g}_2], \tag{13.107}$$

we can rewrite (13.106) as

$$\underbrace{\begin{bmatrix} y_1 \\ y_2^* \end{bmatrix}}_{\tilde{\mathbf{y}}} = \mathbf{G} \begin{bmatrix} x_1 \\ x_2 \end{bmatrix} + \begin{bmatrix} n_1 \\ n_2^* \end{bmatrix} = x_1 \mathbf{g}_1 + x_2 \mathbf{g}_2 + \mathbf{n}. \tag{13.108}$$

Assume $x_i \sim \mathcal{CN}(0, \sigma_x^2)$ and $\mathbf{n} \sim \mathcal{CN}(\mathbf{0}, \sigma_n^2 \mathbf{I})$, and denote $\rho_x = \sigma_x^2 / \sigma_n^2$. Since the conjugation does not change the property of the noise vector, we use the same symbol $\mathbf{n}$ for it. The orthogonality of $\mathbf{X}$ is now converted into the orthogonality of the equivalent channel matrix $\mathbf{G}$, as evidenced by $\mathbf{g}_1^\dagger \mathbf{g}_2 = 0$. Signals from the two transmit antennas are superposed at the receiver, and they are spatially unresolvable if we confine ourselves to a single time slot, say the first row of (13.108). However, with the special space–time code by Alamouti and by introducing an additional temporal dimension, the spatially superposed signals are separable. In particular, the symbols x_1 and x_2 are recoverable by projecting $\tilde{\mathbf{y}}$ onto the vectors $\mathbf{g}_1$ and $\mathbf{g}_2$, respectively. The received SNR for each symbol is identical and given by

$$\gamma = \rho_x \|\mathbf{g}_1\|^2 = \rho_x \|\mathbf{g}_2\|^2 = \rho_x (|h_1|^2 + |h_2|^2). \tag{13.109}$$

In other words, the 2×1 MISO system with Alamouti codes achieves the diversity order of 2, which is the same effect as an MRC on a 1×2 channel. This diversity gain is achieved at the cost of lower data rate of one symbol per channel use.

This coding scheme operates on two time slots; its per-channel-use mutual information can be determined as

$$C_{\text{Alam}} = \frac{1}{2} \log \det \left(\mathbf{I} + \frac{\rho_x}{N_t} \mathbf{G}^\dagger \mathbf{G} \right) = \log \left(1 + 0.5 \rho_x [|h_1|^2 + |h_2|^2] \right), \tag{13.110}$$

which is identical to the channel capacity.

13.13.2 Two receive antennas

We next move to the case with two receive antennas. Suppose that the 2×2 Alamouti coding matrix, given in (13.105), is transmitted over a 2×2 channel $\mathbf{H} = [\mathbf{h}_1 \ \mathbf{h}_2]$, producing channel output

$$\mathbf{Y} = \mathbf{HX} + \mathbf{N}. \tag{13.111}$$

The two columns of $\mathbf{Y} = [\mathbf{y}_1 \ \mathbf{y}_2]$ represent the temporal vectors at the two receive antennas, respectively, and noise matrix $\mathbf{N} = [\mathbf{n}_1 \ \mathbf{n}_2]$ is defined similarly. Thus, it follows that

$$\mathbf{y}_1 = x_1 \mathbf{h}_1 + x_2 \mathbf{h}_2 + \mathbf{n}_1, \tag{13.112}$$

$$\mathbf{y}_2 = x_1^* \mathbf{h}_2 - x_2^* \mathbf{h}_1 + \mathbf{n}_2. \tag{13.113}$$

Taking $\mathbf{h}_1^\dagger \times (13.112)$ and $(\mathbf{h}_1^\dagger \times (13.113))^\dagger$, we obtain

$$\mathbf{h}_1^\dagger \mathbf{y}_1 = x_1 \|\mathbf{h}_1\|^2 + x_2 \mathbf{h}_1^\dagger \mathbf{h}_2 + \mathbf{h}_1^\dagger \mathbf{n}_1, \tag{13.114}$$

$$\mathbf{y}_2^\dagger \mathbf{h}_2 = x_1 \|\mathbf{h}_2\|^2 - x_2 \mathbf{h}_1^\dagger \mathbf{h}_2 + \mathbf{n}_2^\dagger \mathbf{h}_2, \tag{13.115}$$

which, when summed, leads to the decision variable for x_1:

$$z_1 = \mathbf{h}_1^\dagger \mathbf{y}_1 + \mathbf{y}_2^\dagger \mathbf{h}_2 \tag{13.116}$$

$$= x_1 (\|\mathbf{h}_1\|^2 + \|\mathbf{h}_2\|^2) + (\mathbf{h}_1^\dagger \mathbf{n}_1 + \mathbf{n}_2^\dagger \mathbf{h}_2). \tag{13.117}$$

Similarly, we can obtain the decision variable for x_2, as shown by

$$z_2 = \mathbf{h}_2^\dagger \mathbf{y}_1 - \mathbf{y}_2^\dagger \mathbf{h}_1 \tag{13.118}$$

$$= x_2 (\|\mathbf{h}_1\|^2 + \|\mathbf{h}_2\|^2) + (\mathbf{h}_2^\dagger \mathbf{n}_1 - \mathbf{n}_2^\dagger \mathbf{h}_1). \tag{13.119}$$

It is easy to show that the SNR in both z_1 and z_2 is equal to

$$\gamma = \rho_x (\|\mathbf{h}_1\|^2 + \|\mathbf{h}_2\|^2), \tag{13.120}$$

implying that the Alamouti coding scheme achieves the full diversity order of 4 and the symbol rate of 1 symbol/slot.

In fact, we can arrive at the SNR of (13.120) alternatively by writing (13.112) and (13.113) as

$$\begin{bmatrix} \mathbf{y}_1 \\ \mathbf{y}_2^* \end{bmatrix} = \underbrace{\begin{bmatrix} \mathbf{h}_1 & \mathbf{h}_2 \\ \mathbf{h}_2^* & -\mathbf{h}_1^* \end{bmatrix}}_{\mathbf{G} = [\mathbf{g}_1 \ \mathbf{g}_2]} \begin{bmatrix} x_1 \\ x_2 \end{bmatrix} + \underbrace{\begin{bmatrix} \mathbf{n}_1 \\ \mathbf{n}_2^* \end{bmatrix}}_{\mathbf{n}} = x_1 \mathbf{g}_1 + x_2 \mathbf{g}_2 + \mathbf{n}. \tag{13.121}$$

Note $\mathbf{g}_1^\dagger \mathbf{g}_2 = 0$. This indicates that the 2×2 Alamouti space–time coding system is *mathematically* equivalent to transmitting two symbols over a 4×2 block-diagonal MIMO, thus completely eliminating the IAI and enabling the detection of x_1 and x_2 separately. It is thus easy to determine the SINR for both x_1 and x_2, yielding

$$\gamma = \rho_x (\|\mathbf{h}_1\|^2 + \|\mathbf{h}_2\|^2). \tag{13.122}$$

Property 13.13.1 *In Alamouti codes, the orthogonality of code matrices in the time domain is converted into the orthogonal equivalent channels for each transmitted symbol.*

Let us determine the mutual information of a 2×2 MIMO channel with Alamouti codes. From (13.121), it follows that the mutual information is given by

$$C_{\text{Alam}} = \frac{1}{2} \log \det(\mathbf{I} + \frac{\rho_x}{N_t} \mathbf{G} \mathbf{G}^\dagger)|_{N_t = 2}$$

$$= \log (1 + 0.5 \rho_x (\|\mathbf{h}_1\|^2 + \mathbf{h}_2\|^2)). \tag{13.123}$$

We are interested in the theoretical mutual information of the 2×2 MIMO channel, which, by denoting the eigenvalues of $\mathbf{HH}^\dagger$ by λ_1 and λ_2, is equal to

$$C_{\text{theo}} = \log \det(\mathbf{I} + \frac{\rho_x}{2} \mathbf{HH}^\dagger)$$

$$= \log (1 + 0.5 \rho_x \lambda_1)(1 + 0.5 \rho_x \lambda_2)$$

$$= \log (1 + 0.5 \rho_x (\lambda_1 + \lambda_2) + 0.25 \rho_x^2 \lambda_1 \lambda_2)$$

$$> \log (1 + 0.5 \rho_x (\|\mathbf{h}_1\|^2 + \mathbf{h}_2\|^2)). \tag{13.124}$$

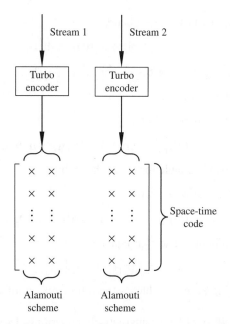

Figure 13.6   *Space–time coding scheme recommended for LTE-A*

The inequality holds if we recognize that $(\lambda_1 + \lambda_2) = \text{tr}(\mathbf{HH}^\dagger) = \|\mathbf{h}_1\|^2 + \|\mathbf{h}_2\|^2$. The last line implies that the data rate achieved by the Alamouti transceiver is less than the channel capacity. This is easy to understand since the Alamouti scheme is not a full-rate system. The maximum rate is 2, but the orthogonal scheme transmits 1 bit per channel use. The advantage of the orthogonal scheme is to use the repetitive transmission in time to trade for the orthogonality in a virtual space–time channel, thereby simplifying the detection and completely removing IAI.

Property 13.13.2   *The mutual information of Alamouti systems is equal to that of the corresponding MIMO channel only for 2×1 MISO channel, and is generally less than the corresponding MIMO channel capacity with $N_t \geq 2$ transmit antennas.*

Given its simplicity and good performance, Alamouti's space–time coding scheme is recommended for use in the standard LTE-A. It can be used alongside turbo codes to fully exploit the advantages of spatial coding and algebraic coding, as illustrated in Figure 13.6.

13.14   General space–time codes

Suppose a matrix symbol $\mathbf{X} \in \mathcal{C}^{N_t \times \ell}$ is transmitted over the $N_r \times N_t$ channel $\mathbf{H} \sim \mathcal{CN}_{N_t \times N_r}(\mathbf{0},\ \mathbf{R}_t \otimes \mathbf{R}_r)$, producing the received signal matrix $\mathbf{Y} \in \mathcal{C}^{N_r \times \ell}$

$$\mathbf{Y} = \underbrace{\sqrt{E_s/N_t}}_{\alpha_s}\, \mathbf{HX} + \mathbf{N}, \tag{13.125}$$

where $\mathbf{N} \in \mathcal{C}^{N_r \times \ell} \sim \mathcal{CN}(\mathbf{0}, N_0\mathbf{I}/2)$ is the white Gaussian noise matrix. On carefully examining the generation of IAI inherent in various parallel transmission schemes, we find that the source comes from the separate treatment of all the parallel transmitted data streams. The IAI problem automatically vanishes if we treat all the parallel data streams as a single matrix symbol, as done in space–time codes. In space–time codes, message symbols are encoded with algebraic structures to form a transmitted matrix symbol $\mathbf{X}$. Only those matrices with a given coding structure constitute the set of feasible space–time codes. Given the transmission of a code matrix $\mathbf{X}_1$ and the received matrix $\mathbf{Y}$, we would like to see the probability that a decision is made in favor of another matrix symbol $\mathbf{X}_2$.

13.14.1   Exact pairwise error probability

To proceed, denote $\text{vec}(\mathbf{X}_i) = \mathbf{x}_i, i = 1, 2$ and represent $\mathbf{Y}$ in vector form, yielding

$$\underbrace{\text{vec}(\mathbf{Y})}_{y} = \underbrace{\alpha_s(\mathbf{I}_n \otimes \mathbf{H})}_{\mathbf{K}}\underbrace{\text{vec}(\mathbf{X})}_{x} + \underbrace{\text{vec}(\mathbf{N})}_{n}, \tag{13.126}$$

where α_s represents the average power at each transmit antenna, and $\mathbf{n} \sim \mathcal{CN}(0, \frac{N_0}{2}\mathbf{I})$ is the AWGN. Suppose that $\mathbf{X} = \mathbf{X}_1$ is sent, and an erroneous decision event occurs when $\|\mathbf{Y} - \mathbf{HX}_1\|^2 > \|\mathbf{Y} - \mathbf{HX}_2\|^2$, or equivalently

$$\|\mathbf{y} - \mathbf{Kx}_1\|^2 > \|\mathbf{y} - \mathbf{Kx}_2\|^2, \tag{13.127}$$

which can be simplified to

$$\|\mathbf{n}\| > \|\mathbf{K}(\mathbf{x}_1 - \mathbf{x}_2) + \mathbf{n}\|^2. \tag{13.128}$$

By denoting

$$\mathbf{D} = \mathbf{X}_1 - \mathbf{X}_2, \tag{13.129}$$

we can further write the erroneous event as

$$\xi \triangleq \underbrace{\mathbf{n}^\dagger\mathbf{K}\,\text{vec}(\mathbf{D}) + \text{vec}^\dagger(\mathbf{D})\mathbf{K}^\dagger\mathbf{n}}_{\text{noise: } w} + \underbrace{\text{vec}^\dagger(\mathbf{D})\mathbf{K}^\dagger\mathbf{K}\,\text{vec}(\mathbf{D})}_{\text{signal: } s} < 0. \tag{13.130}$$

Conditioned on $\mathbf{H}$, the first two terms of the decision variable ξ constitute the Gaussian noise component w. Note that the two noise terms are mutually uncorrelated since they are complex conjugates. It is then easy to find the mean and variance of the noise component w, whereby we can write

$$w \sim \mathcal{N}(0, N_0\text{vec}^\dagger(\mathbf{D})\mathbf{K}^\dagger\mathbf{K}\,\text{vec}(\mathbf{D})). \tag{13.131}$$

The problem described in (13.130) is a binary detection one in Gaussian noise with SNR

$$\gamma = \frac{\text{vec}^\dagger(\mathbf{D})\mathbf{K}^\dagger\mathbf{K}\,\text{vec}(\mathbf{D})}{N_0}, \tag{13.132}$$

whereby the conditional pairwise error probability (PEP) is determined as

$$P_2(\gamma|\mathbf{D}) = Q(\sqrt{\gamma}). \tag{13.133}$$

We need to average $P_2(\gamma)$ over the channel matrix $\mathbf{H}$ to obtain the average error performance. To this end, we need to represent ξ in quadratic form in $\text{vec}(\mathbf{H})$. We use the operational rules of the Kronecker product described in Chapter 2 to rewrite

$$\mathbf{K}\,\text{vec}(\mathbf{D}) = \alpha_s(\mathbf{I}_n \otimes \mathbf{H})\text{vec}(\mathbf{D}) = \alpha_s(\mathbf{D}^T \otimes \mathbf{I}_n)\text{vec}(\mathbf{H}), \tag{13.134}$$

and defining

$$\rho = \frac{\alpha_s^2}{N_0},$$ (13.135)

we obtain the conditional SNR γ as

$$\gamma = \frac{\rho}{2}[(\mathbf{I} \otimes \mathbf{H})\text{vec}(\mathbf{D})]^\dagger[(\mathbf{I} \otimes \mathbf{H})\text{vec}(\mathbf{D})].$$ (13.136)

The CHF of γ equals

$$\phi_\gamma(z) = \mathbb{E}[e^{z\gamma}] = \det\left[\mathbf{I} - \frac{1}{2}\rho\, z(\mathbf{D}^\dagger \mathbf{R}_t \mathbf{D}) \otimes \mathbf{R}_r\right]^{-1},$$ (13.137)

which, when combined with the Gaussian-integral representation of the Q-function in (13.133), allows us to obtain the average pairwise error probability as

$$E_\gamma[P_2(\gamma|D)] = \frac{1}{\pi}\int_0^{\pi/2} \phi_\gamma(z)|_{z=-\frac{1}{2\sin^2\theta}}\, d\theta.$$ (13.138)

This PEP expression, though exact, lacks intuition. To get a more insightful understanding, we note that the CHF in (13.137) implies that γ is a linear combination of a set of independent chi-square variables and, thus, can be well approximated by a single chi-square variable.

Lemma 13.14.6 *A random variable with CHF $\phi_\xi(z) = \det(\mathbf{I} - z\mathbf{A})^{-1}$ can be approximated by a single chi-square variable $\xi \sim c\chi^2(\upsilon)$, for which*

$$\upsilon = \frac{2\,[\text{tr}(\mathbf{A})]^2}{\text{tr}(\mathbf{A}\mathbf{A}^\dagger)},\quad c = \frac{tr(\mathbf{A}\mathbf{A}^\dagger)}{2\,tr(\mathbf{A})}.$$ (13.139)

To apply this lemma to (13.138), we can set $\mathbf{A} = (\rho/2)(\mathbf{D}^\dagger \mathbf{R}_t \mathbf{D}) \otimes \mathbf{R}_r$, and then determine the corresponding scale factor c and the degrees of freedom υ. The resulting geometric intuition is very interesting. The average PEP starts to drop at a certain SNR value of ρ; the earlier it drops, better the performance. The value of c represents the coding gain due to space–time coding, which pushes the start-dropping point to the left along the horizontal SNR axis. The dropping slope depends on the value of υ; a larger υ implies a faster dropping. As such, υ represents the diversity gain. The situation is intuitively illustrated in Figure 13.7. An exact PEP formula for AWGN MIMO channels, in closed form, can be found in (62) of Refs [26, 27].

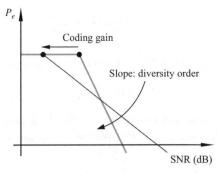

Figure 13.7 *Illustrating coding gain and diversity gain, and their influence on the pairwise error performance*

13.15 Information lossless space–time codes

General orthogonal codes of the Alamouti type introduce loss in information rate compared to the corresponding MIMO channels. To achieve a full channel rate, consider a MIMO system

$$\mathbf{Y} = \rho_t \mathbf{H} \mathbf{B} + \mathbf{N}, \tag{13.140}$$

with a total transmit SNR of ρ_t, $M = 2$ transmit antennas, and a 2×2 transmit space–time input matrix

$$\mathbf{B}_{2,\phi} = \frac{1}{\sqrt{M}} \begin{bmatrix} x_1 + \phi x_2 & \theta(x_3 + \phi x_4) \\ \theta(x_3 - \phi x_4) & x_1 - \phi x_2 \end{bmatrix}, \tag{13.141}$$

where $\phi = e^{j\alpha}$, $\theta^2 = \phi$, and α is a real parameter to be optimized [28]. Denote $\mathbf{n} = \text{vec}(\mathbf{N})$, $\mathbf{y} = \text{vec}(\mathbf{Y})$, and $\mathbf{x} = [x_1, x_2, x_3, x_4]^T$. It is straightforward to rewrite $\mathbf{Y}$ in vector form as

$$\mathbf{y} = \underbrace{\begin{bmatrix} \mathbf{H} & \mathbf{0} \\ \mathbf{0} & \mathbf{H} \end{bmatrix}}_{\mathbf{G}} \mathbf{\Phi} \mathbf{x} + \mathbf{n}, \tag{13.142}$$

where $\mathbf{G}$ is the equivalent virtual MIMO channel matrix, and $\mathbf{\Phi}$ is defined by

$$\mathbf{\Phi} = \frac{1}{\sqrt{2}} \begin{pmatrix} 1 & \phi & 0 & 0 \\ 0 & 0 & \theta & -\theta\phi \\ 0 & 0 & \theta & \theta\phi \\ 1 & -\phi & 0 & 0 \end{pmatrix}. \tag{13.143}$$

It is easy to examine that $\mathbf{\Phi}^\dagger \mathbf{\Phi} = \mathbf{I}$. Thus, the system mutual information can be obtained as

$$C_{B,\phi}(\rho_t) = \frac{1}{2} \log \det \left(\mathbf{I} + \frac{\rho_t}{M} \mathbf{G} \mathbf{G}^\dagger \right)$$

$$= \log \det(\mathbf{I} + \frac{\rho_t}{M} \mathbf{H} \mathbf{H}^\dagger). \tag{13.144}$$

The last line represents the MIMO channel capacity. It is clear that the MIMO system with code matrix $\mathbf{B}_{2,\phi}$ does not suffer from information loss. Intuitively, this scheme achieves the full rate of two symbols per channel use. The drawback of using $\mathbf{B}_{2,\phi}$ is the increase of decoding complexity.

13.16 Multiplexing gain versus diversity gain

In fact, we can trade multiplexing gain for diversity gain, and vice versa. The definition of multiplexing gain is diverse in the literature. For example, for a MIMO system that transmits N distinct symbols over T time slots, Hammons defines his multiplexing gain as the ratio N/T [29]. Clearly, the multiplexing gain has the range $0 \le r_s \le m$. The MIMO system is said to be in the spatial multiplexing mode if $r_s = m$ and in the diversity mode if $r_s \le 1$. The diversity gain for a particular space–time codebook can be intuitively defined as the slope of the PEP curve with respect to SNR. According to Tarokh et al. [30], the diversity gain is defined as the minimum rank, L, of the pairwise difference of all the codeword matrices. The case of $L = m$ is called *full diversity*.

In an $N_r \times N_t$ MIMO fading channel, there are a total of $N_t N_r$ paths linking the transmitter to the receiver. These $N_t N_R$ paths mathematically represent up to $N_t N_r$ degrees of freedom (DFs) that can be used either for parallel transmission (multiplexing) to increase the data rate, in terms of the number of parallel data streams, or to improve the transmission reliability in the form of diversity order. Intuitively, the *diversity gain* corresponds to the number of independent faded paths that a symbol passes through, while the number of parallel channels that can be used for transmission is called *multiplexing*. The allocation of this precious channel resource for multiplexing and diversity is usually implemented either by precoders or by STBC. There is a tradeoff between them, called the diversity-multiplexing tradeoff (DMT).

For a given MIMO channel, there is a tradeoff between the diversity and multiplexing gains. To get some idea of DMT, let us begin with the asymptotic behavior of the MIMO channel capacity as $\rho \to \infty$:

$$C = m \, \log \rho + \prod_{i=1}^{m} \chi_i^2 (2M - 2i + 1) + o(1) \stackrel{\rho \to \infty}{=} m \log \rho, \tag{13.145}$$

implying that the maximum number of parallel streams (i.e., multiplexing) is $m = \min\{N_t, N_r\}$ if all the DFs resource is used for capacity. On the other extreme, if all the DFs resource is invested for transmission reliability, the maximum diversity is $N_t N_r$; see Problem 13.11 for details. The two extremes are called "full rate" and "full diversity", respectively, which were once the design targets of STBC in the early literature. For cases in between, we can imagine that the more parallel data streams are transmitted, creating higher cross-stream interference, the poorer the error performance. Poorer error performance means a drop in diversity order. The DMT can be visualized alternatively from the perspective of transmission reliability. To increase the transmission reliability over a MIMO channel, a common practice is to introduce certain constraints on transmit symbols in the form of code structures or precoding. These constraints increase the minimum distance of the STBC and, thus, the detection reliability on one hand, but reduce the transmission data rate on the other. Two frameworks are employed to characterize the multiplexing, coding, and diversity gains in the literature.

Given the complicated connections within a MIMO channel, there is in general no simple relationship between multiplexing and diversity gains. Tse and Zheng therefore tried to find a DMT relationship in the asymptotic framework of capacity outage by passing the SNR to infinity [31].

13.16.1 Two frameworks

Tarokh et al. [30] investigated the problem in the framework of PEP and found that there exists two exponents in its upper bound, with one representing the diversity order and the other representing the coding gain. The multiplexing gain in Tarohk's framework is simply defined as the number of fixed-rate parallel data streams, which are independent of the channel SNR. Inspired by the observation from (13.145) that the multiplexing gain m is always associated with the scalar $\log \rho$, Zheng and Tse [31] defined the multiplexing gain r as the number of data streams with each of rate on the order of magnitude $\log \rho$ to cope with the channel SNR.

Therefore, they formulated the DMT problem in the framework of the capacity outage $P_{\text{out}} = \Pr\{C < r \log \rho\}$ with an outage threshold defined by $r \log \rho$ [31]. Geometrically, C represents the volume of an m-dimensional geometry, while ρ^r represents an r-dimensional sphere. Their derivation of the key formula $d(r)) = (N_t - r)(N_r - r)$ was based on the *asymptotic* behavior of the capacity outage probability for mathematical tractability. The justification of this methodology is the observation that "diversity gain is an asymptotic concept." The crucial issue is the explanation of the outage threshold $R = r \log \rho_x$. The understanding of multiplexing gain is of duality. It is sometimes interpreted as the number of parallel data streams in transmission, but sometimes as the data rate of $r \log \rho$ bits/per channel time that occupies all the eigenbeams. In determining the theoretical $d(r)$, the transmit symbols are treated as Gaussian random variables; in determining $d(r)$ for a specific scheme, the symbols are simply treated as symbols. The detection error event is dominated by the channel outage. In other words, an outage event will almost surely lead to a detection error, thus intuitively allowing us to assert that the outage probability is a lower bound on the error probability. The resulting outage probability is given in Theorem 4 of Ref. [31].

In summary, Zhang and Tse formulated the DMT problem in the framework of capacity outage, yielding a simple asymptotic DMT relationship. The SNR-dependent multiplexing gain implies the implementation with adaptive modulation. In Tarokh's framework, on the contrary, multiplexing is simply defined as the number

Table 13.2   *Comparison of the methods by Tse and Tarokh*

Method	Definition of multiplexing	Implementation
Tse	$r \log \rho$	Rate-adaptive modulation
Tarokh	Fixed-rate transmission	Fixed-rate modulation

of fixed-rate parallel data streams, which are independent of the channel SNR. The difference in philosophy between the two is compared in Table 13.2.

13.16.2   Derivation of the DMT

Let us outline the two steps of Zheng and Tse in deriving their DMT. First, they treat codewords $\mathbf{X} \in \mathcal{C}^{N_t \times \ell}$ as random matrix symbols chosen from a complex Gaussian alphabet, and derives their basic formula $d(r) = (m - r)(M - r)$ in the framework of capacity outage. In the second step, to prove that $d(r)$ so obtained is indeed the diversity order, they determine the upper and lower bounds of the PEP and show that, as the SNR approaches infinity, both bounds approach the same order of $\rho^{-d(r)}$.

Here we only sketch the basic idea of the first step. We write the probability of capacity outage as

$$P_{\text{out}}(R) = \Pr\{\log \det(\mathbf{I} + \rho \mathbf{H}\mathbf{H}^\dagger) < r \log \rho\}$$

$$= \Pr\{\prod_{i=1}^{m}(1 + \rho \lambda_i) < \rho^r\}. \tag{13.146}$$

Geometrically, $\det(\mathbf{A})$ represents the volume of the manifold $\mathbf{A}$; the second line above implies that $\det(\mathbf{I} + \rho \mathbf{H}\mathbf{H}^\dagger)$ is an m-dimensional supercube with each dimension of a random length $(1 + \rho \lambda_i)$. Outage occurs when this supercube volume is less than that of the r-dimensional cube. As $\rho \to \infty$, the only possibility is that there are $m - r$ dimensions of order $O(1)$ or $\lambda_i = \rho^{-1}$; this is the so-called typical outage event.

The key step is to represent the eigenvalues λ_i of $\mathbf{H}^\dagger \mathbf{H}$ in terms of their singularity levels α's, defined by $\lambda_i = \rho^{-\alpha_i}$, whereby to further write $(1 + \rho \lambda_i) = \rho^{(1-\alpha_i)^+}$ with $(x)^+ = x$ if $x \geq 0$ and equals 0 if $x < 0$. As such, (13.146) becomes

$$P_{\text{out}}(R) = \Pr\{\sum_{i=1}^{m}(1 - \alpha_i)^+ < r\}, \text{ (as } \rho \to \infty). \tag{13.147}$$

Clearly, the joint PDF of α's we need for this calculation is obtainable from the joint PDF of ordered sample eigenvalues $\lambda_1 < \cdots < \lambda_m$. In determining $P_{\text{out}}(R)$, we note that it is *typical* outage events that dominate the outage probability. The typical events are those of the r largest eigenvalues and $(m - r)$ negligibly small eigenvalues. Among all these typical outage events, the one with $\lambda_1 = \cdots = \lambda_{m-r} = 0$ and $\lambda_{m-r+1} = \cdots = \lambda_m = 1$ dominates the outage probability. This corresponds to setting $\alpha_i = 1$ for $i = 1, \cdots, m - r$ and setting the remainder to zero. As such, the capacity-outage probability equals

$$\max_{\alpha} \left\{ \rho^{-\sum_{i=1}^{m}(M-m+2i-1)\alpha_i} \right\} = \rho^{-\sum_{i=1}^{m-r}(M-m+2i-1)}$$

$$= \rho^{-(m-r)(M-r)}. \tag{13.148}$$

The diversity order $d(r)$ is determined by the exponent. It follows that

$$d(r) = (m - r)(M - r). \tag{13.149}$$

13.16.3   Available DFs for diversity

Intuitively, the multiplexing gain is equal to the number of parallel transmitted data streams. From the geometrical perspective, the best way to transmit these data streams is to convey them along the orthogonal eigen directions. Suppose that we want to transmit r parallel data streams, and the question of interest is the number of DFs left for diversity.

To answer this question, we need to find the DFs required for supporting the r eigenbeams. To proceed, represent the channel matrix in its singular decomposition form

$$\mathbf{H} = \sum_{i=1}^{m} \alpha_i \mathbf{u}_i \mathbf{v}_i^\dagger \tag{13.150}$$

in which we need r eigenbeams for transmission, as defined by

$$\mathbf{H}_r = \sum_{i=1}^{r} \alpha_i \mathbf{u}_i \mathbf{v}_i^\dagger. \tag{13.151}$$

We check how many DFs are needed to support this multiplexing. Each pair of $\mathbf{u}_i$ and $\mathbf{v}_i$ requires N_r and N_t complex DFs, respectively, while the singularity values need r complex DFs, implying that there are $(N_r + N_t)r + r$ in total. However, not all these DFs are independent and, in fact, there are constraints among them. In particular, the orthogonality among $\mathbf{u}$'s poses $r(r-1)/2$ complex constraints, plus r constraints due to normalization, adding to a total of $r(r+1)/2$. The same number of constraints exist among $\mathbf{v}$'s. Thus, the true DFs required for multiplexing are given by

$$DF_{\text{mux}} = [(rN_r k + rN_t) + r] - 2\left[\frac{r(r+1)}{2}\right] = (N_t + N_r - r)r, \tag{13.152}$$

which, when subtracted from the total DFs $N_t N_r$ in $\mathbf{H}$, represents the DFs available for diversity:

$$DF_{\text{div}} = N_t N_r - (N_t + N_r - r)r = (N_t - r)(N_r - r) = (m - r)(M - r). \tag{13.153}$$

The result is identical to the diversity gain given in (13.149) obtained as $\rho \to \infty$.

But, it does not mean that all these DFs are completely convertible to the diversity order. In fact, subject to the special connectivity in the MIMO, only a part of them are transformed to the diversity.

✎ **Example 13.5** _____

Determine a DMT expression for an i.i.d. Gaussian MISO channel with m transmit antennas.

Solution

From the previous study in diversity reception, we know that $\gamma \sim \chi^2(2m)$; hence

$$P_{\text{out}} = \Pr\{\log(1 + \rho\gamma) < r \log \rho\}$$
$$= \Pr\{\gamma < \rho^{r-1}\}$$
$$= \int_0^{\rho^{r-1}} x^{m-1} e^{-x} dx$$
$$= \int_0^{\rho^{r-1}} x^{m-1} dx \quad (\text{note}: \ e^{-x} \to 1 \text{ for large } \rho)$$
$$= m^{-1} \rho^{-m(1-r)}. \tag{13.154}$$

Hence, we obtain $d(r) = m(1 - r)$ for $0 \le r \le 1$.   ○

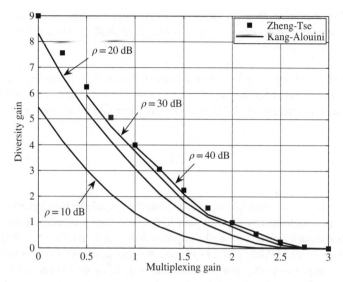

Figure 13.8   *Exact and asymptotic DMT: Nearly i.i.d MIMO channels*

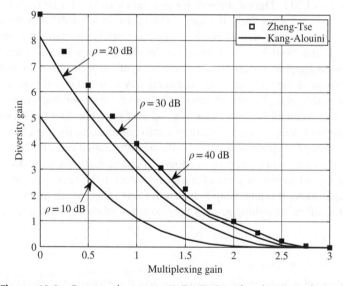

Figure 13.9   *Exact and asymptotic DMT: Correlated MIMO channels*

Property 13.16.1   *For MIMO channels with the simplest topology such as MISO and SIMO, the available DFs for diversity is completely converted to the diversity gain, irrespective of ρ.*

To see the deviation of the asymptotic DMT from its exact counterpart, comparison is made in Figures 13.8 and 13.9. In these figures, the exact DMT is obtained from the exact capacity outage probability via

$$d(r, \rho) = -\frac{\partial \ln P_{\text{out}}(\rho)}{\partial \ln \rho},$$

where $P_{\text{out}}(\rho)$, in turn, is calculated by using (13.32) and (13.146). The asymptotic DMT is computed from $d(r) = (m - r)(M - r)$. Indeed, as ρ approaches 40 dB, the asymptotic DMT is nearly identical to its exact counterpart. A big gap is observed for $\rho < 20$ dB.

Property 13.16.2   *The available DFs for diversity is completely converted into the diversity gain, as described in the DMT formula, only when $\rho \to \infty$. For any finite SNR, however, it is merely partly transferred to the latter.*

General MIMO channels are of complicated connectivity. Given this constraint, there is no simple rule to describe the DMT behavior except for the asymptotic case as $\rho \to \infty$. Though not reflecting the behavior in the SNR regime of practical interest, the DMT formula does provide us with an insightful understanding of MIMO channels.

13.17   Summary

Multiple-antenna technology is a powerful means to exploit the spatial electromagnetic resources in a random scattering field. Random scattering causes multipath fading and thus a significant drop in channel capacity relative to the AWGN channel. Nevertheless, the capacity loss due to multipath fading can be remedied by using collocated or distributive MIMO antenna systems. A main challenge in the implementation of MIMO is IAI. If we focus on a *single* snap shot at a time, the only way to avoid IAI caused by cross-antenna correlation is to transmit symbols in parallel along the *spatially* orthogonal eigenbeams of the channel covariance matrix. The requirement is CSIT. The mechanism for generating IAI is similar to that responsible for ISI in a bandlimited channel except for the lack of a toeplitz structure in a MIMO channel matrix. Thus, for the case without CSIT, as often encountered in practice, one can retrieve the parallelly transmitted symbols by using the method of ZF and MMSE, or that based on the QR decomposition. When focusing on a single snap shot, signal collision happens simply because parallel symbols transmitted by multiple antennas clash in a one-dimensional space at the receiver. Thus, a natural way to avoid signal collision without CSIT is to increase the signal dimension at each transmit antenna so as to form a *temporally* orthogonal transmit signal space. This is the idea of Alamouti's codes. The third scheme is to consider the transmitted entries over a space–time block as a single matrix symbol, thereby totally eliminating the foundation for IAI generation. The problem with the method of space–time codes is its implementation complexity.

A MIMO channel has a number of degrees of freedom that can be converted to multiplexing gain to increase system data rate or converted to diversity gain for transmission reliability. Asymptotically as SNR tends to infinity, there is a simple tradeoff relationship between the two gains, which provides certain insight into a MIMO system. The tradeoff relationship becomes very complex when the SNR is in a normal region.

Problems

Problem 13.1   A classical scheme to use a space–time repetitive transmission to trade for receive diversity with a single antenna was originated by Alamouti. The 2×1 temporal received vector is given by

$$\mathbf{x} = \begin{pmatrix} s_0 & s_1 \\ -s_1^* & s_0^* \end{pmatrix} \begin{bmatrix} h_1 \\ h_2 \end{bmatrix} + \mathbf{n}.$$

Design a receiver to achieve the effect of order-2 diversity reception [25].

Problem 13.2   Assume an $M \times N$ MIMO channel $\mathbf{H}$ of i.i.d. zero-mean and unit-variance Gaussian entries. A widely used formula for featuring its ergodic channel capacity C is given by

$$C \propto \min(M, N).$$

Prove this assertion by assuming that M or N tends to infinity.

Problem 13.3   In Section 13.10, the MMSE receivers were simultaneously derived for all the receive antennas. Show that the same result can be derived by focusing on each single receive antenna.

Problem 13.4   Employ the result obtained in (13.95) to show that the conditional PDF of γ in the example can be expressed as

$$f(\gamma|\mathbf{c}) = \frac{e^{-\gamma/\rho_s}(\gamma/\rho_s)^{m-1}(1+\rho_c)}{\rho_s(m-2)!} \int_0^1 \exp\left(-\frac{\gamma\rho_c}{\rho_s}t\right)(1-t)^{m-2}d\,t.$$

This result was obtained by Winters [22] by invoking a lemma by V.M. Bogachev [32].

Problem 13.5   Suppose a MIMO system with $M = 3$ transmit antennas employs a rate-$\frac{3}{4}$ space–time code matrix of the form

$$\mathbf{X} = \sqrt{\frac{4}{3}}\begin{pmatrix} x_1 & x_2 & x_3 \\ -x_2^* & x_1^* & 0 \\ -x_3^* & 0 & x_1^* \\ 0 & -x_3^* & x_2^* \end{pmatrix}.$$

Find its equivalent virtual channel and corresponding mutual information, and show that it is information lossless [33].

Problem 13.6   Consider another rate-$\frac{3}{4}$ space–time code matrix

$$\mathbf{X} = \sqrt{\frac{4}{3}}\begin{pmatrix} x_1 & x_2 & \frac{x_3}{\sqrt{2}} \\ -x_2^* & x_1^* & \frac{x_3}{\sqrt{2}} \\ \frac{x_3^*}{\sqrt{2}} & \frac{x_3^*}{\sqrt{2}} & y \\ \frac{x_3^*}{\sqrt{2}} & -\frac{x_3^*}{\sqrt{2}} & z, \end{pmatrix}.$$

where $y = (-x_1 - x_1^* + x_2 - x_2^*)/2$ and $z = (x_1 - x_1^* + x_2 + x_2^*)/2$. Find its equivalent virtual channel and the corresponding mutual information [34].

Problem 13.7   In deriving his design criteria for space–time codes, Tarokh employs the upper bound for the average probability of pairwise error on $n \times m$ i.i.d zero-mean unit-variance complex Gaussian MIMO channels with AWGN at the receiver side, as shown by

$$\Pr\{\mathbf{C}_0 \to \mathbf{C}_1\} \le \frac{1}{2}\det[\mathbf{I} + (\rho_t/4)(\mathbf{C}_0 - \mathbf{C}_1)(\mathbf{C}_0 - \mathbf{C}_1)^\dagger]^{-m},$$

where ρ_t denotes the average SNR per transmit antenna. Prove this bound by simply using the integral representation of the Q-function [30].

Problem 13.8   An approximation to the ergodic capacity of i.i.d. Rayleigh $N_t \times N_r$ channels is often used in the literature

$$\overline{C} = m\log\rho + \sum_{i=M-m+1}^{M} \mathbb{E}[\log \chi_{2i}^2] + O(1).$$

Derive this expression [35].

Problem 13.9   In the literature, the ergodic capacity of the $M \times M$ i.i.d. MIMO channel is sometimes lower bounded by

$$\mathbb{E}[C] > \sum_{k=1}^{M} \log[1 + (\rho/M)\chi^2(2k)] \text{ b/s/Hz}.$$

Use the case of 2×2 MIMO as a counter-example to show the incorrectness of this assertion [36].

Problem 13.10   Consider an $n_r \times n_t$ MIMO channel $\mathbf{H}$ of i.i.d. entries $h_{i,j} \sim \mathcal{CN}(0, 1)$. Denote the transmit SNR by ρ_t and denote $m = \min(n_t, n_r)$.

(a) Show that the mutual information can be upper-bounded by C^+ such that

$$C = \log \det(\mathbf{I} + \frac{\rho_t}{n_t}\mathbf{HH}^\dagger) \leq C^+$$

with C^+ defined by

$$C^+ = m \log\left(1 + \frac{\rho_t}{2m \, n_t}\chi^2(2n_t n_r)\right).$$

(b) Further show that

$$C^+ \rightarrow \begin{cases} m \log(1 + \rho_t), & n_t > n_r \text{ and } n_t \rightarrow \infty; \\ m \log[1 + (n_r/n_t)\rho_t], & n_r > n_t \text{ and } n_r \rightarrow \infty. \end{cases}$$

The first line reveals that the MIMO channel is equivalent to m parallel, independent AWGN channels. The second line implies that there is an additional SNR gain if $n_r > n_t$.

Problem 13.11   Given an $N_t \times N_r$ i.i.d. Rayleigh fading MIMO channel, design a transceiver to achieve the maximum degrees of freedom and study its performance.

Problem 13.12   The capacity of an $n \times n$ Gaussian MIMO channel matrix $\mathbf{H}$ with total transmit SNR of ρ is given by

$$C = \log \det[\mathbf{I} + (\rho/n)\mathbf{HH}^\dagger].$$

Assume that the column vectors of $\mathbf{H}$ are i.i.d. with distribution $\mathbf{h}_i \sim \mathcal{CN}(0, \mathbf{R})$, where $\mathbf{R}$ is an exponential correlation matrix defined by $r < 1$, such that its (i, j)th entry equals $r_{ij} = r^{|j-i|}$ for $i, j = 1, 2, \cdots, n$. Using the concavity of the $\log \det(\cdot)$ function, wecan upper-bound the channel ergodic capacity $\mathbb{E}[C]$ by

$$\mathcal{C} = \log \det(\mathbf{I} + \rho\mathbf{R}).$$

Show that

$$C = \sum_{i=1}^{n} \log\left[1 + \frac{\rho(1 - r^2)}{1 - 2r \cos \theta_i + r^2}\right],$$

where θ_i are the values that satisfy one or the other of the equations

$$\sin[(n + 1)\theta/2] = r \sin[(n - 1)\theta/2],$$

$$\cos[(n + 1)\theta/2] = r \cos[(n - 1)\theta/2].$$

Hint: Find the properties of the exponentially correlated matrix using Google or a reference book.

Problem 13.13 Assume that an $n \times L$ STBC is transmitted over an $n \times m$ Rician MIMO channel with i.i.d. Rician link of Rician factor $K_{i,j}$ between transmitter i and receiver j. Show that the pairwise error probability between codewords $\mathbf{C}$ and $\mathbf{C}'$ can be upper-bounded by

$$\Pr\{\mathbf{C} \to \mathbf{C}'\} \leq \prod_{j=1}^{m} \left(\prod_{i=1}^{n} (1 + \rho\lambda_i)^{-1} \exp\left(-\frac{\rho\, K_{i,j}\lambda_i}{1 + \rho\lambda_i}\right) \right),$$

where $\rho = E_s/(4N_0)$. Hint: use the relation

$$Q(x) = \frac{1}{\pi} \int_0^{\pi/2} \exp\left[-\frac{x^2}{2\sin^2\phi}\right] d\phi, \; x \geq 0$$

[30].

Problem 13.14 Suppose that an $n \times n$ matrix $\mathbf{Z}$ follows the complex Wishart distribution $\mathbf{Z} \sim \mathcal{CW}_n(m, \mathbf{\Sigma})$ with $m > n$. The eigenvalues of $\mathbf{\Sigma}$, in descending order, are denoted by $\sigma_1 > \cdots > \sigma_n > 0$. Then, the CDF of the largest eigenvalue (denoted here by λ_1) of $\mathbf{Z}$ is given by Ordonez et al. [11], as

$$F_{\lambda_1}(x) = \prod_{i=1}^{n} \frac{1}{\sigma_i^m (m-i)!} \prod_{i<j}^{n} \frac{\sigma_i \sigma_j}{\sigma_j - \sigma_i} |\mathbf{Y}(x)|, \tag{13.155}$$

where the $n \times n$ matrix $\mathbf{Y}$ is defined as

$$[\mathbf{Y}(x)]_{u,v} = \sigma_u^{m-n+v} \gamma^*(x/\sigma_u, m-n+v), \; u, v = 1, \cdots, n, \tag{13.156}$$

in terms of the lower incomplete gamma function defined by

$$\gamma^*(\varphi, a) = \int_0^{\varphi} e^{-t} t^{a-1} dt, \tag{13.157}$$

which is related to the standard incomplete gamma function by $\gamma^*(\varphi, a) = \Gamma(a)\gamma(\varphi, a)$ [12].

Problem 13.15 Determine a DMT expression for the i.i.d. SIMO channel with m receive antennas.

Problem 13.16 Consider transmitting a coded sequence with a minimum distance $d_{\min}$ over an SIMO channel. Use the chi-square approximation method to determine the slope and coding gain in the pairwise error probability.

Problem 13.17 Design a decoder for the detection of the following 4-transmit antenna rate-$1/2$ orthogonal block code:

$$\mathbf{X} = \begin{pmatrix} x_1 & x_2 & x_3 & x_4 \\ -x_2 & x_1 & -x_4 & x_3 \\ -x_3 & x_4 & x_1 & -x_2 \\ -x_4 & -x_3 & x_2 & x_1 \\ x_1^* & x_2^* & x_3^* & x_4^* \\ -x_2^* & x_1^* & -x_4^* & x_3^* \\ -x_3^* & x_4^* & x_1^* & -x_2^* \\ -x_4^* & -x_3^* & x_2^* & x_1^* \end{pmatrix},$$

whose ith column, denoted by $\mathbf{x}_i$, is transmitted by transmit antenna i. Derive your decoder for the receiver with (1) one receive antenna, and (2) two receive antennas [34].

Problem 13.18   Derive a closed-form expression for the pairwise error probability for a MIMO OFDM system operating on a block quasi-static fading channel. By block quasi-static fading, we mean that the MIMO channel remains unchanged over the duration of each fading block but can independently vary from one block to another [37].

Problem 13.19   Consider a MIMO channel with N transmit and M receive antennas. The received signal is given by

$$\mathbf{y} = \mathbf{Hx} + \mathbf{n},$$

where $\mathbf{x} \sim \mathcal{CN}(\mathbf{0}, \mathbf{R}_x)$ and $\mathbf{n} \sim \mathcal{CN}(\mathbf{0}, \mathbf{R}_n)$ is the AWGN component. The mutual information for this channel is

$$C = \log \det(\mathbf{I} + \underset{n}{\mathbf{R}}^{-1} \mathbf{HR}_x \mathbf{H}^\dagger).$$

Show that the channel capacity can be alternatively expressed as

$$C = \log \frac{\det(\mathbf{R}_x)}{\det(\mathbf{R}_{\mathrm{MMSE}})},$$

where $\mathbf{R}_{\mathrm{MMSE}} = \mathbb{E}[(\hat{\mathbf{x}} - \mathbf{x})(\hat{\mathbf{x}} - \mathbf{x})^\dagger]$ is the covariance matrix of the MMSE estimator $\hat{\mathbf{x}}$ s[38].

Problem 13.20   Consider a MIMO system with M receive and N transmit antennas. The system employs the VBLAST architecture for transmission and linear ZF equalizer for reception. Show that it can achieve a diversity–multiplexing tradeoff given by

$$d(r) = (M - N + 1)(1 - r/N),$$

where r is the total multiplexing gain [39].

Problem 13.21   For a finite SNR ρ, the diversity gain for multiplexing rate r is defined as

$$d(r, \rho) = -\rho \frac{\partial}{\partial \rho} \log P_{\mathrm{out}}(r, \rho). \tag{13.158}$$

For a SIMO system with M receive antennas and employing maximum ratio combining, calculate its finite diversity–multiplexing tradeoff [40].

Problem 13.22   The CDF $F_X(x)$ for the output SINR of the MIMO MMSE receiver is given in Property 13.10.3. Use this CDF to show that, as $\rho \to \infty$, the diversity order of the MIMO MMSE receiver with N_t transmit and N_r receive antennas is given by $(N_r - N_t + 1)$.

References

1. M. Kang and M.-S. Alouini, "Capacity of correlated MIMO Rayleigh channels," *IEEE Trans. Wireless Commun.*, vol. 5, no. 1, pp. 143–155, 2006.

2. M. Kang and M.-S. Alouini, "Capacity of MIMO Rician channels," *IEEE Trans. Wireless Commun.*, vol. 5, no. 1, pp. 112–122, 2006.

3. M. Chiani, M. Win, and A. Zanella, "On the capacity of spatially correlated MIMO Rayleigh-fading channels," *IEEE Trans. Inf. Theory*, vol. 49, no. 10, pp. 2363–2371, 2003.

4. A.T. James, "Distributions of matrix variates and latent roots derived from normal samples," *Ann. Math. Stat.*, vol. 35, pp. 475–501, 1964.

5. M. Kiessling, "Unifying analysis of ergodic MIMO capacity in correlated Rayleigh fading environments," *Eur. Trans. Telecommun.*, Vol. 16, no. 1, pp. 17–35, 2005.

6. I.E. Telatar, "Capacity of multi-antenna Gaussian channels," Bell Labs Technical Memorandum, June 1995.

7. J.W. Silverstein, "Strong convergence of the empirical distribution of eigenvalues of large dimensional random matrices," *J. Multivariate Anal.*, vol. 55, pp. 331–339, 1995.

8. C.R. Rao, *Linear Statistical Inference and its Applications*, 2nd ed., New York: John Wiley & Sons, Inc., 1973.

9. M.R. McKay, A.G. Grant, and I.B. Collings, "Performance analysis of MIMO-MRC in double-correlated Rayleigh environments," *IEEE Trans. Commun.*, vol. 55, no. 3, pp. 497–507, 2007.

10. M.R. McKay, X. Gao, and I.B. Collings, "MIMO multi-channel beaming SER and outage using new eigenvalue distributions of complex noncentral Wishart matrices," *IEEE Trans. Commun.*, vol. 58, no. 3, pp. 424–434, 2008.

11. L.G. Ordonez, D.P. Palomar, and J.R. Fonollosa, "Ordered eigenvalues of a general class of Hermitian random matrices with application to the performance analysis of MIMO systems," *IEEE Trans. Signal Process.*, vol. 57, no. 2, pp. 672–689, 2009.

12. M. Abramowitz and I.A. Stegun, *Handbook of Mathematical Functions*, New York: Dover Publications, 1970, p. 946.

13. C.G. Khatri, "Distribution of the largest or smallest characteristic root under the null hypothesis concering complex multivariate normal populations," *Ann. Math. Stat.*, vol. 35, pp. 1807–1810, 1964.

14. M. Kang and M.-S. Alouini, "Largest eigenvalue of complex Wishart matrices and performance analysis of MIMO MRC systems," *IEEE J. Sel. Areas Commun.*, vol. 21, no. 3, pp. 418–426, 2003.

15. P.A. Dighe, R.K. Mallik, and S.S. Jamuar, "Analysis of trnamsit-receive diversity in Rayleigh fading," *IEEE Trans. Commun.*, vol. 51, no. 4, pp. 694–703, 2003.

16. A. Maaref and S. Aissa, "Closed-form expressions for the outage and ergodic Shannon capacity of MIMO MRC systems," *IEEE Trans. Commun.*, vol. 53, no. 7, pp. 1092–1095, 2005.

17. H. Gao, P.J. Smith, and M.V. Clark, "Theoretical reliability of MMSE linear diversity combining in Rayleigh-fading additive interference channels," *IEEE Trans. Commun.*, vol. 46, no. 5, pp. 666–672, 1998.

18. J.H. Winters, J.S. Salz, and R.D. Gitlin, "The impact of antenna diversity on the capacity of wireless communication systems," *IEEE Trans. Commun.*, vol. 2/3/4, pp. 1740–1751, 1994.

19. D. Gore, B.W. Heath, and A. Paulraj, "On performance of the zero forcing receiver in the presence of transmit correlation," in ISIT 2002, Lausanne Switzerland, June 30–July 5, 2002.

20. R.J. Muirhead, *Aspects of Multivariate Statistical Statistical Analysis*, New York: John Wiley & Sons, Inc., 1982.

21. N. Hedayat and A. Nosratinia, "Outage and diversity of linear receivers in flat-fading MIMO channels," *IEEE Trans. Siganl Process.*, vol. 55, no. 12, pp. 5868–5873, 2007.

22. J.H. Winters, "Optimum combining in digital mobile radio with cochannel interference," *IEEE Trans. Veh. Technol.*, vol. 33, no. 3, pp. 144–155, 1984.

23. E.J. Kelly, "An adaptive detection algorithm," *IEEE Trans. Aerosp. Electron. Syst.*, vol. 22, no. 2, pp. 5–127, 1986.

24. Y. Jiang, X. Zheng, and J. Li, "Asymptotic performance analysis of V-BLAST," in *ICC?*.

25. S.M. Alamouti, "A simple transmit diversity technique for wireless communications," *IEEE J. Sel. Areas Commun.*, vol. 16, no. 8, pp. 1451–1458, 1998.

26. M.K. Simon, "A moment-generating function (MGF)-based approach for performance evaluation of space-time code communication systems," *Wireless Commun. Mobile Comput.*, vol. 2, no. 7, pp. 667–692, 2002.

27. I.-M. Kim, "Exact BER analysis of OSTBCs in spatially correlated MIMO channels," *IEEE Trans. Commun.*, vol. 54, no. 8, pp. 1365–1373, 2006.

28. M.O. Damen, A. Tewfik, and J.-C. Belfiore, "A construction of a space-time code based on number theory," *IEEE Trans. Inf. Theory*, vol. 48, no. 3, pp. 753–760, 2002.

29. A.R. Hammons and H. El Gamal, "On the theory of space-time codes for PSK modulation," *IEEE Trans. Inf. Theory*, vol. 46, no. 2, pp. 524–542, 2000.

30. V. Tarokh, N. Seshadri, and A.R. Calderbank, "Space-time codes for high data rate wireless communication: performance criterion and code construction," *IEEE Trans. Inf. Theory*, vol. 44, no. 2, pp. 744–765, 1998.

31. L. Zheng and D.N.C. Tse, "Diversity and multiplexing: a fundamental tradeoff in multiple-antenna channels," *IEEE Trans. Inf. Theory*, vol. 49, no. 5, pp. 1073–1096, 2013.

32. V.M. Bogachev, "Optimal combining of signals in space diversity reception," *Telecommun. Radio Eng.*, vols. 34/35, p. 83, 1980.

33. B. Hassibi and B.M. Hochwald, "High-rate codes that are linear in space and time," *IEEE Trans. Inf. Theory*, vol. 48, no. 7, pp. 1804–1824, 2002.

34. V. Tarokh, H. Jafarkhani, and A.R. Calderbank, "Space-time block coding for wireless communications: performance results," *IEEE J. Sel. Areas Commun.*, vol. 17, no. 3, pp. 451–460, 1999.

35. G.J. Foschini and M.J. Gan, "On the limits of wireless communication in a fading environment when using multiple antennas," *Wireless Pers. Commun.*, vol. 6, no. 3, pp. 311–335, 1998.

36. G.J. Foschini, "Layered space-time architecture for wireless communication in a fading environment when using multi-element antennas," *Bell Labs Tech. J.*, vol. 1, no. 2, pp. 41–59, Autumn 1996.

37. J. Li and A. Stefanov, "Exact pairwise error probability for block-fading MIMO OFDM systems," *IEEE Trans. Veh. Technol.*, vol. 57, no. 4, pp. 2607–2611, 2008.

38. P. Stoica, Y. Jiang, and J. Li, "On MIMO channel capacity: an intuitive discussion," *IEEE Signal Process. Mag.*, vol. 22, no. 3, pp. 83–84, 2005.

39. Y. Jiang, M. Varanasi, and J. Li, "Performance analysis of ZF and MMSE equalizers for MIMO systems: an in-depth study of the high SNR regime," *IEEE Trans. Inf. Theory*, vol. 57, no. 4, pp. 2008–2026, 2011.

40. N. Lavanis and D. Jalihal, "Finite-SNR diversity multiplexing tradeoff of SIMO diversity combining schemes," in Proceedings of IEEE VTC'2009, Spring 2009, pp. 1–5.

14

Cooperative Communications

Abstract

Physical layer technologies such as MIMO and OFDM have been proved effective in providing a higher data rate for wireless systems. However, such improvements are far from sufficient to meet the increasing demand raised by the explosive growth of mobile devices. Given the limited spectrum, a promising solution is to significantly increase the spectral efficiency of wireless networks. The large amount of distributed wireless nodes, fortunately, also provides a new platform to achieve this goal through cooperation among wireless nodes. The simplest version of node cooperation to assist data transmission can be traced back to the conventional microwave relay system. Cooperative communications, however, is different from the conventional relay system in the sense that multiple nodes will coordinate to transmit the signal and avoid interference. As a result, besides the SNR gain, cooperative communications can also help achieve diversity gain, manage interference, and, thus, greatly improve the aggregate capacity of wireless networks. Various types of node cooperation, such as multi-hop and multi-relay communication, multi-cell MIMO, and two-way relaying, have found their applications in the infrastructured (cellular) and ad hoc (sensor) networks.

14.1   A historical review

Just as any physical phenomenon, information exchange is also governed by a fundamental law. Recall that the maximum mutual information (capacity) of a single point-to-point AWGN channel is governed by Shannon's capacity law [1]

$$C = B \log(1 + \text{SNR}), \tag{14.1}$$

where B and SNR represent the channel bandwidth and the receive SNR, respectively. We also recall that a powerful means to exploit spatial electromagnetic resource is MIMO technology, which increases the link capacity by creating additional channels in the space domain so that the capacity for an $M \times N$ MIMO channel can reach

$$C = \sum_{i=1}^{\min\{M,N\}} B_i \log(1 + \text{SNR}_i). \tag{14.2}$$

However, the continued demand for ubiquitous connectivity, as well as the expected explosive growth of mobile access to the Internet and other applications (e.g., 3-D Internet, augmented reality, social networking, Internet of Things), cannot be fully satisfied by the above physical layer solutions. Specifically, mobile data traffic has been particularly doubling every year due to the explosive growth of wireless devices and this trend is forecast to continue. As a result, the major concern is no longer the link capacity but instead the aggregate capacity of a whole wireless network. Fortunately, the deployment of such high-density devices/nodes also provides the opportunity to apply a promising method to increase network capacity, namely cooperative communications. In this chapter, we shall discuss the advantages of cooperative communications by starting from the conventional relay network. We will then investigate the cooperation of distributed nodes within more general frameworks, including the general multi-relay system, the two-way relaying network, and the multi-cell MIMO network.

14.2   Relaying

The utilization of other nodes for help to exchange information between the source and the destination can be traced back to the conventional microwave relay system. Similar ideas have also found their applications in modern wireless communications systems and been applied in the LTE-A system. In particular, relay nodes (RNs) are utilized in the LTE-A network to relay information between the user equipment (UE) and the evolutional NodeB (eNB). The motivation of using relays can also be found from the Shannon capacity law (14.1), whose purpose is to increase the receive SNR by reducing the pathloss effects.

Consider a simple example where a source (S) wants to transmit information to a destination (D) located $2d$ meters away from the source with a transmit power P_t, as shown in Figure 14.1. To illustrate the benefits

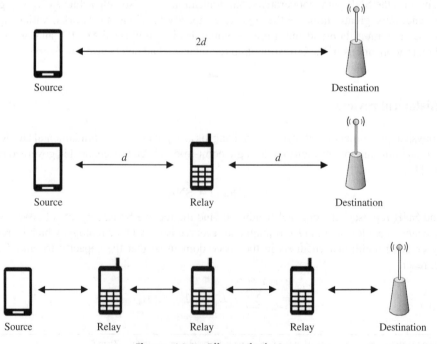

Figure 14.1   *Effects of relaying*

of relaying, we do not consider the small-scale fading here. With the logarithm pathloss model, the receive power at the destination is given by

$$P_d = P_t G_0 \left(\frac{2d}{d_0}\right)^{-\alpha},$$
(14.3)

where G_0 is the power gain at the reference distance (d_0), and α denotes the pathloss factor. Without loss of generality, we assume that the noise power and bandwidth are normalized to 1. Thus, the mutual information between the source and the destination can be given by

$$C = \log\left(1 + 2^{-\alpha} P_t G_0 \left(\frac{d}{d_0}\right)^{-\alpha}\right).$$
(14.4)

Now, let us investigate the effects of relaying. For this purpose, we put a relay (R) at the middle point between the source and the destination. For a fair comparison, the transmit power at the source and the relay is given by $P_t/2$. Under such circumstances, the receive power at the relay and the destination can be expressed as

$$P_r = P_d = \frac{P_t G_0}{2} \left(\frac{d}{d_0}\right)^{\alpha}.$$
(14.5)

Let C_1 and C_2 denote the mutual information of the S–R and R–D links. We can then obtain

$$C_r = \frac{1}{2}\min(C_1, C_2) = \frac{1}{2}\log\left(1 + \frac{1}{2}P_t G_0 \left(\frac{d}{d_0}\right)^{-\alpha}\right).$$
(14.6)

It can be observed by comparing (14.6) with (14.4) that relaying will provide an SNR gain when $\alpha > 1$, but cause a loss of multiplexing gain due to the half-duplex relaying. By comparing (14.6) with (14.4), we can conclude that relaying outperforms single-hop transmission when the SNR gain from less pathloss is greater than the loss of multiplexing gain. On the other hand, relaying is necessary for the case where the direct link is blocked.

A straightforward question to ask is whether deploying more relays can further help increase the capacity, as shown in Figure 14.1. In other words, how will the capacity scale when we have infinite number of relays while the distance between the source and the destination is fixed. Will the capacity keep increasing as the number of relays (n) increases to infinity? Unfortunately, the answer is "No" and the achievable rate will get saturated as n approaches infinity. A basic conclusion is that relaying is beneficial for low-density network or when the pathloss factor is high, say greater than 6. Obviously, relaying itself is not enough to increase the network capacity. Then, how can we fully exploit the space freedoms provided by the large amount of wireless nodes? The answer is "cooperative communications".

14.3   Cooperative communications

Different from the conventional relaying, where only one node is decoding and encoding the message, and signals from other nodes are treated as interference, cooperative communications requires the cooperation among nodes [2, 3]. Specifically, multiple nodes decode and encode messages, and may also exchange information with each other to coordinate transmission and manage interference. Basically, the distributed nodes can be utilized to form a virtual MIMO system, where the major difference from conventional MIMO systems lies in the distributed nature of the node-antennas that cause many issues for channel estimation, synchronization, and joint signal design. The analysis for cooperative networks began from the classical work of Cover and El

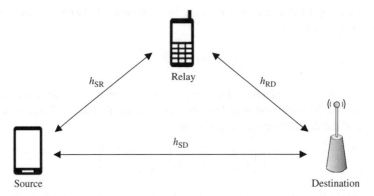

Figure 14.2   *System model for the three-node relay network*

Gamal [4] and has attracted much more attention since the proposal of cooperative diversity [5]. Extensive works have been done on various aspects of systems using relays, including optimization of relay amplification factors [6, 7], power allocation [8], duality analysis [9], transmission over multipath channels [10], and design of codes [11]. Unfortunately, the characterization of the capacity for relay networks is still an open problem even for the simplest three-node case [4].

14.3.1   Cooperation protocols

We now introduce some common cooperation protocols within the framework of the classic three-node relay networks. Consider the three-node relay network as shown in Figure 14.2, where the source S transmits information to the destination D with the help of one relay node R. We further assume that the S–D and R–D transmissions are orthogonal to each other by time division, implying a two-stage transmission scheme. Specifically, the received signals at R and D in the first stage can be expressed as

$$y_r = h_{sr}s + n_r,$$

$$y_{sd} = h_{sd}s + n_{sd}, \tag{14.7}$$

where $s \sim CN(0, P_s)$ is the transmit symbol, $n_r \sim CN(0, \sigma_n^2)$ and $n_{sd} \sim CN(0, \sigma_n^2)$ denote the additive white Gaussian noise (AWGN) at R and D, respectively. In the second stage, the relay R transmits signal to the destination with different protocols.

14.3.1.1   *Amplify-and-forward*

With the amplify-and-forward (AF) scheme, the relay will amplify its received signal in the first stage and then forward it to the destination in the second stage. Specifically, the transmit signal at R is given by

$$x_r = \beta y_r = \beta(h_{sr}s + n_r), \tag{14.8}$$

where β denotes the AF factor. Given the constraints on the transmit power at the source and the relay as P_s and P_r, respectively, the AF factor can be determined as

$$\beta \leq \sqrt{\frac{P_r}{|h_{sr}|^2 \sigma_s^2 + \sigma_n^2}}. \tag{14.9}$$

Note that here we calculate the transmit power for a given channel realization h_{sr}. Another type of power constraint is the average power constraint calculated by averaging over all possible realizations of the channel [12]. Accordingly, the received signal at the destination in the second stage is given by

$$y_{rd} = h_{rd}\beta(h_{sr}s + n_r) + n_d, \tag{14.10}$$

where h_{rd} and n_d denote, respectively, the channel coefficient for the R–D link and the AWGN at the destination.

By combining y_{sd} and y_{rd} together, we can obtain the received signal at the destination as

$$\begin{bmatrix} y_{sd} \\ y_{rd} \end{bmatrix} = \begin{bmatrix} h_{sd} \\ h_{rd}\beta h_{sr} \end{bmatrix} s + \begin{bmatrix} 0 & 1 & 0 \\ h_{rd}\beta & 0 & 1 \end{bmatrix} \begin{bmatrix} n_r \\ n_{sd} \\ n_{rd} \end{bmatrix}, \tag{14.11}$$

where we can further rewrite the received signal in the vector form as

$$\mathbf{y} = \mathbf{h}s + \mathbf{H}\mathbf{n}. \tag{14.12}$$

By treating the vector model as that for a conventional MIMO system, we can obtain the capacity of the AF scheme as

$$C_{AF} = \frac{1}{2}\log\det(I + P_s\mathbf{h}\mathbf{h}^\dagger(\mathbf{H}\mathbb{E}[\mathbf{z}\mathbf{z}^\dagger]\mathbf{H}^\dagger)^{-1}), \tag{14.13}$$

which, after some mathematical manipulations, further gives

$$C_{AF} = \frac{1}{2}\log\left(1 + \gamma_s|h_{sd}|^2 + \frac{\gamma_s\gamma_r|h_{sr}|^2|h_{rd}|^2}{1 + \gamma_s|h_{sr}|^2 + \gamma_r|h_{rd}|^2}\right) \tag{14.14}$$

with

$$\gamma_s = \frac{P_s}{\sigma_n^2}$$

$$\gamma_r = \frac{P_r}{\sigma_n^2}. \tag{14.15}$$

14.3.1.2   *Decode-and-forward*

With the decode-and-forward (DF) scheme, the relay will try to decode its received signal in the first stage and then forward it, with or without encoding, to the destination in the second stage. Accordingly, the received signal at the destination in the second stage is given by

$$y_{rd} = h_{rd}x_r + n_d. \tag{14.16}$$

Combining y_{sd} and y_{rd} together, we can obtain the received signal at the destination as

$$\begin{bmatrix} y_{sd} \\ y_{rd} \end{bmatrix} = \begin{bmatrix} h_{sd} & 0 \\ 0 & h_{rd} \end{bmatrix} \begin{bmatrix} s \\ x_r \end{bmatrix} + \begin{bmatrix} n_{sd} \\ n_{rd} \end{bmatrix}. \tag{14.17}$$

By following a similar procedure as for the AF case, we can obtain the capacity of the DF scheme as

$$C_{DF} = \frac{1}{2}\log(1 + \gamma_s|h_{sd}|^2 + \gamma_r|h_{rd}|^2), \tag{14.18}$$

which, however, only holds when the relay can correctly decode its received signal. Thus, the capacity of the DF scheme is given by

$$C_{\mathrm{DF}} = \frac{1}{2} \min \left\{ \frac{1}{2} \log \left(1 + \gamma_s |h_{sr}|^2\right), \frac{1}{2} \log(1 + \gamma_s |h_{sd}|^2 + \gamma_r |h_{rd}|^2) \right\}.$$ (14.19)

14.3.1.3 *Adaptive relaying*

It can be observed from the above analysis that, with the DF scheme, the cooperation from the relay is meaningful only when the relay can correctly decode the message from the source. By taking this fact into consideration, adaptive relaying was proposed, where the relay only works when the channel gain between the source and the relay is greater than a certain threshold, that is, $|h_{sr}|^2 \geq \eta$. If $|h_{sr}|^2 < \eta$, the source will retransmit the signal in the second stage. As a result, the capacity of the adaptive relaying is given by

$$C_{\mathrm{AR}} = \begin{cases} \frac{1}{2} \log \left(1 + 2\gamma_s |h_{sd}|^2\right) & |h_{sr}|^2 < \eta \\ \frac{1}{2} \log(1 + \gamma_s |h_{sd}|^2 + \gamma_r |h_{rd}|^2) & |h_{sr}|^2 \geq \eta \end{cases}.$$ (14.20)

14.3.1.4 *Incremental relaying*

The adaptive relaying scheme avoids the wastage of energy when the S–R link is bad. However, if the S–D link is good enough to transmit the source information, cooperation from the relay is also a waste of energy. Incremental relaying was proposed to solve this issue. Specifically, after the first stage transmission, the destination will broadcast an ACK/NACK to the source and the relay, depending on whether the source information is correctly received or not. If an ACK is received by the source, it will transmit new information in the second stage. On the other hand, if a NACK is received by the source and the relay, the relay will relay information to the destination using AF or DF scheme. It can be observed that incremental relaying avoids wastage of energy at the relay when the source–destination link is good enough to support the transmission.

14.3.2 Diversity analysis

Given a source and a destination, the multiplexing gain of the concerned link is fixed. As a result, although significant SNR gain can be achieved by cooperation, the major benefit of cooperation among nodes is to achieve the diversity gain, which governs the reliability of a communication link and comes from the independent paths created by the distributed nodes. The diversity gain is formally defined as the slope of the outage probability in the high SNR region with SNR (γ) in log scale. Here, we adopt the DMT measure, where the diversity gain for a target rate $R = r \log \gamma$ is defined as [13]

$$d(r) = - \lim_{\gamma \to \infty} \frac{\log P_e(\gamma)}{\log \gamma},$$ (14.21)

where $P_e(\gamma)$ is the error probability for a given transmit SNR γ and r represents the multiplexing gain. It has been proved in Ref. [13] that the error probability is lower-bounded by the outage probability with

$$\lim_{\gamma \to \infty} \frac{\log P_e(\gamma)}{\log \gamma} \geq \lim_{\gamma \to \infty} \frac{\log P_{\mathrm{out}}(\gamma)}{\log \gamma}.$$ (14.22)

This can be explained by the fact that error-free communications can always be achieved with a rate lower than the capacity of a channel (no outage). Thus, we will follow [13] by utilizing P_{out} to lower-bound P_e so that the optimal diversity gain can be given by

$$d(r) = -\lim_{\gamma \to \infty} \frac{\log P_{\text{out}}(\gamma)}{\log \gamma}. \tag{14.23}$$

14.3.2.1 Single-hop example

Consider the single-hop case where the channel coefficient between the source and destination is denoted as h_{sd}. The capacity of the link is given by

$$C_{\text{SH}} = \log(1 + \gamma |h_{sd}|^2). \tag{14.24}$$

As a result, the outage probability is determined as

$$\begin{aligned} P_{\text{out}}(\gamma, R) &= \Pr(C_{\text{SH}} < R) \\ &= \Pr(\log(1 + \gamma |h_{sd}|^2) < R) \\ &= \Pr\left(|h_{sd}|^2 < \frac{2^R - 1}{\gamma} \right). \end{aligned} \tag{14.25}$$

With a Rayleigh fading channel, $|h_{sd}|^2$ follows the exponential distribution with PDF

$$f_{|h_{sd}|^2}(x) = \frac{1}{\sigma^2} \exp\left(\frac{-x}{\sigma^2} \right). \tag{14.26}$$

It follows that

$$\begin{aligned} P_{\text{out}}(\gamma, R) &= 1 - \exp\left(-\frac{2^R - 1}{\gamma \sigma^2} \right) \\ &= \sum_{i=1}^{\infty} \frac{2^R - 1}{\gamma \sigma^2}, \end{aligned} \tag{14.27}$$

where the second line comes from the series expansion of the exponential function. It then follows from (14.23) that the diversity gain is determined by the first term in the summation, indicating a diversity gain of 1.

In the following, we show the diversity gain of different relaying protocols, where we assume that all links follow i.i.d. Rayleigh fading with different parameters.

14.3.2.2 AF relaying

By following a similar procedure as for the single-hop case, we can obtain the outage probability for AF relaying as

$$P_{\text{out}}(\gamma) = \Pr\left(|h_{sd}|^2 + \frac{\gamma |h_{sr}|^2 |h_{rd}|^2}{1 + \gamma |h_{sr}|^2 + \gamma |h_{rd}|^2} < \frac{2^{2R} - 1}{\gamma} \right), \tag{14.28}$$

which can be further approximated as

$$P_{\text{out}}(\gamma) = \left(\frac{1}{2\sigma_{sd}^2} \frac{\sigma_{sr}^2 + \sigma_{rd}^2}{\sigma_{sr}^2 \sigma_{rd}^2} \right) \left(\frac{2^{2R} - 1}{\gamma} \right)^2, \tag{14.29}$$

where σ_{sd}^2, σ_{sr}^2, and σ_{rd}^2 denote the Rayleigh distribution parameters for the S–D, S–R, and R–D links, respectively. It can be observed that AF relaying can achieve a diversity gain of 2, which makes sense because two independent paths are created by AF relaying.

14.3.2.3   DF relaying

The outage probability for DF relaying can be determined as

$$P_{\text{out}}(\gamma) = \Pr\left(\min\left\{|h_{sr}|^2, |h_{sd}|^2 + |h_{rd}|^2\right\} < \frac{2^{2R} - 1}{\gamma}\right). \tag{14.30}$$

It follows that the outage probability of DF relaying is similar to that of the single-hop case with

$$P_{\text{out}}(\gamma) = \frac{1}{\sigma_{sr}^2}\frac{2^{2R} - 1}{\gamma}. \tag{14.31}$$

Obviously, DF relaying can only achieve a diversity gain of "1", which is restricted by the S–R link.

14.3.2.4   Adaptive relaying

With the adaptive relaying scheme, both AF and DF are available in the second stage. With DF relaying, the outage event is equivalent to

$$(\{|h_{sr}|^2 < \eta_T\}\bigcap\{2|h_{sd}|^2 < \eta_T\})\bigcup(\{|h_{sr}|^2 \geq \eta_T\}\bigcap\{|h_{sd}|^2 + |h_{rd}|^2 < \eta_T\}),$$

where η_T denotes a preselected threshold. It follows that the outage probability can be approximately determined as

$$P_{\text{out}}(\gamma) = \left(\frac{1}{2\sigma_{sd}^2}\frac{\sigma_{sr}^2 + \sigma_{rd}^2}{\sigma_{sr}^2\sigma_{rd}^2}\right)\left(\frac{2^{2R} - 1}{\gamma}\right)^2. \tag{14.32}$$

Thus, the adaptive DF relaying can achieve a diversity gain of "2", which comes from the conditional use of the relay.

It is easy to check that the incremental relaying scheme can also achieve a diversity gain of 2. More importantly, such diversity gain is achieved with nearly twice the spectral efficiency.

14.3.3   Resource allocation

The above analysis compares different cooperation protocols in terms of the achievable diversity gain. For a given scheme, we can optimize the performance by resource allocation, which includes the allocation of power, bandwidth, time slots, and so on. We take the power allocation for the three-node relay system as an example to show the benefits of resource allocation.

14.3.3.1   Single AF-relay

Given the capacity for the AF-based relay network as

$$C_{\text{AF}} = \frac{1}{2}\log\left(1 + \gamma_s|h_{sd}|^2 + \frac{\gamma_s\gamma_r|h_{sr}|^2|h_{rd}|^2}{1 + \gamma_s|h_{sr}|^2 + \gamma_r|h_{rd}|^2}\right), \tag{14.33}$$

we want to maximize C_{AF} by allocating power between γ_s and γ_r with a joint power constraint

$$\gamma_s + \gamma_r = \gamma. \tag{14.34}$$

It can be shown that when $|h_{sr}|$ and $|h_{sd}|$ are sufficiently greater than $|h_{rd}|$, the optimal ratio between γ_s and γ_r can be approximated as

$$\frac{\gamma_s}{\gamma_r} \approx \frac{|h_{sr}|^2|h_{rd}|^2\gamma + |h_{rd}|^2|h_{sd}|^2\gamma + |h_{sd}|^2}{|h_{sr}|^2|h_{rd}|^2\gamma - |h_{sr}|^2|h_{sd}|^2\gamma - |h_{sd}|^2}. \tag{14.35}$$

14.3.3.2   Single DF-relay

Consider a similar problem for the DF-relay case. Given the capacity for the DF-based relay network as

$$C_{DF} = \frac{1}{2}\min\left\{\frac{1}{2}\log\left(1 + \gamma_s|h_{sr}|^2\right), \frac{1}{2}\log(1 + \gamma_s|h_{sd}|^2 + \gamma_r|h_{rd}|^2)\right\}, \tag{14.36}$$

we want to maximize C_{DF} by allocating power between γ_s and γ_r with a joint power constraint

$$\gamma_s + \gamma_r = \gamma. \tag{14.37}$$

It can be shown that the optimal γ_s and γ_r are given by

$$\gamma_s = \gamma\frac{|h_{rd}|^2}{|h_{sr}|^2 + |h_{rd}|^2 - |h_{sd}|^2} \tag{14.38}$$

and

$$\gamma_r = \gamma\frac{|h_{sr}|^2 - |h_{sd}|^2}{|h_{sr}|^2 + |h_{rd}|^2 - |h_{sd}|^2}. \tag{14.39}$$

In the above power allocation, the average power was calculated for given channel realizations. A similar power allocation problem can be done with the average power constraint, where the power is calculated by averaging over all channel realizations.

14.4   Multiple-relay cooperation

The above analysis can be generalized to the case with multiple (M) relays where all relay links can be either orthogonal or nonorthogonal. For the orthogonal case, different relays transmit in turn by time or frequency divisions. The destination will combine the signals coming from different relays where a diversity gain of M is achieved. Unfortunately, such orthogonal transmission will cause a loss of multiplexing gain. With nonorthogonal transmission, relays transmit at the same time over the same frequency band. By utilizing space–time code or coherent combining, a diversity gain of M is also achievable with a much higher spectral efficiency. However, this comes at the cost of synchronization and channel estimation issues. In fact, the same diversity gain can be achieved with a much simpler scheme, that is, relay selection. In particular, the best relay is selected out of the M candidates by following the criterion:

$$\max_m \min\{|h_{sm}|^2, |h_{mr}|^2\}, \tag{14.40}$$

where h_{sm} denotes the coefficient for the channel between the source and the mth relay, and $|h_{mr}|^2$ represents that between the mth relay and the destination.

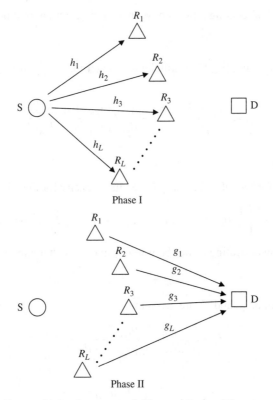

Figure 14.3 *System model for multi-relay AF systems*

14.4.1 Multi-relay over frequency-selective channels

Consider the design of the multiple-relay system over frequency-selective channels, as illustrated in Figure 14.3. The system consists of one source node S, one destination node D, and L relay nodes $R_i, i = 1, \ldots, L$. Let $h_i(t)$ and $g_i(t)$ denote, respectively, the channel impulse response of the frequency-selective S–R_i and R_i–D links with block-constant Rayleigh path coefficients, such that

$$h_i(t) = \sum_{j=1}^{J_i} h_{ij}\delta(t - \tau_{hi} - jT_c),$$

$$g_i(t) = \sum_{k=1}^{K_i} g_{ik}\delta(t - \tau_{gi} - kT_c), \tag{14.41}$$

where J_i and K_i denote the number of paths for $h_i(t)$ and $g_i(t)$, τ_{hi} and τ_{gi} signify the time delay for the first tap of $h_i(t)$ and that of $g_i(t)$, respectively, and T_c stands for the length of a resolution bin. We further assume that there is no direct link between S and D.

Refer to the upper part of Figure 14.3. In Phase I, S broadcasts its information $s(t)$ to $\{R_i\}$ producing the received baseband signal at R_i, given by

$$r_i(t) = \sum_{j=1}^{J_i} h_{ij} s(t - \tau_{hi} - jT_c) + n_i(t), \tag{14.42}$$

where $n_i(t)$ denotes the AWGN at R_i. In AF systems, relays perform a linear operation, say $F_i(\cdot)$, on its received signal [5], thus allowing us to represent the transmit signal of R_i in Phase II as

$$x_i(t) = F_i(r_i(t)) = F_i \left(\sum_{j=1}^{J_i} h_{ij} s(t - \tau_{hi} - jT_c) + n_i(t) \right).$$

The relay signals go through the channels $g_i(t)$ to produce the received signal at D:

$$y(t) = \sum_{i}^{L} \sum_{k=1}^{K_i} g_{ik} x_i(t - \tau_{gi} - kT_c) + n(t) \tag{14.43}$$

with $n(t)$ denoting AWGN at D.

The main problem with frequency-selective channels is inter-symbol interference (ISI), which can be eliminated by using the idea of OFDM, as described in Chapter 10. To be specific, given a length-M Gaussian transmit signal vector $\mathbf{s} \sim \mathcal{CN}(0, \mathbf{Q}_s)$, we can insert a prefix approximately of length l_s to $\mathbf{s}$ to form an equivalent channel matrix of circulant Toeplitz structure. The continuous-time received signal (14.42) at R_i can be expressed in vector form [14–16], as shown by

$$\mathbf{r}_i = \mathbf{H}_i \mathbf{s} + \mathbf{n}_i, \tag{14.44}$$

where the $M \times M$ circulant channel matrix $\mathbf{H}_i$ is constructed from $\{h_{ij}\}$. Correspondingly, the transmit signal of R_i in phase II is expressible as

$$\mathbf{x}_i = \mathbf{A}_i(\mathbf{H}_i \mathbf{s} + \mathbf{n}) \tag{14.45}$$

with $\mathbf{A}_i$ representing the linear operation performed at R_i. By the same argument, another prefix of length l_r is added at relay R_i to combat ISI introduced in the R_i–D link. For a special case without relative transmission delays among different S–R_i–D links, the received signal at the destination, in vector form, can be written as

$$\mathbf{y} = \sum_{i}^{L} \mathbf{G}_i \mathbf{A}_i(\mathbf{H}_i \mathbf{s} + \mathbf{n}_i) + \mathbf{n}, \tag{14.46}$$

where $\mathbf{G}_i$ denotes the $M \times M$ circulant channel matrix constructed from $\{g_{ij}\}$.

Let us show how to extend (14.46) to the general case, in which different S–R_i–D links are subject to different transmission delays. Given the broadcasting nature of the phase I transmission, the delays $\{\tau_{hi}\}$ have no influence on the received signals of relays. Thus, for notational simplicity, we may combine the transmission delays of the S–R_i and R_i–D links as a single one, denoted by $\tau_i = \tau_{hi} + \tau_{gi}$, to account for the total transmission delay of the S–R_i–D link. Further define $\tau_{\min}$ and $\tau_{\max}$ such that

$$\tau_{\min} = \min\{\tau_i\}; \quad \tau_{\max} = \max\{\tau_i\}, \tag{14.47}$$

with which the prefix length at the relays can be determined as

$$l_r \geq \max\{K_i\} + (\tau_{\max} - \tau_{\min})/T_c, \tag{14.48}$$

where we have assumed that the transmission delays are integer times of the resolution bin. Accordingly, we can extend (14.46) to a general form

$$\mathbf{y} = \sum_i^L \mathbf{G}_i(\tau_i)\mathbf{A}_i\mathbf{H}_i\mathbf{s} + \sum_i^L \mathbf{G}_i(\tau_i)\mathbf{A}_i\mathbf{n}_i + \mathbf{n}. \tag{14.49}$$

Here, $\{\mathbf{G}(\tau_i)\}$ result from left-shifting each row of $\mathbf{G}_i$ by τ_i. Clearly, the asynchronous transmission influences both the design of the guard interval and the effective channel response.

Having represented the general signal vector model in (14.49), we are in a position to derive the system mutual information. Denote the channel information of all the links, $\{\mathbf{H}_i, \mathbf{G}_i(\tau_i)\}$, compactly as $\mathbf{HG}$. Then, conditioned on $\mathbf{HG}$, the mutual information between S and D is defined by

$$I(\mathbf{y}, \mathbf{s} \mid \mathbf{HG}) = \mathcal{H}(\mathbf{y} \mid \mathbf{HG}) - \mathcal{H}(\mathbf{y} \mid \mathbf{HG}, \mathbf{s}), \tag{14.50}$$

where $\mathcal{H}(\mathbf{y} \mid \mathbf{HG})$ denotes the entropy of $\mathbf{y}$ conditioned on $\mathbf{HG}$, and the second term on the right is similarly defined. To determine the conditional entropies, we need to find the distribution functions of $\mathbf{y}$ given $\{\mathbf{s}, \mathbf{HG}\}$ and $\mathbf{HG}$, respectively. By defining

$$\mathbf{GAAG} = \sum_{i=1}^L \mathbf{G}_i(\tau_i)\mathbf{A}_i\mathbf{A}_i^\dagger\mathbf{G}_i^\dagger(\tau_i),$$

$$\mathbf{GHQG} = \sum_{i,k=1}^L \mathbf{G}_i(\tau_i)\mathbf{A}_i\mathbf{H}_i\mathbf{Q}_s\mathbf{H}_k^\dagger\mathbf{A}_k^\dagger\mathbf{G}_k^\dagger(\tau_k), \tag{14.51}$$

it is straightforward to determine the two distributions as

$$\mathbf{y}|_{\{\mathbf{HG}, \mathbf{s}\}} \sim \mathcal{CN}\left(\sum_{i=1}^L \mathbf{G}_i(\tau_i)\mathbf{A}_i\mathbf{H}_i\mathbf{s}, \sigma_n^2\left(\mathbf{I} + \mathbf{GAAG}\right)\right),$$

$$\mathbf{y}|_{\mathbf{HG}} \sim \mathcal{CN}(\mathbf{0}, \mathbf{GHQG} + \sigma_n^2(\mathbf{I} + \mathbf{GAAG})), \tag{14.52}$$

from which we can write

$$\mathcal{H}(\mathbf{y} \mid \mathbf{HG}, \mathbf{s}) = \log\det(\pi e \sigma_n^2(\mathbf{I} + \mathbf{GAAG})),$$

$$\mathcal{H}(\mathbf{y} \mid \mathbf{HG}) = \log\det(\pi e(\mathbf{GHQG} + \sigma_n^2(\mathbf{I} + \mathbf{GAAG}))). \tag{14.53}$$

Inserting (14.53) into (14.50) leads to the desired expression for the system mutual information:

$$I(\mathbf{y}, \mathbf{s} \mid \mathbf{HG}) = \log\left(\frac{\det\left(\mathbf{GHQG} + \sigma_n^2\left(\mathbf{I} + \mathbf{GAAG}\right)\right)}{\det(\sigma_n^2(\mathbf{I} + \mathbf{GAAG}))}\right). \tag{14.54}$$

Our purpose is, under the joint power constraint, to design the optimal $\mathbf{Q}_s$ and $\{\mathbf{A}_i\}$ for which the mutual information in (14.54) is maximized. To this end, we assume that all channel information is available at a central point so that global optimization is possible. Given the total power P, we can express the power constraint as

$$P_s + P_r \leq P, \tag{14.55}$$

where P_s and P_r denote the power consumed at the source and that consumed at the relays, respectively. By definition and from (14.45), we can explicitly write

$$P_s = \mathrm{Tr}(Q_s), \quad P_r = \sum_{i=1}^{L} \mathrm{Tr}(\mathbb{E}[\mathbf{x}_i \mathbf{x}_i^\dagger]). \tag{14.56}$$

14.4.2   Optimal matrix structure

To proceed with the global optimization, we need to identify the optimal structure of $\mathbf{Q}_s$ and $\{\mathbf{A}_i\}$. Note that the circulant matrices $\mathbf{H}_i$ and $\mathbf{G}_i(\tau_i)$ have a special expansion eigen decomposition, as shown by Gray [17]:

$$\mathbf{H}_i = \mathbf{F}^\dagger \mathrm{diag}[\phi_{i1}, \ldots, \phi_{iM}]\mathbf{F},$$
$$\mathbf{G}_i(\tau_i) = \mathbf{F}^\dagger \mathrm{diag}[\varphi_{i1}, \ldots, \varphi_{iM}]\mathbf{F}, \tag{14.57}$$

where $\mathbf{F}$ is the unitary DFT (discrete Fourier transform) matrix and $\{\phi_{ij}\}$, $\{\varphi_{ij}\}$ denote the eigenvalues of $\mathbf{H}_i$ and $\mathbf{G}_i(\tau_i)$, respectively. According to Hadamard's inequality [18], we can assert that the transmit covariance matrix $\mathbf{Q}_s$ should have the same eigen decomposition as $\mathbf{H}_i$, taking the form

$$\mathbf{Q}_s = \mathbf{F}^\dagger \mathrm{diag}[p_{s1}, \ldots, p_{sM}]\mathbf{F}. \tag{14.58}$$

Following the same argument, the linear operator $\mathbf{A}_i$ should be chosen such that the transmit covariance matrix at relay R_i has the same unitary matrix as $\mathbf{G}_i$. Namely, there exists a diagonal matrix $\mathbf{D}_i$ such that

$$\mathbb{E}[\mathbf{x}_i \mathbf{x}_i^\dagger] = \mathbf{F}^\dagger \mathbf{D}_i \mathbf{F}. \tag{14.59}$$

Its left-hand side can be directly determined from (14.45), leading to

$$\mathbf{A}_i \mathbf{H}_i \mathbf{Q}_s \mathbf{H}_i^\dagger \mathbf{A}_i^\dagger + \sigma_n^2 \mathbf{A}_i \mathbf{A}_i^\dagger = \mathbf{F}^\dagger \mathbf{D}_i \mathbf{F}, \tag{14.60}$$

which, when solved for $\mathbf{D}_i$ and with the structures of $\mathbf{H}_i$ and $\mathbf{Q}_i$ inserted, gives

$$\mathbf{B}_i \mathrm{diag}[|\phi_{i1}|^2 p_{s1}, \ldots, |\phi_{iM}|^2 p_{sM}]\mathbf{B}_i^\dagger + \sigma_n^2 \mathbf{B}_i \mathbf{B}_i^\dagger = \mathbf{D}_i, \tag{14.61}$$

where $\mathbf{B}_i$ is defined by

$$\mathbf{B}_i = \mathbf{F}\mathbf{A}_i\mathbf{F}^\dagger. \tag{14.62}$$

Thus, we may work on $\mathbf{B}_i$ instead of $\mathbf{A}_i$. It can be shown that $\mathbf{B}_i$ has the following structure:

$$\mathbf{B}_i = \mathbf{D}_i \mathbf{R}_i, \tag{14.63}$$

where $\mathbf{D}_i$ is a diagonal matrix, and $\mathbf{R}_i$ is a permuting matrix having only one nonzero unity entry on each row and each column. The proof is left as an exercise. With this structure in mind, we can rewrite (14.62) as

$$\mathbf{A}_i = \mathbf{F}^\dagger \mathrm{diag}[a_{i1}, \ldots, a_{iM}]\mathbf{R}_i\mathbf{F}. \tag{14.64}$$

The effective channel matrix that links the signal source to the destination, given in (14.49), can now be written as

$$\mathbf{H}_{\text{eff}} = \sum_{i}^{L} \mathbf{G}_i(\tau_i) \mathbf{A}_i \mathbf{H}_i$$

$$= \mathbf{F}^\dagger \sum_{i}^{L}
\begin{bmatrix}
\varphi_{i1} a_{i1} & 0 & \cdots & 0 \\
0 & \varphi_{i2} a_{i2} & \cdots & 0 \\
\vdots & \vdots & \cdots & \vdots \\
0 & 0 & \cdots & \varphi_{iM} a_{iM}
\end{bmatrix}$$

$$\times \mathbf{R}_i
\begin{bmatrix}
\phi_{i1} & 0 & \cdots & 0 \\
0 & \phi_{i2} & \cdots & 0 \\
\vdots & \vdots & \cdots & \vdots \\
0 & 0 & \cdots & \phi_{iM} .
\end{bmatrix}
\mathbf{F} \tag{14.65}$$

Since the transmit covariance matrix should have the same eigen decomposition as the channel matrix, we can assert that $\mathbf{H}_{\text{eff}} \mathbf{H}_{\text{eff}}^\dagger$ should also be decomposed by $\mathbf{F}$; namely

$$\mathbf{H}_{\text{eff}} \mathbf{H}_{\text{eff}}^\dagger = \mathbf{F}^\dagger \mathbf{D}_i \mathbf{F} \tag{14.66}$$

for any arbitrary value of $\{\varphi_{im}, a_{im}, \phi_{im}\}$. Inserting (14.65) into (14.66), it is easy to obtain $\mathbf{R}_i = \mathbf{R}_j = \mathbf{R}$, indicating that all relays should utilize the same switching matrix.

It is observed from (14.65) that, when operating in a multipath environment, each link of a multi-relay system provides a number of virtual subchannels for signal transmission. The operator $\mathbf{A}_i$ has two functions to perform: It coherently amplifies signals on different subchannels with appropriate factors a_{im}, and serves as a switching center in much the same way as that in a telephone network. For example, it may switch the mth subchannel ϕ_{im} of $\mathbf{H}_i$ to an appropriate subchannel of $\mathbf{G}_i$, say φ_{in}, for a better route combination so as to maximize the system mutual information.

14.4.3   Power allocation

Having identified the structure of $\mathbf{Q}_s$ and $\{\mathbf{A}_i\}$, we now turn to the determination of parameters $\{p_{sm}\}$, $\{a_{im}\}$, and $\mathbf{R}$. Matrix $\mathbf{R}$ defines a switching mode between $\mathbf{H}_i$ and $\mathbf{G}_i$. Its design is, in essence, a problem of structure optimization and, thus, differs from parameter optimization in nature. The presence of $\mathbf{R}$ just changes the order of the parameters to be optimized and has no influence on the methodology of parameter optimization. We, therefore, postpone its discussion, and first focus on techniques of optimizing $\{p_{sm}\}, \{a_{im}\}$ for a given $\mathbf{R}$.

Without loss of generality, we assume $\mathbf{R} = \mathbf{I}$ for ease of illustration. Inserting (14.57), (14.58), and (14.64) into (14.54), we obtain

$$I(\mathbf{y}, \mathbf{s}|\mathbf{HG}) = \sum_{m=1}^{M} \log \left(1 + \frac{p_{sm} |\sum_{i=1}^{L} \varphi_{im} \phi_{im} a_{im}|^2}{\sigma_n^2 (1 + \sum_{i=1}^{L} |\varphi_{im}|^2 |a_{im}|^2)} \right), \tag{14.67}$$

whereby the optimization problem becomes

$$\max_{\{p_{sm}\}\{a_{im}\}} \sum_{m=1}^{M} \log \left(1 + \frac{p_{sm} |\sum_{i=1}^{L} \varphi_{im} \phi_{im} a_{im}|^2}{\sigma_n^2 \left(1 + \sum_{i=1}^{L} |\varphi_{im}|^2 |a_{im}|^2 \right)} \right).$$

$$\text{s.t. } P_s + P_r = P, \tag{14.68}$$

For an explicit expression for the power, we insert (14.57) and (14.64) into (14.56), yielding

$$P_s = \text{Tr}(Q_s) = \sum_{m=1}^{M} p_{sm},$$

$$P_r = \sum_{i=1}^{L} \text{Tr}(\mathbb{E}[\mathbf{x}_i \mathbf{x}_i^{\dagger}]) = \sum_{m=1}^{M} p_{rm}, \tag{14.69}$$

where

$$p_{rm} = \sum_{i=1}^{L} |a_{im}|^2 (p_{sm} |\phi_{im}|^2 + \sigma_n^2) \tag{14.70}$$

denotes the power consumed on subchannel m by all relays. Further inserting (14.69) and (14.70) into (14.68), we can rewrite the constraint in (14.68) as

$$P = \sum_{m=1}^{M} P_m$$

$$= \sum_{m=1}^{M} (p_{sm} + p_{rm})$$

$$= \sum_{m=1}^{M} \left(p_{sm} + \sum_{i=1}^{L} |a_{im}|^2 (p_{sm} |\phi_{im}|^2 + \sigma_n^2) \right). \tag{14.71}$$

Clearly, we need to optimize $\{a_{im}\}$ for each linear operator $\mathbf{A}_i$ and find optimal PA of P among $\{p_{sm}\}$ and $\{p_{rm}\}$.

It is observed from (14.68) and (14.71) that, given m, the optimal a_{im} can be uniquely determined from p_{sm} and p_{rm}. It is also observed that, once power P_m is specified to subchannel m, the PA between p_{sm} and p_{rm} can be obtained without the need for any information from other subchannels. These observations suggest that a layered structure can be used for searching the optimal solution. The idea can be implemented through a three-layer procedure. In particular, in Layer 3, we determine the optimal linear operation $\{a_{im}\}$ for given p_{sm} and p_{rm} so that the original problem in (14.68) is simplified to be only a function of p_{sm} and p_{rm}. Then, in Layer 2 we optimally allocate power P_m between the source and relay of subchannel m so that the objective is only a function of P_m, whose optimization is finally obtained in Layer 1. We address these layers in the subsequent sections, in a reverse order, since the results of lower layer (Layer 3) is utilized by the upper layer (Layer 2).

14.4.3.1   Layer 3: Power allocation for different relays

Suppose we are given p_{rm} and p_{sm} for subchannel m. Finding the optimal $\{a_{im}\}$ that maximize

$$I_m = \log\left(1 + \frac{1}{\sigma_n^2}\frac{p_{sm}\left|\sum_{i=1}^{L}\varphi_{im}\phi_{im}a_{im}\right|^2}{1 + \sum_{i=1}^{L}|\varphi_{im}|^2|a_{im}|^2}\right) \tag{14.72}$$

is equivalent to determining $\{a_{im}\}$ that maximize

$$y = \frac{p_{sm}\left(\left|\sum_{i=1}^{L}\varphi_{im}\phi_{im}a_{im}\right|\right)^2}{1 + \sum_{i=1}^{L}|\varphi_{im}|^2|a_{im}|^2}. \tag{14.73}$$

It can be shown that the optimal a_{im} is of the form

$$a_{im} = \frac{c\varphi_{im}^*\phi_{im}^*}{p_{sm}|\phi_{im}|^2 + p_{rm}|\varphi_{im}|^2 + \sigma_n^2}, \tag{14.74}$$

where c is a constant that needs to be adjusted to meet the power constraint. The proof is left as an exercise. Thus, given p_{sm} and p_{rm}, we are able to determine the optimal linear operation $\{a_{im}\}$ and the corresponding mutual information. An interesting observation from (14.72) is that the resulting SNR can be regarded as the summation from L relays, resembling the conventional maximum ratio combining (MRC). Accordingly, for given a power, one can use more relays to increase the system's mutual information without any power penalty. This is attributed to the broadcasting transmission in Phase I and the coherent combining on each relay.

14.4.3.2   Layer 2: Power allocation between source and relays

As a result of Layer 3 optimization, the original problem in (14.68) is simplified to (14.72). Let us move back to Layer 2 to tackle the PA between the source and relays in (14.72). For a given P_m, we utilize $p_{sm} = \alpha_m(P_m)P_m$ to denote the power allocated to the source, where $0 < \alpha(P_m) < 1$. Accordingly, the power allocated to the relays is given by $p_{rm} = (1 - \alpha_m(P_m))P_m$. We then insert the above relations into (14.72) to obtain

$$I_m^{\max} = \log\left(1 + \frac{z_m}{\sigma_n^2}\right), \tag{14.75}$$

where we have defined

$$z_m = \sum_{i=1}^{L}\frac{\alpha_m(P_m)(1 - \alpha_m(P_m))P_m|\varphi_{im}\phi_{im}|^2}{\alpha_m(P_m)|\phi_{im}|^2 + (1 - \alpha_m(P_m))|\varphi_{im}|^2 + \sigma_n^2/P_m}. \tag{14.76}$$

It is easy to examine that

$$\frac{\partial^2 z_m}{\partial\alpha_m(P_m)^2} < 0. \tag{14.77}$$

Thus, the optimal PA between the source and relay can be determined as the solution of

$$\frac{\partial z}{\partial\alpha_m(P_m)} = 0. \tag{14.78}$$

For example, the optimal solution for the case with $L = 1$ is expressible as

$$\alpha_m(P_m) = \frac{\sqrt{P_m|\varphi_{im}|^2 + \sigma_n^2}}{\sqrt{P_m|\varphi_{im}|^2 + \sigma_n^2} + \sqrt{P_m|\phi_{im}|^2 + \sigma_n^2}}, \tag{14.79}$$

where the PA is almost linearly correlated with the ratio of the channel gain for the S–R and R–D links and nearly independent of P_m especially for high SNR environments. The solution for a large number of L can be evaluated efficiently by the Newton method.

To simplify the optimization for a large number of L, we propose the following two properties of $\alpha_m(P_m)$:

Property 14.4.1 *As $L \to \infty$, the optimal power allocation between the source and relays is $\alpha^o = 0.5$.*

The proof is left as an exercise.

This property indicates that, for a large number of relays, equal power allocation between the source and relays is nearly optimal.

Property 14.4.2 *Under high SNR environment, the optimal power allocation between source and the relays at one subchannel is independent of the power allocated to that subchannel.*

The proof can be easily obtained by rewriting (14.76) with high SNR approximation as

$$z_m = P_m \sum_{i=1}^{L} \frac{\alpha_m(P_m)(1 - \alpha_m(P_m))|\varphi_{im}\phi_{im}|^2}{\alpha_m(P_m)|\phi_{im}|^2 + (1 - \alpha_m(P_m))|\varphi_{im}|^2}, \tag{14.80}$$

indicating the independence of optimal $\alpha_m(P_m)$ with P_m.

14.4.3.3 Layer 1: Power allocation for different subchannels

We move back to the first layer of power optimization over different subchannels. Inserting (14.75) into (14.68), we can rewrite the objective function as

$$I = \sum_{m=1}^{M} \log(1 + z_m). \tag{14.81}$$

We need to choose $\{P_m\}$ such that I is maximized under the constraint $P = \sum_{m=1}^{M} P_m$. As a result of AF operations, the objective function (14.81) is no longer convex about P_m; this is in contrast to the classical PA for parallel channels. At this point, it is insightful to elaborate the difference between MR–AF systems and conventional parallel channels in PA.

Recall that the mutual information of M parallel channels with channel gain $\{h_m\}$ usually takes the form

$$I = \sum_{m=1}^{M} \log\left(1 + \frac{P_m}{\sigma_n^2}|h_m|^2\right). \tag{14.82}$$

Its derivative with respect to P_m, denoted here by $L_m(P_m)$, is given by

$$L_m(P_m) = \frac{\partial I}{\partial P_m}. \tag{14.83}$$

It represents the ability of the subchannel m in converting signal energy into mutual information. It is easy to examine the convexity of I in (14.82) about P_m and to verify that $L_m(P_m)$ is a monotonically decreasing function of P_m. Thus, the optimal PA always occurs at the point [18]

$$P_m = (L_m^{-1}(u))^+, \qquad (14.84)$$

where a^+ denotes $\max\{0, a\}$, $L_m^{-1}(\cdot)$ represents the inverse function of $L_m(\cdot)$, and u is chosen to meet the power constraints. This algorithm is also iteratively utilized in power allocation for multiple-access vector channels [14–16].

Consider a typical example where the total power $P = P_{AB}$ is partitioned into P_A and P_B for allocation to subchannels 1 and 2, respectively. In the conventional water-filling for parallel channels, the optimum solution is defined by the stationary points P_A and P_B, for which

$$L_1(P_A) = L_2(P_B), \quad P = P_A + P_B. \qquad (14.85)$$

If we plot $L_1(P)$ and $L_2(P)$ versus P to find the optimal partitioning for a given total power P_{AB}, we can draw a horizontal line and adjust its altitude so that its crossing points with the two curves P_A and P_B satisfy the constraint $P_A + P_B = P$. This corresponds to the optimum PA. Given the monotonically decreasing property of the L_m versus P curves, there always exists a unique solution, and the variation in total power only changes its sharing among the two subchannels. In other words, the power sharing among different subchannels in the traditional water-filling demonstrates the property of continuity.

The situation in MR–AF systems is different. The presence of P_m in the denominator of z_m in (14.81), which is caused by the amplification and forwarding of noise component, makes the objective function (14.81) no longer a convex function of P_m and its derivative L_m no longer a monotone function. A typical situation with two subchannels is illustrated in Figure 14.4. In this situation, if we draw a horizontal line, there can be

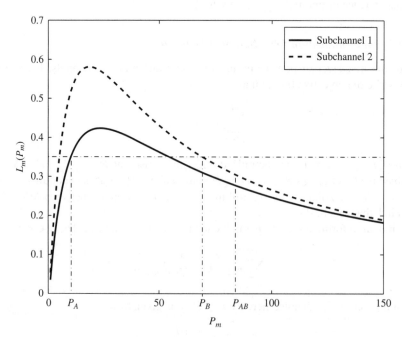

Figure 14.4 *Power allocation for multi-relay AF systems and conventional water-filling: a heuristic comparison*

four crossing points. By extreme value theorem [19], the optimum can be the stationary points or occurs at the boundaries leading to three possible operational schemes: that is, allocating all the power P_{AB} to sub-channel 1; allocating all the power P_{AB} to subchannel 2; or allocating P_A to subchannel 1 and P_B to subchannel 2.

For the setting shown in Figure 14.4 and total power indicated therein, the second scheme is better than the first one since $L_1(P_m) < L_2(P_m)$. More intuitively, the energy to mutual information conversion efficiency of subchannel 2 is higher than that of subchannel 1. For the same setting and power P_B allocated to subchannel 2, scheme 2 is also superior to scheme 3 since the former can convert the remaining power P_A to a higher mutual information than the latter. Mathematically, we have

$$\int_0^{P_A} L_1(P_m)dP_m < \int_{P_B}^{P_{AB}} L_2(P_m)dP_m. \tag{14.86}$$

However, as P_{AB} in Figure 14.4 is increased beyond a certain point, scheme 3 will outperform the other two, implying that the best strategy is appropriately sharing the total power between the two subchannels. In other words, the PA strategy for MR–AF systems can switch abruptly from one operational mode to another, demonstrating a discontinuity property that totally differs from the conventional water-filling.

We conclude this section by summarizing the proposed layered optimization procedure as follows:

(a) Use (14.78) to determine $\alpha_m(P_m)$.
(b) Insert the resulting $\alpha_m(P_m)$ into (14.81) and calculate the derivative $L_m(P_m)$ by virtue of (14.83).
(c) Check all stationary and boundary points to determine the optimal P_m for each subchannel by using the technique described in Section 14.4.3.
(d) Allocate the power P_m obtained above to the source and relays on subchannel m by using $p_{sm} = \alpha(P_m)P_m$ and $p_{rm} = (1 - \alpha(P_m))P_m$.
(e) Finally, allocate power among different relays of subchannel m by using (14.74) to calculate a_{im}.

14.5 Two-way relaying

As a promising candidate to increase the spectral efficiency of the one-way relay (OWR) system [8, 10], two-way relaying (TWR) has attracted much research interest. The idea of TWR is conceptually sketched in Figure 14.5. Several protocols, originally proposed for OWR, have been applied to TWR, including the

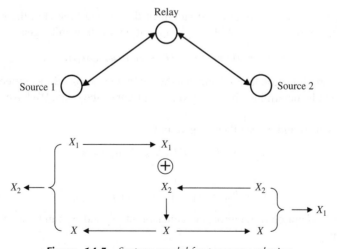

Figure 14.5 *System model for two-way relaying*

decode-and-forward (DF), compress-and-forward (CF) and amplify-and-forward (AF) schemes. Among these protocols, AF-based TWR is quite promising due to its ease of implementation. Many aspects of AF-based TWR have been investigated. Specifically, Ref. [20] investigated the capacity limit of AF-based TWR with linear processing at multi-antenna relays. Han et al. [21] analyzed the average sum rate of AF-based TWR with orthogonal space time block coding scheme. Li et al. [22] considered the outage probability of adaptive TWR scheme, and [23–25] investigated the power allocation of AF-based TWR in order to increase the spectral efficiency.

By reducing the number of time slots needed to communicate between two nodes from 4 to 2, TWR has the potential to increase the throughput of a three-node relay system. However, it is not clear whether TWR is always better than OWR, due to the interference between the two methods of relaying. Related comparisons between TWR and OWR are available in the literature. In particular, Ref. [26] compared the average sum rate of TWR and OWR with the instantaneous power constraint (IPC), where the instantaneous transmit power at each node was constrained. It was shown that, with IPC and the optimal power allocation, TWR always outperforms OWR.

Besides IPC, we also have an average power constraint (APC), where the average transmit powers of three nodes are constrained. With both instantaneous and average power constraints, there are two types of power allocation schemes, namely, joint power allocation and separate power allocation. With joint power allocation, the total transmit power P is allocated to three nodes (two sources and one relay) jointly with $p_{s1} + p_{s2} + p_r \leq P$, where p_{s1} and p_{s2} represent the transmit powers of two source nodes, and p_r denotes the transmit power of the relay. With separate power allocation, the power allocation for the source and the relay is performed separately with $p_{s1} + p_{s2} \leq P_s$ and $p_r \leq P_r$.

Consider a two-way relaying system where two single-antenna sources S_1 and S_2 communicate with the help of a single-antenna AF relay R. The information symbols transmitted from S_1 and S_2 are denoted as s_1 and s_2, respectively. In the first time slot, the received signal at R is given by

$$r_R = h_1 s_1 + h_2 s_2 + n_R, \tag{14.87}$$

where $n_R \sim \mathcal{CN}(0, \sigma_n^2)$ represents the AWGN. Here, h_1 and h_2 denote the channel coefficients for the $S_1 - R$ and $S_2 - R$ channels with

$$h_1 \sim \mathcal{CN}(0, {\sigma_h}^2),$$
$$h_2 \sim \mathcal{CN}(0, {\sigma_h}^2), \tag{14.88}$$

where ${\sigma_h}^2$ denotes the large-scale fading information. In the second time slot, the relay will amplify and forward its received signal to S_1 and S_2. With the AF operation, the transmit signal of R is given by

$$t_R = \beta \cdot r_R = \beta \cdot (h_1 s_1 + h_2 s_2 + n_R), \tag{14.89}$$

where β represents the amplification factor. Note that the coefficient β has a direct influence on the throughput of the system and should be determined by the power constraint of the relay. We have more to say about this in later sections.

The received signals at S_1 and S_2 can then be given by

$$r_1 = h_1 \beta \cdot (h_1 s_1 + h_2 s_2 + n_R) + n_R \tag{14.90}$$

and

$$r_2 = h_2 \beta \cdot (h_1 s_1 + h_2 s_2 + n_R) + n_R, \tag{14.91}$$

respectively. With the assumption of reciprocal channels, both S_1 and S_2 can remove their own signals and obtain the desired signals as

$$r_1 = h_1 \beta \cdot (h_2 s_2 + n_R) + n_R \tag{14.92}$$

and

$$r_2 - h_2\beta \cdot (h_1 s_1 + n_R) + n_R, \tag{14.93}$$

respectively. Thus, the received SNRs at S_1 and S_2 are given by

$$\text{SNR}_1 = \frac{\beta^2 |h_1|^2 |h_2|^2 p_{s2}}{\beta^2 |h_1|^2 \sigma_n^2 + \sigma_n^2} \tag{14.94}$$

and

$$\text{SNR}_2 = \frac{\beta^2 |h_1|^2 |h_2|^2 p_{s1}}{\beta^2 |h_2|^2 \sigma_n^2 + \sigma_n^2}, \tag{14.95}$$

where p_{s1} and p_{s2} denote the transmit power of S_1 and S_2, respectively. In the above derivation, we have assumed that the noise variances at different nodes are identical, and, for ease of illustration, we shall assume $\sigma_n^2 = 1$ in the following derivation. The mutual information of the two links can then be given by

$$R_1 = \log_2(1 + \text{SNR}_1) \tag{14.96}$$

and

$$R_2 = \log_2(1 + \text{SNR}_2), \tag{14.97}$$

respectively. Thus, the sum rate of this system is given by

$$R = R_1 + R_2 = \log_2((1 + \text{SNR}_1)(1 + \text{SNR}_2)). \tag{14.98}$$

In the following sections, we shall investigate the optimal power allocation to maximize the sum rate R with average and instantaneous power constraints, respectively. Based on these results, we shall investigate whether TWR always outperforms OWR.

14.5.1   Average power constraints

With APC, the average transmit power is constrained. Specifically, we can determine the average transmit power at the relay as

$$p_r = \mathbb{E}\left[|\beta_a (h_1 s_1 + h_2 s_2 + n_R)|^2\right], \tag{14.99}$$

where β_a denotes the amplification factor with average power constraint. According to (14.88) and noting that the signal power of s_i is p_{si}, we can further obtain

$$p_r = |\beta_a|^2 (\sigma_h^2 p_{s1} + \sigma_h^2 p_{s2} + 1). \tag{14.100}$$

As a result, the amplification factor is given by

$$\beta_a = \sqrt{\frac{p_r}{\sigma_h^2 p_{s1} + \sigma_h^2 p_{s2} + 1}}. \tag{14.101}$$

Note that β_a is independent of the instantaneous channel information.

14.5.1.1   Separate power allocation

With separate power constraints for the source and the relay, the optimization problem can be expressed as

$$f = \max_{p_{s1}, p_{s2}, p_r} \{R\}$$

$$s.t. \quad p_{s1} + p_{s2} \le P_s$$

$$p_r \le P_r.$$

By substituting (14.94), (14.95), and (14.101) in (14.98), we can obtain the sum rate for the network as

$$R = \log_2 \left[\left(1 + \frac{p_r p_{s2} |h_1|^2 |h_2|^2}{p_r |h_1|^2 + p_{s1} \sigma_h{}^2 + p_{s2} \sigma_h{}^2 + 1} \right) \right.$$

$$\left. \left(1 + \frac{p_r p_{s1} |h_1|^2 |h_2|^2}{p_r |h_2|^2 + p_{s1} \sigma_h{}^2 + p_{s2} \sigma_h{}^2 + 1} \right) \right]. \tag{14.102}$$

By taking the derivative of R with respect to p_r, we can show that R is an increasing function of p_r. As a result, the relay node should utilize the highest power with $p_r = P_r$.

Next, we consider the optimal power allocation between the two source nodes. For this purpose, we denote the transmit powers from the two source nodes as

$$p_{s1} = \alpha p_s, \tag{14.103}$$

$$p_{s2} = (1 - \alpha) p_s, \tag{14.104}$$

where $p_s \leq P_s$ denotes the utilized transmit power, and α denotes the power allocation ratio. By substituting (14.103) and (14.104) into (14.102) and taking the derivative of R with respect to p_s, we can show that the sum rate R is an increasing function of p_s. As a result, the source nodes should utilize the highest power with $p_{s1} + p_{s2} = p_s = P_s$ in order to maximize R.

By substituting (14.94), (14.95), (14.96), and (14.103) into (14.96) and (14.97), we can determine the two rates as

$$R_1 = \log_2 \left(1 + \frac{P_r |h_1|^2 |h_2|^2 (1 - \alpha) P_s}{P_r |h_1|^2 + \sigma_h{}^2 P_s + 1} \right) \tag{14.105}$$

and

$$R_2 = \log_2 \left(1 + \frac{P_r |h_1|^2 |h_2|^2 \alpha P_s}{P_r |h_2|^2 + \sigma_h{}^2 P_s + 1} \right), \tag{14.106}$$

respectively.

It can be shown that the optimal α is given by

$$\alpha_o = \frac{|h_1|^2 - |h_2|^2}{2 P_s |h_1|^2 |h_2|^2} + \frac{1}{2}, \tag{14.107}$$

where c is a constant to meet the power constraints. The proof is left as an exercise. If TWR outperforms OWR, we shall have $0 < \alpha_o < 1$, which requires

$$P_s \geq \frac{||h_1|^2 - |h_2|^2|}{|h_1|^2 |h_2|^2}. \tag{14.108}$$

Given the complex Gaussian distributions for h_1 and h_2, we know the above condition will not hold with probability 1. Thus, we can conclude that TWR is not always optimal in this case. In particular, we have the following conclusion:

Property 14.5.1 *With average power constraint and separate power allocation between the source and relay nodes, TWR is not always better than OWR. The optimality of TWR holds when*

$$P_s > \frac{||h_1|^2 - |h_2|^2|}{|h_1|^2 |h_2|^2}. \tag{14.109}$$

14.5.1.2 Joint power allocation

We now consider the case with joint power allocation, where the optimization problem is given by

$$f = \arg \max_{p_{s1}, p_{s2}, p_r} \{R\}$$

$$s.t. \quad p_{s1} + p_{s2} + p_r \leq P.$$

Note that the difference between joint and separate power allocation lies in the fact that, with joint power allocation, we can further allocate power between the source and relay nodes.

Let P_s denote the power consumed by the two source nodes with $p_{s1} + p_{s2} = P_s$, and let $P_r = P - P_s$ denote the power utilized by the relay. By following a similar procedure as for the separate power allocation case, we define $p_{s1} = \alpha P_s$ and $p_{s2} = (1 - \alpha)P_s$. The optimal P_s that maximizes the sum rate is determined by setting the derivative of $(1 + \mathrm{SNR}_1)(1 + \mathrm{SNR}_2)$ with respect to P_s to zero with

$$\frac{\partial}{\partial P_s}(1 + \mathrm{SNR}_1)(1 + \mathrm{SNR}_2) = 0. \tag{14.110}$$

Similar to the separate power allocation case, the optimality of TWR holds with the condition

$$P_s > \frac{\left| |h_1|^2 - |h_2|^2 \right|}{|h_1|^2 |h_2|^2}, \tag{14.111}$$

$$P_r = P - P_s > 0. \tag{14.112}$$

Given the additional freedom with joint power allocation to allocate power between the source and relay nodes, we can rewrite the above condition as

$$P > \frac{\left| |h_1|^2 - |h_2|^2 \right|}{|h_1|^2 |h_2|^2}. \tag{14.113}$$

Note that the condition for TWR to outperform OWR with joint power allocation is looser than that with separate power allocation. However, the possibility of $P \leq \frac{||h_1|^2 - |h_2|^2|}{|h_1|^2|h_2|^2}$ still exists, thus enabling the following assertion:

Property 14.5.2 *With average power constraint and joint power allocation among all three nodes, TWR is not always better than OWR. The optimality of TWR holds when*

$$P > \frac{\left| |h_1|^2 - |h_2|^2 \right|}{|h_1|^2 |h_2|^2}. \tag{14.114}$$

14.5.2 Instantaneous power constraint

With instantaneous power constraint, the transmit power of the relay is given by

$$p_r = |\beta_i|^2 \left(|h_1|^2 p_{s1} + |h_2|^2 p_{s2} + 1 \right), \tag{14.115}$$

where the amplification factor can be determined as

$$\beta_i = \sqrt{\frac{p_r}{|h_1|^2 p_{s1} + |h_2|^2 p_{s2} + 1}}. \tag{14.116}$$

With joint power allocation among three nodes, the optimal power allocation with instantaneous constraints for TWR is given in Ref. [25], as

$$p_1 = \frac{\overline{p}}{2 + 2\sqrt{\frac{|h_1|^2\overline{p}+1}{|h_2|^2\overline{p}+1}}}, p_2 = \frac{\overline{p}}{2 + 2\sqrt{\frac{|h_2|^2\overline{p}+1}{|h_1|^2\overline{p}+1}}}, p_R = \frac{\overline{p}}{2} \tag{14.117}$$

and TWR is shown to be better than OWR. However, with instantaneous power constraint and separate power allocation between the source and relay nodes, TWR is not always better than OWR [12].

14.6   Multi-cell MIMO

Given the limited spectrum resource, an effective way to meet the increasing demand for wireless networks is to significantly increase the spectral efficiency. Reducing the size of cells has been proved very effective for this purpose, as can be seen from the evolution of cellular system from the first-generation FDMA system to the fourth-generation LTE system. However, small-cell networks need to deal with interference, which will become even more serious as we further decrease the size of the cells. In this area, cooperation communications provides a promising method for interference management. By coordinating the transmission/reception of multiple base stations (BSs), multi-cell processing (MCP) is a revolutionary technique to suppress inter-cell interference and improve the performance of multi-cell networks [27, 28]. With sufficient number of antennas, MCP with full BS coordination can eliminate all the inter-cell interference and transform the multi-cell system from an interference-limited regime to a noise-limited one.

With cooperation among multiply distributed BSs, MCP can be regarded as a distributed multiple-input multiple-output (MIMO) system. However, the distributed nature of those BS antennas brings two major differences from single-cell MIMO systems. First, inter-BS information exchange, which includes sharing of user data and channel state information (CSI), is required and demands high-capacity backhauls connecting different BSs. However, the backhaul capacity in current cellular systems is limited, so inter-BS information sharing can cause a great burden for practical applications. Second, for a given user terminal (UT), the distributed BS antennas make different BS–UT links independent but not identically distributed, whereby the large–scale pathloss and shadowing will have different effects on different links. While the first issue related to the backhaul capacity has attracted much attention and has been treated to some extent, the second one related to nonidentical BS–UT links has not been investigated.

In MCP, the amount of inter-BS data sharing grows with the network size, the number of users in each cell, and the number of antennas at each BS. This has motivated the analysis regarding the impact of limited backhaul capacity from an information-theoretic point of view by constraining the information exchange between BS and the central unit (CU) [29], limiting the number of coordinating cells [30] or the knowledge of the codebook information about neighboring cells [31]. In Ref. [32], inner capacity bounds were derived for different BS cooperation schemes by considering constrained backhaul and imperfect channel knowledge. In Ref. [33], the bandwidth requirements for sharing information among coordinated BSs were analyzed, and it was found that CSI occupies a negligible fraction of the overall required backhaul bandwidth for moderate Doppler speeds. To relieve the burden on the backhaul capacity, different MCP schemes have been proposed with reduced inter-BS data-sharing. Cluster-based coordination is one approach [34–36] that performs local coordination and only requires data-sharing among neighboring BSs. In Refs [37, 38], distributed BS coordination strategies were proposed, which are based on the virtual signal-to-interference-and-noise (SINR) framework, which only requires local CSI. A similar distributed scheme with zero-forcing (ZF) beamforming was investigated in Ref. [39]. In Ref. [40], a distributed implementation of coordinated multicell beamforming was proposed via message-passing between neighboring BSs.

There is another category of MCP that does not require sharing user data between different BSs, and the data transmission for each UT comes from a single BS. This essentially forms a type of interference channel. In Ref. [41], the Pareto boundary for the MISO interference channel (MISO-IC) was characterized. A similar problem was investigated in Ref. [42] where the transmitters only have statistical channel knowledge. By considering local CSI and without sharing user data, different distributed beamforming algorithms were proposed in Ref. [43]. In Ref. [44], a multi-cell downlink beamforming scheme, which achieves different rate-tuples on the Pareto boundary of the achievable rate region for the MISO-IC, was proposed. Multi-cell downlink beamforming was investigated in Refs [45, 46], based on the downlink–uplink duality. In Ref. [47], an adaptive inter-cell interference cancellation scheme was proposed that could further reduce the CSI requirement by adaptively switching between ZF and MRT (maximum ratio transmission) based on user locations.

The MCP schemes based on local CSI, such as in Refs [37, 38], reduce the backhaul capacity requirement at the cost of fewer channel freedoms (independent channels in one link). However, they still require sharing user data among different BSs, which demands more backhaul capacity than the sharing of CSI, as demonstrated in Ref. [33]. The MCP schemes based on the interference channel model have a lower requirement on the backhaul capacity, as they do not need to share user data, but will suffer from inter-cell interference. The question is whether we can achieve a better tradeoff between performance and backhaul capacity requirement. Similar issues also exist in the heterogeneous network, which is regarded as the most promising solution for the future wireless networks.

14.7 Summary

The explosive growth of wireless devices demands a 1000× increase for network capacity. Such requirement cannot be fully satisfied by conventional physical layer technologies. Fortunately, the large number of densely deployed wireless nodes also enables cooperative communications technology. By coordinating the transmission among wireless nodes, cooperative communication can help achieve both SNR gain and diversity gain, and has found application in the LTE-A network. On the other hand, cooperative communication also provides an effective way to manage interference. For instance, with the cooperative transmission/reception of multiple base stations, multi-cell processing is a revolutionary technique to suppress inter-cell interference and improve the performance of multi-cell networks. Node cooperation has also been proved effective in resource allocation and interference management for heterogeneous networks, interference channels, and sensor networks. We strongly believe that cooperative communication will play an extremely important role in future wireless systems.

Problems

Problem 14.1 Consider a wireless link where the source (S) wants to transmit information to a destination (D) located $2d$ meters away from the source. For the first case, the source transmits with power P_t through the direct link. For the second case, a relay (R) is put at the middle point between the source and the destination, where the transmit power at the source and the relay is $P_t/2$. With the logarithm pathloss model, determine the throughput of the S–D and S–R–D links. Comment on the results.

Problem 14.2 Prove that the matrix $\mathbf{B}_i$ in (14.61) is of the structure $\mathbf{B}_i = \mathbf{D}_i \mathbf{R}_i$, where $\mathbf{D}_i$ is a diagonal matrix and $\mathbf{R}_i$ is a permuting matrix having only one nonzero unity entry on each row and each column.

Problem 14.3 Prove that the optimal a_{im} in (14.73) is of the form of

$$a_{im} = \frac{c\varphi_{im}^* \phi_{im}^*}{p_{sm}|\phi_{im}|^2 + p_{rm}|\varphi_{im}|^2 + \sigma_n^2}.$$

Problem 14.4   Prove that the optimal $\alpha_m(P_m)$ in (14.76) for the case with $L = 1$ is expressible as

$$\alpha_m(P_m) = \frac{\sqrt{P_m|\varphi_{im}|^2 + \sigma_n^2}}{\sqrt{P_m|\varphi_{im}|^2 + \sigma_n^2} + \sqrt{P_m|\phi_{im}|^2 + \sigma_n^2}}.$$

Problem 14.5   Prove Property 14.4.1.

Problem 14.6   Prove that the optimal α in (14.103) is of the form

$$\alpha_o = \frac{|h_1|^2 - |h_2|^2}{2P_s|h_1|^2|h_2|^2} + \frac{1}{2}.$$

References

1. C.E. Shannon, "A mathematical theory of communication," *Bell Syst. Tech. J.*, vol. 27, pp. 379–423, 623–656, 1948.
2. R. Pabst, B. Walke, D. Schultz, P. Herhold, H. Yanikomeroglu, S. Mukherjee, H. Viswanathan, M. Lott, W. Zirwas, M. Dohler, H. Aghvami, D. Falconer, and G. Fettweis, "Relay-based deployment concepts for wireless and mobile broadband radio," *IEEE Commun. Mag.*, vol. 42, no. 9, pp. 80–89, 2004.
3. P. Gupta and P.R. Kumar, "The capacity of wireless networks," *IEEE Trans. Info. Theory*, vol. 46, pp. 388–404, 2000.
4. T.M. Cover and A.A. El Gamal, "Capacity theorems for the relay channel," *IEEE Trans. Inf. Theory*, vol. 25, no. 5, pp. 572–584, 1979.
5. J.N. Laneman and G.W. Wornell, "Distributed space-time-coded protocols for exploiting cooperative diversity in wireless networks," *IEEE Trans. Inf. Theory*, vol. 49, no. 10, pp. 2415–2425, 2003.
6. I. Maric and R.D. Yates, "Bandwidth and power allocation for cooperative strategies in Gaussian relay networks," in *Conference Record of the 38th Asilomar Conference on Signals, Systems and Computers*, vol. 2, 2004.
7. P. Larsson and H. Rong, "Large scale cooperative relaying network with optimal coherent combining under aggregate relay power constraints," in *Future Telecommunications Conference*, Beijing, China, Dec. 2003.
8. M.O. Hasna and M.S. Alouini, "Optimal power allocation for relayed transmission over Rayleigh-fading channels," *IEEE Trans. Wireless Commun.*, vol. 3, no. 6, pp. 1999–2003, Nov. 2004.
9. S.A. Jafar, K.S. Gomadam, and C. Huang, "Duality and rate optimization for multiple access and broadcast channels with amplify-and-forward relays," *IEEE Trans. Inf. Theory*, vol. 53, no. 10, pp. 3350–3370, 2007.
10. S.A. Jafar, K.S. Gomadam, and C. Huang, "Duality and rate optimization for multiple access and broadcast channels with amplify-and-forward Relays," *IEEE Trans. Inf. Theory*, vol. 53, no. 10, pp. 3350–3370, 2007.
11. S.H. Song and Q.T. Zhang, "Design collaborative systems with multiple AF-relays for asynchronous frequency-selective fading channels," *IEEE Trans. Commun.*, vol. 57, no. 9, pp. 2808–2817, 2009.
12. W. Zhang and K.B. Letaief, "Full-rate distributed space-time codes for cooperative communications," *IEEE Trans. Wireless Commun.*, vol. 7, no. 7, pp. 2446–2451, 2008.
13. H. Li, M.W. Liu, S.H. Song, and K.B. Letaief, "Optimality of amplify-and-forward based two-way relaying," *IEEE Wireless Communications and Networking Conference*, 2013.
14. L.Z. Zheng and D.N.C. Tse, "Diversity and multiplexing: a fundamental tradeoff in multiple-antenna channels," *IEEE Trans. Inf. Theory*, vol. 49, no. 5, pp. 1073–1096, 2003.
15. W. Yu, W. Rhee, J. Cioffi, and S. Boyd, "Iterative water-filling for Gaussian vector multiple-access channels," *IEEE Trans. Inf. Theory*, vol. 50, no. 1, pp. 141–152, 2004.
16. D. Popescu, O. Popescu, and C. Rose, "Interference avoidance for multiaccess vector channels," *IEEE Trans. Commun.*, vol. 55, no. 8, pp. 1466–1471, 2007.
17. S. Ohno, G. Giannakis, and Z. Luo, "Multicarrier multiple access is sum-rate optimal for block transmissions over circulant ISI channels," *IEEE J. Sel. Areas Commun.*, vol. 24, no. 6, pp. 1256–1260, 2006.
18. R.M. Gray, "On the asymptotic eigenvalue distribution of Toeplitz matrices," *IEEE Trans. Inf. Theory*, vol. IT-18, no. 6, pp. 725–730, 1972.

19. I.E. Telatar, "Capacity of multi-antenna Gaussian channels," *Eur. Trans. Telecommun.*, vol. 10, pp. 585–596, 1999.

20. D.A. Pierre, *Optimization Theory with Applications*, New York. Dover Publications, 1986.

21. R. Zhang, Y.C. Liang, C.C. Chai, and S. Cui, "Optimal beamforming for two-way multi-antenna relay channel with analogue network coding," *IEEE J. Sel. Areas Commun.*, vol. 27, no. 5, pp. 699–712, 2009.

22. Y. Han, S.H. Ting, C.K. Ho, and W.H. Chin, "Performance bounds for two-way amplify-and-forward relaying," *IEEE Trans. Wireless Commun.*, vol. 8, no. 1, pp. 432–439, 2009.

23. Q. Li, S.H. Ting, A. Pandharipande, and Y. Han, "Adaptive two-way relaying and outage analysis," *IEEE Trans. Wireless Commun.*, vol. 8, no. 6, pp. 3288–3299, 2009.

24. Y.U. Jang, E.R. Jeong, and Y.H. Lee, "A two-step approach to power allocation for OFDM signals over two-way amplify-and-forward relay," *IEEE Trans. Signal Process.*, vol. 58, no. 4, pp. 2426–2430, 2010.

25. X.J. Zhang and Y. Gong, "Adaptive power allocation in two-way amplify-and-forward relay networks," in *Proceedings of IEEE International Conference on Communications*, June 2009 (ICC09).

26. Q. Yuan, Y. Zhou, M. Zhao, and Y. Yang, "Optimal transmission power allocation for two-way relay channel using analog network coding," in *Proceedings of International Conference on Consumer Electronics, Communications and Networks*, Apr. 2011 (CECNet).

27. J.C. Park, I. Song, S.R. Lee, and Y.H. Kim, "Average rate performance of two-way amplify-and-forward relaying in asymmetric fading channels," *J. Commun. Networks*, vol. 13, no. 3, pp. 250–256, 2011.

28. D. Gesbert, S. Hanly, H. Huang, S. Shamai, O. Simeone, and W. Yu, "Multi-cell MIMO cooperative networks: a new look at interference," *IEEE J. Sel. Areas Commun.*, vol. 28, no. 9, 2010.

29. O. Somekh, O. Simeone, Y. Bar-Ness, A.M. Haimovich, U. Spagnolini, and S. Shamai (Shitz), *Distributed Antenna Systems: Open Architecture for Future Wireless Communications*, Auerbach Publications, CRC Press, 2007.

30. P. Marsch and G. Fettweis, "A framework for optimizing the uplink performance of distributed antenna systems under a constrained backhaul," in *Proceedings of the IEEE International Conference on Communications (ICC'07), Glasgow, Scotland*, Jun. 24-28 2007.

31. O. Somekh, B.M. Zaidel, and S. Shamai, "Spectral efficiency of joint multiple cell-site processors for randomly spread DS-CDMA systems," *IEEE Trans. Inf. Theory*, vol. 52, no. 7, pp. 2625–2637, 2007.

32. A. Sanderovich, O. Somekh, and S. Shamai, "Uplink macro diversity with limited backhaul capacity," in *Proceedings of IEEE International Symposium on Information Theory (ISIT'07)*, pp. 11–15, June 24-29, 2007.

33. P. Marsch and G. Fettweis, "On downlink network MIMO under a constrained backhaul and imperfect channel knowledge," in *Proceedings of the IEEE GLOBECOM, Honolulu, Hawaii*, pp. 1–6, Nov. 30-Dec. 4, 2009.

34. D. Samardzija and H. Huang, "Determining backhaul bandwidth requirements for network MIMO," in *17th European Signal Processing Conference (EUSIPCO 2009), Glasgow, Scotland*, pp. 1494-1498, Aug. 2009.

35. S. Venkatesan, "Coordinating base stations for greater uplink spectral efficiency in a cellular network," in *Proceedings of the IEEE International Symposium on Personal Indoor and Mobile Radio Communications, Athens, Greece*, pp. 1–5, Sept. 2007.

36. A. Papadogiannis, D. Gesbert, and E. Hardouin, "A dynamic clustering approach in wireless networks with multi-cell cooperative processing," in *Proceedings of the IEEE International Conference on Communicaitons, Beijing, China*, pp. 4033–4037, May 2008.

37. J. Zhang, R. Chen, J.G. Andrews, A. Ghosh, and R.W. Heath Jr., "Networked MIMO with clustered linear precoding," *IEEE Trans. Wireless Commun.*, vol. 8, no. 4, pp. 1910–1921, 2009.

38. E. Bjornson, R. Zakhour, D. Gesbert, and B. Ottersten, "Distributed multicell and multiantenna precoding: characterization and performance evaluation," in *Proceedings of the IEEE GLOBECOM'09*, 2009.

39. R. Zakhour and D. Gesbert, "Distributed multicell-MISO precoding using the layered virtual SINR framework," *IEEE Trans. Wireless Commun.*, vol. 9, no. 8, pp. 2444–2448, 2010.

40. M. Kobayashi, M. Debbah, and J. Belfiore, "Outage efficient strategies in network MIMO with partial CSIT," in *Proceedings of the IEEE ISIT'09, Seoul, Korea*, pp. 249–253, Jun. 28-Jul. 3, 2009.

41. B.L. Ng, J.S. Evans, S.V. Hanly, and D. Aktas, "Distributed downlink beamforming with cooperative base stations," *IEEE Trans. Inform. Theory*, vol. 54, no. 12, pp. 5491–5499, 2008.

42. E. Jorswieck, E.G. Larsson, and D. Danev, "Complete characterization of the Pareto boundary for the MISO interference channel," *IEEE Trans. Signal Process.*, vol. 56, no. 10, pp. 5292–5296, 2008.

43. J. Lindblom, E. Karipidis, and E.G. Larsson, "Selfishness and altruism on the MISO interference channel: the case of partial transmitter CSI," *IEEE Commun. Lett.*, vol. 13, no. 9, pp. 667–669, 2009.

44. R. Zakhour, K.M. Ho, and D. Gesbert, "Distributed beamforming coordination in multicell MIMO channels," in *IEEE 69th Vehicular Technology Conference, Barcelona, April 26-29*, 2009.

45. R. Zhang and S. Cui, "Cooperative interference management with MISO beamforming," *IEEE Trans. Signal Process.*, vol. 58, no. 10, pp. 5450–5458, 2010.

46. F. Rashid-Farrokhi, K.J.R. Liu, and L. Tassiulas, "Transmit beamforming and power control for cellular wireless systems," *IEEE J. Sel. Areas Commun.*, vol. 16, no. 8, pp. 1437–1450, 1998.

47. H. Dahrouj and W. Yu, "Coordinated beamforming for the multicell multi-antenna wireless system," *IEEE Trans. Wireless Commun.*, vol. 9, no. 5, pp. 1748–1759, 2010.

15

Cognitive Radio

Abstract

Cognitive radio (CR) emerges to address the challenge of inefficient use of the precious electromagnetic spectrum resource, which stems from the widely adopted policy of fixed spectrum allocation. This chapter is focused on the enabling technology of CR, that is, spectrum sensing, although a brief summary of its current status in standardization and commercialization is also included.

♣

15.1 Introduction

The ever-increasing demand for high-data-rate wireless transmission poses a challenge of efficiently utilizing the precious electromagnetic spectrum. The current practice in many countries is a static spectrum allocation policy, whereby governmental agencies allocate wireless spectra to licensed users on a long-term basis for large geographical regions. Licensed users are also known as *primary users* (PU) in the literature . A careful study released by the U.S. Federal Communications Commission (FCC) [1], however, reveals that most of the allocated spectra are underutilized, at less than 25% [2]. A possible solution is the use of dynamic spectrum access (DSA). *Cognitive radio* (CR) [3, 4] is one such emerging technique that provides unlicensed users, usually referred to as *CR users* or *secondary users* (SU), with the capability to share the licensed spectrum in an opportunistic manner. The secondary users usually do not operate individually, but rather work jointly to form a CR network coexisting with the primary network over the same geographical area. The two networks constitute a *heterogeneous* wireless architecture, but with different priority in using the frequency spectrum.

The key technology of CR consists of three essential components: spectrum sensing, dynamic spectrum management, and adaptive transmission [5]. Spectrum sensing is the key enabling function of cognitive radio. CR needs to reliably detect PU signals of possibly unknown types in relatively low signal-to-noise ratio (SNR), and obtain spectrum usage characteristics across multiple dimensions, such as time, space, code, and angle, in real time. Since its introduction in 1999 [3], CR has become one of the most active research areas in communications and signal processing, and is believed to be the next generation of wireless communication systems.

The introduction of the CR concept prompted an outpouring of academic research on related algorithms and systems design, mainly in the physical (PHY), medium access control (MAC), and network layers of the open systems interconnection (OSI) model. Physical layer research efforts mainly focus on adding functionalities

Wireless Communications: Principles, Theory and Methodology, First Edition. Keith Q.T. Zhang.
© 2016 John Wiley & Sons, Ltd. Published 2016 by John Wiley & Sons, Ltd.
Companion Website: www.wiley.com/go/zhang7749

required by the CR concept. One major direction is to provide reliable spectrum sensing results. Examples include using feature detection (e.g., cyclostationary detectors) to improve the accuracy of simple energy detection (ED) [6], or exploring compressive sensing theory to design efficient wideband spectrum sensing algorithms [7]. Another direction is to design and evaluate CR/PU location estimation algorithms to support the geolocation feature required by the CR network [8]. MAC layer research implements most of the major spectrum management and adaptation visions of CR, aiming to provide large spectrum efficiency for the secondary network while maintaining sufficient protection and transparency for the primary system. Numerous interesting topics have been studied in the literature, including sensing time and algorithm optimization, PU traffic modeling, spectrum sharing mechanism (i.e., CR self coexistence), interference management and mitigation, and multiband operation [9]. The spectrum heterogeneity due to PU activity and node mobility may cause the network topology to vary significantly with time and location; therefore, new routing protocols with strong adaptability and fast route recovery capability need to be designed in the network layer [10]. Furthermore, since CR networks require rapid learning and reconfigurability, cross-layer algorithm designs may be necessary to optimize the overall CR performance.

Clearly, the central task of a CR network is twofold: (1) to enable SUs to implement opportunistic sharing of the licensed spectrum, and (2) to ensure its seamless communications. Spectrum sharing, in nature, is the problem of multiple access. Unlike the traditional multiple-access schemes, the SUs are always in a lower priority, in the sense that the SUs can use an unused licensed channel but need to immediately vacate from its occupancy once the PU appears again. The vacating SU needs another available channel to continue its communication, forming a very unique characteristic of CR networks, that is, DSA. The phenomenon of switching SU's channels from one to another is referred to as *spectrum mobility*. Thus, from a spectrum management perspective, the CR networks are confronted with four challenges:

- spectrum sensing
- spectrum decision
- spectrum sharing
- spectrum mobility.

In this chapter, we mainly focus on various detection techniques for spectrum sensing. The issues of standardization of CR networks and the current progress in experimental and commercial CR systems are briefly described in Sections 15.7 and 15.8, respectively.

15.2   Spectrum sensing for spectrum holes

The term *spectrum hole* was originally coined to heuristically indicate the unused frequency portions in the entire spectrum. In a general sense used today, spectrum holes are defined as multidimensional regions within the frequency–time–space domain, in which a particular secondary use is possible without introducing interference.

Spectrum sensing, a key enabling function of CR networks, aims to find a spectrum hole for the use by an SU. It requires the reliable detection of primary signals, possibly of a random/unknown waveform with some unknown parameters, even in relatively low SNR. The detection must be carried out in a continuous manner in order to monitor the target band and activities of PU, so that the CR can vacate the occupied spectrum as soon as the PU is turned on again. Therefore, reliability and rapid reaction are two basic requirements for a sensing algorithm to be adopted in CR networks.

The detection of deterministic and stochastic primary signals relies on different techniques, and is therefore addressed, separately. In this chapter, we are concerned with detectors with a constant false-alarm rate (CFAR). For notational simplicity, we may use the same symbol Λ or its variation to denote the thresholds of different detectors. Hence, the reader should beware of the context and relevant formulas for threshold evaluation.

15.3   Matched filter versus energy detector

In this section, primary signals are treated as *deterministic*. The matched-filter (MF) approach and energy detector (ED) represent two extreme cases in terms of the prior information required at their receivers. The MF receiver requires the knowledge of the transmit signal waveform, whereas the only information required by the energy detector is the noise power. We will treat the two techniques in the same framework so as to unveil their nature.

To begin with, let $s(t), 0 \leq t < T$ denote a possible deterministic primary signal with frequency bandlimited to B Hz. Its energy is given by

$$E_s = \int_0^T |s(t)|^2 dt. \tag{15.1}$$

The received signal at a CR user is then expressible as

$$x(t) = s(t) + n(t), \ 0 \leq t < T, \tag{15.2}$$

where $n(t)$ is a zero-mean complex white Gaussian process with power σ_n^2 known to the receiver. The SNR is then equal to

$$\rho = E_s/\sigma_n^2. \tag{15.3}$$

For performance analysis in this section, we assume signal and noise of *real* values. The physical implication of real and complex signals is studied in Problem 15.2.

15.3.1   Matched-filter detection

In the MF approach, any scaled waveform of $s(t)$ can be used as a local reference. For convenience and without loss of generality, we use the normalized reference $\psi(t) = s(t)/\sqrt{E_s}$ for correlation, which produces a sample output at $t = T$:

$$\xi = \sqrt{E_s} + n. \tag{15.4}$$

In mathematics, this operation is equivalent to projecting the received signal $x(t)$ onto a single basis function $\psi(t)$, resulting a single noise component $n \sim \mathcal{N}(0, \sigma_n^2)$. The advantage of using the normalized reference is that n maintains the same variance as $n(t)$. It is clear that

$$\xi \sim \begin{cases} \mathcal{N}(0, \sigma_n^2), & H_0 \\ \mathcal{N}(\sqrt{E_s}, \sigma_n^2), & H_1 \end{cases}, \tag{15.5}$$

which is compared with a preset threshold Ω for a decision, as shown by

$$\xi \rightarrow \begin{cases} < \Omega, & H_0 \\ > \Omega, & H_1 \end{cases}. \tag{15.6}$$

Thus, for a given probability of false alarm $P_{fa} = \alpha$, we have $P_{fa} = Q(\Omega/\sigma_n) = \alpha$. By denoting $\Lambda = \Omega/\sigma_n$ and $\rho = E_s/\sigma_n^2$, the corresponding detection probability is obtained as

$$P_d = \int_\Omega^\infty \frac{1}{\sqrt{2\pi}\sigma_n} e^{-\frac{(x-\sqrt{E_s})^2}{2\sigma_n^2}} dx = \begin{cases} Q(\Lambda - \sqrt{\rho}), & \sqrt{\rho} < \Lambda \\ 1 - Q(\sqrt{\rho} - \Lambda), & \sqrt{\rho} > \Lambda \end{cases}. \tag{15.7}$$

The MF method requires perfect information of the signal waveform arriving at its receiver. This type of knowledge is nearly impossible for a practical CR user. Thus, the MF method is only used as a benchmark for performance comparison for various CR detectors.

15.3.2 Energy detection

Very often, the secondary user has no prior information about the timing, symbol period, and initial phase of the signal waveform of the primary user, or even does not know the transmission scheme used by the latter. In such a case, ED is an appropriate choice. A block diagram for ED implementation and its comparison with the matched-filter receiver is depicted in Figure 15.1.

A simple and feasible technique to implement ED is to sample the received signal and then accumulate their energy for detection. Recall from the sampling theorem that the band-pass signal $s(t)$ can be represented by its samples $s(kT_s)$ separated by $T_s = \frac{1}{2B}$ in terms of sinc functions, as shown by

$$s(t) = \sum_{k=0}^{N-1} \sqrt{T_s}\, s(kT_s)\psi_k(t), \tag{15.8}$$

where $\phi_k(t) = \sqrt{2B}\mathrm{sinc}(2B(t - kT_s))$ constitute a set of orthonormal bases defined over $t \in (-\infty, \infty)$. The accumulation of sample energy leads to $v = \sum_k |s(k) + n(k)|^2$. When viewed from a multi-dimensional receiver, this is equivalent to projecting $x(t)$ onto a bank of N orthogonal receivers, each matched to a $\psi_k(t)$ function, in much the same way as the noncoherent quadrature reception of a sinusoidal signal with unknown phase. The resulting received vector is given by

$$\mathbf{x} = \sqrt{T_s}\,\mathbf{s} + \mathbf{n}, \tag{15.9}$$

where $\mathbf{s} = [s(0T_s), s(T_s), \cdots, s((N-1)T_s)]^T$, and $\mathbf{n} = [n(0T_s), n(T_s), \cdots, n((N-1)T_s)]^T$. We collect all the energy at different coordinates and normalize it with the noise variance to yield an energy decision variable

$$\xi = \sum_{k=0}^{N-1} |x(kT_s)|^2 / \sigma_n^2 = \begin{cases} < \Lambda, & H_0 \\ > \Lambda, & H_1 \end{cases}. \tag{15.10}$$

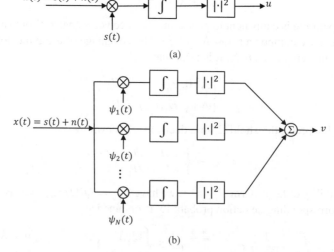

(a)

(b)

Figure 15.1 *Matched-filter versus energy detection approaches in the same framework: a) Matched filter; b) Energy detector*

Just as for the matched filter, here we use the same symbol Λ, again, to denote the threshold. But its determination is different, as derived below. Note that for real-valued data

$$|x(kT_s)|^2/\sigma_n^2 = \begin{cases} \chi^2(1), & H_0 \\ \chi^2(1, T_s|s(kT_s)|^2/\sigma_n^2), & H_1 \end{cases}, \tag{15.11}$$

which, when using the properties of chi-square distribution, enables the assertion

$$\xi = \begin{cases} \chi^2(N), & H_0 \\ \chi^2(N, E_s/\sigma_n^2), & H_1 \end{cases}, \tag{15.12}$$

where $E_s = [|s(0T_s)|^2 + \cdots + |s((N-1)T_s)|^2]T_s$ is the total accumulated energy.

The MF receiver knows the exact received signal waveform and uses it to fully capture all the signal energy, leading to a Gaussian-distributed decision variable. The ED method, on the contrary, has to employ multiple basis functions to grasp all the signal energy, leading to a noncentral chi-square distributed decision variable of N degrees of freedom. As N increases, the accumulated energy increases, but a larger threshold is required to maintain the same false-alarm rate, implying that part of the increased energy must be used to compensate the increase in threshold. The fundamental difference between MD and MF is summarized in Table 15.1.

Let us consider the performance of the ED method. Theoretically, the detection performance can be evaluated by using the PDF of the noncentral $\chi^2(v, \lambda)$ variable given by Johnson and Kotz [11]:

$$\chi^2(v, \lambda) \sim \frac{1}{2}(z/\lambda)^{\frac{1}{4}(v-2)} I_{\frac{1}{2}(v-2)}(\sqrt{\lambda z}) \exp\left[-(\lambda + z)/2\right], \quad z > 0, \tag{15.13}$$

where $I_m(\cdot)$ denotes the modified Bessel function of the first kind and order m. It is common practice to expand I_m as a series (see, e.g., (9.6.10) of Ref. [12]) and represent the noncentral PDF as a summation of central chi-squares. The noncentral chi-square expansion can be integrated term by term in terms of an incomplete gamma function, leading to ([12], p. 942)

$$\begin{aligned} P_d &= \Pr\left\{\frac{1}{2}\chi^2(2m, 2\rho) > \Lambda\right\} \\ &= 1 - \Pr\{\chi^2(2m, 2\rho) < 2\Lambda\} \\ &= 1 - \sum_{k=0}^{\infty} \frac{\rho^k e^{-\rho}}{k!} \Pr\{\chi^2(2m + 2k) < 2\Lambda\} \\ &= 1 - \sum_{k=0}^{\infty} \frac{\rho^k e^{-\rho}}{k!} \gamma(\Lambda, m + k), \end{aligned} \tag{15.14}$$

where $m = 1$ for the MF method and N for ED. The incomplete gamma function, defined as $\gamma(x, a) = \int_0^x \frac{1}{\Gamma(a)} t^{a-1} e^{-t} dt$, can be easily calculated by using the built-in MATLAB function *gammainc*. In the above expression, the weighting coefficients for incomplete gamma functions are easy to memorize and are equal to the Poisson distribution.

Table 15.1 *MF versus ED: A comparison*

Method	MF	ED
Projection onto	Single base	Multiple bases
Decision variable	(15.5)	(15.10)
PDF under H_1	$\mathcal{N}(\sqrt{E_s}, \sigma_n^2)$	$\chi^2(N, \rho)$

The above expression can be also used for calculating the threshold for a given false-alarm probability, say $P_{fa} = \alpha$, by setting $\rho = 0$. However, it is easier to use

$$\Pr\{\chi^2(2m) > 2\Lambda\} = \alpha \rightarrow \gamma(\Lambda, m) = 1 - \alpha. \tag{15.15}$$

More specifically,

$$\text{MF} : \Lambda = -\ln\alpha,$$

$$\text{ED} : \gamma(\Lambda, N) = 1 - \alpha. \tag{15.16}$$

To simplify the calculation of the detection performance, we may approximate ξ under H_1 by $\alpha\chi^2(v)$. To this end, recall that, for $u \sim \chi^2(m, \lambda)$, we have $\mathbb{E}[u] = m + \lambda$ and $\text{var}[u] = 2(m + 2\lambda)$. By equating the first two moments, we obtain

$$\alpha = \frac{1}{2}(N + 2\rho)/(N + \rho); v = 2(N + \rho)^2/(N + 2\rho). \tag{15.17}$$

In practice, the energy detector is easier to implement by using an alternative scheme. In this scheme, the received signal is first applied to a band-pass filter before sampling at the rate $t = kT_s$ and energy accumulation; it yields

$$\xi = \sum_{k=1}^{N} |s(kT_s) + e(kT_s)|^2/(N_0B), \tag{15.18}$$

where $e(t)$ is the noise after band-pass filtering, whose variance is N_0B.

15.4   Detection of random primary signals

In this section and subsequent ones, primary signals are assumed to be *random* in nature and take *complex* values. The treatment of real-valued signals is similar. The presence of a PU can alter the covariance structure of the received signal, as compared to the background additive white Gaussian noise (AWGN). Fully exploiting such changes is central to the construction of a powerful detector for CR. The information used for signal detection usually takes the form of magnitude, correlation, or their combination, up to available prior knowledge in a given operational environment. In this chapter, we are concerned with the detection of random CR signals in AWGN with different types of prior information, assuming the knowledge of exact noise variance, estimated noise variance, or no noise power reference at the receiver, respectively. Different prior information on the noise variance calls for different detection strategies, which are the focus of this chapter.

The most popular detector that exploits magnitude information is the energy-based detector [13–15]. Magnitude itself is a relative quantity, requiring a reference noise power level for comparison to indicate the possible presence of a primary signal. Using the total received signal energy alone, without normalization by noise variance, will generally lead to a detector without constant false-alarm rate (CFAR). Recently, efforts have been made to exploit signal correlation structures for the detection of CR signals in environments without noise variance information [16, 17]. The correlation among data can produce peaks and troughs in its power spectrum. Among all types of spectra produced by unitary transforms, the eigen spectrum is the most spread out, thus motivating the use of the maximum eigenvalue or the ratio of maximum to minimum eigenvalue as a better indicator of signal presence in white noise [16, 17]. Two problems arise. First, using the maximum value alone without appropriate normalization fails to produce a CFAR detector. Second, the eigen ratio test, though enjoys the CFAR property, faces a difficulty in its CFAR threshold determination. Note that this type of eigen tests are intuitively constructed without a rigorous framework. A more solid and systematic procedure for constructing eigen-based tests is the classical likelihood principle, which leads to a detector with a slightly better performance and a much easier threshold determination.

In many practical applications, the noise variance has to be estimated from a finite number of previous noise-alone samples. This inaccurate noise variance estimate has an impact upon the detection performance and threshold design, and is usually treated as uncertainty in maintaining a CFAR for the energy detector [17]. As will be shown subsequently, a better practice is to absorb the noise estimation uncertainty into the total sample energy to form a single test variable.

The features that can be extracted for the detection of a random signal are possible changes in the power level and the data correlation structure. To proceed, suppose we are given mN received samples $\mathbf{x} = [x(1), x(2), \cdots, x(mN)]^T$, which may contain a primary signal $\mathbf{s} = [s(1), s(2), \cdots, s(mN)]^T$. The signal detection in AWGN, denoted here by $\mathbf{n} = [n(1), n(2), \cdots, n(mN)]^T$, is to choose one of the two possible hypotheses: that is, H_0: $\mathbf{x}$ consists of noise $\mathbf{n}$ alone against its general alternative, H_1: $\mathbf{x} = \mathbf{s} + \mathbf{n}$. In this section, we assume that $\mathbf{s} \sim \mathcal{CN}(\mathbf{0}, \sigma_s^2\mathbf{I})$ and $\mathbf{n} \sim \mathcal{CN}(\mathbf{0}, \sigma_n^2\mathbf{I})$ for ease of analysis. To exploit the received data correlation, we partition the received data into N blocks (column vectors), each of m samples, such that $\mathbf{x}_i = [x((i-1)N+1), \cdots, x(iN)]^T$. As such, the covariance matrix of $\mathbf{x}$ can be estimated as

$$\hat{\mathbf{R}}_x = \frac{1}{N} \sum_{i=1}^{N} \mathbf{x}_i \mathbf{x}_i^\dagger. \tag{15.19}$$

Energy detectors only employ the trace information of $\hat{\mathbf{R}}_x$ while other detectors try to exploit its correlation structures as well.

15.4.1 Energy-based detection

In the LRT given in (15.24), if we neglect the correlation information term and just keep the energy term, the resulting test is the well-known energy detector test given by

$$\xi_{en} = \frac{2\mathbf{x}^\dagger\mathbf{x}}{\sigma_n^2} \begin{cases} > \Lambda_2, & H_1 \\ \leq \Lambda_2, & H_0 \end{cases}. \tag{15.20}$$

Here, we have used the fact that $trace(\hat{\mathbf{R}}) = \mathbf{x}^\dagger\mathbf{x}$, with $\mathbf{x}$ denoting the $mN \times 1$ vector containing all data samples. Since the energy test is well known, we only briefly summarize results that may not be available in the literature.

To determine threshold Λ_2, it is common practice to approximate ξ_{en} as a Gaussian variable. In fact, there is a simple, exact expression for the PDF of ξ_{en}. Under H_0, it is easy to use the result of Giri [18] to assert

$$\xi_{en} \sim \chi^2(2mN), \tag{15.21}$$

whereby we can easily determine the threshold Λ_α for $P_{fa} = \alpha$ by using the "gammainc," a built-in incomplete gamma function in MATLAB. Specifically,

$$P_{fa} = \Pr\{\xi_{en} > \Lambda_\alpha | H_0\} = \int_{\frac{1}{2}\Lambda_\alpha}^{\infty} \frac{1}{\Gamma(mN)} y^{mN-1} e^{-y} dy = \alpha, \tag{15.22}$$

which is equivalent to finding Λ_α such that gammainc$\left(\frac{\Lambda_\alpha}{2}, mN\right) = 1 - \alpha$. The solution can be obtained by using the Newton–Raphson iteration algorithm.

Another approach for finding a desired threshold Λ_2 for $P_{fa} = \alpha$ is to relate the chi-square threshold to its Gaussian counterpart, as shown by

$$\Lambda_c = \frac{1}{2}[\Lambda_g + \sqrt{2v - 1}\,]^2, v > 100,$$

$$\Lambda_c = v[1 - a + \Lambda_g\sqrt{a}\,]^3, v > 30, \tag{15.23}$$

where v is the degrees of freedom, $a = 2/(9v)$, Λ_c, and Λ_g are the upper α-point of the chi square distribution and of the Q-function, respectively. These formulae are obtained through nonlinear transforms and much more accurate than the widely adopted approximation based on the central limit theorem.

15.4.2   Maximum likelihood ratio test

After manipulations, and using the notation in (15.19), we can write the log likelihood ratio as

$$\xi_{\ell rt} = 2\log \frac{\max_{\mathbf{R}_x} f(\mathbf{x}_1, \cdots, \mathbf{x}_{mN}|H_1)}{f(\mathbf{x}, \cdots, \mathbf{x}_{mN})}$$

$$= 2\log \underbrace{\left[\frac{(\sigma_n^2)^m}{\det\hat{\mathbf{R}}} \exp\left(\frac{tr\hat{\mathbf{R}}}{\sigma_n^2} - m \right) \right]^N}_{z}$$

$$= 2n\left[\log\left(\frac{(\sigma_n^2)^m}{\det\hat{\mathbf{R}}} \right) + \left(\frac{tr\hat{\mathbf{R}}}{\sigma_n^2} - m \right) \right], \tag{15.24}$$

where on the last line, we use $n = N - 1$ to replace N to remove the bias. In (15.24), the factor of 2 is used for theoretic convenience. The terms inside the square brackets on the right-hand side of (15.24) have clear physical significance. The first term collects the correlation information in the data, whereas the second term accumulates the signal energy information.

For a given threshold Λ_1, the decision rule is given by

$$\xi_{\ell rt} \begin{cases} > \Lambda_1, & H_1 \\ < \Lambda_1, & H_0 \end{cases}. \tag{15.25}$$

We need to determine the threshold for a specified false-alarm rate, say $P_{fa} = \alpha$. There are two ways to achieve this goal. In the first approach, we invoke a theorem of Ref. [19] to obtain the characteristic function (CHF) of $\xi_{\ell rt}$, as shown by

$$\phi_{\xi_{\ell rt}}(jt) = (e/n)^{-j2tmn}(1 - j2t)^{-mn(1-j2t)}$$

$$\times \prod_{i=1}^{m} \frac{\Gamma(n - j2tn - i + 1)}{\Gamma(n - i + 1)}. \tag{15.26}$$

We directly relate this CHF to the false-alarm rate for a given threshold Λ_1 by using the Imhof lemma [20]. It turns out that

$$P_{fa} = \Pr\{\xi_{\ell rt} > \Lambda_1|H_0\}$$

$$= \frac{1}{2} + \frac{1}{\pi} \int_0^\infty \frac{\Im\{\phi_{\xi_{\ell rt}}(jt)e^{-jt\Lambda_1}\}}{t} dt. \tag{15.27}$$

This formula contains one integral, which can be numerically calculated by using the method of Gauss quadrature. The integrand becomes negligible beyond $t \geq 5$. The gamma functions inside the integrand will assume

Table 15.2   *Theoretic thresholds of $\xi_{\ell rt}$*

	Setting: $m = 5$, $N = 500$		
Thresholds	$P_{fa} = 0.1$	0.01	0.001
Theoretic	34.495	44.470	52.96
Simulated	34.389	44.486	52.820

Table 15.3   *Theoretic thresholds for $\xi_{\ell rt}$ with $\chi^2(m^2)$ approximation*

	Setting: $m = 5$, $N = 500$		
Thresholds	$P_{fa} = 0.1$	0.01	0.001
$\xi_{\ell rt}$	34.512	44.645	53.009
$\rho_1 \xi_{\ell rt}$	34.389	44.486	52.820
$\chi^2(25)$	34.382	44.314	52.620

large values because of their large arguments. For numerical accuracy, we calculate their logarithms instead, by using, for example, the continued fraction expansion (6.1.48) of Ref. [12]. After summarizing these logarithms and converting back to the exponential, we obtain the value of the integrand. As an illustration, theoretic thresholds for $m = 5$ and $N = 500$ are given in Table 15.2, where the simulation results obtained over $50,000$ independent runs are also provided for comparison.

The second approach is much simpler and is based on the result of Wilks [21] which asserts that, for a large sample size, the log likelihood ratio approaches a chi-square variable as long as the parametric space under H_0 is interior to the one under H_1. The degrees of freedom (DFs) equal the difference in free parameters under the two hypotheses. In our case, the parametric space under H_1 is defined by the Hermitian matrix $\mathbf{R}_x$, which has $m(m - 1)/2$ complex parameters and m real parameters. Hence, we have DF $= m^2$ and $\xi_{\ell rt} \sim \chi^2(m^2)$. Furthermore, a correction factor ρ_1 can be used for a more accurate fitting, namely $\rho_1 \xi_{\ell rt} \sim \chi^2(m^2)$. The derivation for ρ_1 is quite complicated, but the idea is very simple. We first determine the CHF of $\rho_1 \xi_{\ell rt}$, and then expand it into an appropriate series such that the dominant term corresponds to a chi-square variable. We choose the correction factor ρ_1 such that the $O(n^{-1})$ term vanishes to improve the chi-square approximation accuracy to the order of $O(n^{-2})$. It turns out that [22]

$$\rho_1 = 1 - \frac{2m^2 + 3m - 1}{6n(m + 1)}. \tag{15.28}$$

The theoretic thresholds so obtained for $m = 5$ and $N = 500$ are tabulated in Table 15.3, where thresholds based on the theoretically asymptotic PDF of $\chi^2(m^2)$ with $m = 5$ is also included as a benchmark for comparison.

The relative error for $\xi_{\ell rt}$ for $P_{fa} = 0.1, 0.01$, and 0.001 is 0.38%, 0.75%, and 0.74%, respectively. After using the bias correction factor ρ_1, the corresponding errors reduce to 0.02%, 0.39%, and 0.38%, respectively.

15.4.3   Eigenvalue ratio test

To exploit signal correlation for detection, it is common practice to choose correlation-sensitive parameters as a test statistic. Among all the spectra, the eigen spectrum is the most spreading and is, thus, often used as an indicator for detection of correlated CR signals. Two schemes can be considered. One scheme is based on the maximum eigen value, whereas the other is based on the ratio of two extreme eigenvalues. The first scheme does not have the CFAR property, but the eigen ratio test does. Let $\{\ell_i\}$ denote the sample eigen values, and denote $\ell_{\max} = \max\{\ell_i\}$ and $\ell_{\min} = \min\{\ell_i\}$. The eigen ratio test variable is then defined by

$$\xi_{\text{eig}} = \ell_{\max}/\ell_{\min}. \tag{15.29}$$

At this point, it is interesting to compare the eigen-ratio test with the LRT shown in (15.24). We note that the LRT is also an eigen-based test since, as observed in (15.24), the determinant and trace of the sample correlation matrix $\hat{\mathbf{R}}$ represent the product and the sum of the eigenvalues of $\hat{\mathbf{R}}$, respectively. The difference is that LRT employs the information in all the sample eigenvalues rather than just focusing on the maximum and minimum eigenvalues.

Therefore, we expect that the LRT should perform better than, or equal to, the eigen ratio test in terms of detection performance. More importantly, LRT has an easy way to determine its threshold that is not shared by the eigen ratio test. Theoretically, the PDFs for $\ell_{\max}$, $\ell_{\min}$, and their ratio, which are necessary for threshold determination, are available in the literature [23]. The marginal PDF of the maximum sample eigenvalue has been available since 1963, representable in terms of a confluent hypogeometric function of matrix argument [22], and, recently, progress has been made along this line [24]. Yet, it is still difficult to manage in practice. The PDF of the smallest sample eigenvalue can be found in the literature [25] but in a form that is difficult to handle. It is, therefore, common practice to determine thresholds for the maximum eigenvalue test or eigen ratio test by computer simulations.

15.5   Detection without exact knowledge of σ_n^2

In this section, we consider scenarios in which an inaccurate noise variance estimate, say $\hat{\sigma}_n^2$, is available at the receiver. We assume that this estimate is obtained from L independent noise-only samples, so that we can assert

$$2L\hat{\sigma}_n^2 \sim \chi^2(2L). \tag{15.30}$$

15.5.1   LRT with $\hat{\sigma}_n^2$

We use $\hat{\sigma}_n^2$ to replace σ_n^2 in (15.24) to obtain a modified test variable

$$\xi'_{\ell rt} = 2\log\underbrace{\left[\frac{(\hat{\sigma}_n^2)^m}{\det\hat{\mathbf{R}}}\exp\left(\frac{tr\hat{\mathbf{R}}}{\hat{\sigma}_n^2} - m\right)\right]^N}_{w}. \tag{15.31}$$

To determine the threshold for a specified CFAR, we invoke (15.17), again, to determine the conditional CHF of $\xi'_{\ell rt}$ under H_0, yielding

$$\phi_{\xi'_{\ell rt}}(jt|\hat{\sigma}_n^2) = (e/n)^{-j2tmn}(1 - j2tu^{-1})^{-mn(1-j2t)}u^{j2tmn}$$

$$\times \prod_{i=1}^{m}\frac{\Gamma(n - j2tn - i + 1)}{\Gamma(n - i + 1)}, \tag{15.32}$$

where $u = \hat{\sigma}_n^2/\sigma_n^2 \sim \frac{1}{2L}\chi^2(2L)$. The average CHF can be obtained by averaging over the PDF of u, which is then used to determine P_{fa} by using, again, the Imhof lemma [20].

15.5.2   LRT without noise-level reference

In this case, the only information that can be exploited for the detection of PUs is the correlation structure among the received data. This correlation is either directly introduced by the primary signal or caused by multipath fading channels.

The log likelihood ratio is given by

$$\xi_{\ell rt-} = 2\log \frac{\max_{\mathbf{R}_x} f(\mathbf{x}_1, \cdots, \mathbf{x}_{mN}|H_1)}{\max_{\sigma_n^2} f(\mathbf{x}, \cdots, \mathbf{x}_{mN})}$$

$$= 2n\log \frac{[(tr\hat{\mathbf{R}})/m]^m}{\det\hat{\mathbf{R}}}$$

$$= 2n\log \left[\frac{(\ell_1 + \cdots + \ell_m)/m}{(\ell_1 \cdots \ell_i)^{1/m}} \right]^m, \qquad (15.33)$$

where, again, $n = N - 1$, and $\{\ell_i\}$ denote the eigenvalues of $\hat{\mathbf{R}}_x$. Without noise power-level reference, the LRT examines the spread out of the sample eigenvalues by calculating the ratio of their arithmetic to geometric mean.

Without prior information, the noise variance has to be estimated from the data, thereby losing one degree of freedom. As such, we can assert that under H_0, asymptotically

$$\xi_{\ell rt-} \sim \chi^2(m^2 - 1). \qquad (15.34)$$

By the same token as for $\xi_{\ell rt}$, we can introduce a bias correcting factor

$$\rho_2 = 1 - \frac{2m^2 + 1}{6mn} \qquad (15.35)$$

so that

$$\rho_2 \, \xi_{\ell rt-} \sim \chi^2(m^2 - 1) \qquad (15.36)$$

provides a more accurate approximation under H_0. Simulated thresholds for $\ell_{\ell rt-}$ $m = 5$ with and without bias correction are compared in Table 15.4, where results based on theoretic $\chi^2(24)$ approximation are provided on the last line as a benchmark. Simulation results are obtained from $50,000$ independent computer runs.

Table 15.4   *Thresholds for $\xi_{\ell rt-}$ and $\rho\xi_{\ell rt-}$ $m = 5$: $\chi^2(m^2 - 1)$ approximation versus simulation*

Thresholds	N = 500		
	$P_{fa} = 0.1$	0.01	0.001
$\xi_{\ell rt-}$	33.323	43.234	50.807
$\rho_2\xi_{\ell rt-}$	33.209	43.087	50.634
$\chi^2(24)$	33.196	42.980	51.179

15.6 Cooperative spectrum sensing

When using a single CR user for spectrum sensing, a primary signal is likely to be missing due to channel fluctuation or shadowing during signal transmission, resulting in the so-called hidden primary-user problem. The principle of diversity combining, as studied in Chapter 8, is equally applicable to the hidden primary-user problem. The idea is to exploit the potential diversity gain inherent in spatially distributed CR users, forming a cooperative sensing system.

There are many specific schemes for diverse application scenarios addressed in the literature. Because of space limits, we only discuss the configuration shown in Figure 15.2.

Assume that a baseband PU signal $s(t)$ is transmitted over a flat Rayleigh fading channel $g_k(t)$ and experiences a random time delay τ_k, producing the received signal $x_k(t)$ at CR user k

$$x_k(t) = g_k(t)s(t - \tau_k)e^{j\omega_c t} + n_k(t), \tag{15.37}$$

where the fading gain $g_k(t)$ is a Gaussian process with correlation $R_k(\tau) = \mathbb{E}[g_k(t)g_k^*(t - \tau)]$ specified by Jake's model, and the AWGN $n_k(t)$ is with mean zero and variance σ_n^2. For simplicity, we assume that $s(t)$ is normalized to have unit power, and its waveform is up to the modulation scheme used by the PU which, typically, takes one of the following forms:

$$s(t) = \begin{cases} e^{j\omega_c t}, & \text{Pilot tone} \\ \sum_{m=-\infty}^{\infty} a_{km}\text{rect}(t - mT_s), & \text{QAM} \\ \sum_{m=-\infty}^{\infty} e^{j\theta_m}\text{rect}(t - mT_s), & \text{MPSK} \end{cases} \tag{15.38}$$

where ω_c denotes the carrier frequency, T_s denotes the symbol duration, and a_{km} is the complex amplitude. The receiver of the CR user k samples its received signal to produce an $N \times 1$ vector $\mathbf{x}_k$ and forwards the signal energy

$$\xi_k = \mathbf{x}_k^\dagger \mathbf{x}_k \tag{15.39}$$

to the fusion center for a decision making.

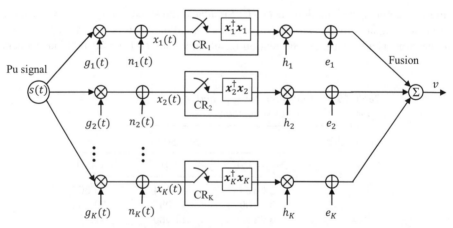

Figure 15.2 *Block diagram of cooperative sensing*

For the case of pilot tone or MPSK signal, the only change in the received signal is its phase, which, however, has no influence on the energy calculation. Since the pilot or MPSK signal is assumed of unit magnitude, we may thus write

$$\mathbf{x}_k = \underbrace{[g_k(1), \cdots, g_k(N)]^T}_{\mathbf{g}_k} + \mathbf{n}_k, \tag{15.40}$$

where $\mathbf{g}_k \sim \mathcal{CN}(\mathbf{0}, \mathbf{R}_k)$ and the AWGN $\mathbf{n}_k \sim \mathcal{CN}(\mathbf{0}, \sigma_n^2 \mathbf{I})$. The case of QAM signal is different, in the sense that both its amplitude and phase can vary and, thus, have an impact upon the correlation structure of $\mathbf{x}_k$. Thus, for the case of QAM, we have

$$\mathbf{x}_k = [a_{k1} g_k(1) \mathbf{1}_1, \cdots, a_{kM} g_k(M) \mathbf{1}_M]^T + \mathbf{n}_k, \tag{15.41}$$

where $\mathbf{1}_m$ is an all-1 row vector, resulting from sampling the mth rectangular pulse $\text{rect}(t - mT_s)$, and its size depends on the sampling rate and timing. The random data a_{km} breaks the correlation of the flat fading channel.

Consider the case of MPSK signals. Each CR receiver accumulates the energy $\xi_k = \mathbf{x}_k^\dagger \mathbf{x}_k$ and forwards it to the fusion center, so that the information can be accumulated to synthesize a decision variable

$$\ell = \sum_{k=1}^{K} |h_k|^2 \xi_k + e, \tag{15.42}$$

where $e \sim \mathcal{CN}(0, \sigma_e^2)$ denotes the AWGN at the fusion center, and $h_k \sim \mathcal{CN}(0, \rho_k)$ denotes the channel gain that links the fusion center to the kth CR. A decision is made according to the rule

$$\ell \begin{cases} > \Lambda, & H_1 \\ < \Lambda, & H_0 \end{cases}, \tag{15.43}$$

where Λ is the threshold determined by a tolerant false-alarm rate. Then, conditioned on $\mathbf{h} = [h_1, \cdots, h_K]$, the detection probability is given by

$$\Pr\{\ell > \Lambda\} = \frac{1}{2} + \frac{1}{\pi} \int_0^\infty \frac{\Im\{\phi_\ell(z) e^{-jz\Lambda}\}}{z} \, dz. \tag{15.44}$$

To determine the CHF $\phi_\ell(z)$ of ℓ, we start from its conditional CHF, $\phi_\ell(z|\mathbf{h})$ which, by definition, can be calculated as

$$\phi_\ell(z|\mathbf{h}) = \mathbb{E}[e^{jz\ell}]$$

$$= \prod_{k=1}^{K} \mathbb{E}[e^{jz|h_k|^2 \mathbf{x}_k^\dagger \mathbf{x}_k}] e^{-z^2 \sigma_e^2 / 2}$$

$$= e^{-z^2 \sigma_e^2 / 2} \prod_{k=1}^{K} \det(\mathbf{I} - jz|h_k|^2 (\sigma_n^2 \mathbf{I} + \mathbf{R}_k))]^{-1}. \tag{15.45}$$

It is straightforward to average the conditional CHF over h's to obtain $\phi_\ell(z)$. In what follows, we consider a special case. Under the i.i.d. assumption on $g_k(t)$ such that $\mathbf{R}_k = \sigma_k^2 \mathbf{I}$, the above expression is simplified to

$$\phi_\ell(z|\mathbf{h}) = e^{-z^2 \sigma_e^2 / 2} \prod_{k=1}^{K} [1 - jz|h_k|^2 (1 + \sigma_k)]^{-N}. \tag{15.46}$$

Here, we assume that the noise variance has been normalized to be $\sigma_n^2 = 1$, so that σ_k^2 stands for the SNR γ_k at CR user k. Averaging $\phi_\ell(z|\mathbf{h})$ over $|h_k|^2 \sim \frac{\rho_k}{2}\chi^2(2)$ and using (3.353(2)) of Ref. [26], we obtain for $N > 2$,

$$
\phi_\ell(z) = e^{-z^2\sigma_e^2/2} \prod_{k=1}^{K} \beta^N \int_0^\infty \frac{e^{-u}}{(u+\beta)^N} du
$$

$$
= e^{-z^2\sigma_e^2/2} \prod_{k=1}^{K} \frac{(-1)^{N-1}\beta^N}{(N-1)!} \left[\sum_{m=1}^{N-1} (m-1)!(-1)^m \beta^{-m} + e^\beta \, E_1(\beta) \right], \tag{15.47}
$$

where $\beta = -jz\rho_k(1+\gamma_k)$ and the exponential integral can be calculated by (5.1.11) of Ref. [12]:

$$
E_1(\beta) = -\psi + \ln\beta - \sum_{n=1}^{\infty} \frac{(-1)^n \beta^n}{n\, n!}, \ |\arg\ \beta| < \pi \tag{15.48}
$$

with $\psi = 0.57721\,56649\ldots$ denoting the Euler's constant. We can use this CHF and (15.44) to determine the false-alarm rate and detection performance. For false-alarm rate, we need to set all $\gamma_k = 0$.

15.7   Standardization of CR networks

The booming of CR technology in the academia expedites its standardization process. The IEEE has developed two families of standards for DSA/CR, coordinated by its IEEE Standard Coordinating Committee (SCC) 41 and IEEE 802 Working Groups (WG), respectively. The IEEE SCC 41 worked on a series of standards, that is, the P1900.x standards, defining the basic architectures, interfaces, protocols, and policies for DSA/CR. Specifically, by standardizing logical interfaces and data structure required for the exchange of sensing-related information between sensors and their clients (i.e., the spectrum management unit), the DYSPAN-P1900.6 standard facilitates interoperability between independently developed devices and thus allows for separate evolution of spectrum sensors and other system functions [27]. While the P1900.x standard series is more high-level and conceptual, the IEEE 802 family of standards is more practical. The IEEE 802.22 Wireless Regional Area Networks (WRAN) standard and its two amendments IEEE 802.22a and IEEE 802.22b specify the air interface (including MAC and PHY) for CR operation in high-VHF/low-UHF TV white space (TVWS) with a potential coverage of 100 km [28]. A recent evolvement of the commercially successful IEEE 802.11 standard series, namely the IEEE 802.11af Wireless Local Area Networks (WLAN) standard, extends the most advanced WLAN standard in mess production (i.e., 802.11ac which uses OFDM and MIMO) to include operation in the TVWS as a CR network with maximum coverage of 1 km [9]. Significant standardization advancements on DSA/CR have also been made by several other major regulatory organizations, including the International Telecommunication Union (ITU), the European Telecommunications Standards Institute (ETSI), and the European Association for Standardizing Information and Communication Systems (ECMA) [9, 29].

15.8   Experimentation and commercialization of CR systems

As part of the research and development effort, abundant experimental and trial studies of CR prototype systems for its feasibility have been performed. Popular choices of hardware/software platforms to implement CR functionalities by researchers include GNU radio, Universal Software Radio Peripheral (USRP), Berkeley Emulation Engine (BEE), and Wireless Open-Access Research Platform (WARP) [30]. In 2006, the Defense Advanced Research Projects Agency (DARPA) demonstrated the capabilities of the XG radios to work similar to the CR concept, which is considered the first private CR system trail. Demonstrations of DSA/CR have been presented continuously at dedicated CR conferences such as the IEEE DySPAN and SDR Forum (now Wireless Innovation Forum) since 2007. To verify the feasibility and commercial potential of CR, several regulation bodies have introduced key experiments and trials. The Office of Engineering and Technology (OET) of FCC in the US conducted a hardware trail for spectrum sensing, in 2008, in the TVWS channels the FCC opened for secondary access. Five companies (Adaptrum, I2R Singapore, Microsoft, Motorola, Philips) participated in the trial, which includes sensing of both TV signals and wireless microphone signals. Recently, the UK communication regulator Ofcom has also conducted coexistence test on CR-related technologies [31]. Compared to research, standardization, and experiments of CR networks, its commercialization is still at a very early stage. Several licensed TV channels in VHF/UHF (i.e., the TVWS) have been approved by FCC in the US and ETSI/Ofcom in Europe to be used for secondary access. Since geolocation database is a valid alternative to spectrum sensing in the IEEE 802.22 standard, commercial white space databases (e.g., from Google, Microsoft, SpectrumBridge) are already available for wireless devices to get information on what channels are vacant. However, the response delay, data accuracy, and energy consumption due to frequent querying are still open issues to be addressed. Furthermore, only a small number of companies, such as Adaptrum [32], are producing TVWS devices, resulting in limited choices of hardware and software products. In terms of attempts on commercial deployment of CR networks, the first urban White Space network was deployed by Microsoft in 2009 [33], and the company later on has shown some impressive development in Ghana, South Africa, Philippines, Singapore, and so on [34]. However, to obtain global commercial success and adoption of CR/DSA network, the CR community and industry still have a long way to go. Challenges such as adaptive RF front-ends, cost- and energy-efficient platforms, network coexistence, and security need to be well studied and addressed.

There is abundant material on cognitive radio in the literature. The reader is referred to [37–53] for additional reading.

Problems

Problem 15.1 Draw a simple circuit diagram of an energy detector for the detection of an MPSK primary signal with unknown symbol period, unknown symbol timing, and unknown carrier frequency.

Problem 15.2 Complex-valued baseband signal models are widely used in the context of spectrum sensing. What are the physical requirements and implications of using such complex signal models?

Problem 15.3 Energy detection is used for sensing a zero-mean random signal $x(t)$ with variance σ_x^2 in AWGN $n(t)$ with mean zero and unknown variance σ_n^2, so that the received signal is given by $y(t) = x(t) + n(t)$. By sampling $y(t)$, we acquire L noisy samples $\mathbf{y}_1 = [y_1, \cdots, y_L]^T$, and acquire M noise-alone samples $\mathbf{y}_2 = [n_{L+1}1, \cdots, n_{L+M}]^T$ from $y(t)$ during a quiet period of the primary user. Assume that $\mathbf{y}_1 \sim \mathcal{N}(\mathbf{0}, \sigma_n^2\mathbf{I} + \sigma_x^2\mathbf{C}_x)$, and $\mathbf{y}_2 \sim \mathcal{N}(\mathbf{0}, \sigma_n^2\mathbf{I})$, where $\mathbf{C}_x$ denotes the signal correlation matrix. The test statistic is defined by $\xi = (\mathbf{y}_1^T\mathbf{y}_1)/(\mathbf{y}_2^T\mathbf{y}_2)$.

(a) Determine the PDFs of ξ under H_1 and H_0.
(b) Determine the threshold for a given false-alarm rate, say $P_{fa} = \alpha$.
(c) Investigate the influence of the sample size M on the detection performance.

Problem 15.4 Detection of correlated primary signals in AWGN can be equivalently done in the frequency domain. Fourier transformation projects the received signal vector into the point-frequency vector, whereas the multi-taper method of spectral (MTS) estimation projects the same vector onto a frequency subspace. Investigate the MTS method through computer experiments [35].

Problem 15.5   A discrete complex i.i.d. primary signal $s_k \sim \mathcal{CN}(0, \sigma_s^2)$ is transmitted over an AWGN channel with $n_k \sim \mathcal{CN}(0, \sigma_n^2)$, so that the received signals under the two hypotheses are given by

$$H_0 : \ x_k = n_k,$$

$$H_1 : \ x_k = s_k + n_k.$$

Suppose that an N-sample-based energy detector $\xi = (1/\sigma_n^2) \sum_{k=1}^N |x_k|^2$ is used for the detection of the primary user. (1) Determine the PDF of ξ under H_0 and H_1, respectively. (2) Given a detection threshold Λ, determine the corresponding false alarm rate and the probability of detection [36].

References

1. Federal Communication Commission, "Spectrum-Policy Task Force," Rep. ET Docket no. 02-135, Nov. 2002.
2. M.A. McHenry, "NSF Spectrum Occupancy Measurements Summary," Shared Spectrum Co. report, Aug. 2005.
3. J. Mitola III and G.Q. Maguire Jr., "Cognitive radio: making software radios more personal," *IEEE Pers. Commun.*, vol. 6, no. 4, pp. 13–18, 1999.
4. J. Mitola, "Cognitive radio: an integrated agent architecture for software defined radio," Doctor of Technology, Royal Institute of Technology (KTH), Stockholm, Sweden, 2000.
5. S. Haykin, "Cognitive radio: brain-empowered wireless communications," *IEEE J. Sel. Areas Commun.*, vol. 23, no. 2, pp. 201–220, 2005.
6. E. Rebeiz, P. Urriza, and D. Cabric, "Optimizing wideband cyclostationary spectrum sensing under receiver impairments," *IEEE Trans. Signal Process.*, vol. 61, no. 15, pp. 3931–3943, 2013.
7. Z. Tian and G.B. Giannakis, "*Compressed sensing for wideband cognitive radios*," in *IEEE International Conference on Acoustics, Speech and Signal Processing*, Apr. 2007.
8. J. Wang, J. Chen, and D. Cabric, "Cramer-Rao bounds for joint RSS/DoA-based primary-user localization in cognitive radio networks," *IEEE Trans. Wireless Commun.*, vol. 12, no. 3, pp. 1363–1375, 2013.
9. L. Gavrilovska, D. Denkovski, V. Rakovic, and M. Angjelichinoski, "Medium access control protocols in cognitive radio networks: overview and general classification," *IEEE Commun. Surv. Tutorials*, vol. 16, no. 4, pp. 2092–2124, 2014.
10. M. Youssef, M. Ibrahim, M. Abdelatif, L. Chen, and A.V. Vasilakos, "Routing metrics of cognitive radio networks: a survey," *IEEE Commun. Surv. Tutorials*, vol. 16, no. 1, pp. 92–109, 2014.
11. N.L. Johnson and S. Kotz, *Continuous Univariate Distributions-2*, New York: John Wiley & Sons, Inc., 1970, pp. 132–134.
12. M. Abramowitz and I.A. Stegun, *Handbook of Mathematical Functions*, New York: Dover Publications, 1972, pp. 504–505, pp. 931–933, and p. 941.
13. H. Urkowitz, "Energy detection of unknown deterministic signals," *Proc. IEEE*, vol. 54, no. 4, pp. 523–531, 1967.
14. J.E. Slat and H.H. Nquyen, "Performance prediction for energy detection of unknown signals," *IEEE Trans. Veh. Technol.*, vol. 57, no. 6, pp. 3900–3904, 2008.
15. F.F. Digham, M.-S. Alouini, and M.K. Simon, "On the energy detection of unknown signals over fading channels," *IEEE Trans. Commun.*, vol. 55, no. 1, pp. 21–24, 2007.
16. Y.H. Zeng, C.L. Koh, and Y.-C. Liang, "Maximum eigenvalue detection: Theory and application," IEEE International Conference on Communications, (ICC'08), 2008.
17. Y.H. Zeng and Y.-C. Liang, "Maximum-minimum eigenvalue detection for cognitive radio," in *IEEE International Symposium on Personal, Indoor and Mobile Radio Communications, (PIMRC'07)*, 2007.
18. N. Giri, "On the complex analysis of T^2- and R^2-tests," *Ann. Math. Stat.*, vol. 36, pp. 665–670, 1965.
19. P.R. Krishnaiah, J.C. Lee, and T.C. Chang, "The distribution of the likelihood ratio statistics for certain covariance structures of complex multivariate normal populations," *Biometrika*, vol. 63, no. 3, pp. 543–549, 1976.
20. J. Gil-Pelaez, "Note on the inversion theorem," *Biometrika*, vol. 38, pp. 481–482, 1951.

21. S.S. Wilks, "The large-sample distribution of likelihood ratio for testing composite hypotheses," *Ann. Math. Stat.*, vol. 9, pp. 60–66, 1938.

22. R.J. Muirhead, *Aspects of Multivariate Statistical Theory*, New York: John Wiley & Sons, Inc., 1982, pp. 353–364.

23. T. Ratnarajah, R. Vaillancourt, and M. Alvo, "Eigenvalues and condition numbers of complex random matrices," *SIAM J. Matrix Anal. Appl.*, also see in Web.

24. I.M. Johnstone, "On the distribution of the largest eigenvalue in principal component analysis," *Ann. Stat.*, vol. 29, no. 2, pp. 295–327, 2001.

25. Z.D. Bai, "Methodologies in spectra analysis of large dimensional random matrices: a review," *Stat. Sin.*, vol. 9, pp. 611–677, 1999.

26. I.S. Gradshteyn and I.M. Ryzhik, *Table of Integrals, Series, and Products*, New York: Academic Press, 1980.

27. K. Moessner, H. Harada, C. Sun, Y.D. Alemseged, H.N. Tran, D. Noguet, R. Sawai, and N. Sato, "Spectrum sensing for cognitive radio systems: technical aspects and standardization activities of the IEEE P1900.6 working group," *IEEE Wireless Commun.*, vol. 18, no. 1, pp. 30–37, 2011.

28. C. Stevenson, G. Chouinard, Z. Lei, W. Hu, S.J. Shellhammer, and W. Caldwell, "IEEE 802.22: the first cognitive radio wireless regional area network standard," *IEEE Commun. Mag.*, vol. 47, no. 1, pp. 130–138, 2009.

29. S. Filin, H. Harada, H. Murakami, and K. Ishizu, "International standardization of cognitive radio systems," *IEEE Commun. Mag.*, vol. 49, no. 3, pp. 82–89, 2011.

30. P. Pawelczak, K. Nolan, L. Doyle, S.W. Oh, and D. Cabric, "Cognitive radio: ten years of experimentation and development," *IEEE Commun. Mag.*, vol. 49, no. 3, pp. 90–100, 2011.

31. Online: http://stakeholders.ofcom.org.uk/market-data-research/other/technology-research/.

32. Online: http://www.adaptrum.com/.

33. Online: http://research.microsoft.com/en-us/projects/KNOWS/deployment.aspx.

34. Online: http://research.microsoft.com/en-us/projects/spectrum/pilots.aspx.

35. Q.T. Zhang, "Theoretical performance and thresholds of the multitaper method for spectrum sensing," *IEEE Trans. Veh. Technol.*, vol. 60, no. 5, pp. 2128–2138, 2011.

36. K.B. Letaief and W. Zhang, "Cooperative communications for cognitive radio networks," *Proc. IEEE*, vol. 97, no. 5, pp. 978–892, 2009.

37. J. Mitola III, "Cognitive radio: architecture evolution," *Proc. IEEE*, vol. 97, no. 4, pp.626–641, 2009.

38. S. Haykin, D.J. Thomson, and J.H. Reed, "Spectrum sensing for cognitive radio," *Proc. IEEE*, vol. 97, no. 5, pp. 849–977, 2009.

39. J. Ma, G.Y. Li, and B. Hwang, "Signal processing in cognitive radio," *Proc. IEEE*, vol. 97, no. 5, pp. 805–823, 2009.

40. M.D. Springer, *Algebra of Random Variables*, New York: John Wiley & Sons, Inc., 1973.

41. T. Yucek and H. Arslan, "A survey of spectrum sensing algorithms for cognitive radio applications," *IEEE Commun. Surv. Tutorials*, vol. 11, no. 1, pp. 116–130, 2009.

42. I.F. Akyildiz, W.-Y. Lee, M.C. Vuran, and S. Mohanty, "A survey on spectrum management in cognitive radio networks," *IEEE Commun. Mag.*, vol. 46, no. 4, pp. 40–48, 2008.

43. W.-Y. Lee and I.F. Akyildiz, "Optimal spectrum sensing framework for cognitive radio networks," *IEEE Trans. Wireless Commun.*, vol. 7, no. 10, pp. 3845–3857, 2008.

44. N. Devroye, M. Vu, and V. Tarokh, "Cognitive radio networks: highlights of information theoretic limits, models and design," *IEEE Signal Process.*, vol. 25, no. 6, pp. 12–23, 2008.

45. A. Jovicic and P. Viswanath, "Cognitive radio: an information-theoretic perspective," *IEEE Trans. Inf. Theory*, vol. 55, no. 9, pp. 3945–3958, 2009.

46. R. Zhang, S. Cui, and Y.-C. Liang, "On ergodic sum capacity of fading cognitive multiple-access and broadcast channels," *IEEE Trans. Inf. Theory*, vol. 55, no. 11, pp. 5161–5178, 2009.

47. A. Goldsmith, S.A. Jafar, and S. Srinivasa, "Breaking spectrum gridlock with cognitive radios: an information theoretic perspective," *Proc. IEEE*, vol. 97, no. 5, pp. 894–914, 2009.

48. J. Unnikrishnan and V.V. Veeravalli, "Cooperative sensing for primary detection in cognitive radio," *IEEE J. Sel. Top. Signal Process.*, vol. 2, no. 1, pp. 18–27, 2008.

49. Z. Quan, S. Cui, and A.H. Sayed, "Optimal linear cooperation for spectrum sensing in cognitive radio networks," *IEEE J. Sel. Top. Signal Process.*, vol. 2, no. 1, pp. 28–40, 2008.

50.   G. Ganesan and Y. Li, "Cooperative spectrum sensing in cognitive radio, Part I: two user networks," *IEEE Trans. Wireless Commun.*, vol. 6, no. 6, pp. 2204–2213, 2007.

51.   Q. Zhang, J. Jai, and J. Zhang, "Cooperative relay to improve diversity in cognitive radio networks," *IEEE Commun. Mag.*, vol. 47, no. 2, pp. 111–117, 2009.

52.   S. Chaudhari, V. Koivunen, and H.V. Poor, "Autocorrelation-based decentralized sequential detection of OFDM signals in cognitive radio," *IEEE Trans. Signal Process.*, vol. 57, no. 7, pp. 2690–2700, 2009.

53.   A.J. Barabell, "Improving the resolution performance of eigenstructure-based direction-finding algorithms," *Proceedings of the IEEE ICASSP-83*, vol. 1, pp. 336–339, Boston, MA, 1983.

Index

Wireless Communications: Principles, Theory and Methodology, First Edition. Keith Q.T. Zhang.
© 2016 John Wiley & Sons, Ltd. Published 2016 by John Wiley & Sons, Ltd.
Companion Website: www.wiley.com/go/zhang7749